AF772138

PALLADACYCLES

PALLADACYCLES
CATALYSIS AND BEYOND

Edited by

ANANT R. KAPDI
Department of Chemistry, Institute of Chemical Technology, Mumbai, India

DEBABRATA MAITI
Department of Chemistry, Indian Institute of Technology, Mumbai, India

ELSEVIER

Elsevier
Radarweg 29, PO Box 211, 1000 AE Amsterdam, Netherlands
The Boulevard, Langford Lane, Kidlington, Oxford OX5 1GB, United Kingdom
50 Hampshire Street, 5th Floor, Cambridge, MA 02139, United States

© 2019 Elsevier Inc. All rights reserved.

No part of this publication may be reproduced or transmitted in any form or by any means, electronic or mechanical, including photocopying, recording, or any information storage and retrieval system, without permission in writing from the publisher. Details on how to seek permission, further information about the Publisher's permissions policies and our arrangements with organizations such as the Copyright Clearance Center and the Copyright Licensing Agency, can be found at our website: www.elsevier.com/permissions.

This book and the individual contributions contained in it are protected under copyright by the Publisher (other than as may be noted herein).

Notices
Knowledge and best practice in this field are constantly changing. As new research and experience broaden our understanding, changes in research methods, professional practices, or medical treatment may become necessary.

Practitioners and researchers must always rely on their own experience and knowledge in evaluating and using any information, methods, compounds, or experiments described herein. In using such information or methods they should be mindful of their own safety and the safety of others, including parties for whom they have a professional responsibility.

To the fullest extent of the law, neither the Publisher nor the authors, contributors, or editors, assume any liability for any injury and/or damage to persons or property as a matter of products liability, negligence or otherwise, or from any use or operation of any methods, products, instructions, or ideas contained in the material herein.

Library of Congress Cataloging-in-Publication Data
A catalog record for this book is available from the Library of Congress

British Library Cataloguing-in-Publication Data
A catalogue record for this book is available from the British Library

ISBN: 978-0-12-815505-9

For information on all Elsevier publications
visit our website at https://www.elsevier.com/books-and-journals

Publisher: Susan Dennis
Acquisition Editor: Emily McCloskey
Editorial Project Manager: Michelle Fisher
Production Project Manager:
Prem Kumar Kaliamoorthi
Designer: Mark Rogers

Typeset by SPi Global, India

ELSEVIER Book Aid International
Working together to grow libraries in developing countries
www.elsevier.com • www.bookaid.org

CONTENTS

CONTRIBUTORS

Prachi Bangde
Department of Pharmaceutical Technology, Institute of Chemical Technology, Mumbai, India

Bhalchandra M. Bhanage
Department of Chemistry, Institute of Chemical Technology, Mumbai, India

Shatrughn A. Bhilare
Department of Chemistry, Institute of Chemical Technology, Mumbai, India

Pui Ying Choy
Department of Chemistry, The Chinese University of Hong Kong, Shatin, Hong Kong

Prajakta Dandekar
Department of Pharmaceutical Technology, Institute of Chemical Technology, Mumbai, India

Ian J.S. Fairlamb
Department of Chemistry, University of York, York, United Kingdom

Tejpalsingh R. Girase
Department of Chemistry, Institute of Chemical Technology, Mumbai, India

Gaurav R. Gupta
Department of Chemistry, Institute of Chemical Technology, Mumbai, India

Xinwei He
Department of Chemistry, The Chinese University of Hong Kong, Shatin, Hong Kong

Anant R. Kapdi
Department of Chemistry, Institute of Chemical Technology, Mumbai, India

Raysa Khan
Department of Chemistry, School of Life Sciences, University of Sussex, Brighton, East Sussex, United Kingdom

Fuk Yee Kwong
Department of Chemistry, The Chinese University of Hong Kong, Shatin, Hong Kong

Fátima Lucio-Martínez
Departamento de Química Inorgánica, Universidad de Santiago de Compostela, Santiago de Compostela, Spain

Debabrata Maiti
Department of Chemistry, Indian Institute of Technology, Mumbai, India

Helena C. Malinakova
Department of Chemistry, The University of Kansas, Lawrence, KS, United States

M. Teresa Pereira
Departamento de Química Inorgánica, Universidad de Santiago de Compostela, Santiago de Compostela, Spain

Dharmendra Prajapati
Department of Chemistry, Institute of Chemical Technology, Mumbai, India

Benudhar Punji
Organometallic Synthesis and Catalysis Group, Chemical Engineering Division, CSIR–National Chemical Laboratory (CSIR–NCL), Pune, India

Francisco Reigosa
Departamento de Química Inorgánica, Universidad de Santiago de Compostela, Santiago de Compostela, Spain

José L. Serrano
Department of Mining, Geological and Cartographic Engineering, Inorganic Chemistry Section, Technical University of Cartagena, Regional Campus of International Excellence "Campus Mare Nostrum", Cartagena, Spain

Jagrut A. Shah
Department of Chemistry, Institute of Chemical Technology, Mumbai, India

Kevin H. Shaughnessy
Department of Chemistry & Biochemistry, The University of Alabama, Tuscaloosa, AL, United States

Vineeta Soni
Organometallic Synthesis and Catalysis Group, Chemical Engineering Division, CSIR–National Chemical Laboratory (CSIR–NCL), Pune, India

John Spencer
Department of Chemistry, School of Life Sciences, University of Sussex, Brighton, East Sussex, United Kingdom

José M. Vila
Departamento de Química Inorgánica, Universidad de Santiago de Compostela, Santiago de Compostela, Spain

CHAPTER 1

Palladacycles as Efficient Precatalysts for Suzuki-Miyaura Cross-Coupling Reactions

José M. Vila, M. Teresa Pereira, Fátima Lucio-Martínez, Francisco Reigosa
Departamento de Química Inorgánica, Universidad de Santiago de Compostela, Santiago de Compostela, Spain

Contents

1. INTRODUCTION

Palladacycles *(1–5)* are a paramount class of organometallic compounds which, over recent years, have arisen as valuable devices in the preparation of organometallic substances, especially in cross-coupling processes leading to the formation of carbon-carbon bonds, namely the Suzuki-Miyaura reaction *(6–10)*. Palladacycles constitute the precatalysts that provide the necessary palladium itself in an oxidation state dependant on the proposed catalytic cycle, i.e., Pd(0)/Pd(II)//Pd(II)/Pd(IV), or alternatively, where palladium intercedes in the course of the reaction, more often than not palladacycles emerge as intermediates. Well known reasons for their ample use in such performances are, among others, that they are fairly stable and that with few exceptions may be readily prepared in reasonably good yields. After the pioneering work by Herman et al. *(11–13)* involving

Palladacycles
https://doi.org/10.1016/B978-0-12-815505-9.00002-0
© 2019 Elsevier Inc.
All rights reserved.

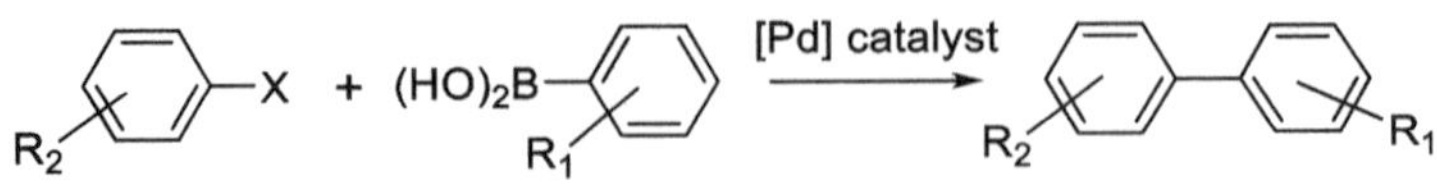

Scheme 1.1 The Suzuki-Miyaura cross-coupling reaction.

phosphine palladacycles, countless amounts of new palladacycles have been tested for this purpose.

The Suzuki-Miyaura process consists in the cross-coupling reaction of organoborated compounds (Ar_1-$B(OR)_2$) with electrophiles (Ar_2-X) catalyzed by palladium *(14, 15)*. This is one of the most efficient methods for making carbon–carbon bonds between aryl rings (Scheme 1.1).

Precisely, one of the main advantages of the Suzuki-Miyaura reaction is that it uses organoborated species as precursors; in particular, various reactions have been carried out with organoborated derivatives, such as esters of boronic acids and alkylboranes. Another important aspect is the possibility of using water as solvent *(16, 17)*. Because water is an excellent solvent for microwave heating, the combined use of water as solvent plus heating by microwave radiation has also been considered *(18)*. It is generally accepted that the catalytic cycle for the Suzuki–Miyaura reaction is initiated with an oxidative addition of the electrophile (Ar_2-X) to the catalyst generating the intermediate {Ar_2-[Pd]-X}. The resulting complex undergoes transmetallation between {Ar_2-[Pd]-X} and (Ar_1-$B(OH)_2$) to give the disubstituted palladium species {Ar_1-[Pd]-Ar_2}, after which reductive elimination of the palladium-disubstituted moiety occurs, releasing the coupling product (Ar_1-Ar_2) and regenerating the catalyst, which is incorporated again into the cycle; the latter step verifies the catalytic efficiency of the palladacycle (Scheme 1.2).

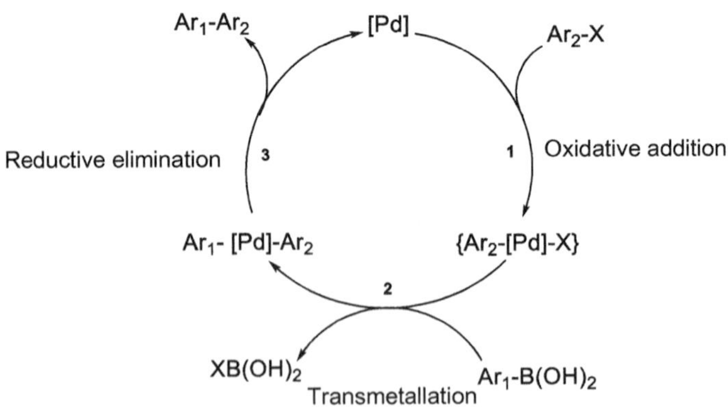

Scheme 1.2 The Suzuki-Miyaura catalytic cycle.

The amount of references, inclusive of reviews, related to palladacycles that have been used in the aforementioned process is overwhelming. So, far from being comprehensive on this issue, the reader may find an account of the most outstanding aspects of this chemistry as from the work of Dupont *(3)* and Bedford (Chapter 8 in *(2)*). Also, despite strict frontiers between the palladacycles sometimes being difficult to define with precision, herein they are mainly classified as a function of the nature of the metallated ligand.

2. IMINE AND AMINE PALLADACYCLES

This is a rather large group of palladacycles owing to the accessibility of the amine ligands and to the ease of the preparation of Schiff bases *aka* imines. Yang et al. showed that Shiff base palladacycles bearing arsines or stibines **1** were good catalysts for the Suzuki-Miyaura cross-coupling between arylboronic acids and aryl bromides, with yields *c.*90%.

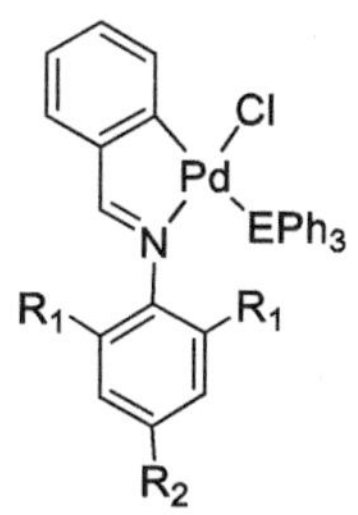

1 E = As, Sb; R_1, R_2 = Me; R_1 = iPr, R_2 = H

However, for the analogous aryl chloride or chloronitro derivatives the result was *c.*30% or lower. Attempts to perform the reaction in aqueous media produced only traces of the coupled products *(19)*. Similar mononuclear palladacyles, albeit coordinated to tertiary phosphines, were successfully tested by Serrano et al. for the reaction between benzyl bromide and arylboronic acids with yields >70%, which proceed with formation of a $Pd^0(PR_3)$ intermediate *(20)*. Nitrogen donor palladacycles have been used in aqueous-biphasic solvent systems in order to obtain a recyclable catalytic system. The hydrophilic palladacycles **2**, prepared from *p*-hydroxybenzylamine, and **3**, from a sulfonated imine produced active catalysts for the cross-coupling of aryl bromides combined with t-Bu-Amphos *(21)*. The **3**/t-Bu-Amphos system was shown to be greatly recyclable, with yields above 86% and a life span of twelve cycles.

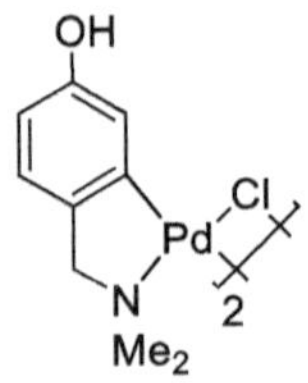

2

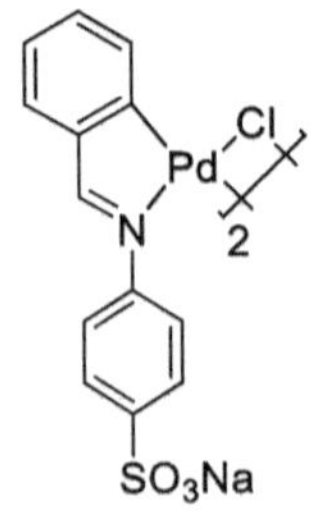

3

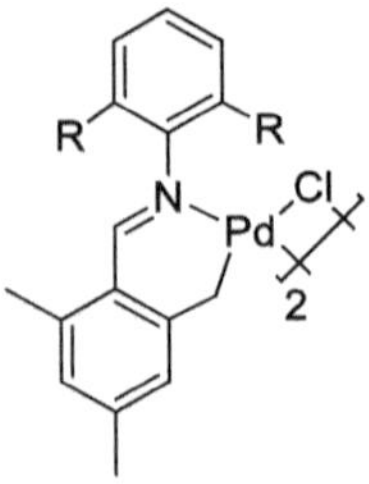

4 R = Me, iPr

Similar Schiff base palladacycles *(22)* showing air and moisture insensitivity were inspected for the Suzuki-Miyaura coupling reaction between aryl halides and arylboronic acids. For example, complex **4** displayed an activity of up to TON *c.* 10^7, it being suggested that palladacyclic catalysis was *via* palladium nanoparticles. Furthermore, the catalytic activity of amine *N*-donor palladacycles may be enhanced by supporting the catalyst over graphene oxide *(23)*; the resulting nanohybrid system **5** shows a TON of 80,000 and a TOF (h^{-1}) of 240,000. Other more sophisticated amine *N*-donor palladacycles have been applied as catalysts. Thus, the bulky pisdene complex **6** (pisdene = 3-benzyl-7-methyl-3,7-diazabicyclo[3.3.1]nonane) displayed an 85% yield under mild conditions for the cross-coupling between 4-bromoanisole and phenylboronic acid *(24)*; while the tetranuclear palladacycle **7** was an air stable catalyst with a TON as high as 228×10^3 with a catalyst mol percentage of only 390×10^{-6} *(25)*. Phenylalanine palladacycles **8** may lead to polymer-type species as in the Suzuki cross-coupling polymerization of dibromides with diboronic acids with shorter reaction times than other conventional palladium catalysts *(26)*. Pyridinium salts related to those in ionic liquid solvents withstand palladation as well to produce palladium-pyrilydene compounds **9** as active catalysts for the Suzuki reaction *(27)*.

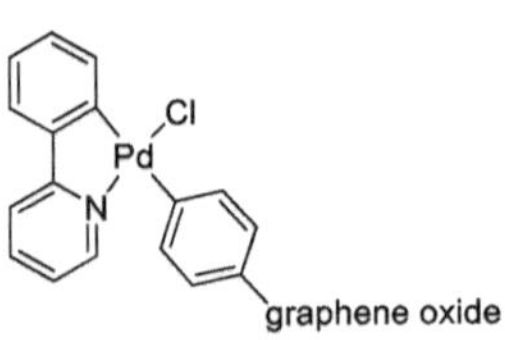

5

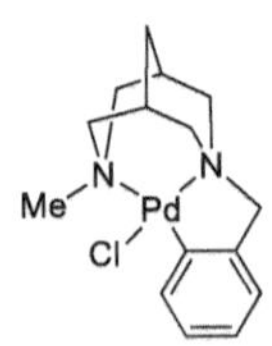

6

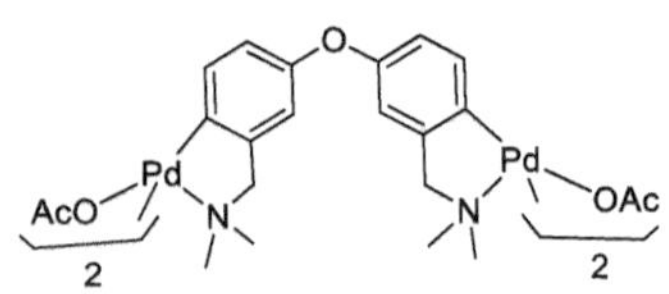

7

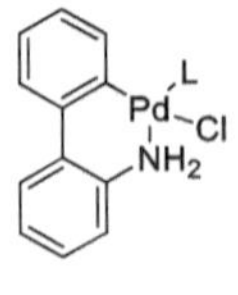

8 L = P(t-Bu)3, PPh3, PCy3

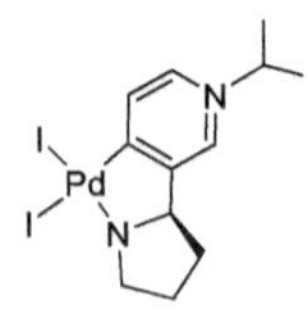

9

3. OXIME PALLADACYCLES

The related oxime palladacycles have been shown to be likewise effective for the Suzuki coupling *(28)*. They furnish the origins of remarkably effective compounds that allow for less rigorous reaction conditions. Oximes of the type **10** have been successfully applied to the cross-coupling reaction of even sterically impeded and deactivated aryl chlorides in water using microwave heating *(29–31)*. The "light fluorine" containing dinuclear oxime palladacycle **11** (fluorine content less than 50%) with two perfluoro octyl groups attached is very active for promoting the Suzuki reaction in aqueous media, with very low palladium leaching, as well as the use of microwave heating *(32)*.

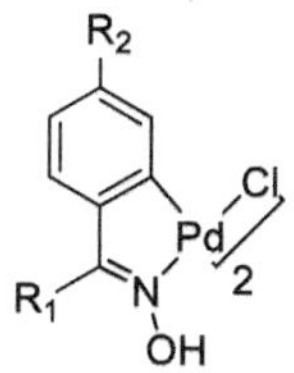

10 R_1 = Me; R_2 = OH
R_1 = 4-ClC_6H_4, R_2 = OH

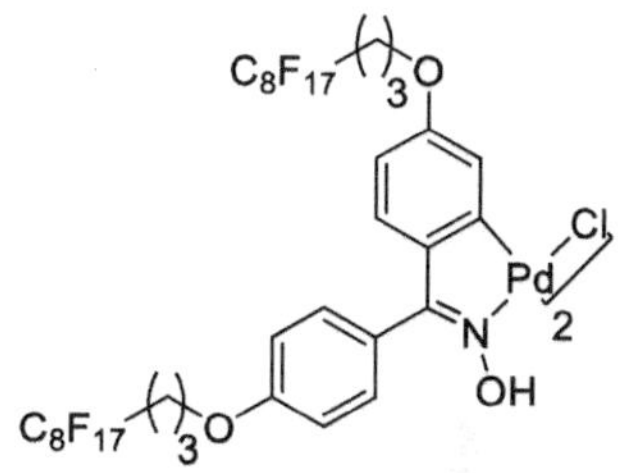

11

Oxime-based palladacycles have also been used as heterogeneous catalysts. An alcoxy oxime palladacycle resin **12** was tested for the Suzuki reaction. Microwave radiation improved its activity for the coupling of 4-bromoanisole with phenyboronic acid under 20 minutes with a yield greater than 94% *(33, 34)*.

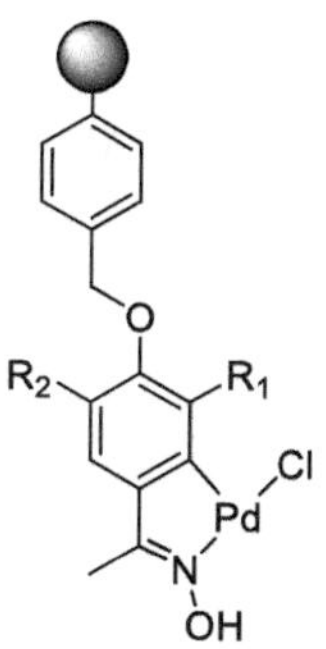

12 a, R_1 = R_2 = H
b, R_1 = H, R_2 = OCH_3
c, R_1 = R_2 = OCH_3

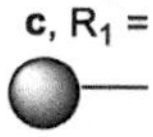
Polymer

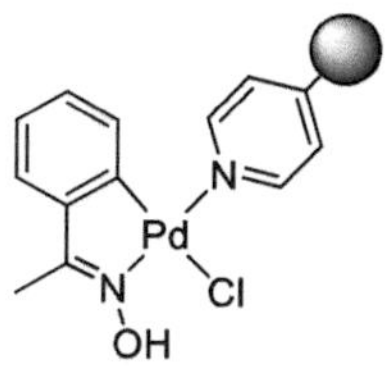

13

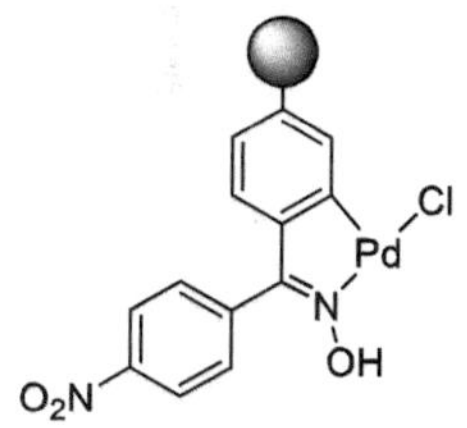

14

Fixing the catalyst on polyvinylpyridine (PVP) resin affords a reusable, air, moisture, and thermally stable palladium precatalyst **13**, which shows low intensity leaching and that is suitable for continuous flow operations *(35)*. Nájera et al. *(36)* have evaluated the effectiveness of the Kaiser oxime-based palladacycle **14** with K_2CO_3 as base in water under reflux for the coupling between phenylboronic acid and several substituted aryls, among others, with yields >90%. Poorer results were obtained for alkyl boronic acids, but diverse styrenes and stilbenes were prepared using this catalyst. Other alternatives include, for example, oxime palladacycles supported on graphene oxide *(37)* (vide supra), which are most convenient for the reaction between aryl bromides and arylboronic acids working with little catalyst quantities in water at *c.*25°C. Other materials available for supporting oxime palladacycle precursors **15** developed by Corma et al. *(38)* include mesoporous silica yielding an heterogeneous catalyst.

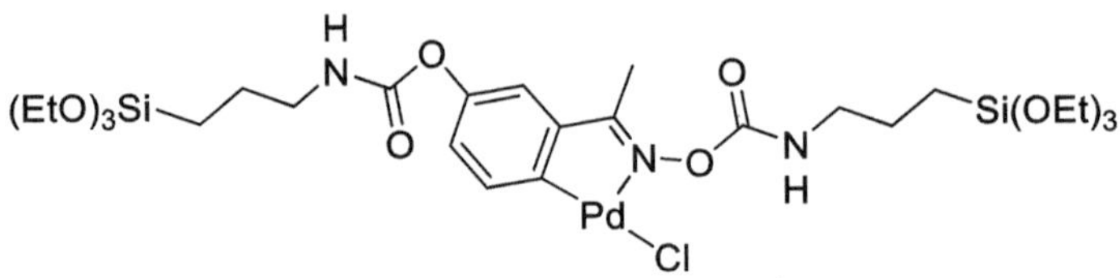

15

Moreover, a mini review concerning oxime palladacycles as catalysts not only in the Suzuki-Miyaura coupling, but in other in carbon-carbon bond formation processes as well, has recently been given by Nájera *(39)*.

4. FERROCENE PALLADACYCLES

A wide variety of ferrocene derivatives has been researched by Wu et al. The ferrocenylimine palladacycle **16** was an advantageous catalyst for the cross-coupling of 3-pyridylboronic pinacol ester with an assortment of iodide, bromide, and the electron-rich chlorides *(40)*. Reasonable to good yields, close to or above 96%, were obtained for catalysis of 3-pyridin boronic pinacol ester with aryl iodides. Lower yields were obtained in general for the analogous bromide derivatives. Also, the catalyst proved to be rather efficient in the cross-coupling of aryl chlorides with yields reaching 80% using cesium carbonate as base. Reduction of the C=N double bond with $NaBH_4$ or with $LiAlH_4$ gives the related ferrocenylamine palladium complexes **17** *(41)* and **18** *(42)* which are also quite suitable catalysts. The latter presents two pairs of racemates R_nR_p/S_nS_p and S_nR_p/R_nS_p. The corresponding carbene adducts have also been examined and they have successfully been employed in the Suzuki coupling *(43)*.

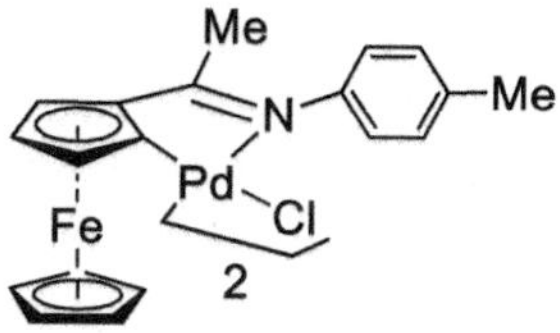

16

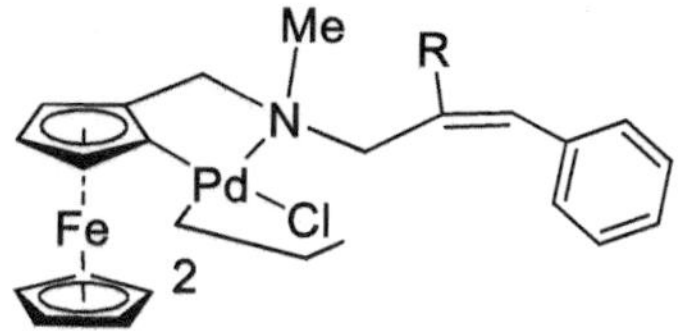

17

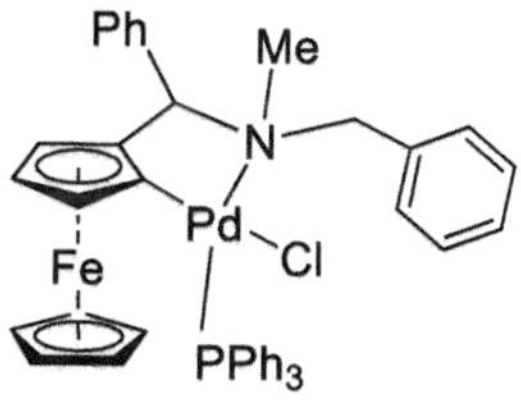

18

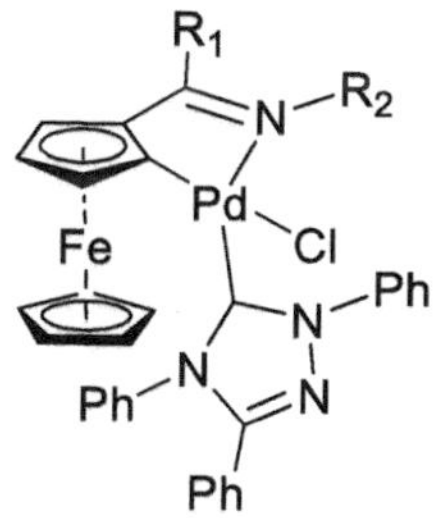

19: R_1 = H, Me; R_2 = substituted aryl, Cy

Triscyclohexylphosphine or 2-(dicyclohexylphosphanyl)-2′-(dimethylamine)biphenyl as ancillary ligand in the mononuclear air-stable palladacycles **20** *(44)* and **21** *(45)*, respectively, enhances the activity of the catalysts. It has been shown that compound **20** with $R_1 = Me$, $R_2 = p\text{-}MeC_6H_4$, is highly efficient for borylation catalysis to give unsymmetrical biaryls *(46)*.

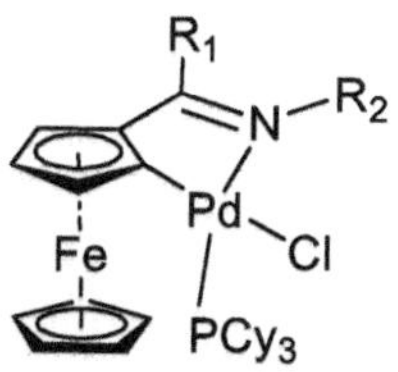

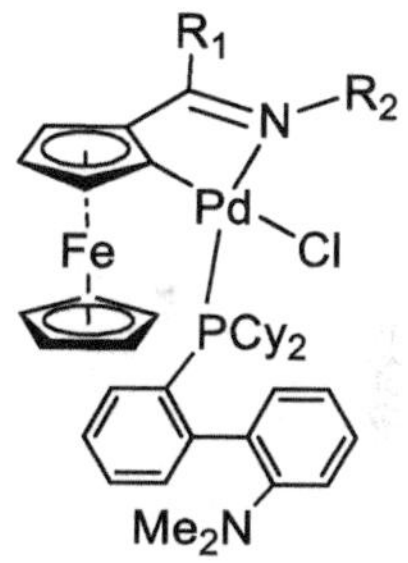

20: R_1 = H, Me; R_2 = ^{i}Pr, Cy substituted aryl

21: R_1 = H, Me; R_2 = ^{i}Pr, Cy substituted aryl

Analogues of **20** are quite adequate for binding the CF_3CH_2 moiety to an aromatic ring *(47)*, albeit that the yields are far from high ranging roughly *ca.* 40%–50%, as well as for couplings between boronic acids and carboxylic anhydrides or acyl chlorides *(48)*.

In order to avoid the use of cumbersome bulky phosphines, these may be substituted by pyridine to produce phosphine-free catalysts **22** *(49)* which display quite high yields in the Suzuki reaction. Other phosphine-free palladacycles *(50)* prepared from ferrocenylimidazolines have been investigated as catalyst showing good to excellent results. Complexes of type **23** may give yields >95% for cross-coupling of aryl bromides with phenyl boronic acid, and for those cases where the choice of the halide derivative was the corresponding chloride, the resulting yields were in the range 51%–88%, with a peak at 97% when **24** *(51)* was employed in the coupling with 4-nitrochlorobenzene. The Suzuki reaction may be used to synthesize polydentate palladacycles. This may be achieved by cross-coupling on the ligand itself, which is then metallated; or alternatively, the ferrocene palladacycle **25** is made first and then it undergoes the Suzuki coupling with the appropriate biphenyl boronic acid *(52)*. Whichever is the case, the final product is **26**; however, by the second pathway the reaction can proceed without addition of a second palladium source as catalyst.

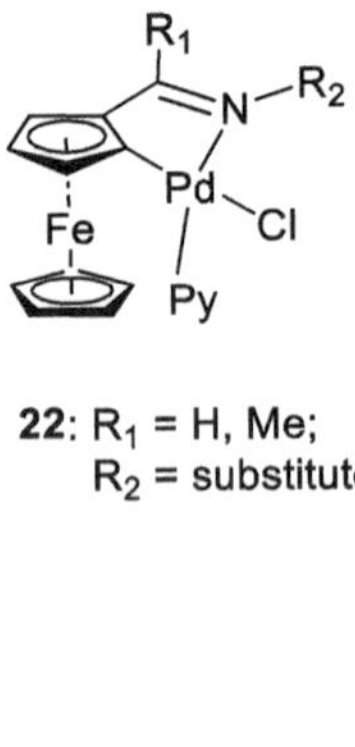

22: R_1 = H, Me;
R_2 = substituted aryl

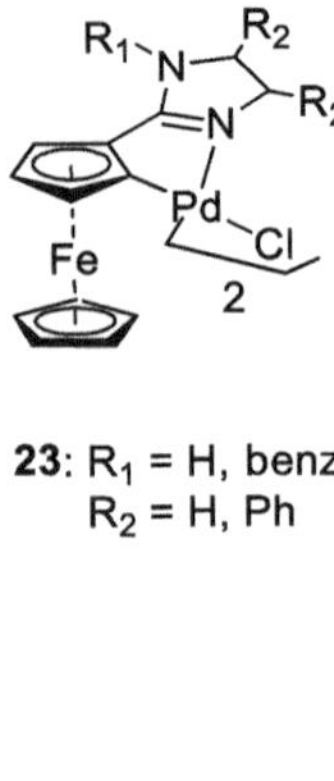

23: R_1 = H, benzyl, acetyl
R_2 = H, Ph

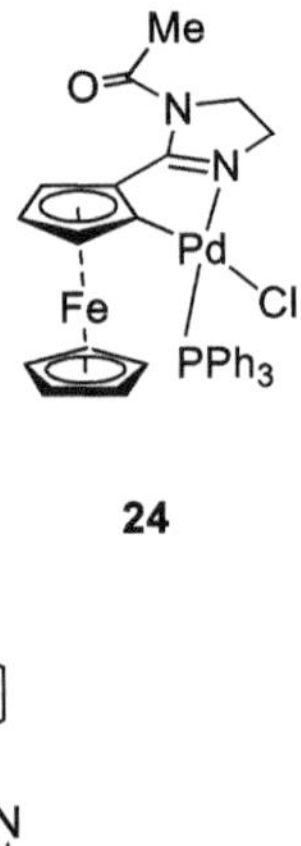

24

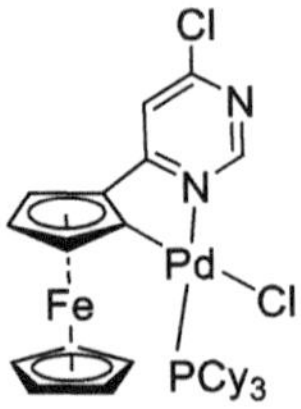

25

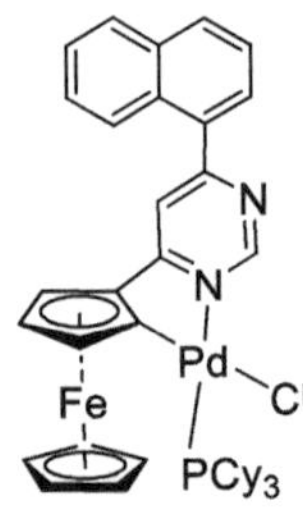

26

Terdentate [*C*,*N*,*E*] (E = S, Se) ferrocenylimine palladacycles **27** *(53)* have been explored for the Suzuki-Miyaura reaction between arylboronic acids and an array of substituted aryl chlorides and bromides yielding satisfactory results, with the selenium derivative being somewhat more powerful than the sulfur analogues. Half-sandwich cyclopentadienyl metal moieties as a fundamental part of the palladacycle have been invoked as potential catalysts in the Suzuki cross-coupling reaction, for example the dinuclear Re/Pd complexes described by Gladysz et al. *(54)* **28** and **29**, albeit conclusive experiments, still remain outstanding.

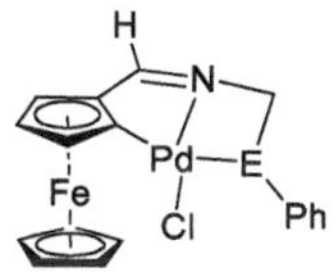

27: E = S, Se

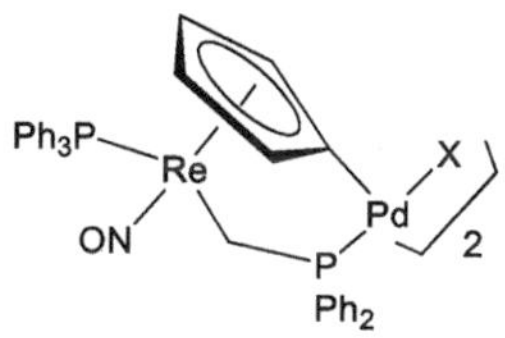

28: X = Cl, Br

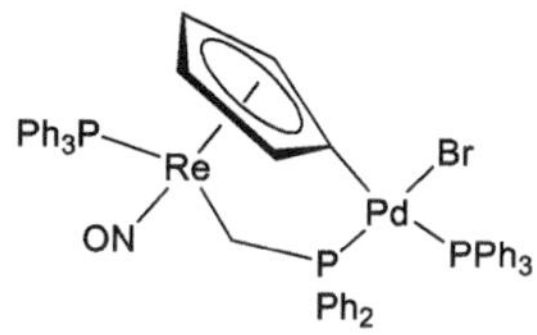

29

5. PINCER PALLADACYCLES

Pendant benzamidinate palladacycles are good catalysts for the Suzuki coupling, with those having the oxazoline group **30** showing greater activities than those with pyridine or amine group *(55)*. In this case, lower loadings may be used and the conversion, as followed by ^{1}H NMR, is up to 99% with an isolated yield of 92%. Palladacycles with phosphaalkene ligands **31** *(56)*, or with nitrogen and selenium or sulfur donors **32** *(57, 58)* give air stable catalysts with TONs of 93,000 and the conversion to nano-sized particles $Pd_{17}Se_{15}$, is believed to be the real active species.

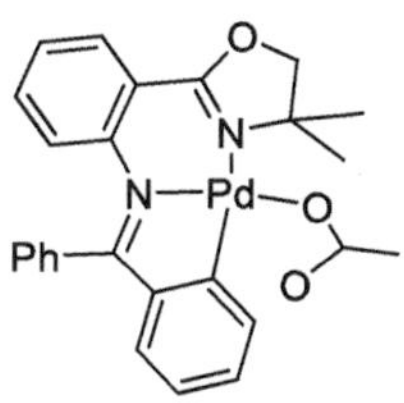

30

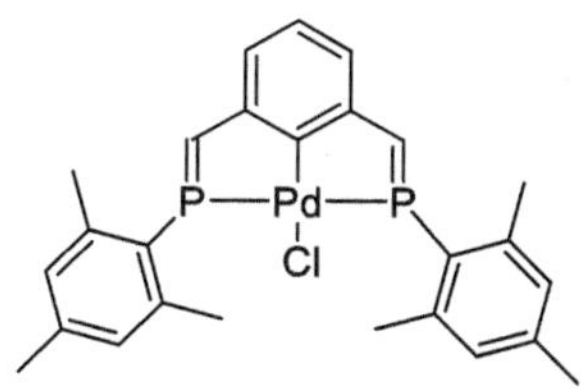

31

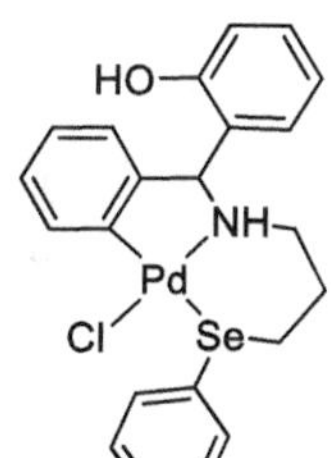

32

The critical role of nanoparticles has also been invoked in the case of phosphine-free palladacycles with selenium donors such as **33** where Beletskaya et al. *(59)* provide a critical point of view related to this issue, while pursuing evidence on the nanoparticle-enabled route as opposed to the mononuclear route. Nonsymmetrical [N,C,N], [N,C,P] and

[N,C,S] palladium pincer complexes such as **34** *(60)*, **35** *(61)*, **36** *(62)*, proved to be highly efficient catalysts with yields of nearly 100% in the cases of **34** and **36**. Other unsymmetrical [C,N,N] high performing palladacycles containing iminopyridyl **37** *(63)* have been developed by Hu et al. for the synthesis of 1,1-diarylalkanes for coupling of benzylic bromides with arylboronic acids with substantial selectivity and TON of 47,000.

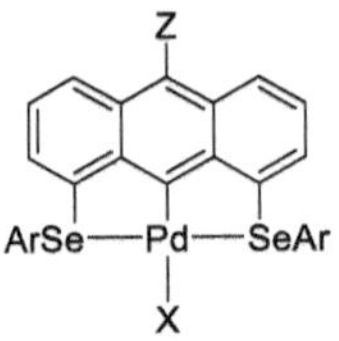

33: X = OAc, Cl
Z = H, NHAc, OMe

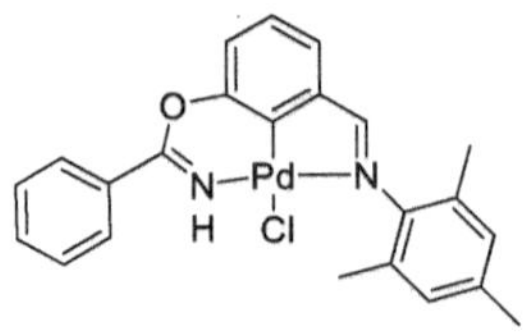

34

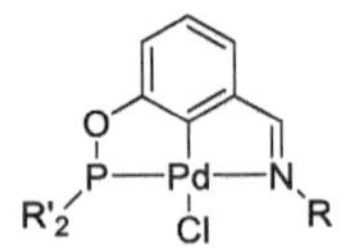

35: R = Ph, *i*-Pr, *m*-ClC_6H_4;
(*S*)-1-phenylethyl;
R' = Ph, *i*-Pr

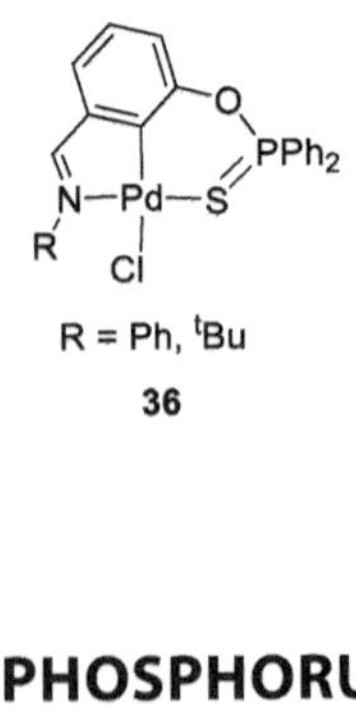

R = Ph, ^{t}Bu

36

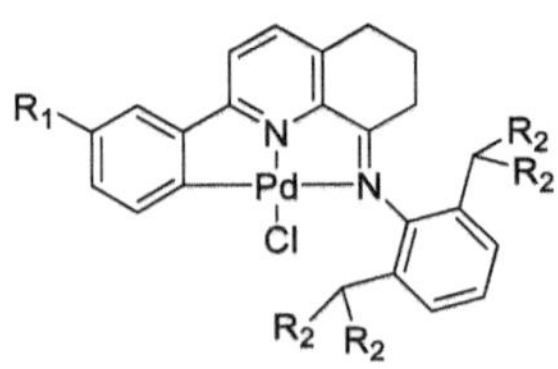

37, R_1 = H, CF_3, Me, OMe;
R_2 = Me, Ph

6. PHOSPHORUS DONOR ATOM PALLADACYCLES

Polystyrene supported adducts of amine and phosphite palladacycles have been used as catalysts in the Suzuki reaction for aryl chlorides *(64)*. Confinement of the catalyst in the polystyrene base may lead to enhanced performance when matched against their homogeneous counterparts. However, the supported catalysts may not be recovered. Better conversion in terms of TON was obtained in one reaction for **38**. In contrast, the carbene and phosphine adducts of similar phosphite and phosphinite palladacycles **39** for coupling of alkyl bromides with phenyl boronic acid showed poor results *(65)*. No improvement with respect to other palladium sources, such as palladium acetate, was found. Furthermore, it seems at this point that the oxidative addition of the alkyl halide is not the rate determining step.

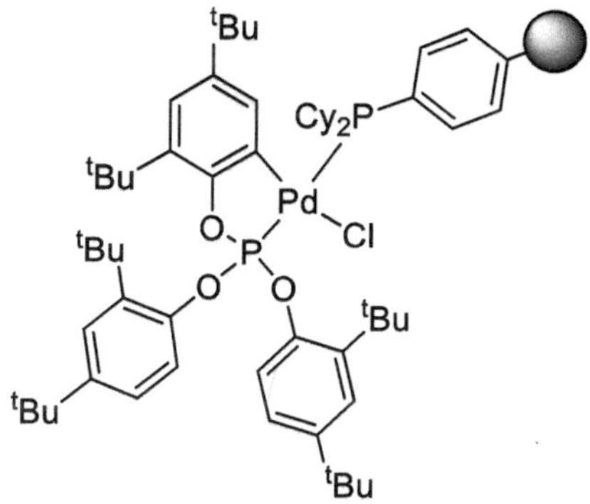

38: — Polystyrene

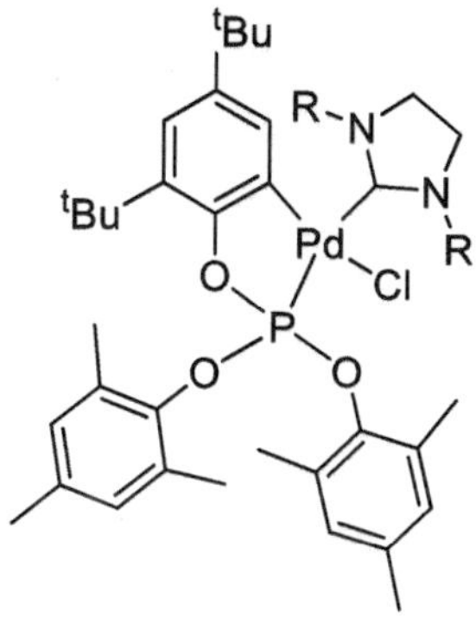

39: R =2,4,6-$Me_3C_6H_2$

A palladacycle giving very high activity in aqueous media at ambient temperature **40** has been devised using an upgraded protocol, selected from amongst a series of related phosphorus donor compounds and tested to give a 91% yield *(66)*. An improved protocol employing **41** in aqueous media was achieved avoiding use of organic solvents, with isolated yields of ≥99% in some cases *(67)*. Initial treatment of the catalyst with boronic acid shortens the reaction times.

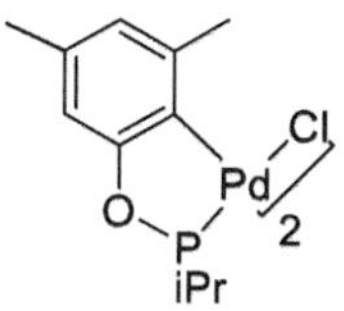

40

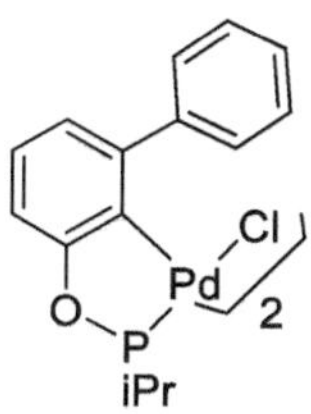

41

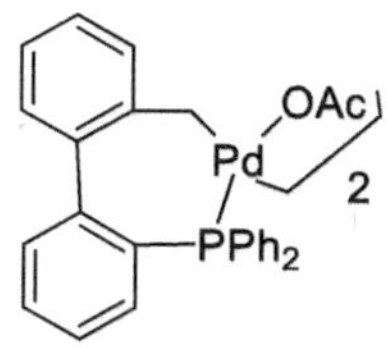

42

Dinuclear phosphorus donor palladacycles are also useful catalysts *(68, 69)*. The palladation of the biphenylphosphine $Ph_2P\{2\text{-}(2\text{-}MeC_6H_4)C_6H_4\}$ gave **42** as a water soluble complex. This palladacycle was remarkably valid for the coupling between aryl halides and phenylboronic acids producing yields of 100% with TON of 100,000. Phosphite palladacycles may be applied to the preparation of fluorenones *(70)*. Thus, compound **43** catalyzes the addition reaction between arylboronic acids and halogen containing aromatic aldehydes that, after cyclization, renders fluorenones and indenofluorenediones. The related palladated phosphites **44** can be utilized for the carbonylative Suzuki-Miyaura cross-coupling of aryl iodides with boronic acids to give biaryl ketones *(71)*. The numerous coupling reactions show turnover numbers and turnover frequencies spreading over 10^6–10^7 and 10^5–10^6 h^{-1}, respectively.

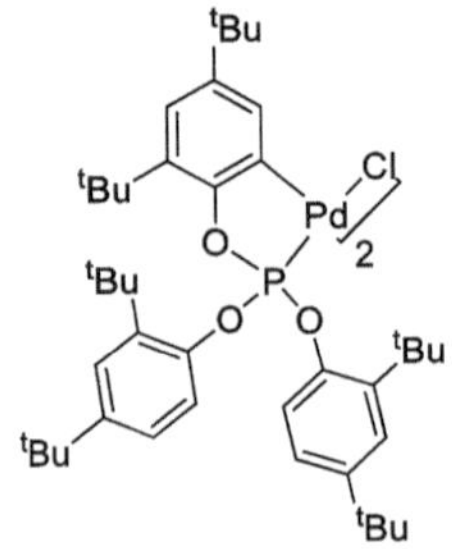

43

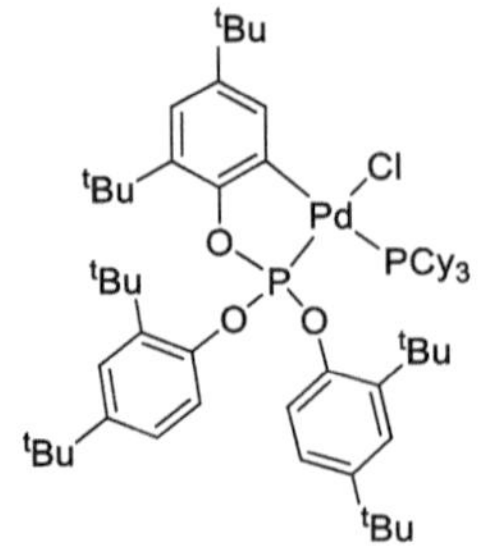

44

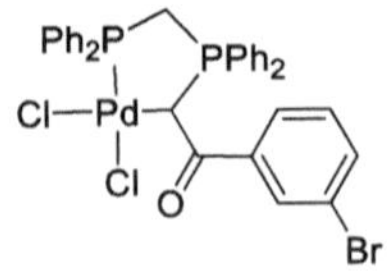

45

Finally, palladacycle ylides **45** *(72, 73)* have been submitted to the Suzuki coupling reaction of aryl chlorides and bromides with good to excellent results; yields ranging *c.*70%–85%.

7. CARBENE PALLADACYCLES

Carbene palladacycles as catalysts for the Suzuki-Miyaura coupling have emerged in the past as a strong challenge to the widely employed phosphine complexes. The carbene ligand is a firm σ donor and the resulting complexes enjoy flexibility and firmness of the molecular architecture. Nolan *(74)* has given a general overview regarding palladium-carbene compounds suitable for cross-coupling reactions, inclusive of a section dedicated to the palladacycles. Since then, a good number of papers have appeared in relation to these catalysts, which we will briefly comment on herein.

Complexes of type **46**, **47** *(75)* and **48** *(76)* have been shown to be first-rate catalysts for the Suzuki coupling with more or less activated aryl bromides against phenylboronic acids even at low catalyst loadings; **46** being more active than **47**, consequent on the greater stability of the former. The methyl groups on the benzimidazole in **48** play a determining role in enhancing the activity of the catalyst. In all cases, yields ranging *c.*85%–95% were achieved.

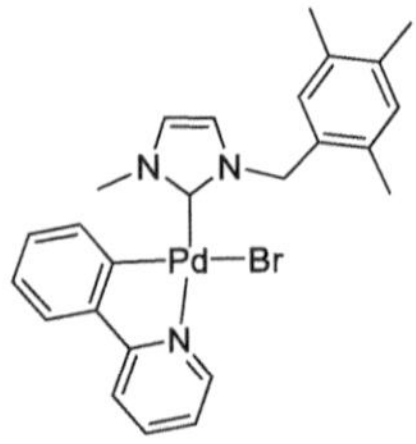

46

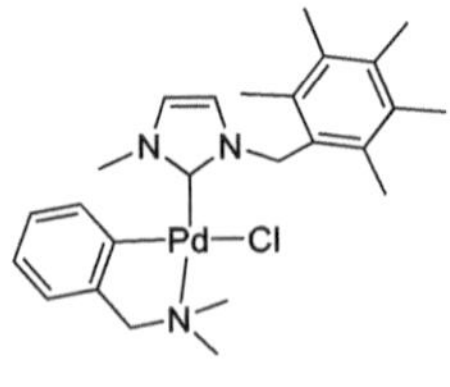

47

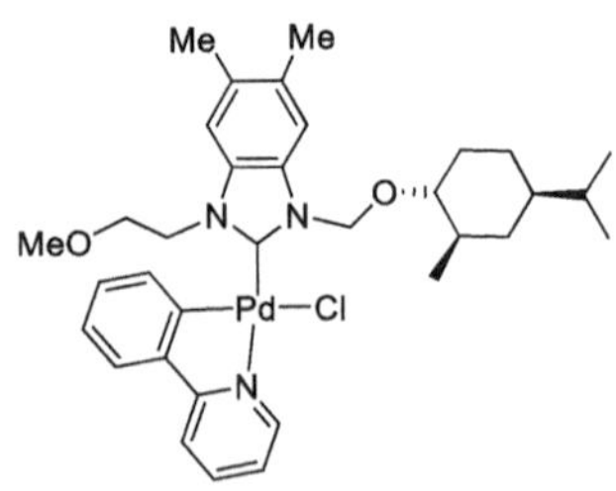

48

Ying et al. *(77)* have described a differing library of palladacycles, for example **49**, prepared in a novel fashion, consisting in the heating of imidazolium salts with adequate acetate palladacycles, for the cross-coupling between deactivated aryl chlorides and phenylboronic acid under smooth reaction conditions. The bulky carbene substituents are paramount in expanding the catalytic ability. Later, Zhang et al. *(78)* implemented this and other palladium complexes on the surface of hydrophobic mesoporous ethane-silica gel **50** making a solid phosphine-free catalyst applicable to the Suzuki coupling even in the case of the less activated aryl chlorides. The inclusion of the azide ligand in the palladacycle **51** *(79)* seems to improve the catalytic quality of the corresponding species, as opposed to the analogous chlorides, with yields *c.*99% in some cases.

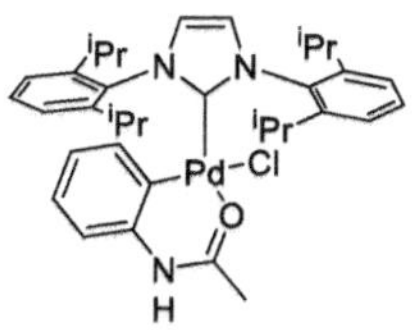

49

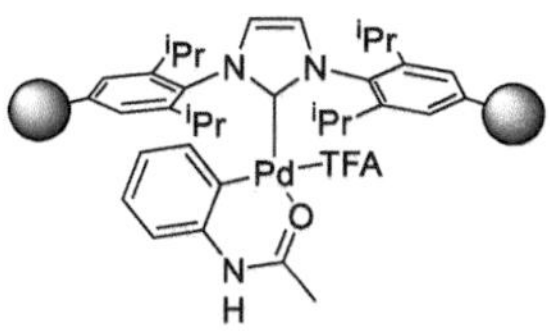

50 — Polymer

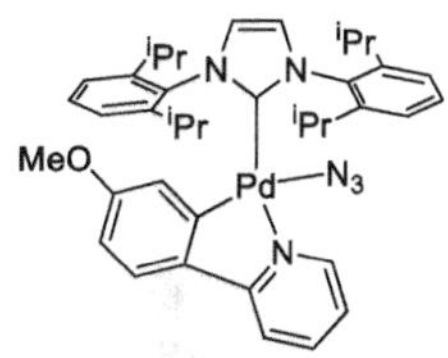

51

Kapdi et al. *(80)* described the preparation of carbene palladacycles **52**, **53** derived from dinuclear hydroxy-bridged palladium complexes, that are most suitable for the arylation of 9-bromophenanthrene with a wide variety of arylboronic acids. The yields in the pertinent conditions are up to 96%.

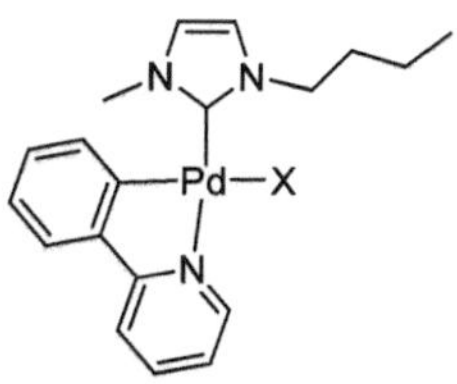

52: X = Cl, Br

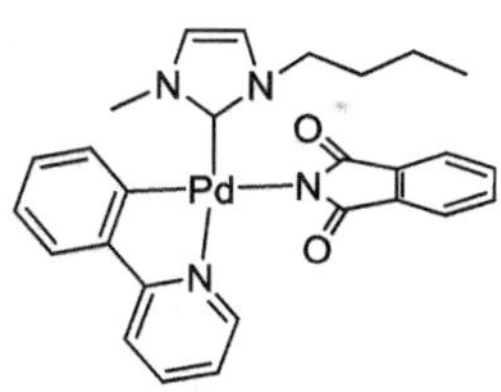

53

8. ALTERNATIVE PALLADACYCLES

Further examples of palladacycles that could well fit into several of the categories described above will be now mentioned. Mono- and dinuclear palladacycles exhibiting [*S*,*C*] **54** and [*S*,*C*,*C*] **55** coordination have been found to be active catalysts *(81)*; the latter present σC^2_{sp}-Pd and σC^3_{sp}-Pd bonds on the same compound. The mononuclear [*S,C*] bonded complexes display greater activity in the Suzuki coupling tests. The related palladacycles derived from benzamidines with pendant thioether groups *(82)*, showing [*C,N,S*] coordination **56,** have also displayed a significant activity for the Suzuki coupling reaction.

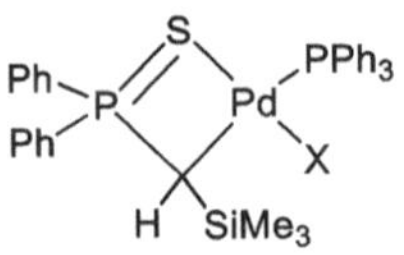

54

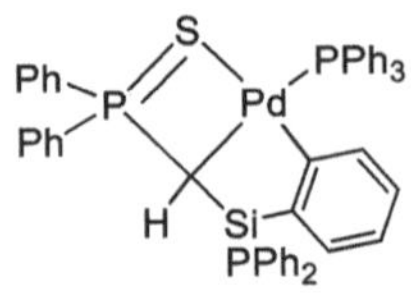

55

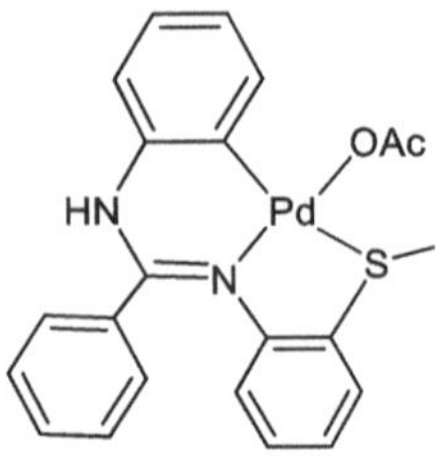

56

Treatment of dinuclear six-membered [*C,N*] guanidine palladacycles with bridging halide ligands (Cl, Br) with dimethyl pyrazole gave mono-nuclear complexes after halide substitution to render active catalysts for the cross-coupling reaction between phenylboronic acid and 4-bromotoluene; namely **57** showed a yield of 100% and TON of 100 *(83)*. Dunina et al. *(84)* studied the competition of two catalytic cycles in the Suzuki coupling process by probing various palladacycles in protic and aprotic media, especially in the cases of metallated cyclophane Schiff base ligands. For instance, complex **58** concurs mainly with the Pd(II)/Pd(0) pathway and leaves the palladacycle framework practically intact.

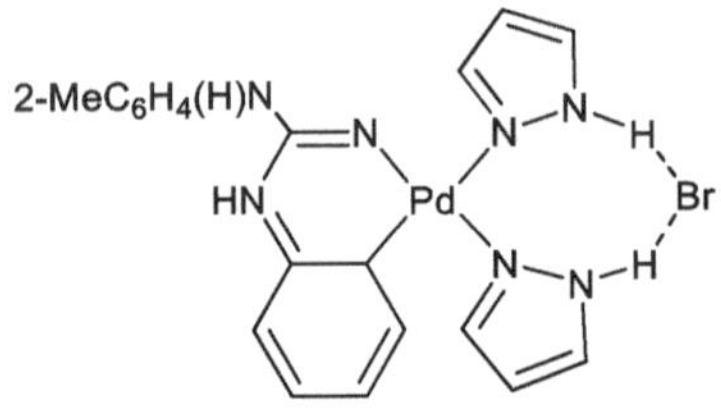

57

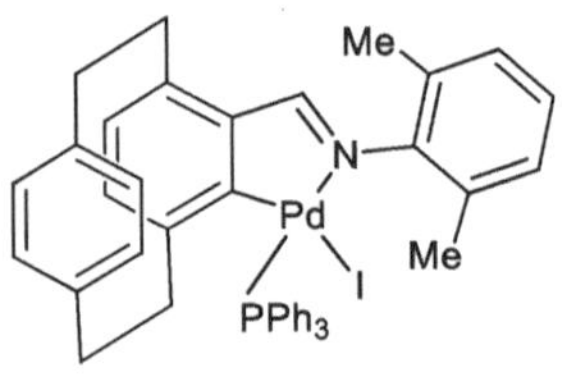

58

Indole based [*C,N,N*] palladacycles *(85)* with a pincer-type arrangement **59** or with spiro chelating rings **60**, possessing one or two metal centers, respectively, or alternatively the terdentate [*C,N,N*] triphenylpyrazole derivatives **61** *(86)*, have been shown to be highly active catalysts.

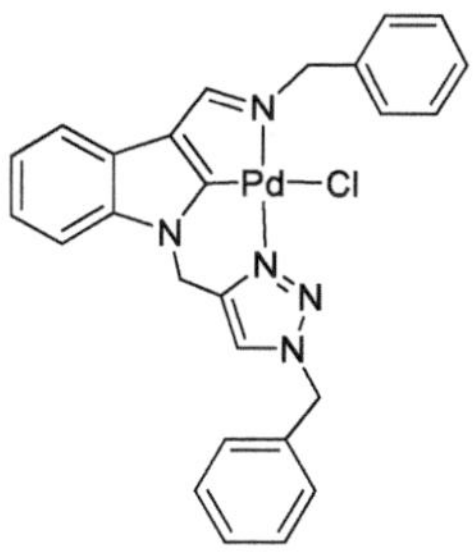

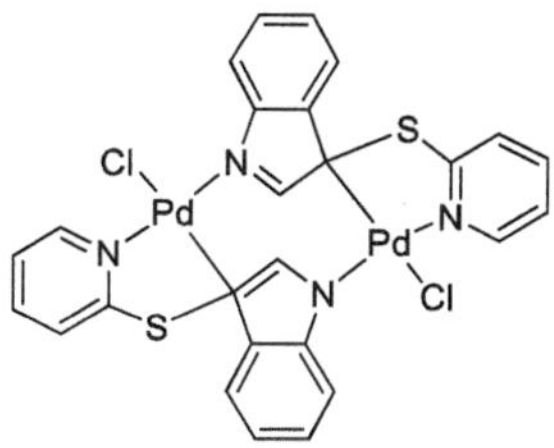

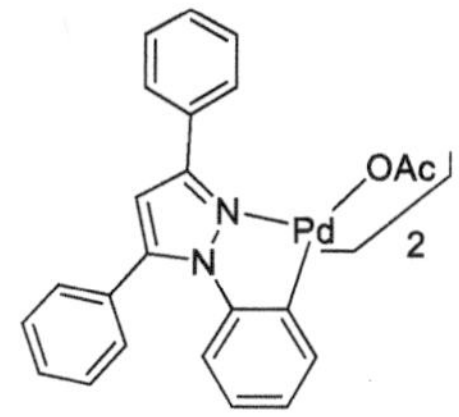

59 60 61

From the foregoing, it can be stated that we are not far from the truth if we say that actually any palladacycle may be used as catalyst more or less effectively in the Suzuki–Miyaura reaction. After all, palladacycles are really precatalysts which liberate Pd(0), or in any case "active palladium" in the media. The question stands as, can we perform this reaction without addition of a palladium source as catalyst? Recently, the Vila group *(87)* has answered this by showing that if the Suzuki-Miyaura carbon-carbon bond formation reaction takes place solely on the palladacycle, it may proceed in the absence of addition of a "traditional" palladium catalyst, such as for example $Pd(PPh_3)_4$, or of an appropriate palladacycle. This suggests that it is the functionalized palladacycle itself **62** that plays the role of the catalyst; it simultaneously holds the precatalyst and the boronic function. The adequately functionalized palladacycle auto-directs an aryl ketone halide producing an autocatalytic operation with formation of the carbon–carbon bond to give **63**, Scheme 1.3.

The catalytic cycles may involve either Pd(0)/Pd(II) or Pd(II)/Pd(IV) mechanism, with the former being more often than not invoked as it implies release of Pd(0) in the reaction medium. However, in this case the initial

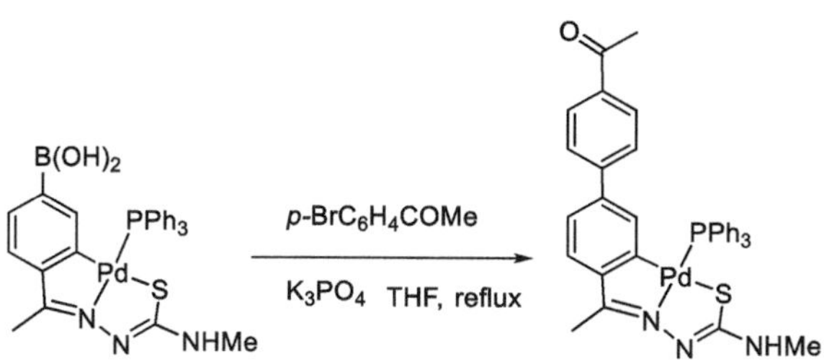

62 63

Scheme 1.3 Formation of **63** via autocatalysis from **62**.

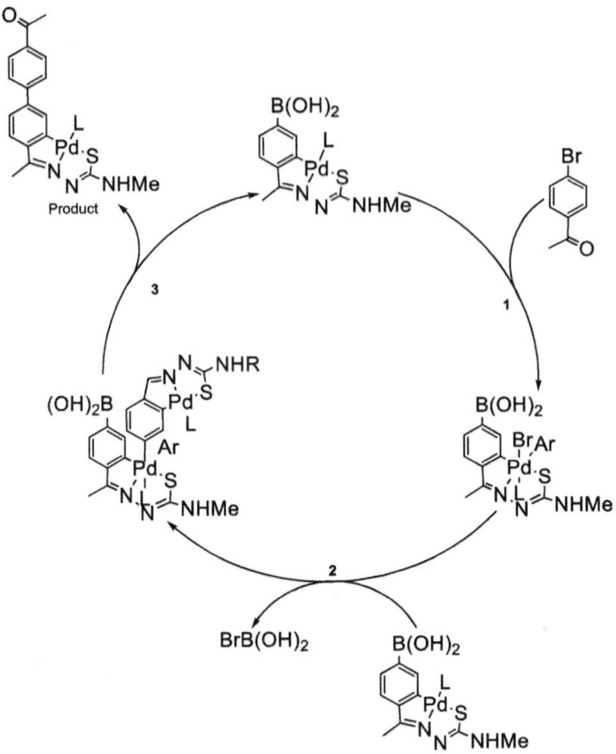

Scheme 1.4 Proposed Pd(II)/Pd(IV) mechanism for the Suzuki-Miyaura cross-coupling of **62** to **63**. 1: Oxidative addition; 2: Transmetallation; 3: Reductive elimination. L = PPh_3.

palladium oxidation state, Pd(II), on the one hand, and the strength of the bonds at the metal imposed by the strongly chelating thiosemicarbazone ligand, on the other, is surely in favor of the latter mechanism. This is what was proposed by Vila et al., and further confirmed by DFT calculations on species in the cycle (Scheme 1.4).

9. CONCLUSIONS

In recent publications that include palladacycles, or other metallic compounds, for that matter, that act as catalysts when cross-coupling reactions are considered, it can be seen that the number of species is increasingly numerous. The figures depicted in this chapter are only the tip of the iceberg in what relates to the tremendous variability of complexes that have been applied to the Suzuki-Miyaura coupling reaction, an endless myriad of more or less efficient catalysts. It would seem, however, that there are much more simple compounds, such as $PdCl_2$, $Pd(AcO)_2$ or $Pd(PPh_3)_4$, which can act as excellent catalysts and which are commercially available, thus avoiding the often tedious and low-yield work of preparing the more complex species mentioned; it would not make much sense to continue researching to obtain compounds of the type discussed above. One author of such species, Beletskaya *(8)* stated some time ago that, in relation to the particular case of palladacycles as catalysts "*after a survey over the application of palladacycles in catalysis it becomes evident that the initial are often announced as outstanding because of very high catalytic activity in several test reactions, very rarely find applications in preparative chemistry. Neither enantioselectivity nor recyclability has been realized. Dozens of palladacycles of all imaginable classes have been studied in various cross-coupling reactions, but none appeared to be the "well-defined catalyst", which has been the initial leitmotiv of this story.*"

The literature illustrates that the search for new catalysts continues to resolve aspects such as the improvement of the effectiveness of the catalyst in terms of its performance and TON, its recoverability, lower working temperature, water soluble catalysts, possibility of application to multiple processes, industrial uses ..., in short, an ideal catalyst that compensates for the synthesis of sophisticated metal compounds, in particular the palladacycles.

REFERENCES

1. Albrecht, M. *Chem. Rev.* **2010**, *110*, 576–623.
2. Dupont, J.; Pfeffer, M. *Palladacycles.* Wiley-VCH: Weinheim, 2008.
3. Dupont, J.; Consorti, C. S.; Spencer, J. *Chem. Rev.* **2005**, *105*, 2527–2571.
4. Ghedini, M.; Aiello, I.; Crispini, A.; Golemme, A.; La Deda, M.; Pucci, D. *Coord. Chem. Rev.* **2006**, *250*, 1373–1390.
5. Omae, I. *Cyclometallation Reactions. Five-Membered Ring Products as Universal Reagents*; Springer-Verlag: New York, 2014.
6. Bedford, R. B. *Chem. Commun.* **2003**, 1787–1796.
7. Beletskaya, I. P.; Cheprakov, A. V. *Chem. Rev.* **2000**, *100*, 3009–3066.
8. Beletskaya, I. P.; Cheprakov, A. V. *J. Organomet. Chem.* **2004**, *689*, 4055–4082.
9. Dupont, J.; Pfeffer, M.; Spencer, J. *Eur. J. Inorg. Chem.* **1917–1927**, *2001.*
10. Littke, A. F.; Fu, G. C. *Angew. Chem. Int. Ed. Eng.* **2002**, *41*, 4176–4211.
11. Weskamp, T.; Böhm, V. P. W.; Herrmann, W. A. *J. Organomet. Chem.* **2000**, *600*, 12–22.
12. Beller, M.; Fischer, H.; Herrmann, W. A.; Öfele, K.; Brossmer, C. *Angew. Chem. Int. Ed. Eng.* **1995**, *34*, 1848–1849.
13. Herrmann, W. A.; Brossmer, C.; Öfele, K.; Reisinger, C. P.; Priermeier, T.; Beller, M.; Fischer, H. *Angew. Chem. Int. Ed. Eng.* **1995**, *34*, 1844–1848.
14. Suzuki, A. *Chem. Commun.* **2005**, 4759–4763.
15. Miyaura, N. A. S. *J. Chem. Soc. Chem. Commun.* **1979**, 866–867.

16. Röhlich, C.; Wirth, A. S.; Köhler, K. *Chem. Eur. J.* **2012**, *18*, 15485–15494.
17. Fihri, A.; Luart, D.; Len, C.; Solhy, A.; Chevrin, C.; Polshettiwar, V. *Dalton Trans.* **2011**, *40*, 3116–3121.
18. Leadbeater, N. E. *Chem. Commun.* **2005**, 2881–2902.
19. Wang, H.; Yang, J. *Appl. Organomet. Chem.* **2016**, *30*, 262–267.
20. Serrano, J. L.; García, L.; Pérez, J.; Pérez, E.; García, J.; Sánchez, G.; Sehnal, P.; De Ornellas, S.; Williams, T. J.; Fairlamb, I. J. S. *Organometallics* **2011**, *30*, 5095–5109.
21. Huang, R.; Shaughnessy, K. H. *Organometallics* **2006**, *25*, 4105–4112.
22. Chen, C. L.; Liu, Y. H.; Peng, S. M.; Liu, S. T. *Organometallics* **2005**, *24*, 1075–1081.
23. Hashemi Fath, R.; Hoseini, S. J. *J. Organomet. Chem.* **2017**, *828*, 16–23.
24. Bulgiyina, L. A.; Khrushcheva, N. S.; Peregudov, A. S.; Sokolov, V. I. *Russ. Chem. Bull. Int. Ed.* **2016**, *65*, 2479–2484.
25. Puls, F.; Richter, N.; Kataeva, O.; Knölker, H. J. *Chem. Eur. J.* **2017**, *23*, 17576–17583.
26. Zhang, H. H.; Xing, C. H.; Tsemo, G. B.; Hu, Q. S. *ACS Macro Lett.* **2013**, *2*, 10–13.
27. Albrecht, M.; Stoeckli-Evans, H. *Chem. Commun.* **2005**, *5*, 4705–4707.
28. Alonso, D. A.; Nájera, C. *Chem. Soc. Rev.* **2010**, *39*, 2891–2902.
29. Alonso, D. A.; Cívicos, J. F.; Nájera, C. *Synlett* **2009**, 3011–3015.
30. Cívicos, J. F.; Alonso, D. A.; Nájera, C. *Adv. Synth. Catal.* **2011**, *353*, 1683–1687.
31. Cívicos, J. F.; Alonso, D. A.; Nájera, C. *Eur. J. Org. Chem.* **2012**, *1*, 3670–3676.
32. Susanto, W.; Chu, C. Y.; Ang, W. J.; Chou, T. C.; Lo, L. C.; Lam, Y. *J. Organomet. Chem.* **2012**, 77, 2729–2742.
33. Cho, H. J.; Jung, S.; Kong, S.; Park, S. J.; Lee, S. M.; Lee, Y. S. *Adv. Synth. Catal.* **2014**, *356*, 1056–1064.
34. Park, S. J.; Cho, H. J.; Lee, S. M.; Lee, Y. S. *Tetrahedron Lett.* **2017**, *58*, 2670–2674.
35. Solodenko, W.; Mennecke, K.; Vogt, C.; Gruhl, S.; Kirschning, A. *Synthesis (Stuttg)* **2006**, 1873–1881.
36. Alacid, E.; Nájera, C. *J. Organomet. Chem.* **2009**, *694*, 1658–1665.
37. Gómez-Martínez, M.; Baeza, A.; Alonso, D. A. *Catalysts* **2017**, 7, 94.
38. Corma, A.; Das, D.; García, H.; Leyva, A. *J. Catal.* **2005**, *229*, 322–331.
39. Nájera, C. *ChemCatChem* **2016**, *8*, 1865–1881.
40. Zhang, J.; Zhao, L.; Song, M.; Mak, T. C. W.; Wu, Y. *J. Organomet. Chem.* **2006**, *691*, 1301–1306.
41. Qian, H.; Zhang, T.; Yin, Z.; Wang, Q.; Yuan, Q.; Jiang, R.; Yan, S.; Zhao, X. *J. Organomet. Chem.* **2016**, *824*, 124–130.
42. Wang, H. X.; Wu, H. F.; Yang, X. L.; Ma, N.; Wan, L. *Polyhedron* **2007**, *26*, 3857–3864.
43. Li, J. Y.; Yu, A. J.; Wu, Y. J.; Zhu, Y.; Du, C. X.; Yang, H. W. *Polyhedron* **2007**, *26*, 2629–2637.
44. Xu, C.; Gong, J. F.; Yue, S. F.; Zhu, Y.; Wu, Y. J. *Dalton Trans.* **2006**, 4730–4739.
45. Xu, C.; Gong, J. F.; Guo, T.; Zhang, Y. H.; Wu, Y. J. *J. Mol. Catal. A Chem.* **2008**, *279*, 69–76.
46. Wang, L.; Cui, X.; Li, J.; Wu, Y.; Zhu, Z.; Wu, Y. *Eur. J. Org. Chem.* **2012**, 595–603.
47. Leng, F.; Wang, Y.; Li, H.; Li, J.; Zou, D.; Wu, Y.; Wu, Y. *Chem. Commun.* **2013**, *49*, 10697–10699.
48. Yu, A.; Shen, L.; Cui, X.; Peng, D.; Wu, Y. *Tetrahedron* **2012**, *68*, 2283–2288.
49. Xu, C.; Gong, J. F.; Zhang, Y. H.; Zhu, Y.; Wu, Y. J. *Aust. J. Chem.* **2007**, *60*, 190–195.
50. Ma, J.; Cui, X.; Zhang, B.; Song, M.; Wu, Y. *Tetrahedron* **2007**, *63*, 5529–5538.
51. Ma, J.; Cui, X.; Gao, L.; Wu, Y. *Inorg. Chem. Commun.* **2007**, *10*, 762–766.
52. Xu, C.; Wang, Z. Q.; Zhang, Y. P.; Dong, X. M.; Hao, X. Q.; Fu, W. J.; Ji, B. M.; Song, M. P. *Eur. J. Inorg. Chem.* **2011**, 4878–4888.

53. Sharma, A. K.; Joshi, H.; Bhaskar, R.; Kumar, S.; Singh, A. K. *Dalton Trans.* **2017**, *46*, 2485–2496.
54. Friedlein, F. K.; Kromm, K.; Hampel, F.; Gladysz, J. A. *Chem. Eur. J.* **2006**, *12*, 5267–5281.
55. Wu, K. M.; Huang, C. A.; Peng, K. F.; Chen, C. T. *Tetrahedron* **2005**, *61*, 9679–9687.
56. Deschamps, B.; Le Goff, X.; Ricard, L.; Le Floch, P. *Heteroat. Chem.* **2007**, *18*, 363–371.
57. Rao, G. K.; Kumar, A.; Ahmed, J.; Singh, A. K. *Chem. Commun.* **2010**, *46*, 5954–5956.
58. Rao, G. K.; Kumar, A.; Kumar, S.; Dupare, U. B.; Singh, A. K. *Organometallics* **2013**, *32*, 2452–2458.
59. Sigeev, A. S.; Peregudov, A. S.; Cheprakov, A. V.; Beletskaya, I. P. *Adv. Synth. Catal.* **2015**, *357*, 417–429.
60. Avila-Sorrosa, A.; Jiménez-Vázquez, H. A.; Reyes-Arellano, A.; Pioquinto-Mendoza, J. R.; Toscano, R. A.; González-Sebastián, L.; Morales-Morales, D. *J. Organomet. Chem.* **2016**, *819*, 69–75.
61. Zhang, B. S.; Wang, C.; Gong, J. F.; Song, M. P. *J. Organomet. Chem.* **2009**, *694*, 2555–2561.
62. Kozlov, V. A.; Aleksanyan, D. V.; Nelyubina, Y. V.; Lyssenko, K. A.; Vasilev, A. A.; Petrovskii, P. V.; Odinets, I. L. *Organometallics* **2010**, *29*, 2054–2062.
63. Jin, L.; Wei, W.; Sun, N.; Hu, B.; Shen, Z.; Hu, X. *Org. Chem. Front.* **2018**, *5*, 2484–2491.
64. Bedford, R. B.; Coles, S. J.; Hursthouse, B.; Scordia, J. M. *Dalton Trans.* **2005**, 991–995.
65. Bedford, R. B.; Betham, M.; Coles, S. J.; Frost, R. M.; Hursthouse, M. B. *Tetrahedron* **2005**, *61*, 9663–9669.
66. Marziale, A. N.; Faul, S. H.; Reiner, T.; Schneider, S.; Eppinger, J. *Green Chem.* **2010**, *12*, 35–38.
67. Marziale, A. N.; Jantke, D.; Faul, S. H.; Reiner, T.; Herdtweck, E.; Eppinger, J. *Green Chem.* **2011**, *13*, 169–177.
68. Joshaghani, M.; Daryanavard, M.; Rafiee, E.; Nadri, S. *J. Organomet. Chem.* **2008**, *693*, 3135–3140.
69. Karami, K.; Ghasemi, M.; Naeini, N. H. *Tetrahedron Lett.* **2013**, *54*, 1352–1355.
70. Liu, T. P.; Liao, Y. X.; Xing, C. H.; Hu, Q. S. *Org. Lett.* **2011**, *13*, 2452–2455.
71. Gautam, P.; Bhanage, B. M. *J. Organomet. Chem.* **2015**, *80*, 7810–7815.
72. Sabounchei, S. J.; Sedghi, A.; Hashemi, A.; Hosseinzadeh, M.; Bayat, M.; Gable, R. W. *Appl. Organomet. Chem.* **2017**, *31*, 1–15.
73. Sabounchei, S. J.; Ahmadi, M.; Panahimehr, M.; Bagherjeri, F. A.; Nasri, Z. *J. Mol. Catal. A Chem.* **2014**, 383–384, 249–259.
74. Marion, N.; Nolan, S. P. *Acc. Chem. Res.* **2008**, *41*, 1440–1449.
75. Emin Günay, M.; Gümüşada, R.; Özdemir, N.; Dinçer, M.; Çetinkaya, B. *J. Organomet. Chem.* **2009**, *694*, 2343–2349.
76. Kilinçarslan, R.; Günay, M. E.; Firinci, R.; Denizalti, S.; Çetinkaya, B. *Appl. Organomet. Chem.* **2016**, *30*, 268–272.
77. Peh, G. R.; Kantchev, E. A. B.; Er, J. C.; Ying, J. Y. *Chem. Eur. J.* **2010**, *16*, 4010–4017.
78. Li, G.; Yang, H.; Li, W.; Zhang, G. *Green Chem.* **2011**, *13*, 2939–2947.
79. Kim, Y. J.; Lee, J. H.; Kim, T.; Ham, J.; Zheng, Z. N.; Lee, S. W. *Eur. J. Inorg. Chem.* **2012**, 6011–6017.
80. Serrano, J. L.; Perez, J.; Garcia, L.; Sanchez, G.; Garcia, J.; Lozano, P.; Zende, V.; Kapdi, A. *Organometallics* **2015**, *34*, 522–533.
81. Molitor, S.; Schwarz, C.; Gessner, V. H. *Organometallics* **2015**, *35*, 159–167.
82. Wang, W. C.; Peng, K. F.; Chen, M. T.; Chen, C. T. *Dalton Trans.* **2012**, *41*, 3022–3029.
83. Agarwal, P.; Thirupathi, N.; Nethaji, M. *Organometallics* **2016**, *35*, 3112–3123.

84. Gorunova, O. N.; Grishin, Y. K.; Ilyin, M. M.; Kochetkov, K. A.; Churakov, A. V.; Kuz'mina, L. G.; Dunina, V. V. *Russ. Chem. Bull.* **2017**, *66*, 282–292.
85. Singh, M. P.; Saleem, F.; Pal, R. S.; Singh, A. K. *New J. Chem.* **2017**, *41*, 11342–11352.
86. Mamidala, R.; Mukundam, V.; Dhanunjayarao, K.; Venkatasubbaiah, K. *Dalton Trans.* **2015**, *44*, 5805–5809.
87. Lucio-Martínez, F.; Adrio, L. A.; Polo-Ces, P.; Ortigueira, J. M.; Fernández, J. J.; Adams, H.; Pereira, M. T.; Vila, J. M. *Dalton Trans.* **2016**, *45*, 17598–17601.

CHAPTER 2

Palladacycles as Precatalysts for Heck and Sonogashira Cross-Coupling Reactions

Pui Ying Choy, Xinwei He, Fuk Yee Kwong
Department of Chemistry, The Chinese University of Hong Kong, Shatin, Hong Kong

Contents

1. Introduction

The palladium-catalyzed Heck reaction was firstly discovered in the late 1960s and remains one of the most valuable and prominent strategies to synthesize complex carbon molecules *(1)*. This reaction aims to connect arene and olefin fragments together via a transition metal-catalyzed reaction

Palladacycles
https://doi.org/10.1016/B978-0-12-815505-9.00003-2
© 2019 Elsevier Inc.
All rights reserved.

pathway. In 2010, Prof. Richard F. Heck was recognized by the Nobel Prize in Chemistry for his pioneering invention and the exceedingly significant contribution of this reaction *(2)*. In fact, the versatility of this cross-coupling process has been demonstrated in the pharmaceutical, agricultural, and electronics synthesis *(3)*. This chemistry offers excellent opportunity in developing new fields of research, and provides a solid platform for convergent organic synthesis *(4)*. Sonogashira cross-coupling reactions is the direct coupling of terminal alkyne C—H bond with electrophilic aryl halides *(5,6)*. This method features a modular approach in preparing an array of diversified aryl alkynes and conjugated enynes, which are important synthetic precursors for a range of pharmaceutically attractive intermediates and subunits of materially valuable organic compounds *(7)*. These two coupling reactions have been extensively studied and their related topics such as the applications, techniques, and mechanism have also been reported *(8)*. Apart from those reviews, there are a few literatures that illustrate a general overview of the catalyst design toward these reactions.

Palladacycles have emerged as a complimentary family of organopalladium catalysts (or catalyst precursors) for many C—C and C—X (where X = N, O, S, P) cross-coupling reactions. The first cyclopalladation was discovered in the 1960s by Cope *(9)* and the palladacyclic catalyst has become one of the most popular catalysts used for the Mizoroki-Heck type reactions in recent years. The general structures of the palladacycles can be categorized into two main classes with two subareas: 1st class) anionic four/six-electron donor palladacycles—(A) the simple palladacycles with one ligand coordination to the palladium species in a κ^2-*L,C* manner to form a stable, five/six membered ring (C-L type) and (B) "pincer" palladacycles in which the metallated carbon is supported by two donors in a κ^3-*L,C-L* manner (L-C-L type); 2nd class) the palladacycles exist as halogen or acetate bridged dimers, with *cis*- or *trans*-conformations.

In the past decade, palladacycles have received increasing attention because of their unique structure, chirality in the metallocene series, and their superior catalytic activities in many organic transformations *(10)*. The landmark employment of palladacycles in cross-coupling reactions occurred in the late 1990s with the introduction of Herrmann-Beller's catalyst. Since then, a wide variety of palladacycles have been reported, such as phosphorus-, ferrocenyl-, nitrogen-, NHC-, pincer-, and sulfur-based palladacycles. For brevity, we herein focus on the recent developments in the past decade with these palladacycles in the Heck and Sonogashira cross-coupling reactions. Not surprisingly, there is overlap between sections;

many publications reported more than one class of palladacycles and many palladacycles contain more than one class of structures (e.g., both P and N-donor atom exist in the same palladacycle). Numerous reviews and book chapters have been written covering the literature, and a selection of the publications will be appraised.

2. Phosphine-Based Palladacycles

Palladacycles serving as the catalysts in cross-coupling reactions appeared in 1995 *(11)*. The Herrmann-Beller's catalyst (**1**) is a very effective catalyst/promoter for the Heck and Sonogashira coupling reactions. With the successful examples of **1**, different phosphorous-derived palladacycles were developed and named "phosphapalladacycles" *(12)*. There are two major approaches for modifying phosphapalladacycles: (1) modification of the electronic and steric properties; and (2) alternation of the nature of the auxiliary ligands in the coordination environment of palladium *(13)*. In general, the palladium is bound directly to one or two phosphorus(III) atom(s) and to one carbon-anionic center (bi- or tridentate ligand chelation). They are simply illustrated as PC- and PCP-type cyclopalladated complexes, respectively.

2.1 Heck Cross-Coupling Reaction

2.1.1 Herrmann-Beller-Type Palladacycles

The representative "Herrmann-Beller's catalyst" exists in the form of acetate-bridged dimer. It is capable of dealing with the Heck reactions between a wide variety of aryl halides and styrene/acrylate (Table 2.1). TBAB was added to further promote the coupling reaction when difficult aryl chlorides were used as the coupling partners. LiBr, $[AsPh_4]Cl$, $[PPh_4]Cl$ and $Bu_4NCl{\cdot}H_2O$ are other alternatives of additives *(16)*. Although the catalyst system was able to facilitate this reaction, high reaction temperature (130–160°C) was found to be necessary.

Detailed mechanistic studies and kinetic modeling investigations were carried out by Böhm, Herrmann *(18)* and Blackmond, and Pfaltz *(19)*. The reaction rate is highly dependent on the catalyst preactivation step. The kinetic model of the Heck coupling between bromobenzaldehyde and *n*-butyl acrylate showed that the catalytic efficiency increased at lower catalyst concentration. Later, de Vries further indicated that Pd(0) black was formed at high Pd concentration. A kinetic study was carried out to compare the kinetic behavior of palladacycles and a homeopathic ligand-free palladium catalytic system. The results showed that the resting state of the catalyst within the

Table 2.1 Herrmann-Beller's catalyst (**1**) in Heck reaction

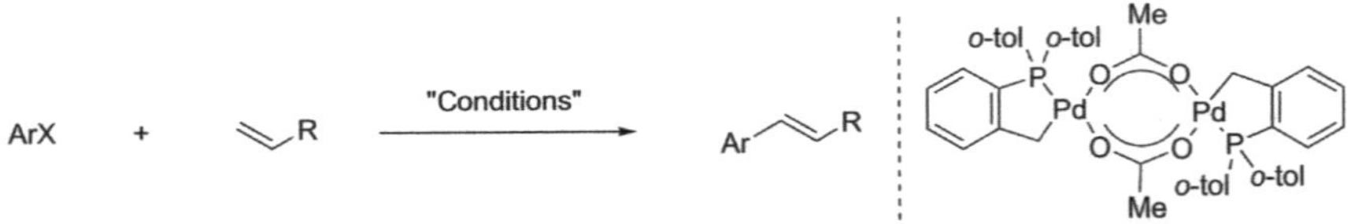

1 (Herrmann-Beller's catalyst)

Entry	1 (mol% Pd)	ArX	R	Time (h)	Solvent	Additive	Yield (%)	Ref.
1	2.0	C_6H_5Br	CO_2*n*-Bu	48	DMA	—	96	11
2	1.0×10^{-3}	*p*-$CHOC_6H_4Br$	CO_2*n*-Bu	12	DMA	—	>99	11
3	2.0	*o*-MeC_6H_4Br	CO_2*n*-Bu	48	DMA	—	92	14
4	0.1	*p*-*n*-BuC_6H_4Br	Ph	15	TBAB	—	83	15
5	0.2	*p*-$CHOC_6H_4Cl$	CO_2*n*-Bu	24	DMA	TBAB	81	11
6	0.1	*p*-CNC_6H_4Cl	Ph	68	DMA	TBAB	48	14
7	1.0	*p*-$CF_3C_6H_4Cl$	Ph	15	TBAB	LiBr	91	15
8	2.0	C_6H_5Cl	Ph	16	TBAB	$[AsPh_4]Cl$	96	16
9	4.0	*p*-$MeOC_6H_4Cl$	Ph	14	TBAB	$[PPh_4]Cl$	18	16
10	0.2	*o*-CNC_6H_4Cl	Ph	24	DMA	$Bu_4NCl{\cdot}H_2O$	94	17

catalytic cycle is the intermediate derived from oxidative addition, and the proposed catalytic cycle is depicted in Fig. 2.1 *(20)*. In general, the bromoarenes undergo an acetate/bromide exchange at the beginning of the coupling reaction and a Pd(II)–Pd(IV) cycle was proposed *(11)*. In 2001, a detailed investigation was carried out about the influence of the reaction conditions in the Heck coupling of 1-chloro-4-trifluormethylbenzene using palladacycle **1** *(17)*. The basicity, additive, and solvent are all important factors to the success of this reaction. Other kinetic studies were further performed and they were in good agreement with the mechanism suggested by Blackmond and Pfaltz *(21)*.

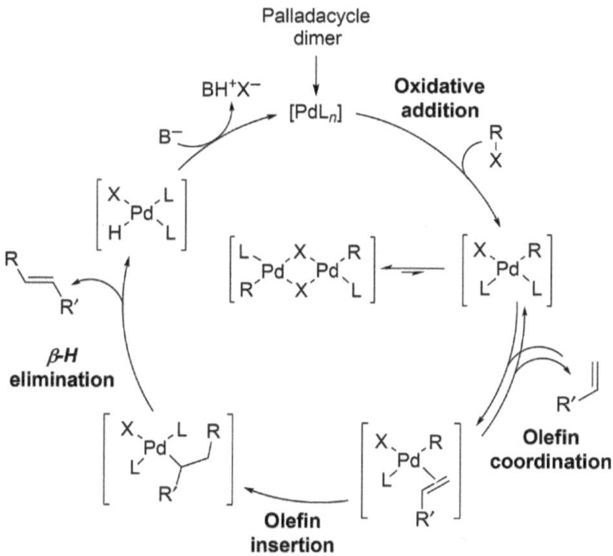

Fig. 2.1 Proposed catalytic cycle by Blackmond and Pfaltz.

Methacrylic acid derivatives are common reagents for the Heck reactions. Beller and co-workers have pioneered the use of **1** for the coupling of aryl bromides with the alkyl methacrylate. With the standard reaction conditions, a mixture of products was formed including the C=C double bond isomers as well as the doubly arylated products. By detailed screening, the selectivity of the reaction is highly influenced by the nature of the base *(22)*. The same group further tested the selectivity of Heck reactions of aryl bromides with cyclohexene in 1999 (Scheme 2.1) *(23)*. Mixture of isomers was obtained and it showed that the double-bond migration was catalyzed slowly by the base but not by HPdX complex.

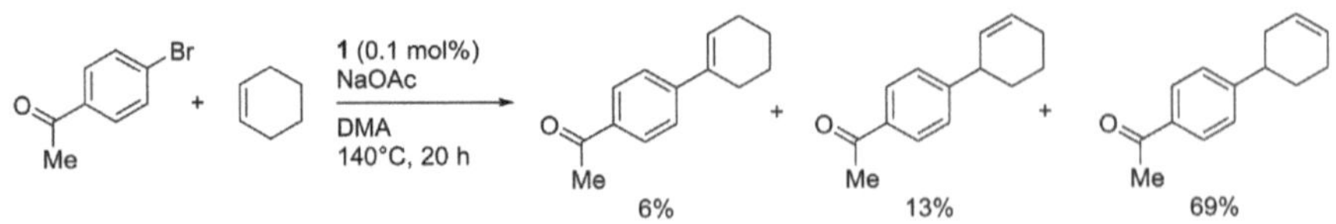

Scheme 2.1 Mixture of isomers formed by Heck reaction between aryl bromide and cyclohexene.

Herrmann-Beller's catalyst is a well-established precatalyst for the general Heck coupling reactions. Recently, chemists moved forward to employ this notable catalyst to other interesting Heck-related reactions. An inter- and intramolecular double Heck coupling of bridged *o,o′*-dibromobiaryls with ethyl acrylate was demonstrated in the presence of Herrmann-Beller catalyst (Scheme 2.2) *(24)*.

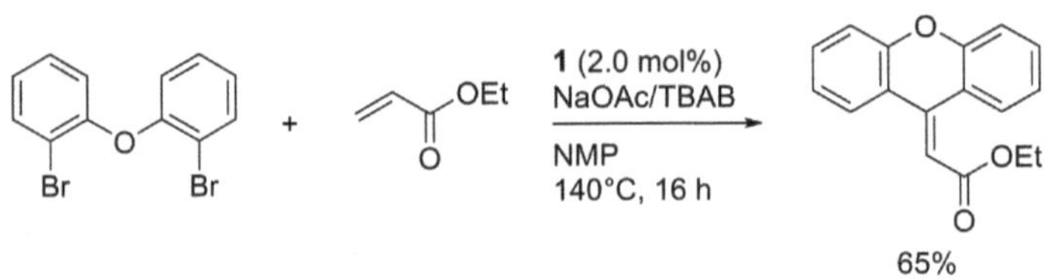

Scheme 2.2 Double Heck reaction of bridged *o,o′*-dibromobiaryls.

Alkyl acrylates are usually employed as the model substrates for initial investigation of the Heck process. Djakovitch group expanded the scope of this reaction to acrolein *(25)*. Selective cinnamyl derivatives or double-arylated products could be afforded depending on the catalyst, the base, and the solvent used. In the presence of **1**, NaOAc as the base and DMA as solvent, 33% of the mono-arylated products in conjunction with 25% of the β,β-diarylated products were given. However, 89% of the mono-arylated products and only 5% of diarylated products resulted when NMP solvent was used. Djakovitch further applied palladacycle **1** in the Heck arylation of diethyl vinylphosphonate. It is an important exploration that polymerization of acrolein often results instead of the desired products by using common palladium catalyst system. Acrolein diethyl acetal is alternatively used instead of acrolein to avoid possible polymerization problems. Saturated esters or α,β-unsaturated aldehydes could be given as the products (Scheme 2.3). Interestingly, when aryl iodide was coupled with acrolein diethyl acetal using the same catalytic system, except water was used as the solvent, no saturated esters but only α,β-unsaturated aldehyde was afforded *(26)*. In 2014, Larhed demonstrated a microwave-heated continuous-flow system for the Heck coupling of aryl bromides and substituted *n*-butyl acrylate using palladacycle **1** *(27)*. It was noted even for large-scale synthesis.

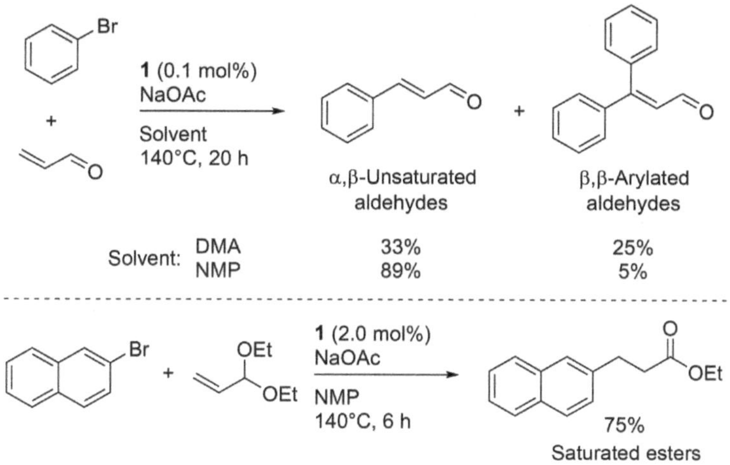

Scheme 2.3 Herrmann-Beller's catalyst in Heck arylation of acrolein and its derivatives.

Herrmann-Beller's catalyst is a representative precatalyst which can be used for genuine application of organic synthesis. Through an intra- or intermolecular or double Heck coupling, intermediates of some useful materials can be accessed for further transformations (Scheme 2.4) *(28)*. Fluorous ponytail-substituted aromatics are useful building blocks for the construction of functionalized soluble ligands. In order to synthesize these fluorous ligands with ethylene spacers, the Heck reaction using Hermann-Beller's catalyst is one of the possible methods *(29)*. Sjöberg and co-workers employed phosphapalladacycle **1** in the vinylation of 2-iodo-*p*-carborane which can be used as hydrophobic pharmacophores *(30)*. The catalyst **1** allowed the Heck arylation at C-8 of adenine nucleosides which is an important medicinal and biological active unit *(31)*. A small amount of *cis*-isomer was detected and the *trans*-olefin was afforded as the major product. An intramolecular Heck cyclization of protected amines was successfully achieved with **1** for synthesizing useful phenanthridine-incorporated alkaloid frameworks *(32)*. A highly diastereoselective construction of tetracyclic steroid core was established with a two-fold Heck reaction and the unnatural *cis*-product can be further converted into orally contraceptive desogestrel by common methods *(33)*. The Stewart group reported the synthesis of *N*-heterocycles by a series of domino reactions having the key Heck reaction with Herrmann-Beller's catalyst system *(34)*.

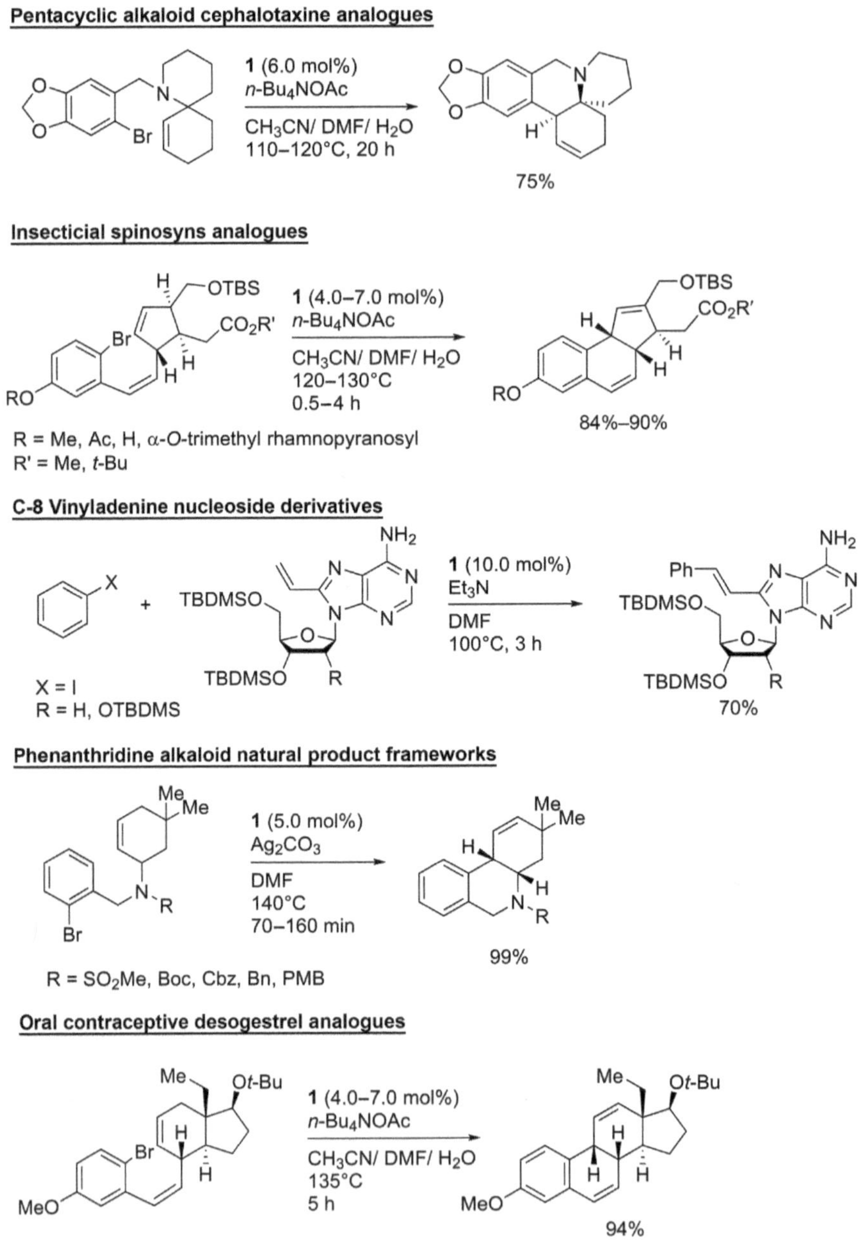

Scheme 2.4 Selected applications of Herrmann-Beller's catalyst in organic synthesis.

Some closely related analogues of **1**, such as dimeric phosphapalladacycle **2–14** were also reported and applied in the Heck reactions (Table 2.2; Fig. 2.2) *(11,14,35)*. In 2008, Fu and Guo carried out a DFT study and proposed two main pathways for the precatalyst activation. The computational results indicated that the phosphapalladacycle **2** was activated via an olefin insertion into a Pd-C bond followed by the β-hydride elimination, and subsequent HBr abstraction under basic medium (Scheme 2.5) *(36)*.

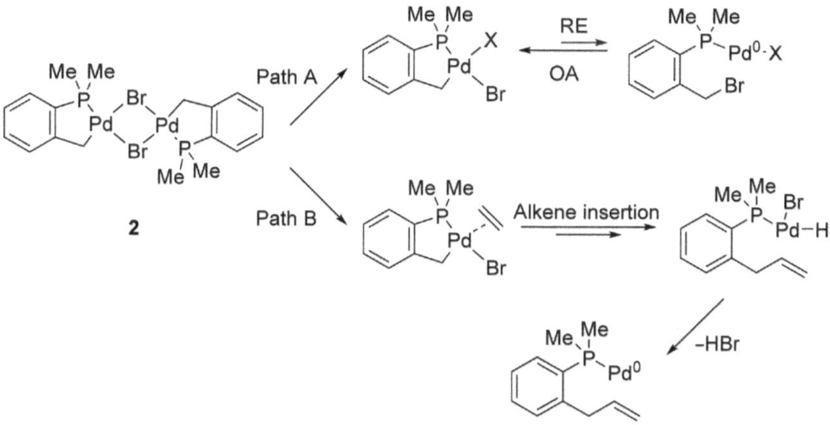

Scheme 2.5 Two possible mechanisms for the activation of palladacycle **2** in Heck coupling.

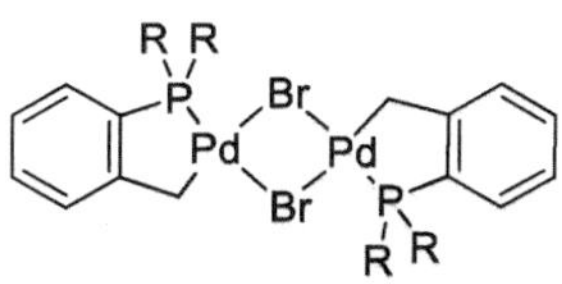

2 R = Me
3 R = *o*-tolyl
4 R = mesityl
5 R = Ph
6 R = *t*-Bu

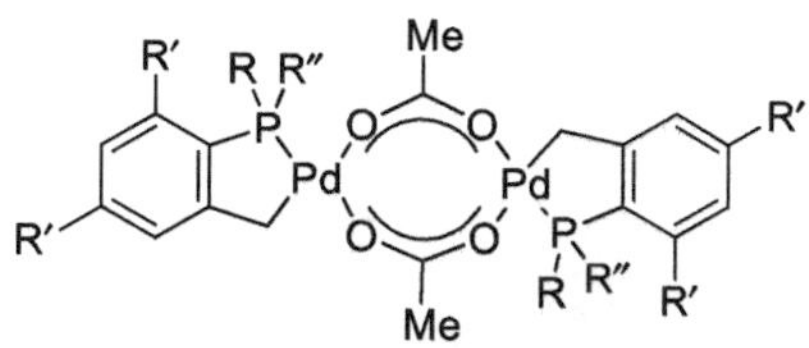

7 R = R″ = mesityl, R′ = Me
8 R = *o*-tolyl, R″ = *t*-Bu, R′ = H
9 R = R″ = *t*-Bu, R′ = H
10 R = *o*-Tolyl, R″ = Cy, R′ = H
11 R = R″ = Cy, R′ = H

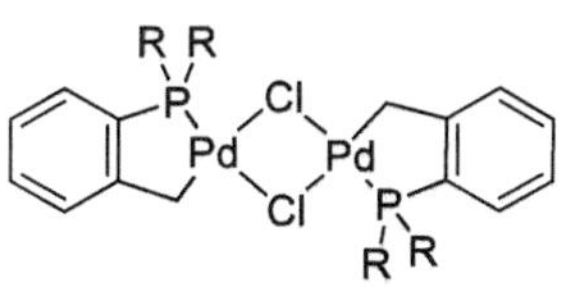

12 R = *o*-tolyl
13 R = mesityl

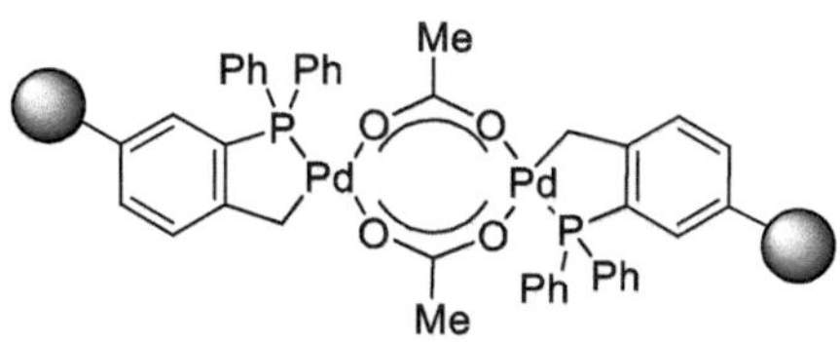

14 ● = polystyrene-supported

Fig. 2.2 Examples of dimeric phosphapalladacycles.

Table 2.2 Dimeric Herrmann-Beller-type palladacycles in Heck reaction

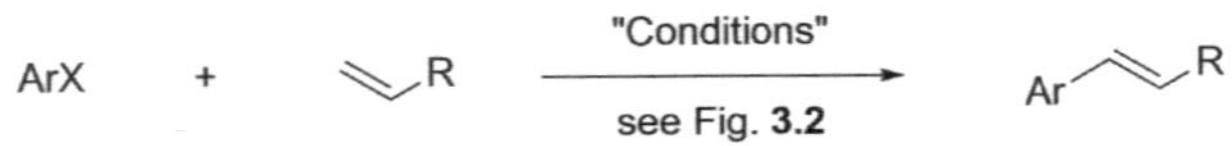

Entry	Palladacycle (mol% Pd)	ArX	R	Solvent	Time (h)	Temp. (°C)	Yield (%)	Ref.
1	**4** (2.0)	p-$MeOC_6H_4Br$	CO_2n-Bu	DMA	44	138	74	11, 14
2[a]	**4** (0.2)	p-$CHOC_6H_4Cl$	CO_2n-Bu	DMA	24	130	40	11, 14
3	**5** (2.0×10^{-4})	p-$MeC(O)C_6H_4Br$	CO_2n-Bu	DMA	20	170	85	35
4	**6** (2.0×10^{-2})	p-$MeC(O)C_6H_4Br$	CO_2n-Bu	DMA	24	130	98	35
5	**14** (10.0)	C_6H_5Br	CO_2Me	DMA	48	100	97	37
6	**14** (0.4)	o-$CHOC_6H_4Br$	CO_2Me	DMA	10	100	99	37
7	**15** (0.1)	p-FC_6H_4Br	Ph	DMA	20	140	3	38

[a]LiBr was added as an additive.

Recoverable and hence reusable catalytic systems are attractive means for cross-coupling reactions. Herrmann-Beller's catalyst anchored on a polystyrene-supported matrix **14** *(37)* or entrapped in zeolites NaY **15** *(38)* were prepared and examined as the possible catalysts for Heck and Sonogashira reactions of aryl iodides and bromides. For the Herrmann-Beller's catalyst entrapped in the zeolites NaY, a very low activity was obtained with only 3% GC yield. It is believed that the steric hindrance of the complexes in the zeolite super cage limited the diffusion of the educts and/or products and the accessibility of the metal center for the educts.

Multicomponent reactions are an attractive protocol for synthesizing complex molecules. In 2012, the Malinakova group reported an investigation of phosphapalladacycles **16–18** in the three-component coupling of aryl-boronic acids, allenes and imines for accessing the homoallylic amines (Scheme 2.6) *(39)*. The P-heteroatom offers back-bonding via the d-orbitals and diminished the electron density at the Pd(II) center in ^{1}-bonded allyl-Pd(II) intermediate and thus resulted in low product yield. Phosphapalladacycle **18** gave a very low catalytic activity when compared to the traditional $Pd(OAc)_2/P(t\text{-}Bu)_3$ catalytic system. It is expected that the four-membered cyclopalladated ligands formation should be facile.

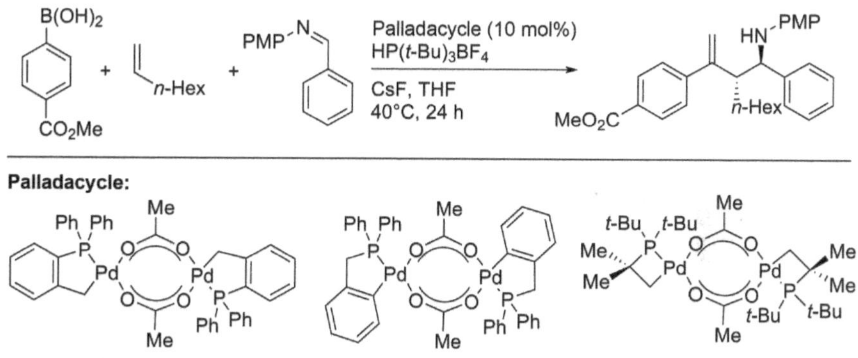

16 16% **17** 34% **18** 29%

Scheme 2.6 Three-components Heck reaction of arylboronic acids, allenes, and imines.

Monomeric form of Herrmann-Beller's catalyst was also effective for the Heck reaction (Table 2.3). The acetylacetonate palladacycles were prepared and they showed excellent performance in the Heck coupling of aryl bromides *(40,41)*. Unfortunately, only 1%–2% product yield could be obtained when chlorobenzene was used.

Table 2.3 Monomeric-form Herrmann-Beller-type palladacycles in Heck reaction

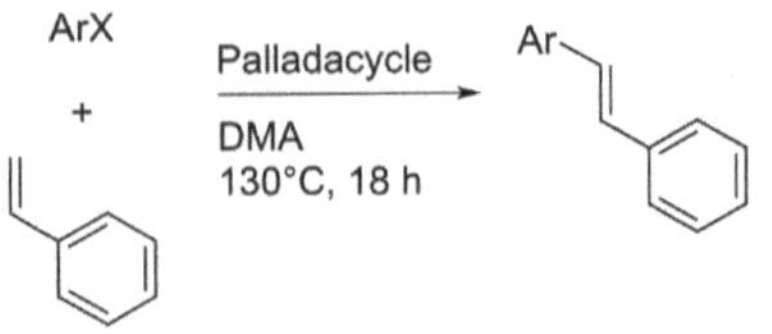

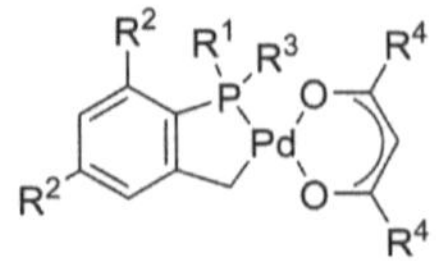

19 $R^1 = R^3$ = *o*-tolyl, R^2 = H, R^4 = Me
20 $R^1 = R^3$ = *o*-tolyl, R^2 = H, $R^4 = CF_3$
21 $R^1 = R^3$ = mesityl, R^2 = Me, R^4 = Me
22 R^1 = *o*-tolyl, R^2 = H, R^3 = Cy, R^4 = Me
23 R^1 = *o*-tolyl, R^2 = H, R^3 = *t*-Bu, R^4 = Me
24 $R^1 = R^3$ = *t*-Bu, R^2 = H, R^4 = Me
25 $R^1 = R^3$ = Cy, R^2 = H, R^4 = Me

Entry	Palladacycle (mol% Pd)	ArX	Yield (%)	Ref.
1	**19** (0.35)	C_6H_5I	83	40
2	**20** (0.1)	C_6H_5I	90	40
3	**21** (1.0)	C_6H_5Br	80	41
4	**22** (1.0)	C_6H_5Br	94	41
5	**23** (1.0)	C_6H_5Br	86	41
6	**24** (1.0)	C_6H_5Br	94	41
7	**25** (1.0)	C_6H_5Br	90	41
8	**21** (0.1)	$MeC(O)C_6H_4Cl$	51	41
9	**22** (0.1)	$MeC(O)C_6H_4Cl$	8	41
10	**23** (0.1)	$MeC(O)C_6H_4Cl$	50	41
11	**24** (0.1)	$MeC(O)C_6H_4Cl$	34	41
12	**25** (0.1)	$MeC(O)C_6H_4Cl$	38	41

N-Heterocyclic carbenes (NHC) are well-known electron-rich, neutral σ-donor ligands in transition metal chemistry. By combining the features of NHC into the phosphapalladacycle, an improved catalyst system was developed. The mono-carbene-substituted complex with various bulky group attached **26–30** gave different effects to the catalyst efficacy in the Heck reaction of activated aryl bromide/chloride with styrene (Scheme 2.7) *(42)*.

2.1.2 Other Dimeric and Monomeric Phosphapalladacycles

Inspired by the initial groundwork of the Herrmann-Beller's catalyst, subsequent development of other related phosphapalladacycles for Heck coupling reaction continues. Several monomeric and dimeric palladacycles were designed and are depicted in Fig. 2.3.

Phosphapalladacycle **46** can be easily synthesized from 2-diphenylphosphino-2′-methylbiphenyl **(L1)** by Pd-catalyzed reaction. It is designed as a "mixed catalyst" in which it combines Herrmann's framework with a sterically demanding biphenyl moiety. The performance of **46** and the conventional

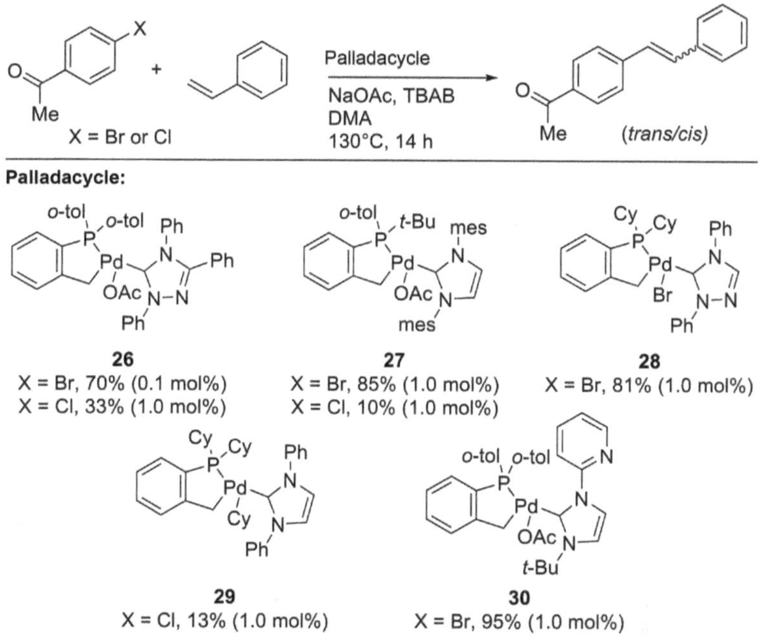

Scheme 2.7 NHC-phosphapalladacycles with Herrmann's catalyst skeleton in Heck reaction.

biphenyl phosphine ligand **L1** were compared in Heck coupling of different substituted olefin. Catalyst **46** gave both terminal and internal products when 1,1-disubtituted olefins were coupled with aryl bromides, yet **L1** gave terminal products as the main product *(46)* (Table 2.4, Entry 17).

In 2011, the Matano group reported examples of porphyrin-appended phosphapalladacycles **40–43,** and studied the effect of the central metals on the catalytic activity in Heck reactions (Fig. 2.3, Table 2.4, Entries 11–14) *(44)*. When the central metal was Ni, its catalytic activity was better than the Zn- and the free-base-one. The results were found not to be simply correlated with the σ-donating ability of the P-donor and the electron-richness of the porphyrin π-systems.

A water-soluble palladacycle aqua complex **51** was designed for the Heck coupling ofbenzoic acid derivatives in water (Scheme 2.8) *(48,49)*. The reaction was found to be highly dependent on the pH environment. When the pH value was larger than 11.5, a decrease in the catalytic reactivity was observed due to the deprotonation of the H_2O ligand on **51**. The reaction also proceeded smoothly when aryl bromides were used (51% for Br, 70% for I).

Kwong and co-workers developed an indolyl-phosphine ligand **CM-phos** for cross-coupling reactions. The palladium complex derived from **CM-phos** and $Pd(OAc)_2$ was found in a dimeric form with two acetate

Monomeric palladacycles

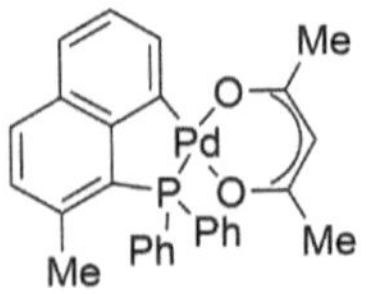

31

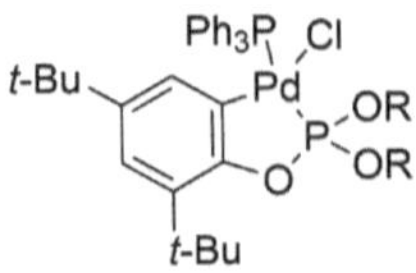

32 R = 2,4-*t*-Bu$_2$Ph

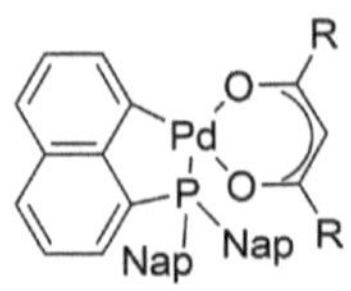

33 R = CF_3
34 R = Me

Dimeric palladacycles

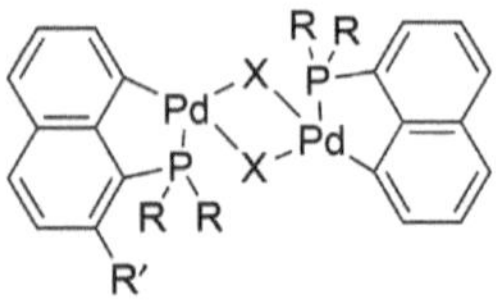

35 R = naphthyl, R′ = H, X = OAc
36 R = naphthyl, R′ = H, X = I
37 R = naphthyl, R′ = H, X = Br
38 R = Ph, R′ = Me, X = OAc

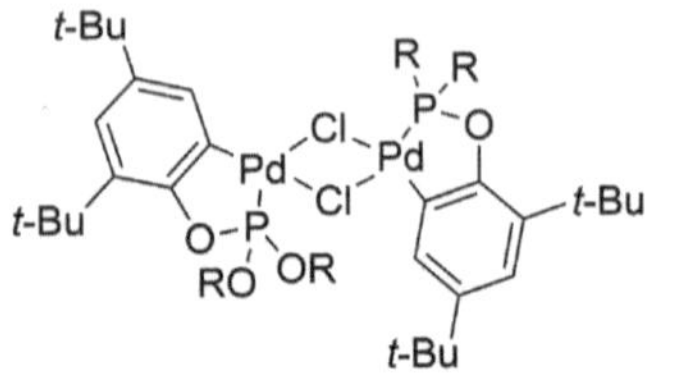

39 R = 2,4-*t*-Bu$_2$Ph

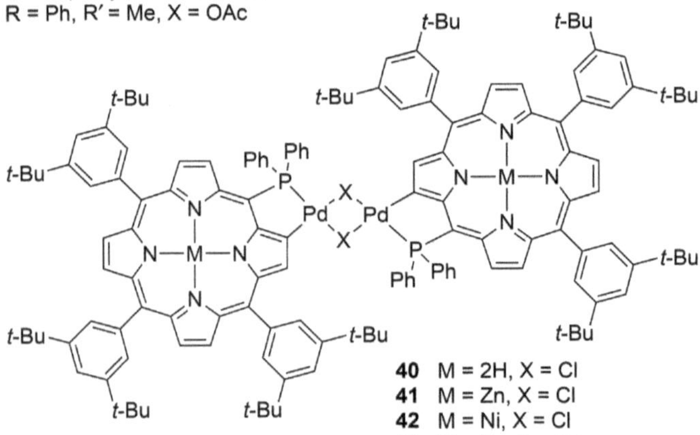

40 M = 2H, X = Cl
41 M = Zn, X = Cl
42 M = Ni, X = Cl
43 M = Zn, X = OAc

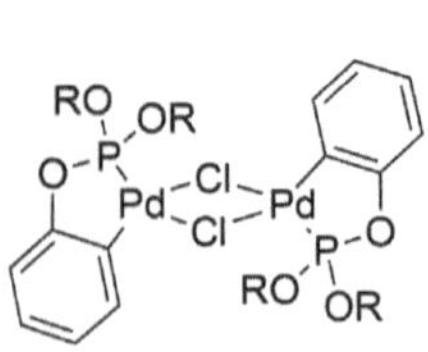

44 R = Ph
45 R = 2,4-*t*-BuPh

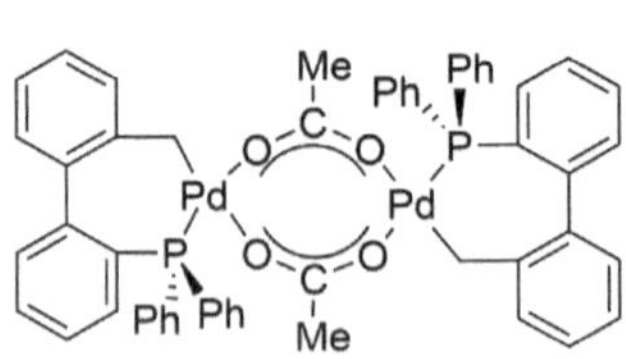

46

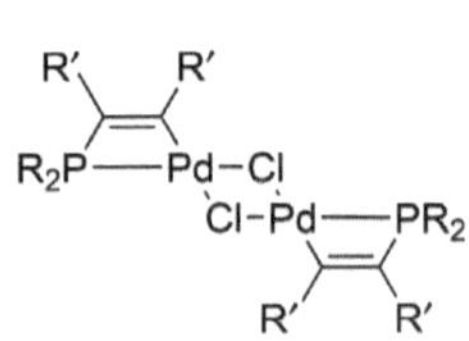

47 R = R′ = Ph
48 R = Ph, R′ = *p*-tol
49 R = *i*-Pr, R′ = Ph
50 R = Ph, R′ = *n*-Bu

Fig. 2.3 Examples of phosphapalladacycles.

Table 2.4 Other phosphapalladacycles in Heck reaction

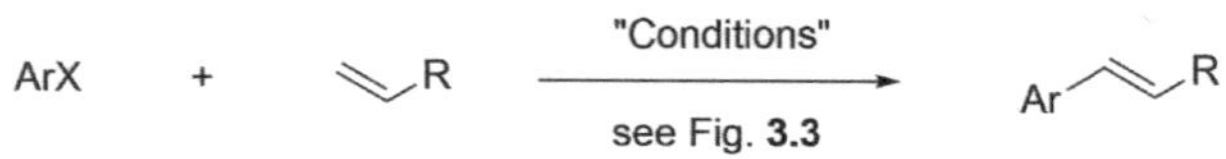

Entry	Palladacycle (mol% Pd)	ArX	R	Solvent	Time (h)	Temp. (°C)	Yield (%)	Ref.
1	**31** (0.1)	C_6H_5Br	Ph	DMA	19	130	79	41
2	**32** (0.2)	*p*-MeC(O)C_6H_4Br	CO_2*n*-Bu	DMA	13	140	100	43
3	**33** (0.1)	C_6H_5Br	Ph	DMF	30	115	77	40
4	**34** (0.5)	C_6H_5I	Ph	DMF	24	95	56	40
5	**35** (0.26)	*p*-MeC(O)C_6H_4Br	Ph	DMF	7	125	94	40
6	**35** (0.26)	*p*-CNC_6H_4Br	Ph	DMF	7	125	85	40
7[a]	**38** (0.2)	*p*-MeOC_6H_4Br	Ph	DMA	18	130	85	41
8	**38** (0.2)	*p*-MeC(O)C_6H_4Cl	Ph	DMA	14	130	29	41
9	**39** (1.0×10^{-4})	*p*-MeC(O)C_6H_4Br	CO_2*n*-Bu	DMA	6	180	100	43
10	**39** (1.0)	C_6H_5Cl	Ph	TBAB	17	150	45	16
11	**40** (0.1)	*p*-CHOC_6H_4Br	CO_2*n*-Bu	DMA	7	135	100	44
12	**41** (0.1)	*p*-CHOC_6H_4Br	CO_2*n*-Bu	DMA	8	135	<20	44
13	**42** (0.1)	*p*-CHOC_6H_4Br	CO_2*n*-Bu	DMA	8	135	<20	44
14	**43** (0.1)	*p*-CHOC_6H_4Br	CO_2*n*-Bu	DMA	9	135	93	44
15	**44** (1.0)	C_6H_5Cl	Ph	DMF	4	130	52	45
16	**44** (1.0)	*p*-$NO_2C_6H_4Cl$	Ph	DMF	4	130	95	45
17	**46** (0.2)	C_6H_5Br	Ph	DMA	1	130	60	46
18	**47** (1.0×10^{-2})	C_6H_5Br	Ph	DMF	10	140	89	47
19	**48** (0.1)	C_6H_5Br	Ph	DMF	5	140	89	47
20	**49** (0.1)	C_6H_5Br	Ph	DMF	5	140	92	47
21	**50** (0.1)	C_6H_5Br	Ph	DMF	5	140	63	47

[a]TBAB was added as an additive.

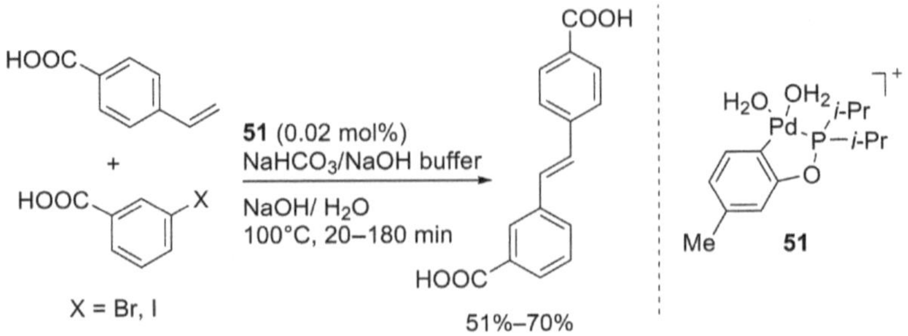

Scheme 2.8 Water-soluble phosphapalladacycle **51** in Heck reaction.

bridging groups (Fig. 2.4) *(50)*. The six-membered cyclometallated ring is puckered and it was employed in various cross-coupling reactions with excellent catalytic efficacy.

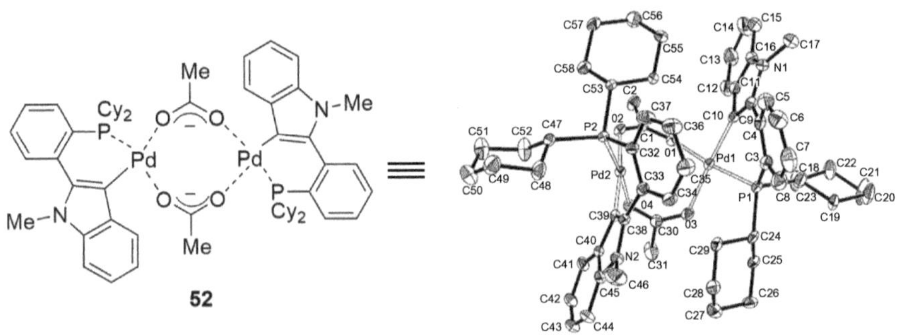

Fig. 2.4 Indolyl-phosphine palladacycle **52** derived from **CM-phos**.

2.1.3 Diphosphine-Palladacycles

Chelating bisphosphine-Pd-dihalide complexes are good catalysts for Heck reaction. Different diphosphine-containing palladacycles were developed (Fig. 2.5; Table 2.5) *(51)*. Phosphapalladacycle **56–59** were designed with different length of the carbon chain between the P-donor *(52)*. Interestingly, the longer the carbon chain, the lower the reactivity of the catalyst in the Heck reaction. For the diphosphine palladacycle **61–62**, the Heck coupling of aryl halides and ethyl acrylate underwent smoothly with NMP and Cs_2CO_3 *(53)*. When styrene was used in place of ethyl acrylate, DMF and K_2CO_3 were necessary to promote this reaction. Mono-ylide phosphine palladacycle **63**, **64**, and **71** were also successful precatalysts for Heck reaction of aryl bromides and chlorides *(56,57)*. The chelation of ylide through the ylidic carbon atom and phosphino group afforded six-membered ring where the P,C coordination was confirmed by the single crystal X-ray crystallography. The Sabounchei group further reported a bis-ylide palladacycle in comparing the catalytic activity of its mono-ylide form in Heck coupling *(54)*.

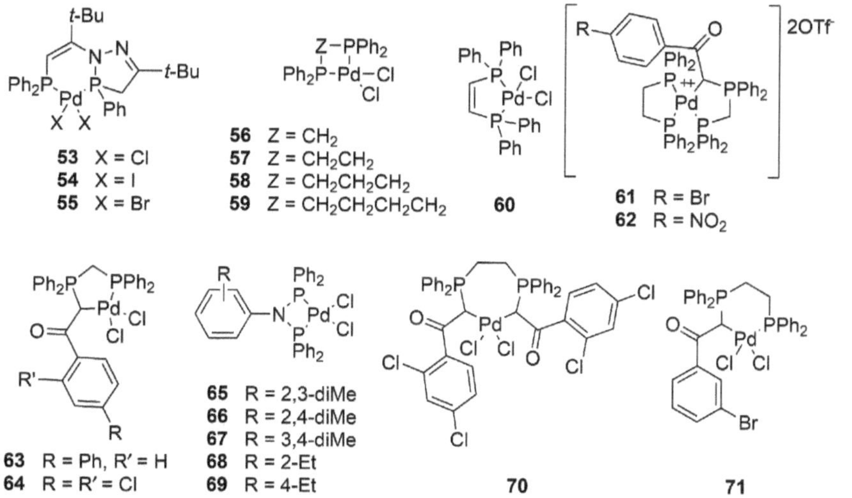

Fig. 2.5 Examples of diphosphine-palladacycles.

2.2 Sonogashira Cross-Coupling Reaction

Nowadays, the mostly employed reaction conditions for the Sonogashira coupling still involve simple phosphine-based palladium complexes such as [Pd(PPh_3)$_2$$Cl_2$], [Pd($PPh_3$)$_4$], [Pd(dppe)$Cl_2$], [Pd(dppp)$Cl_2$], and [Pd (dppf)$Cl_2$] as the catalysts, *(58)* in which a catalytic amount of copper(I) salt is required *(59)*. Excellent product yields were often obtained by using these phosphine-based palladium complexes, under a medium of excess amine (or inorganic base). High catalyst and co-catalyst loadings are required in more demanding situations as in the case of nonactivated aryl chlorides. Considering the drawback of high-cost palladium catalysts in these C_{sp}^{2}-C_{sp} cross-coupling reactions, *(60)* the search for new phosphine-types of ligands amendable to improve the catalytic efficiency is necessary.

Herrmann-Beller's catalyst **1** served as the landmark catalyst for most of the cross-coupling reactions. In 1996, Herrmann demonstrated the Sonogashira coupling which proceeded well with a wide range of aryl bromides *(61)*. Later investigation showed that even without the addition of CuI, aryl bromides could also be coupled smoothly with phenylacetylene to give the desired product. However, when aryl chlorides were used as the electrophilic partner, no satisfactory results were obtained. It was synthetically notable that the tolane products synthesized by Sonogashira coupling could be further coupled with 2-bromobenzaldehyde to afford 2,3-diphenylindenone *(62)*.

Table 2.5 Diphosphine palladacycles in Heck reaction

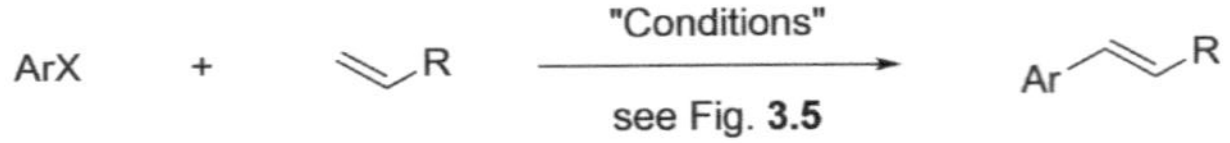

Entry	Palladacycle (mol% Pd)	ArX	R	Solvent	Base	Time (h)	Temp. (°C)	Yield (%)	Ref.
1	**53** (3.3×10^{-3})	4-MeC(O)C_6H_4Br	Ph	DMF	TBA	66	125	91	52
2	**55** (0.23)	4-CNC_6H_4Br	Ph	DMF	TBA	12	125	75	52
3	**56** (5.8×10^{-2})	4-MeC(O)C_6H_4Br	Ph	DMF	TBA	24	125	88	52
4	**57** (6.0×10^{-2})	4-MeC(O)C_6H_4Br	Ph	DMF	TBA	24	125	87	52
5	**58** 0.465	4-MeC(O)C_6H_4Br	Ph	DMF	TBA	24	125	75	52
6	**59** (1.5×10^{-2})	4-MeC(O)C_6H_4Br	Ph	DMF	TBA	24	125	79	52
7	**60** (5.2×10^{-2})	4-MeC(O)C_6H_4Br	Ph	DMF	$KHCO_3$	24	125	82	52
8	**61** (1.0×10^{-3})	C_6H_5Cl	Ph	DMF	K_2CO_3	4	130	77	53
9	**62** (1.0×10^{-3})	C_6H_5Cl	Ph	DMF	K_2CO_3	4	130	69	53
10	**64** (1.0×10^{-3})	C_6H_5Cl	CO_2Et	NMP	K_2CO_3	24	130	71	54
11	**65** (1.0×10^{-3})	4-MeOC_6H_4Br	Ph	dioxane	K_2CO_3	16	80	75	55
12	**66** (1.0×10^{-3})	4-MeOC_6H_4Br	Ph	dioxane	K_2CO_3	16	80	70	55
13	**67** (1.0×10^{-3})	4-MeOC_6H_4Br	Ph	dioxane	K_2CO_3	16	80	84	55
14	**68** (1.0×10^{-3})	4-MeOC_6H_4Br	Ph	dioxane	K_2CO_3	16	80	68	55
15	**63** (1.0×10^{-3})	C_6H_5Cl	CO_2Et	DMF	K_2CO_3	24	180	74	56
16	**69** (1.0×10^{-3})	4-MeOC_6H_4Br	Ph	dioxane	K_2CO_3	16	80	80	55
17	**70** (1.0×10^{-3})	C_6H_5Cl	CO_2Et	NMP	K_2CO_3	24	130	76	54
18	**71** (1.0)	C_6H_5Cl	Ph	DMF/H_2O	K_2CO_3	6	reflux	76	57

The polymer-supported diphenylphosphinoethane palladium(0) complex **72** was recently reported and used as the precatalyst in the copper-free Sonogashira coupling of aryl iodides with terminal alkynes (Scheme 2.9) *(63)*. Excellent product yields (85%–98%) resulted even at room temperature conditions. This catalyst could be recycled and reused up to five times without considerable loss of activity.

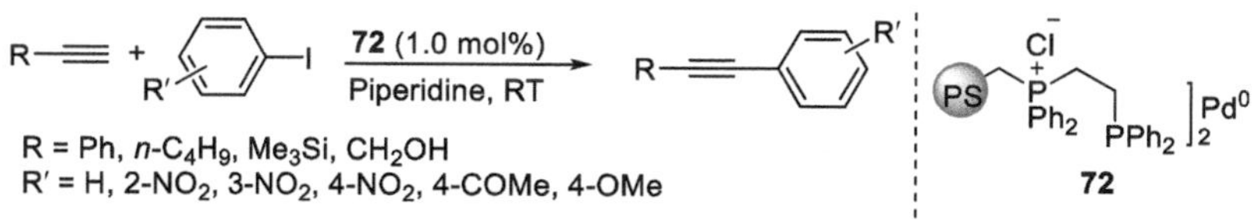

Scheme 2.9 Sonogashira reaction catalyzed by a polymer-supported Pd(0) diphenylphosphinoethane complex.

Tamami and Ghasemi reported the immobilization of Pd(0) NPs into a modified cross-linked polyacrylamide containing phosphinite **73** *(64)*. The catalytic performance of **73** was studied in the Sonogashira reaction of various aryl iodides (80%–90%), bromides (75%–87%), and chlorides (70%–78%) as illustrated in Scheme 2.10. According to ICP analysis, 4.8% leaching of palladium was observed during the reaction with iodobenzene, yet without significant loss of activity after five runs.

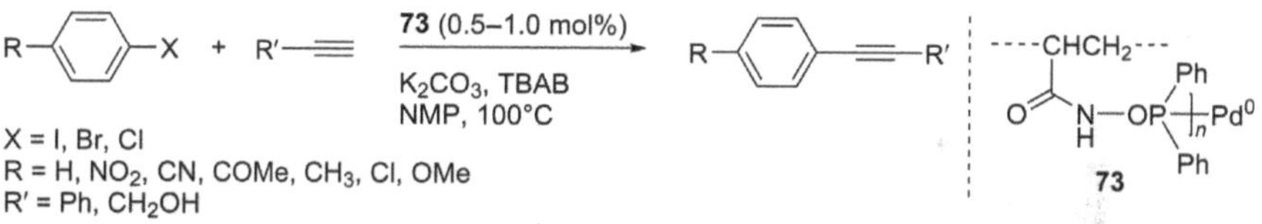

Scheme 2.10 Sonogashira coupling reaction catalyzed by a crosslinked polyacrylamide containing phosphinite complex.

In addition to the general mechanism of copper-free Sonogashira reaction, a mechanism particularly referring to the catalyst **72** was proposed (Fig. 2.6). The catalytic cycle is initiated, as usual, by the oxidative addition of the aryl halide in step 2. Subsequent insertion of the carbon-carbon triple bond to Ar-Pd-X generates the alkenylpalladium species (step 3), which then undergoes β-hydride elimination (step 4) to afford the disubstituted alkyne. Finally, Pd(0) is regenerated by reductive elimination of the Pd(II) compound in the presence of a base in step 5.

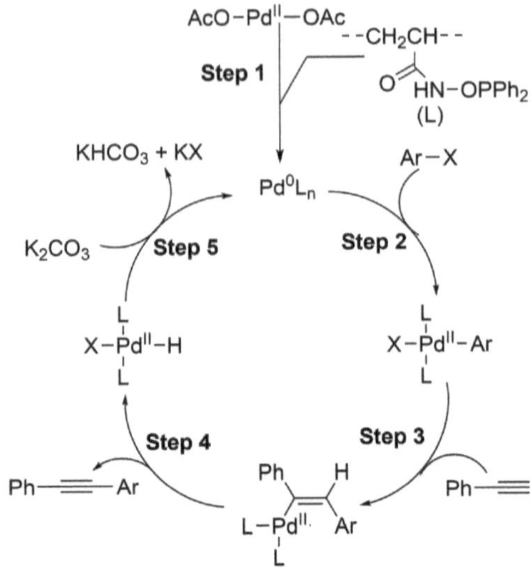

Fig. 2.6 Proposed mechanism for the polyacrylamide-phosphinite Pd(0) complex catalyzed Sonogashira reaction.

P,N-Donor bidentate ligand exhibits hemilabile behavior when it is coordinated to a palladium center. The soft phosphorus atom coordinates strongly whereas the hard nitrogen donor is weakly bound. The palladium(II) complex **74**, having a ferrocenyl phosphinimine-phosphine ligand, was tested in the copper- and amine-free Sonogashira coupling of aryl iodides and aryl bromides with terminal alkynes (Table 2.6) *(65)*. One equivalent of tetrabutylammonium acetate (which serves as an activator) was necessary to facilitate this reaction. The corresponding disubstituted alkynes were obtained in high yields even when only 0.1 mol% of the Pd catalyst was employed. In addition, monomeric phosphine-derived palladacycle **75** was assayed in this cross-coupling of aryl bromides and activated aryl chlorides with various terminal alkynes under both copper- and amine-free conditions *(66)*. The corresponding products were obtained in excellent yields at 80°C with 2.5–3.0 mol% of palladacycle **75** used as catalyst and the presence of TBAB as the additive in which it was found necessary for the coupling of aryl chlorides (Table 2.6). The P,O-donor bidentate palladacycles **76–78**, containing 3-oxo-1,3-diphosphapropene ligands (Fig. 2.7), were employed as precatalysts for this coupling of iodobenzene and phenylacetylene at room temperature, with yields up to 68% (R = *p*-MeOC$_6$H$_4$) *(67)*.

Table 2.6 A ferrocene-based phosphinimine-phosphine ligand for copper- and amine-free alkynylation of aryl iodides and bromides

ArX + R—≡ —"Conditions" (see Fig. **3.7**)→ Ar—≡—R

Entry	Palladacycle (mol% Pd)	ArX	R	Solvent	Base	Time (h)	Temp. (°C)	Yield (%)	Ref.
1	**74** (0.1)	*o*-$MeOC_6H_4I$	Ph	NMP	*n*-Bu_4NOAc	1	110	96	65
2	**74** (0.1)	*o*-$MeOC_6H_4Br$	Ph	NMP	*n*-Bu_4NOAc	1	110	70	65
3	**74** (0.1)	*o*-$MeOC_6H_4I$	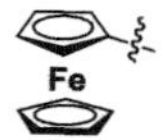	NMP	*n*-Bu_4NOAc	1	110	97	65
4	**74** (0.1)	*p*-CNC_6H_4Br	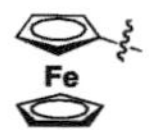	NMP	*n*-Bu_4NOAc	1	110	64	65
5	**74** (0.1)	*p*-MeC_6H_4I	*i*-Pr_3Si	NMP	*n*-Bu_4NOAc	1	110	93	65
6	**75** (2.5)	*p*-$MeOC_6H_4Br$	Ph	CH_3CN	Cs_2CO_3	24	80	100	66
7	**75** (2.5)	*p*-MeC_6H_4Br	*t*-Bu	CH_3CN	Cs_2CO_3	24	80	100	66
8	**75** (2.5)	2,4,6-$Me_3C_6H_3Br$	*n*-Hex	CH_3CN	Cs_2CO_3	24	80	45	66
9	**75** (3.0)	*p*-$MeOC_6H_4Cl$	Ph	CH_3CN	Cs_2CO_3	24	80	80	66
10	**75** (3.0)	*p*-CNC_6H_4Cl	*p*-NC	CH_3CN	Cs_2CO_3	24	80	100	66
11	**75** (3.0)	C_6H_5Cl	*t*-Bu	CH_3CN	Cs_2CO_3	24	80	70	66

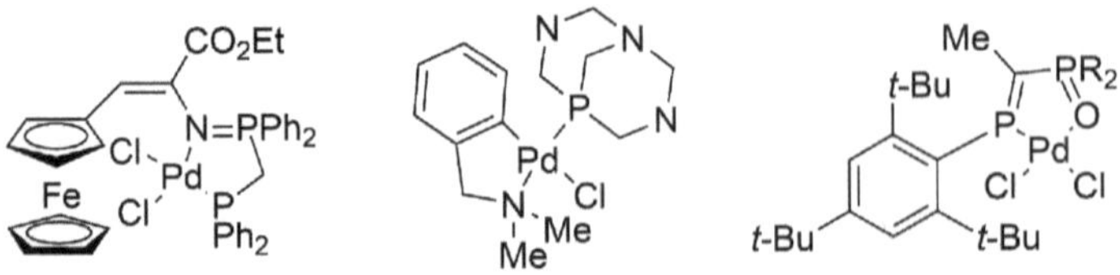

74 **75**

76 R = *p*-MePh
77 R = *p*-MeOPh
78 R = *n*-Bu

Fig. 2.7 Examples of P,N-bidentate palladacycles.

An air-stable Pd(I) dimer complex **79** was developed as a highly active catalyst in direct arylation of terminal alkynes, in which it was found necessary to use a substoichiometric amount of $ZnCl_2$ as additive (Scheme 2.11) *(68)*. This system allowed the reaction to proceed at room temperature without generating diyne byproduct. A more robust palladacycle phosphine-ylide **80** was invented for both additive- and amine-free Sonogashira coupling reaction of various aryl halides with phenylacetylene (Scheme 2.12) *(69)*. Remarkably, the corresponding coupling products were obtained in good-to-excellent yields in the presence of only 0.001 mol% Pd.

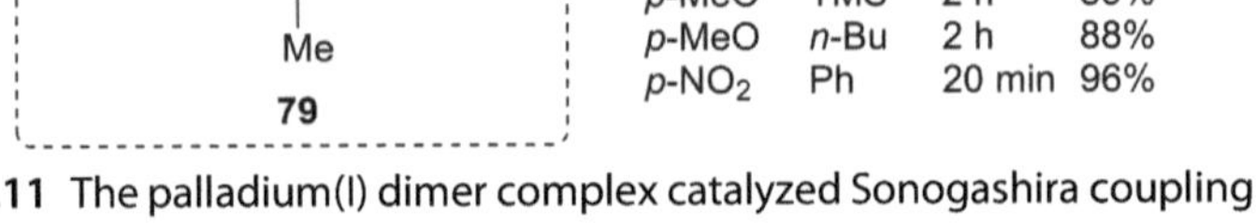

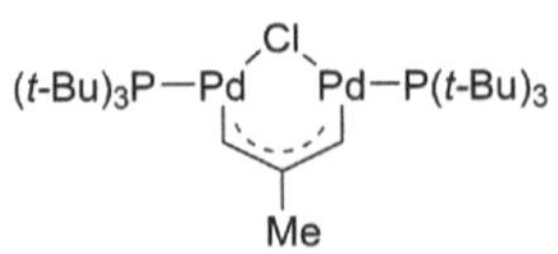

79

R	R′	time	yield
H	Ph	30 min	93%
p-MeO	TMS	2 h	89%
p-MeO	*n*-Bu	2 h	88%
p-NO_2	Ph	20 min	96%

Scheme 2.11 The palladium(I) dimer complex catalyzed Sonogashira coupling reaction.

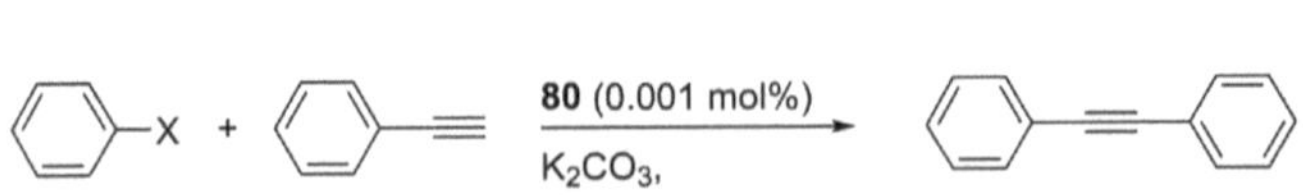

X = I, 81%
X = Br, 78%
X = Cl, 67%

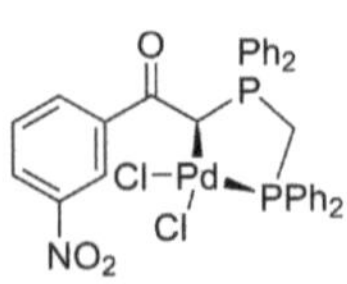

80

Scheme 2.12 The coupling of aryl halides with phenylacetylene by phosphine-ylide palladacycle.

Palladacycles **61** and **62** (Fig. 2.5) were found to be efficient in facilitating Heck reaction. Sabounchei later employed these mononuclear containing bidentate phosphine ligands in the copper- and amine-free Sonogashira coupling *(70)*. 0.001 mol% of palladacycles allowed the coupling to proceed with moderate-to-good yields. Notably, aryl chlorides were able to be coupled when 0.1 mol% of catalyst was used.

The reactivity of dimeric palladacycles bearing substituted diaryl/alkyl phosphito ligands was tested in a copper-free Sonogashira reaction with aryl iodides and phenylacetylene as substrates (Fig. 2.8) *(71)*. Eppinger, Ziółkowski and co-workers reported this reaction to be carried out in the presence of triethylamine as base in water or in various ionic liquids, such as [bmim][PF_6], [bmim][BF_4] and [emim][SO_4Et]. Thus, some of the coupling products could be easily isolated in excellent purity without tedious work-up, such as column chromatographic purification. The best results were obtained by the orthometallated complex **83** in all ionic liquids, which permitted the recycling of catalysts with only slightly diminished the catalyst activity.

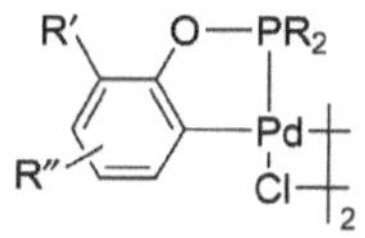

81 R = *i*-Pr, R′ = Ph, R″ = H
82 R = Ph, R′ = R″ = H
83 R = *m*-MeC$_6$H$_4$, R′ = H, R″ = *m*-Me
84 R = *o*-MeC$_6$H$_4$, R′ = H, R″ = *o*-Me

Fig. 2.8 The dimeric palladacycles bearing diaryl/alkyl phosphito ligands.

The soluble polymer supported palladium catalyst was explored as the catalyst in Sonogashira coupling reactions. The recovery of the catalyst from such a system involves a simple filtration process. The soluble supports may be either precipitated from solution or isolated by membrane filtration. Recently, more readily accessible polymers (such as polystyrenes or polyethylene glycol) were found as suitable supports for catalyst recycling.

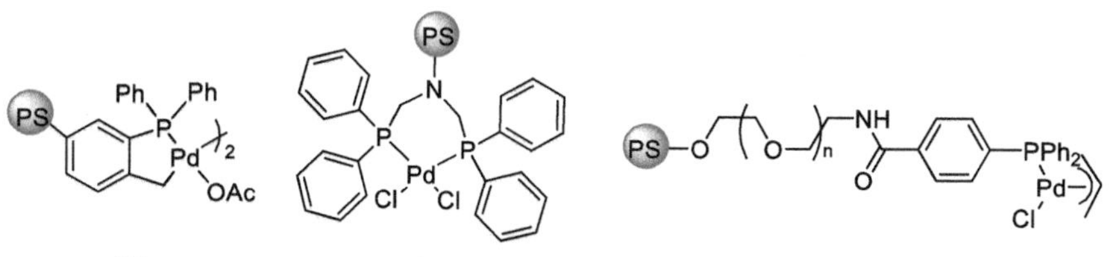

85 **86** **87**

Fig. 2.9 Examples of linear polystyrene-supported palladacycles.

The soluble linear polystyrene-supported palladacycle **85** was used in the alkynylation of activated 4-bromoacetophenone with phenylacetylene (Fig. 2.9) *(37)*. The reaction took place at 90°C with only 0.2 mol% of precatalyst, although a reaction time of 72 h was required. This supported catalyst was allowed to precipitate by addition of diethyl ether. This catalyst could be reused up to four cycles, without significant loss of activity. In addition, the supported diphosphine-based palladacycle **86** (Fig. 2.9), derived from aminomethyl-functionalized polystyrene beads, was utilized as an effective precatalyst of the Sonogashira coupling of aryl iodides with either aromatic or aliphatic terminal alkynes (Scheme 2.13) *(72)*. Almost quantitative product yields were afforded in the presence of 2 mol% CuI in dioxane/piperidine solvent mixtures at 60°C. The catalyst **86** could be recycled four times. Moreover, an amiphiphilic polystyrene-poly(ethylene glycol) resin-supported palladium-phosphine complex **87** was developed (Fig. 2.9) *(73)*. This catalyst system allowed only aryl iodides or activated aryl bromides to proceed well, yet deactivated aryl bromides and activated aryl chlorides showed low conversion.

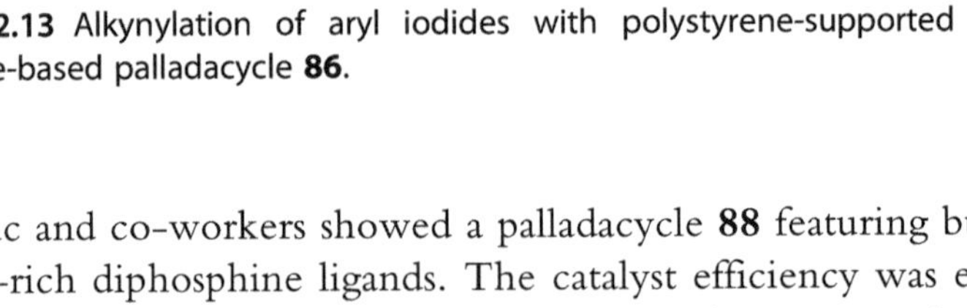

Scheme 2.13 Alkynylation of aryl iodides with polystyrene-supported chelating phosphine-based palladacycle **86**.

Astruc and co-workers showed a palladacycle **88** featuring bulky and electron-rich diphosphine ligands. The catalyst efficiency was evaluated in Sonogashira reaction of iodobenzene, bromobenzene, and activated aryl chlorides (Scheme 2.14) *(74)*. Quantitative product yields were obtained at 1.0 mol% Pd loading for iodobenzene and bromobenzene at room temperature conditions. However, chlorobenzene was found to be nonapplicable as only 9% conversion was observed. Nevertheless, there was up to 30% yield when activated aryl chlorides were used as the substrates.

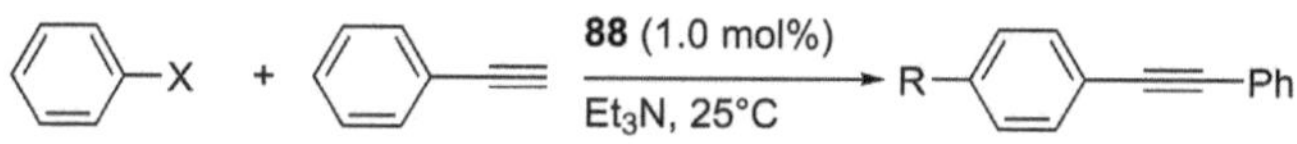

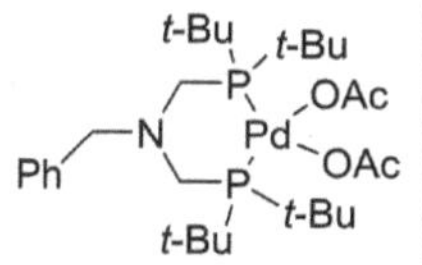

88

X	Time	Conversion
I	30 min	100%
Br	1 h	100%
Cl	3 h	9%

Scheme 2.14 Alkynylation of aryl halides using palladium and an electron-rich aminobisphosphine.

The same research group later prepared dendritic supported electron-rich phosphine-based palladacycles (metallodendrimers with 4-, 8-, or 16- palladium nuclei), and put into Sonogashira coupling for initial investigation (Fig. 2.10) *(75)*. The metallodendrimers having P*t*-Bu_2 moieties (1.0 mol %, employed at 25–120°C) were found to be efficient for the alkynylation of iodobenzene (97%–100% yields) and bromobenzene (93%–100% yields). However, the corresponding catalysts bearing dicyclohexylphosphino groups were less efficient under similar reaction conditions (iodobenzene conversion: 46%–79%, bromobenzene conversion: 15%–20%).

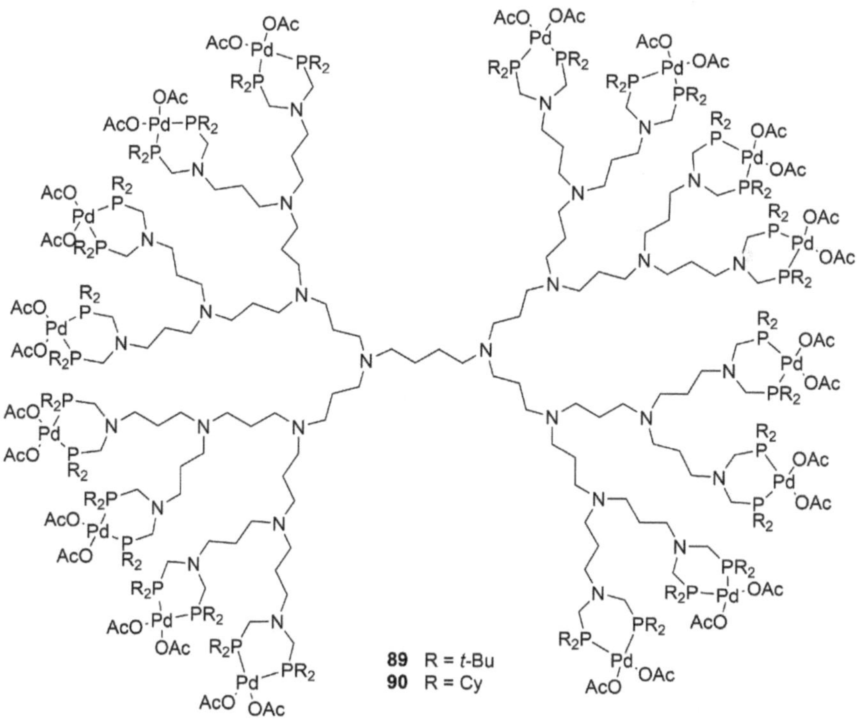

89 R = *t*-Bu
90 R = Cy

Fig. 2.10 Dendritic electron-rich phosphine-based palladacycles (sixteen pairs of chelating P donors).

An interesting dendritic palladium(II) complex **91** was prepared by a covalent grafting via a reaction between the carboxylic acid groups of core-shell γ-Fe_2O_3/polymer superparamagnetic nanoparticles and the amino groups at the focal of an amino-terminated dendron (Fig. 2.11) *(76)*. The catalytic activity of the palladacycle **91** was investigated in the copper-free Sonogashira coupling of aryl iodides and bromides with phenylacetylene in methanol or water at 70°C, using Triton X-405 as a surfactant and lithium hydroxide as the base.

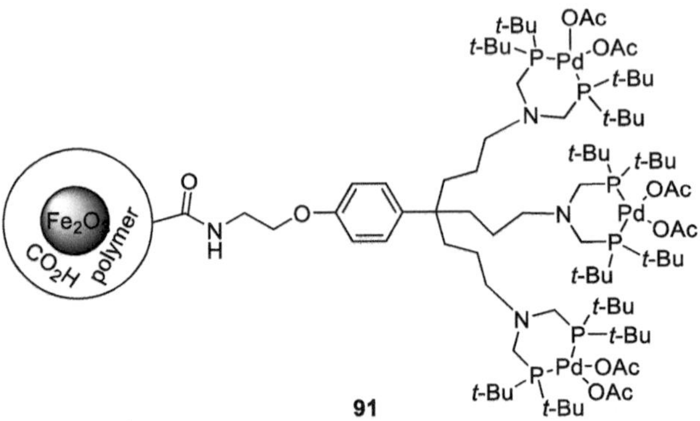

91

Fig. 2.11 Dendritic palladium(II) complex supported by the core-shell γ-Fe_2O_3/polymer superparamagnetic nanoparticles.

Mesoporous silicate MCM-41 was found applicable to serve as the support for anchoring a bidentate phosphine palladium(II) complex **92** (Fig. 2.12) *(77)*. Only 0.05 mol% of Pd was sufficient to promote the coupling between aryl iodides and terminal alkynes at room temperature. This catalyst could be reused up to ten cycles without any diminishing of the catalytic activity. The capability of this catalyst was shown even in the carbonylative Sonogashira coupling of terminal alkynes with aryl iodides under atmospheric pressure of carbon monoxide *(78)*. There was no observable decrease of activity after ten times recycling. Furthermore, another palladacycle catalyst supported on commercial silica gel **93** was used in the copper-free Sonogashira coupling of aryl iodides and activated aryl bromides with terminal alkynes (Fig. 2.12). The reaction was carried out in the presence of triethylamine as the base in ethylene glycol as the solvent at 100°C, and the catalyst retained its activity after being reused for six times *(79)*.

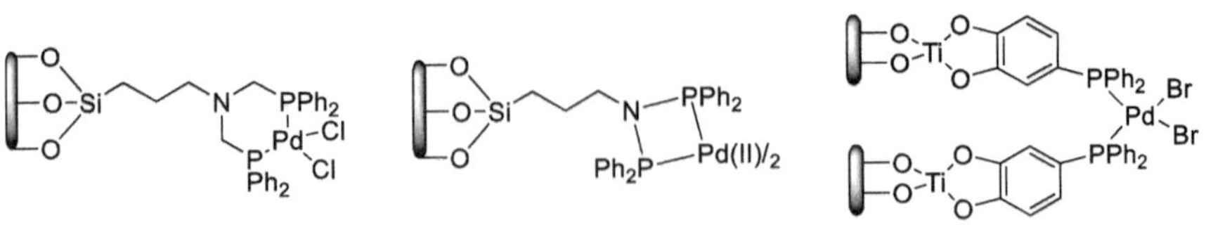

92 **93** **94**

Fig. 2.12 Examples of solid-supported palladacycles.

Apart from the silica-immobilized palladium catalysts, McDonagh and co-workers proposed the incorporation of palladium complex to titanium dioxide bearing a strongly bound catecholate ligand to form a TiO_2-immobilized palladacycle **94** (Fig. 2.12) *(80)*. This catalyst was evaluated in the model coupling of 4-iodonitrobenzene with phenylacetylene at room temperature.

3. Ferrocene-Based Palladacycles

Cyclopalladated ferrocenyl complexes have been successful as efficient catalysts in many cross-coupling reactions. They are often found to be stable, easy to access and handle, and they can potentially be used as alternatives to common palladium salts. The ferrocene derived cyclopalladated at the substituted cyclopentyl ring structure and the intramolecular coordination of donor atom stabilized the Pd–C bond. The donor atoms are usually nitrogen, phosphorus, oxygen, or sulfur atom(s). Among them, nitrogen atoms often serve as the electron-rich atom for the coordination of a positively charged metal center for ferrocene-based palladacycles. The nitrogen atom can be either sp^3 (amine) or sp^2 (imine and oxime) *(81)*. Correspondingly, *N*-cyclopalladated ferrocenes are classified into two sections in this chapter according to the state of the nitrogen atom (Fig. 2.13). The environment of the palladium atom can be depicted as $Pd(C_{(Fc)}N)$ or $Pd(C_{(Fc)}\ NP)$ and they will be further discussed: (1) ferrocenyl-palladacycle dimers in which the metal center is stabilized by the five-membered ring (Fig. 2.13, upper); and (2) phosphine adducts of palladacycles (Fig. 2.13, lower). The solution of halide-bridged palladacyclic dimer was treated with triaryl/trialkylphosphine per palladium in dichloromethane to afford the corresponding monomeric phosphine adducts of palladacycles.

3.1 Heck Cross-Coupling Reaction

3.1.1 Ferrocene-Based Palladacycles With N(sp^3) Atom

Dimethylaminomethylferrocene, as the simplest achiral amine model, was first introduced as the main backbone of the palladacycles to afford **95** and **96** (Fig. 2.14). These cyclopalladated complexes of tertiary arylamines were served as catalysts for the Heck reaction of aryl iodides (Table 2.7) *(82)*. It is highly efficient with turnover numbers 9.6×10^6 and turnover frequencies up to 8.7×10^5 with **96** *(83)*. By increasing the catalyst loading, 1.0×10^{-3} of **95** catalyzed the coupling with 99% yield and TON

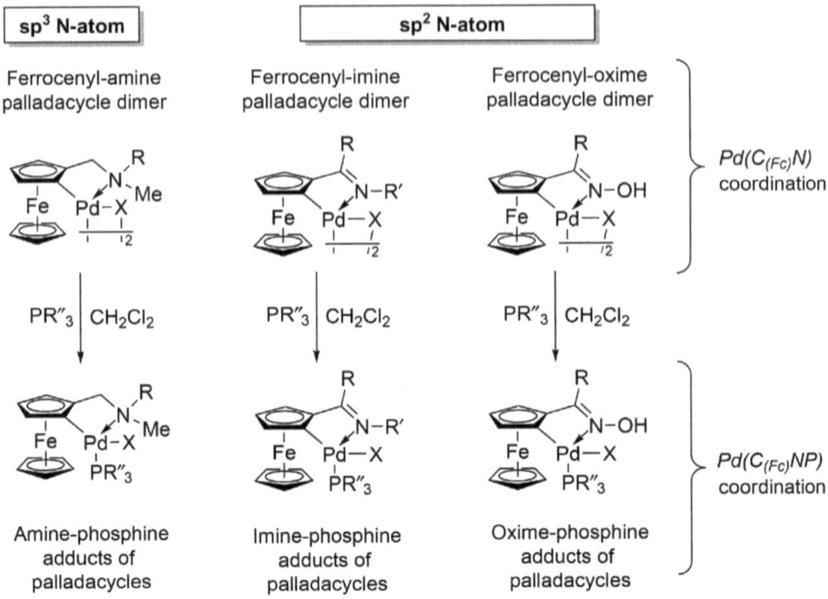

Fig. 2.13 Type of Ferrocene-based palladacycles.

9.9 × 10⁴. However, only 23% yield were obtained when aryl bromides were used as the coupling partner under the typical reaction system *(84)*. Sokolov and co-workers investigated the stability of **95** by heating up the reaction mixture and it was found that the dimer **95** completely decomposed at 100°C. It is worth noting that in most of the cases known, the substituents at the nitrogen atom are two methyl groups. Few exceptions are reported. In 2007, racemic palladacycles **97–99** based on ferrocenylmethylamines were synthesized consisting of R_NR_p and S_NS_p configurations and they successfully achieved the Heck cross-coupling of aryl bromides with styrene *(85)*.

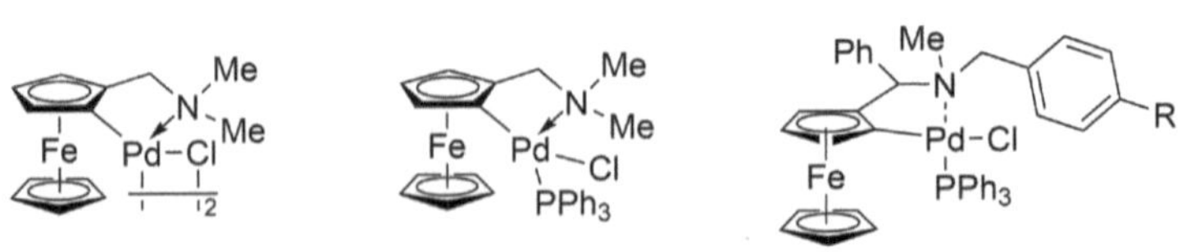

95 **96** **97** R = H **98** R = OMe **99** R = Me

Fig. 2.14 Examples of ferrocenyl-amine palladacycles.

Table 2.7 Ferrocenyl-amine palladacycles in Heck reaction

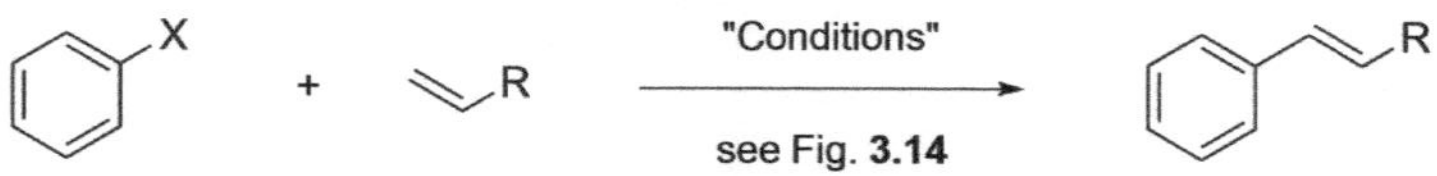

Entry	Palladacycle (mol% Pd)	X	R	Base	Solvent	Temp. (°C)	Yield (%)	Ref.
1	**95** (1.0×10^{-5})	I	CO_2Me	Et_3N	DMA	140	33	81, 82
2	**96** (1.0×10^{-5})	I	CO_2Me	Et_3N	DMA	140	96	81, 82
3	**95** (0.2)	Br	CO_2Et	K_3PO_4	DMF	140	23	84
4	**97** (0.1)	Br	Ph	K_2CO_3	DMF	140	96	85
5	**98** (0.1)	Br	Ph	K_2CO_3	DMF	140	100	85
6	**99** (0.1)	Br	Ph	K_2CO_3	DMF	140	98	85

3.1.2 *Ferrocene-Based Palladacycles With N(sp^2) Atom*

3.1.2.1 Imine-Derived Ferrocenyl Palladacycles

Imine is often complimentarily used as a part of the cyclopalladated ferrocenyl compounds. They were easily synthesized by the condensation of ferrocenecarbaldehyde or acetylferrocene with amine. In the last decade, a large portion of the works with ferrocenyl imines was developed by the group of Wu. In 2001, the dimeric complexes **100** and **101** were synthesized and applied in the Heck reaction. The metal center was stabilized by a five-membered ring which was easily prepared by treatment of ferrocenyl imines with Li_2PdCl_4 or K_2PdI_2 in methanol *(86)*. The palladacycles performed well in the coupling of iodobenzene with ethyl acrylate with comparably high product yields and turnover numbers (Table 2.8). However, when styrene was used as the coupling partners, **100** gave a poor result with 38% product yield and **101** afforded the desirable product in 99% yield. Activated aryl bromides and chlorides were feasible coupling partners with palladacycle **101** as the catalyst. High catalyst loading and high reaction temperature were required to afford the desirable product in good-to-excellent yield. Later, Wu successfully achieved the Heck reaction in neat water in the presence of palladacycle **101** *(87)*. Notably, not only aryl iodides but also aryl bromides could be coupled with the olefins (Fig. 2.15).

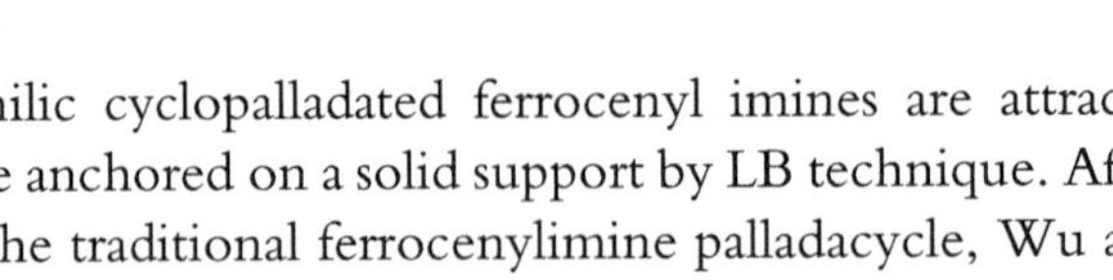

100 **101**

Fig. 2.15 Examples of ferrocenyl-imine palladacycles.

Amphiphilic cyclopalladated ferrocenyl imines are attractive catalysts which can be anchored on a solid support by LB technique. After the investigation on the traditional ferrocenylimine palladacycle, Wu and Li turned their attention to amphiphilic cylopalladated ferrocenylimines with long-chain hydrocarbon in 2007 *(88)*. A series of novel long alkyl ferrocenylimines were synthesized in which the alkyl chain can be up to 16 carbons (Fig. 2.16). Ethyl acrylate and bromobenzene were used as the coupling partners for the Heck reaction. Palladacycles **104–111** all exhibited high activity with 0.1 mol% catalyst loading to give 94%–99% product yields. The catalyst loading (**106**) can even go down to 0.0001 mol% for the coupling of styrene with aryl iodides while to 0.01 mol% for aryl bromides.

Table 2.8 Ferrocenyl-imine palladacycles in Heck reaction

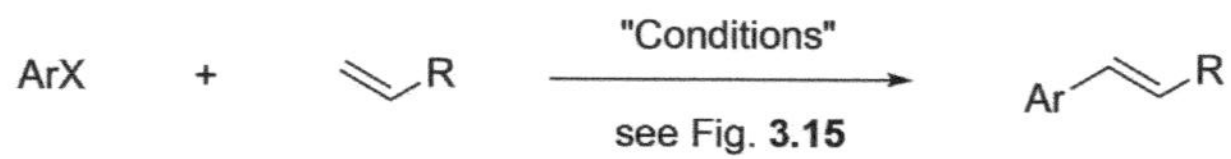

Entry	Palladacycle (mol% Pd)	ArX	R	Time (h)	Solvent	Temp. (°C)	Yield (%)	Ref.
1	**100** (6.54×10^{-2})	C_6H_5I	CO_2Et	2.5	dioxane	100	98	86
2	**101** (5.46×10^{-2})	C_6H_5I	CO_2Et	2.5	dioxane	100	95	86
3	**100** (6.54×10^{-2})	C_6H_5I	Ph	8	dioxane	100	38	86
4	**101** (5.46×10^{-2})	C_6H_5I	Ph	8	dioxane	100	99	86
5[a]	**101** (5.0×10^{-3})	C_6H_5I	Ph	10	H_2O	100	92	87
6	**101** (5.46×10^{-5})	*p*-$NO_2C_6H_4Br$	CO_2Et	55	DMF	140	64	86
7[a]	**101** (5.0×10^{-2})	*p*-$NO_2C_6H_5Br$	CO_2Et	20	H_2O	100	99	87
8	**101** (0.36)	*p*-$NO_2C_6H_5Cl$	CO_2Et	10	DMF	140	73	86

[a]TBAB was used as additive.

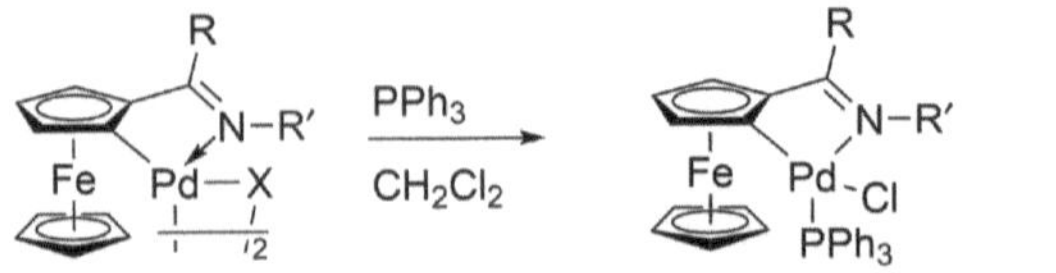

104 R = H, R′ = *n*-$C_{12}H_{25}$
105 R = H, R′= *n*-$C_{16}H_{33}$
106 R = Me, R′ = *n*-$C_{12}H_{25}$
107 R = Me, R′= *n*-$C_{16}H_{33}$

108 R = H, R′ = *n*-$C_{12}H_{25}$
109 R = H, R′= *n*-$C_{16}H_{33}$
110 R = Me, R′ = *n*-$C_{12}H_{25}$
111 R = Me, R′= *n*-$C_{16}H_{33}$

Fig. 2.16 Examples of amphiphilic cyclopalladated ferrocenylimines.

Later, Wu and Li continued their works on the development of new cyclopalladated ferrocenylimine with ester groups (Table 2.9, **112–115**) *(89)*. A wide range of aryl or heteroaryl bromides could be coupled with styrene successfully in a common organic solvent and even neat water to afford moderate-to-excellent product yield. The imine-phosphine adducts of palladacycles **114** and **115** were easily synthesized by mixing **112** or **113** with triphenylphosphine in dichloromethane for 1 hour. The relative activities of these palladacycles in the coupling of bromobenzene and styrene were investigated. Under the same reaction conditions, **112** and **113** performed slightly better than their phosphine adducts **114** and **115** in the Heck reaction.

Table 2.9 Ferrocenyl-imine palladacycles in Heck reaction[89]

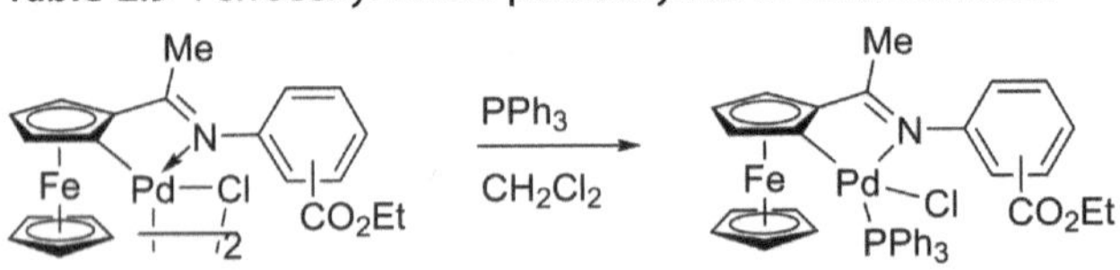

112 3-CO_2Et
113 4-CO_2Et

114 3-CO_2Et
115 4-CO_2Et

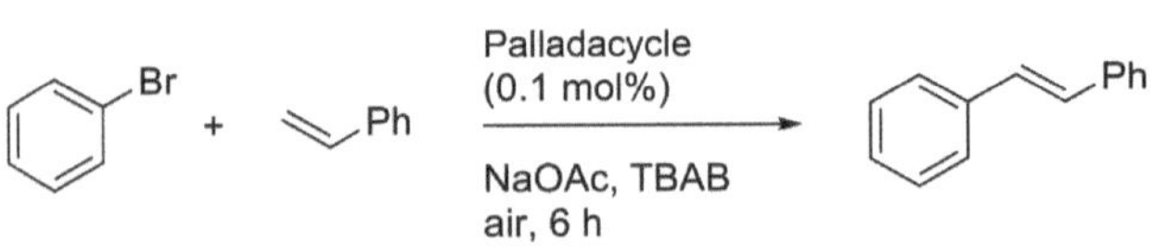

Entry	Palladacycle	Solvent	Temp. (°C)	Yield (%)
1	**112**	DMF	130	98
2	**113**	DMF	130	>99
3	**112**	water	100	91
4	**113**	water	100	98
5	**114**	DMF	130	95
6	**115**	DMF	130	94
7	**114**	water	100	91
8	**115**	water	100	93

To further modify the ferrocenylimine palladacycles, Csámpai and Kotschy developed the ferrocenecarboxaldehyde hydrazones palladacycles **116** and **117** (Fig. 2.17) *(90)*. The kinetic study showed that palladacycle **117** exhibited a longer induction phase and gave a lower reactivity than palladacycle **116**.

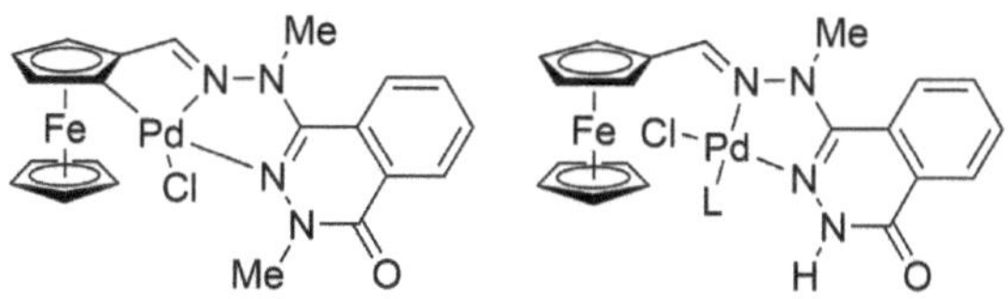

116
N-Pd-C(Fc) coordination

117
N-Pd-N coordination

Fig. 2.17 Examples of ferrocenecarboxaldehyde hydrazones palladacycles.

In 2006, Wu and Gong developed a series of novel tricyclohexylphosphine-cyclopalladated ferrocenylimine complexes **118–124** that displayed good activity in the Heck reaction of aryl bromides with acrylic acid ethyl ester (Scheme 2.15) *(91)*. Only *m*-NO_2PhCl could be activated with 1.0 mol% catalyst loading to afford the coupling product in moderate yield (47%). The single-crystal X-ray analysis showed that there are altogether six types of intermolecular hydrogen bonds in the crystals of **118**·CH_2Cl_2.

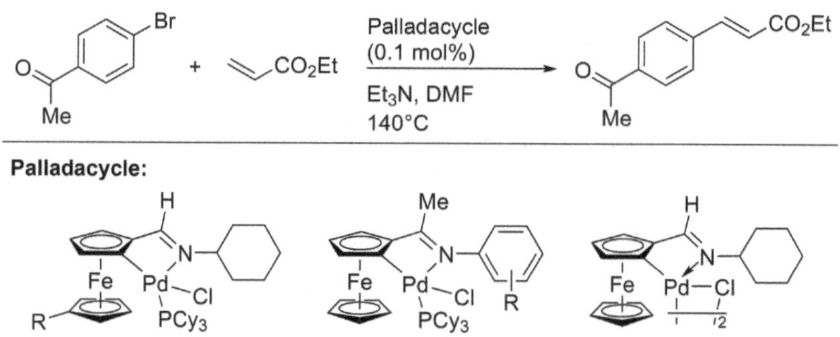

118 R = H, 90%
119 R = CHO, 85%

120 R = 4-OEt, 75%
121 R = 4-Br, 81%
122 R = 4-Cl, 80%
123 R = 2-Cl, 79%

124 53%
(0.05 mol%)

Scheme 2.15 Ferrocenyl-imine palladacycles in Heck reaction.

Apart from the imine-phosphine adducts of palladacycles, Wu also demonstrated the carbene adduct of cyclopalladated ferrocenylimine **125** in the Heck reaction (Scheme 2.16) *(92)*. Kinetic studies were carried out by the in situ ^{13}C NMR investigation and Hg poisoning experiments. The result showed that the palladacycle proceeded via a classical Pd(0)/Pd(II) catalytic cycle. Using higher catalyst loading, activated *p*-nitrochlorobenzene could be coupled to give the desired products in good yield.

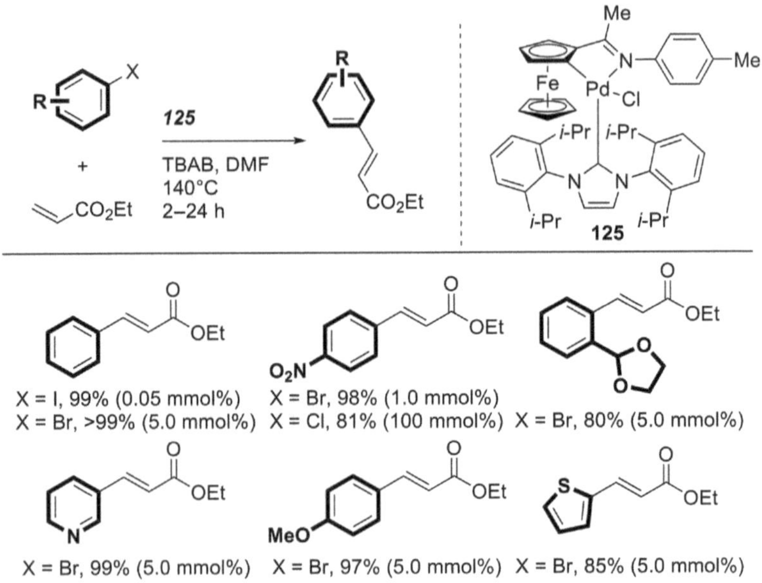

Scheme 2.16 Ferrocenyl-Imine-carbene palladacycles in Heck reaction.

Palladacycles are seldom used in oxidative Heck reaction. In 2010, Wu reported the use of ferrocenylimine palladacycles **100** in the oxidative Heck reaction of olefins with arylboronic acids (Table 2.10) *(93)*. The catalytic system is remarkable and simple. Low catalyst loading (0.5 mol%) was employed with O_2 as the only oxidant and mild reaction conditions were used (50°C). The reaction is also called "deborylative Heck reaction". Both steric and electronic effects of olefin give no critical influence on the reaction.

2.1.2.2 Oxime-Derived Ferrocenyl Palladacycles

Oxime, which is also known as hydroxylimines, can also be coordinated as part of the ferrocenyl palladacycles. Acetylferrocenyloxime **126** was one of

Table 2.10 Oxidative Heck coupling of olefins with arylboronic acids

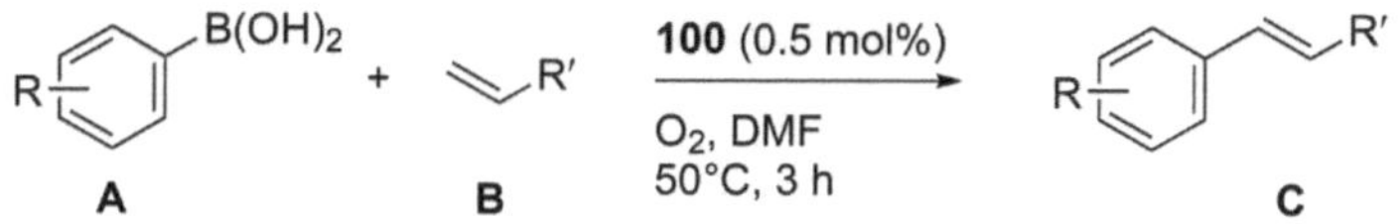

Entry	A	B	C	Yield/ %
1	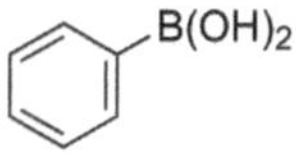	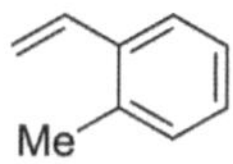	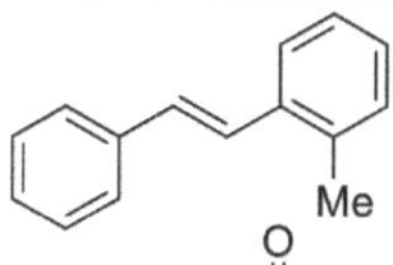	76
2	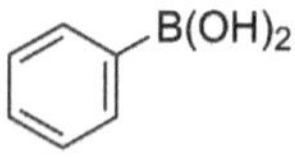	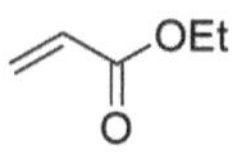	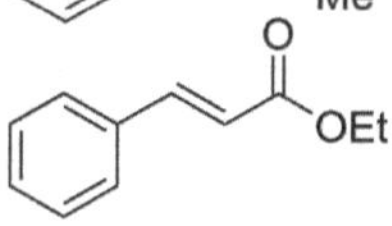	82
3	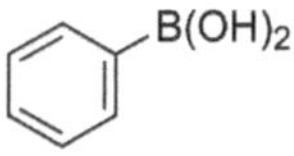	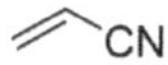	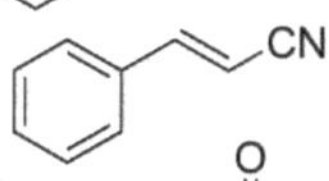	78
4	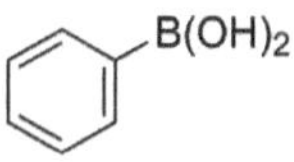	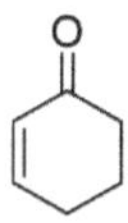	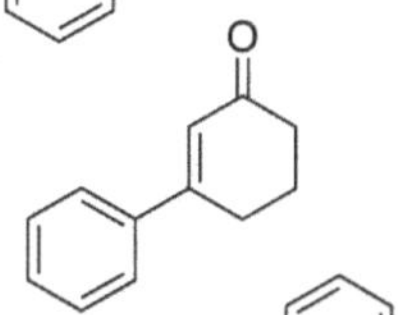	47
5	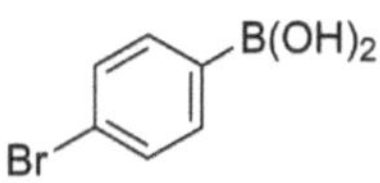	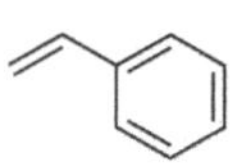	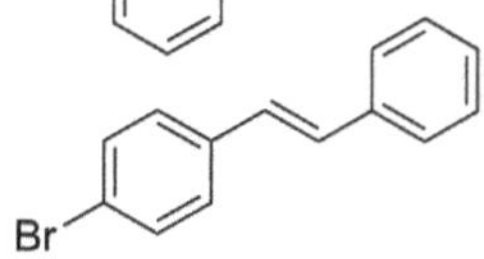	73

the examples that served as the catalyst for Heck reactions by Iyer in 2001 *(94)*. Its phosphine adducts were composed of PPh_3 (**127**) or $P(OEt)_3$ (**128**). Activated *p*-chloroacetophenone was shown to be coupled with styrene under these catalyst systems although poor yields were obtained (Fig. 2.18; Table 2.11). Later, Iyer further demonstrated a large-scale coupling (2 vs 25 mmol) using these acetylferrocenyloximes without deleterious effect *(95)*.

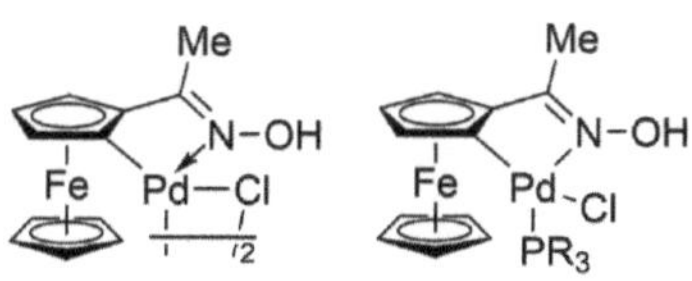

126 **127** R = Ph **128** R = OEt

Fig. 2.18 Examples of ferrocenyl-oxime palladacycles.

Table 2.11 Ferrocenyl-oxime palladacycles in Heck reaction

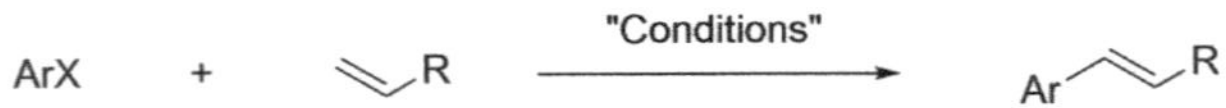

Entry	Palladacycle (mol%)	ArX	R	Base	Solvent	Temp. (°C)	Yield (%)	Ref.
1	**126** (0.5)	C_6H_5Br	Ph	NaOAc	NMP	140–150	84	94
2	**127** (0.5)	C_6H_5Br	Ph	NaOAc	NMP	140–150	79	94
3	**128** (0.5)	C_6H_5Br	Ph	NaOAc	NMP	140–150	83	94
4	**126** (0.5)	*p*-COC_6H_4Cl	Ph	NaOAc	NMP	140–150	30	94
5	**128** (0.5)	*p*-COC_6H_4Cl	Ph	NaOAc	NMP	140–150	53	94
6	**127** (1.0×10^{-2})	C_6H_5I	CO_2Me	Et_3N	DMF	110	94	96

3.1.3 Other Ferrocenyl Palladacycles

Nitrogen heterocyles could be palladated together with the ferrocene to give a five-membered stable structure *(97)*. A series of tridentate (C,N,N) coordinated systems was presented by Chupakhin for Heck reactions *(98)*. Sokolov demonstrated a fused bispalladacycle **129** with a chelating ligand of C,N, N system in the Heck reactions *(84)*. With the same reaction conditions (K_2CO_3 as the base) of coupling of bromobenzene with ethyl acrylate, the C,N,N coordinated **129** (52%) performed better than the C,N coordinated **95** (3%).

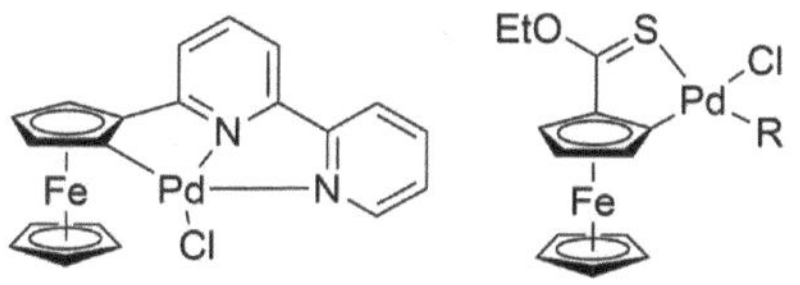

129

130 R = PPh_3
131 R = $P(o\text{-}Tol)_3$
132 R = PMe_3
133 R = $SbPh_3$

Fig. 2.19 Other ferrocenyl palladacycles.

Apart from the nitrogen-containing ferrocenyl palladacycles, organosulfur ligands can also be a part of the ferrocenyl palladacycles. In 2016, López-Cortés and Ortega-Alfaroa demonstrated four new sulfur-containing palladacycles **130–133** with different phosphine adducts for Heck cross-coupling (Fig. 2.19) *(99)*. 4-Iodotoluene and 4-bromotoluene could be coupled with methyl acrylate smoothly using 0.05 mol% of the palladacycles. Different heat sources such as conventional heating, microwave, and infrared irradiation (Osram lamp with 250 W, 125 V) were applied. Surprisingly, no coupling products could be obtained when microwave irradiation was used.

3.2 Sonogashira Cross-Coupling Reaction

The palladium complexes containing ferrocenyl fragment are widely used in carbon-carbon and carbon-heteroatom bond formation, owing to their structural versatility, stability, easily synthetic accessibility, and high catalytic activity in many organic reactions. Wu and co-workers developed cyclopalladated ferrocenylimine **134** for the Sonogashira reaction of ArX (X = I, Br, and Cl) with various terminal alkynes to give moderate-to-excellent yields (Table 2.12) *(100)*. TBAB was added as an additive and some activated aryl chlorides could be coupled with a lower product yield when compared to aryl iodides and bromides.

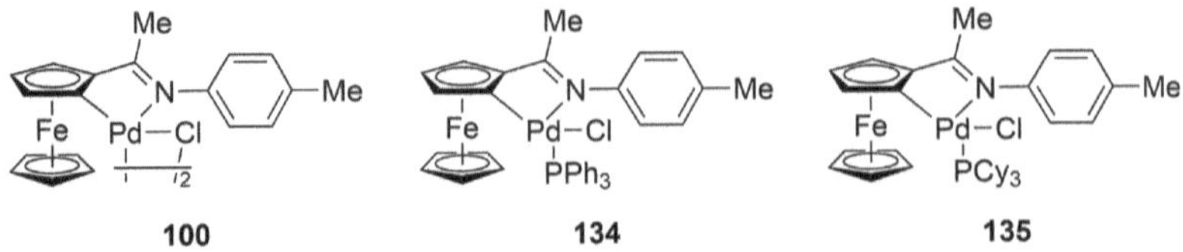

Fig. 2.20 Examples of ferrocenylimine palladacycles in Sonogashira coupling.

Table 2.12 Cyclopalladated ferrocenylimine in Sonogashira reaction[100]

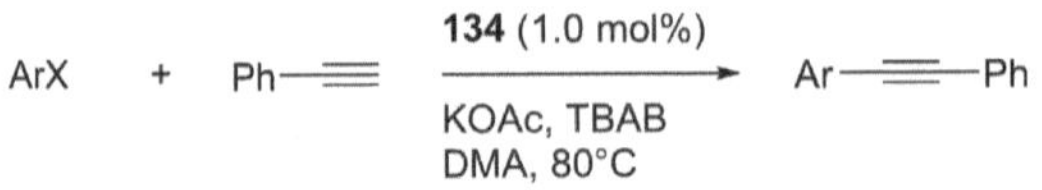

Entry	ArX	Time (h)	Yield (%)
1	C_6H_5I	3	90
2	*p*-$NO_2C_6H_4Br$	6	91
3	*o*-PyBr	20	99
4	*m*-PyBr	20	96
5	*p*-$NO_2C_6H_4Cl$	24	50
6	*o*-PyBr	24	87

Table 2.13 The Sonogashira reaction of bromostyrenes with terminal alkynes[101]

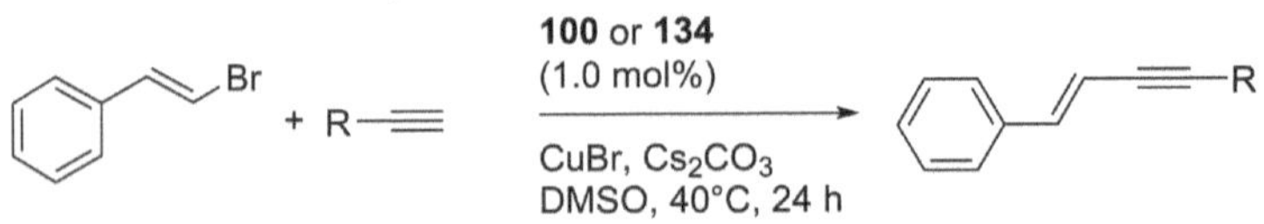

Entry	Palladacycle	R	Yield (%)
1	**100**	Ph	91
2	**100**	*m*-MeC_6H_4	83
3	**100**	*p*-*t*-BuC_6H_4	75
4	**100**	*p*-FC_6H_4	77
5	**134**	*n*-Hexyl	75
6	**134**	*n*-Pentyl	89

The same research group continued their studies on Sonogashira coupling of β-bromostyrenes with terminal alkynes, and deacetonative Sonogashira coupling reaction of aryl propargyl alcohols with aryl chlorides. Satisfying results were observed with the use of cyclopalladated ferrocenylimine **100** as catalyst (Fig. 2.20, Table 2.13) *(101)*. A series of conjugated enynes were successfully obtained via these catalyst systems with moderate-to-excellent yields. A broad range of functional groups could be tolerant, and the products were furnished as specific *E* isomers (Table 2.13).

Later, Wu and co-workers further demonstrated the deacetonative Sonogashira coupling of aryl propargyl alcohols with aryl chlorides with the catalyst system of palladacycle **100/XPhos**. Symmetric and unsymmetric diaryl acetylenes were afforded in good-to-excellent yields (Table 2.14) *(102)*. Electron-poor, electron-neutral, and even inactive sterically hindered electron-rich aryl chlorides were feasible coupling partners in this reaction. Especially, electron-poor alkynes were well-tolerated under this reaction conditions, for which the traditional catalyst system failed to activate such a coupling partner.

Table 2.14 The deacetonative Sonogashira coupling of aryl propargyl alcohols with aryl chlorides

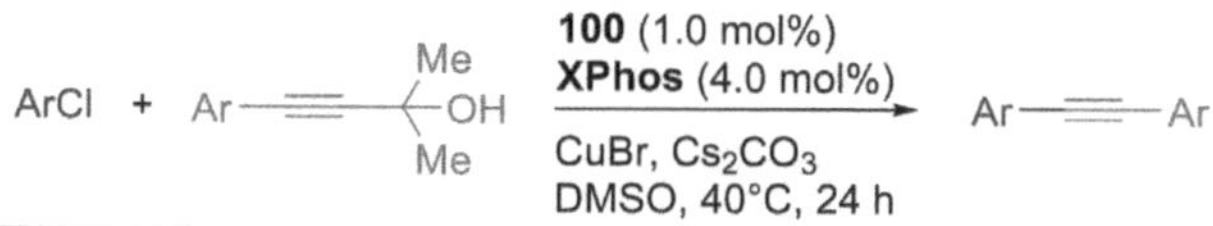

Entry	ArCl	Alkyne	% yield
1	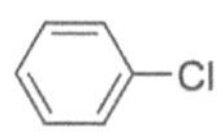		96
2	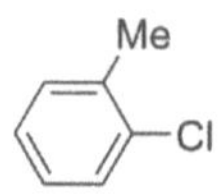		97
3	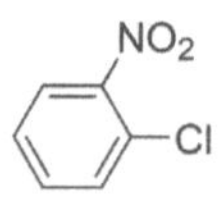		95
4	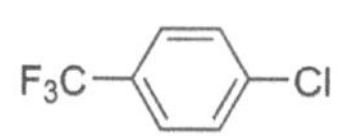	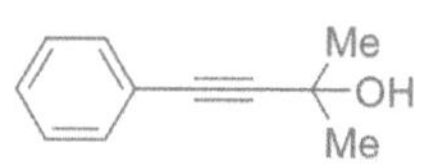	94
5	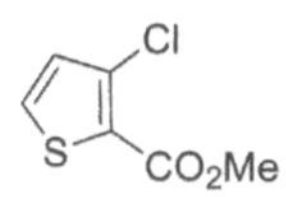	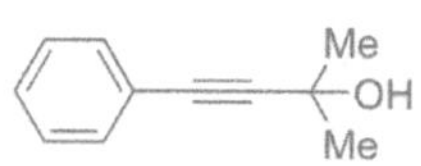	94
6	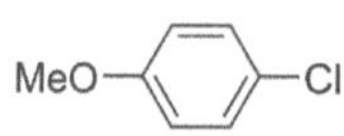	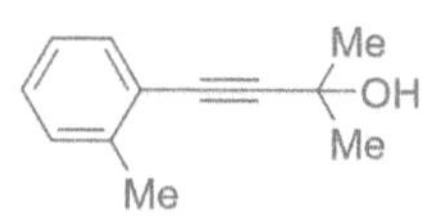	76
7	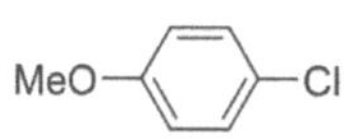	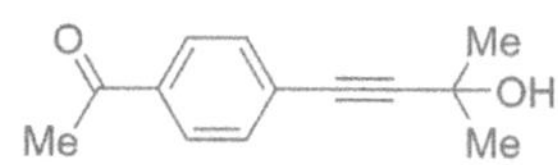	78
8	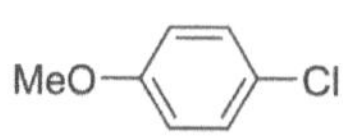	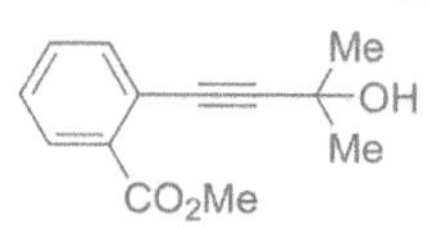	58
9	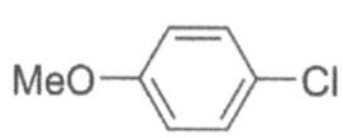	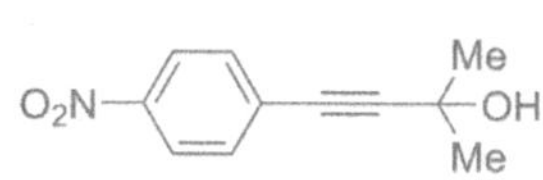	75

Xu and co-workers performed Sonogashira coupling reaction of aryl halides with phenylacetylene using a new triphenylphosphine adduct of cyclopalladated ferrocenylpyridazine containing a halide anion (**136–139**) (Fig. 2.21) *(103)*. The results showed that palladacycles **136** and **137** exhibited a higher catalytic activity than palladacycles **138** and **139** in the Sonogashira reaction (Table 2.15). 0.1 mol% of the catalyst was enough to catalyze the reaction in excellent product yields.

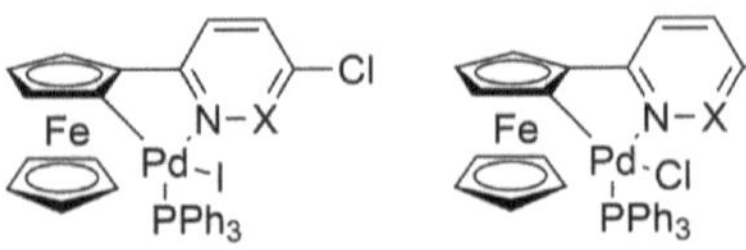

136 X = C
137 X = N
138 X = C
139 X = N

Fig. 2.21 Examples of ferrocenylpyridazine palladacycles.

Table 2.15 Sonogashira reaction of aryl halides and phenylacetylene with palladacycles **136-139**

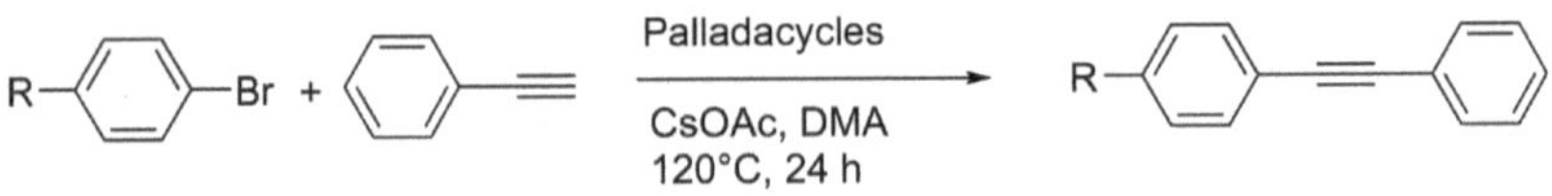

Entry	Palladacycle (mol% Pd)	R	Yield (%)	Ref.
1	**136** (1.0)	H	96	103
2	**137** (1.0)	H	97	103
3	**138** (1.0)	H	88	103
4	**139** (1.0)	H	90	103
5	**136** (0.1)	*p*-MeC(O)	96	103
6	**137** (0.1)	*p*-MeC(O)	98	103
7	**138** (0.1)	*p*-MeC(O)	91	103
8	**139** (0.1)	*p*-MeC(O)	92	103

4. Amine/Imine/Oxime-Based Palladacycles

Nitrogen-containing ligands are often used in the coordination chemistry. There are various types of nitrogen donor ligands employed in the homogeneous catalysis. When they coordinated in the palladacycle, it is well known that the formation of π-back bonds contribute to the total M-L bond strength. Furthermore, the nitrogen donors are part of the chelating systems together with phosphorus- or carbon-donors. Here they will be divided into different sections on the basis of their structures or the hybridization of the *N*-donor atom.

4.1 Heck Cross-Coupling Reaction

4.1.1 Amine-Based Palladacycles

An early discovery of C,N palladacycles **140** was made by Cope and Friedrich in 1968 *(9)*. The application of **140** in the Heck coupling reaction appeared in 1998 in a study by the Bedford group *(43)*. The palladacycle was used to test the coupling of 4-bromoacetophenone with *n*-butylacrylate. 0.2 mol% of Pd could catalyze the reaction to give 53% product yield in 1 hour. Lengthening the reaction time to 20 h gave a complete conversion. It was noted that the monomeric form **142** performed better than the dimeric form **140** (Fig. 2.22; Table 2.16).

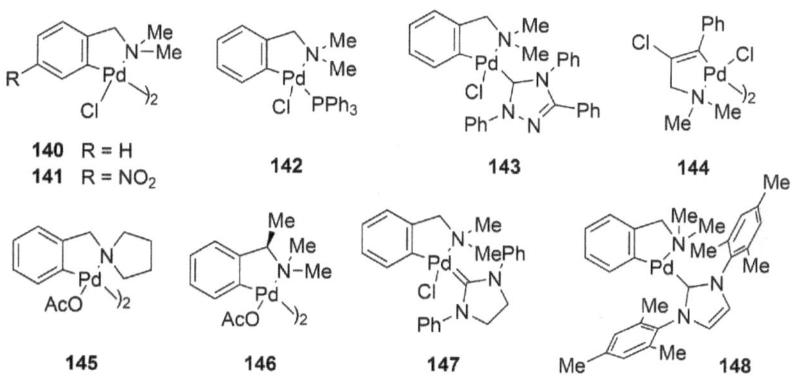

Fig. 2.22 Examples of aliphatic amine-based palladacycles.

A special propargyl amine-palladacycle **144** was developed by Dupont group (Fig. 2.22) *(105)*. It can be easily obtained by the chloropalladation of heterosubstituted alkynes. Under standard reaction conditions, various activated aryl iodides, bromides, and chlorides could be coupled effectively with the *n*-butylacrylate. A kinetic and mechanistic study was carried out which showed that the palladacycles are merely a reservoir of active Pd(0) species that undergo oxidative addition of aryl halide on the surface with subsequent detachment, generating homogenous Pd(II) species *(109)*. It also showed that a slight excess of olefin would give a rapid increase of the Pd(0) concentration.

The Malinakova group further developed the aliphatic amine-based palladacycles **145** and **146** and applied in the three-component coupling of arylboronic acids, allenes, and imines to synthesize the homoallylic amines *(39)*. Moderate yields (58% for **145**, 58% for **146**) were obtained which showed a better catalytic activity than the phosphapalladacycles **16–18** (Scheme 2.6).

Table 2.16 C,N-donor palladacycles in Heck reaction

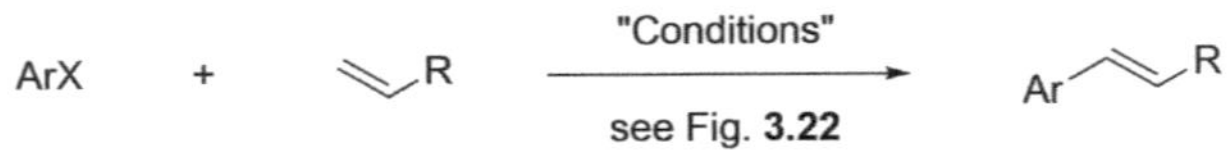

Entry	Palladacycle (mol% Pd)	ArX	R	Time (h)	Solvent	Base	Temp. (°C)	Yield (%)	Ref.
1	**140** (0.2)	*p*-MeC(O)C_6H_4Br	CO_2*n*-Bu	20	DMA	NaOAc	140	100[a]	43
2	**140** (0.2)	C_6H_5I	Ph	35	DMF	*n*-Bu_3N	85	72	104
3	**140** (1.0×10^{-5})	C_6H_5I	CO_2Me	11	DMA	Et_3N	140	34	81, 82
4	**141** (0.2)	C_6H_5I	Ph	9	DMF	*n*-Bu_3N	85	55	104
5	**142** (1.0×10^{-5}) 10^{-5}	C_6H_5I	CO_2Me	11	DMA	Et_3N	140	91	81, 82
6	**143** (1.0)	*p*-$MeOC_6H_4Br$	Ph	14	DMA	NaOAc	130	79[a]	42
7	**143** (1.0)	*p*-MeC(O)C_6H_4Cl	Ph	14	DMA	$Ca(OH)_2$	130	97[a]	42
8[b]	**144** (0.1)	*p*-CNC_6H_4Cl	CO_2*n*-Bu	24	DMA	NaOAc	150	60	105
9[b]	**147** (1.0×10^{-2})	*p*-$MeOC_6H_4Cl$	Ph	24	NMP	NaOAc	150	61	106
10	**148** (0.1)	*p*-$MeOC_6H_4Br$	CO_2*n*-Bu	24	NMP	K_2CO_3	140	86	107
11	**148** (2.0)	*p*-ClC_6H_4I	CN	18	NMP	K_2CO_3	100	88[a]	107
12[c]	**148** (2.0)	*p*-ClC_6H_4I	$PO(OEt)_2$	18	NMP	K_2CO_3	100	61	107
13[c]	**148** (2.0)	*p*-ClC_6H_4I	$SO_2C_6H_4$	18	NMP	K_2CO_3	100	37	107
14[b,c]	**148** (2.0)	*m,m′*-$Me_2C_6H_4OTf$	*p*-$MeOC_6H_4$	18	NMP	K_2CO_3	140	73	108

[a]Yield of mixture of *trans*- and *cis*-product.
[b]TBAB was added as an additive.
[c]4Å Finely ground molecular sieves was added.

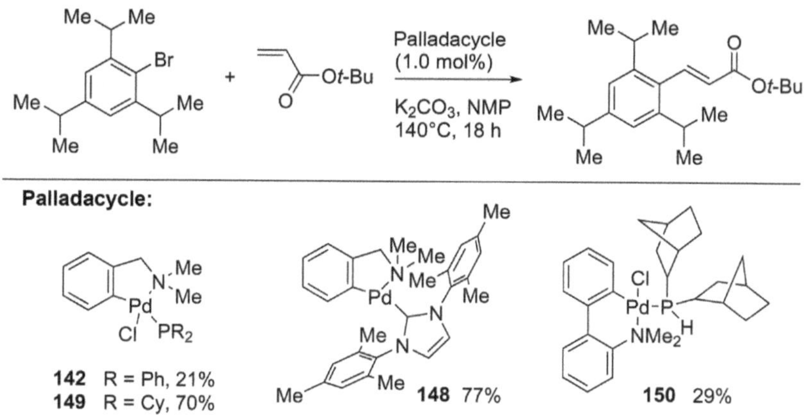

Scheme 2.17 Comparison of catalytic activity of amine-based palladacycles in challenging Heck reaction.

The amine-based palladacycles **148**, **149,** and **150** were used to evaluate the catalyst activity in challenging the Heck coupling reaction (Scheme 2.17) *(107)*. 2,4,6-Diisopropylbromobenzene is a steric congested substrate that is seldom a successful coupling partner in the cross-coupling reaction. This challenging substrate could be coupled with *t*-butyl acrylate with 1.0 mol% Pd (palladacycle **148**) to give 77% product yield. However, when aniline-based palladacycle **150** was used, only 29% product was obtained. Palladacycle **148** was further employed in the Heck coupling of 3,5-dimethylphenyl trifluoromethanesulfonate and 4-methoxystyrene and 73% isolated yield was obtained *(108)*.

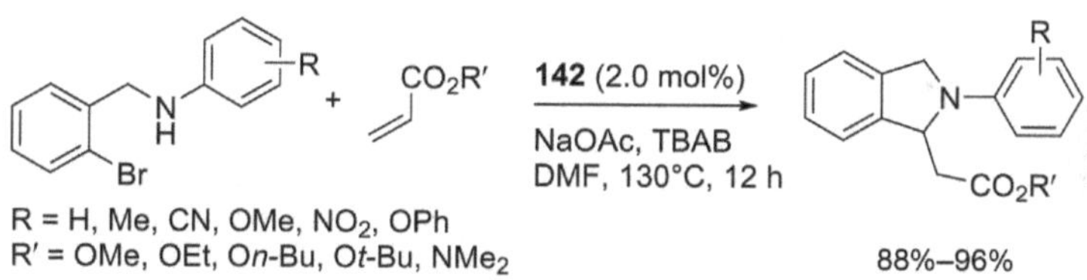

Scheme 2.18 One-pot tandem Heck-aza-Michael addition reaction.

One-pot synthesis is an attractive tool for industrial application. A tandem Heck-intramolecular-aza-Michael addition reaction protocol was demonstrated for one-pot synthesis of isoindolines from unprotected amines with high yields *(110)*. Palladacycles **140** and **142** were a feasible catalyst used in such a reaction (Scheme 2.18).

Apart from the aliphatic amine palladacycles, aromatic amine (aniline) is also a desirable skeleton of the palladacycles. Palladacycles **151–154** were easily synthesized by treating the chloride-bridged palladacycles

Table 2.17 C,N donor palladacycles in Heck cross-coupling reaction

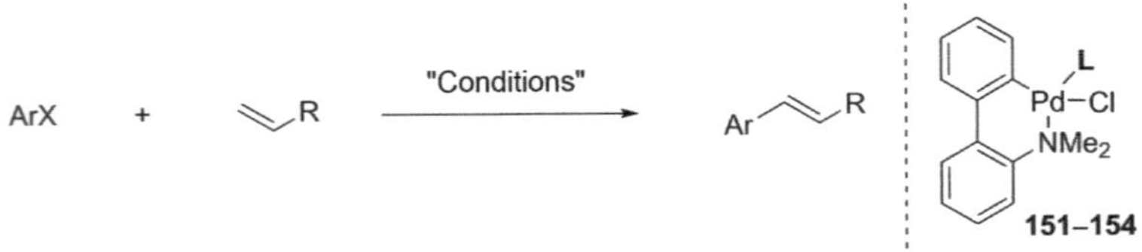

151–154

Entry	Palladacycle (mol% Pd)	L	ArX	R	Solvent	Base	Temp. (°C)	Yield (%)	Ref.
1	**151** (0.25)	$HPNor_2$	*p*-MeC_6H_4Cl	CO_2*n*-Bu	DMA	Na_2CO_3	140	48	111
2	**152** (0.25)	HP(*t*-Bu)$_2$	*p*-MeC_6H_4Cl	CO_2*n*-Bu	DMA	Na_2CO_3	140	46	111
3	**153** (0.25)	$HPCy_2$	*p*-MeC_6H_4Cl	CO_2*n*-Bu	DMA	Na_2CO_3	140	89	111
4	**152** (0.25)	PCy_3	*p*-MeC_6H_4Cl	CO_2*n*-Bu	DMA	Na_2CO_3	140	42	111
5	**153** (2.0)	Pyridine	*p*-$MeOC_6H_4Br$	Ph	MeOH	$K_3PO_4{\cdot}3H_2O$	65	40	112
6	**154** (2.0)	4-Picoline	*p*-$MeOC_6H_4Br$	Ph	MeOH	$K_3PO_4{\cdot}3H_2O$	65	27	112

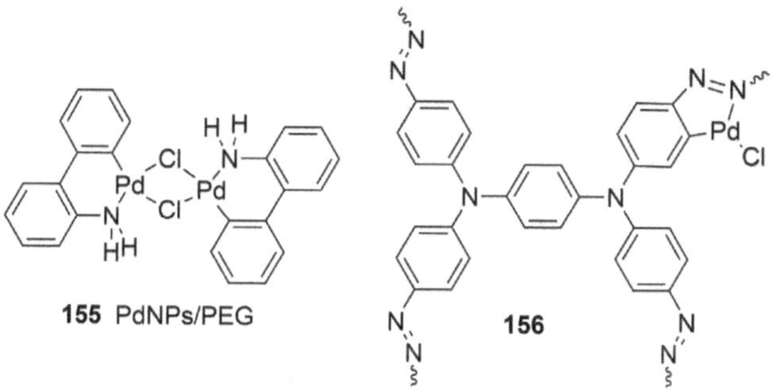

Fig. 2.23 Examples of polymer-supported aromatic amine-based palladacycles.

with the common phosphines and they exhibited a good activity in the Heck reaction (Table 2.17) *(111)*. Surprisingly, PCy_3 often show a superior effect in cross-coupling reaction but **154** (bearing PCy_2) showed a negative result when compared to other phosphine ligands with the palladacycles. However, when the phosphine ligand was replaced by other common ligands, such as pyridine, 4-picoline, phen, and bipy, poor results were obtained even using aryl bromide as a coupling partner with a high catalyst loading *(112)*.

Supported palladacycles were investigated for various organic transformations. The Karami group developed a polyethylene glycol-supported recyclable NC palladacycle **155** for Heck coupling (Fig. 2.23) *(113)*. The dimeric aniline-based palladacycle nanoparticles were added into PEG with different molecular weights to yield solidified palladacycles. Pd/PEG 15000 showed a better defined geometrical shape and it is relatively more active than Pd/PEG 1000 in the Heck reaction between iodobenzene and styrene. A cyclopalladated azo-linked porous polymer **156** was investigated and it was found that the cyclic-form palladacycles decomposed to acyclic-form during the first cycle of the Heck reaction *(114)*.

4.1.2 Imine-Based Palladacycles

Blackmond and co-workers investigated the reactivity of imine palladacycles **157–159** (Fig. 2.24) by the kinetic measurements *(19,115)*. It was showed that the palladacycles with nitrogen-based ligands were more active than the Herrmann's phosphapalladacycles in the Heck coupling of aryl halides with olefins.

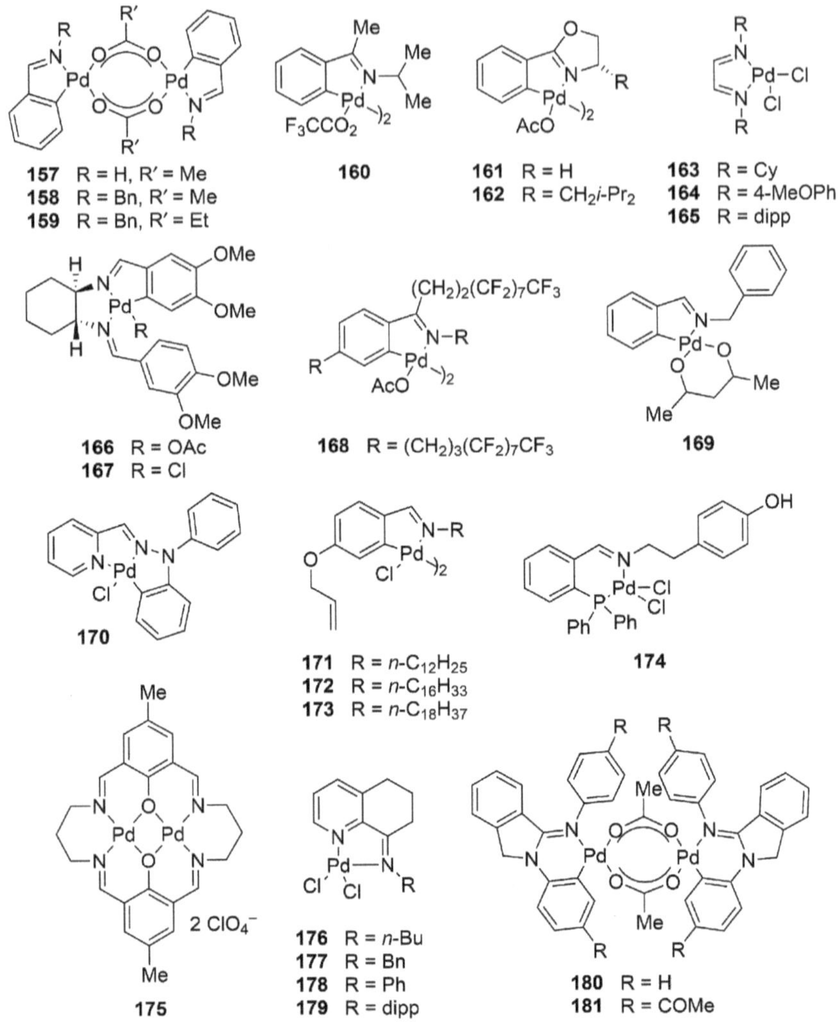

Fig. 2.24 Examples of Imine-based palladacycles (dipp = 2,6-diisopropyl).

Further studies showed that both dimeric and monomeric imine-based Milstein's palladacycles performed well in the Heck coupling of aryl iodides and bromides (Table 2.18). Notably, infrared irradiation and microwave heating with shorter reaction time could be used which gave comparable results to the conventional heating with longer reaction time *(116)*. Recently, the Wu and Li group developed a series of highly ordered amphiphilic cyclopalladated arylimine monolayer compounds **171–173** for Heck reaction (Fig. 2.24) *(117)*. Slightly lower product yield was obtained when the coupling medium changed from DMF to water.

Table 2.18 Imine-based palladacycles in Heck reaction

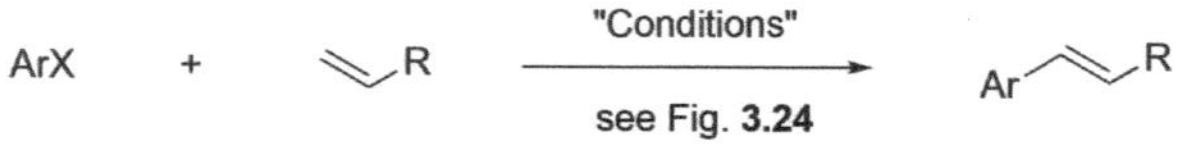

Entry	Palladacycle (mol% Pd)	ArX	R or alkene	Solvent	Base	Time (h)	Temp. (°C)	Yield (%)	Ref.
1	**160** (1.4×10^{-3})	C_6H_5Br	CO_2Me	NMP	Na_2CO_3	43	140	93	118
2	**161** (2.8×10^{-3})	C_6H_5Br	CO_2Me	NMP	Na_2CO_3	130	140	96	118
3	**162** (2.8×10^{-2})	C_6H_5I	Indene	NMP	Na_2CO_3	115	120	72	118
4	**166** (1.0×10^{-3})	C_6H_5I	CO_2Me	DMA	Et_3N	24	reflux	84	119
5	**167** (1.0)	C_6H_5Br	CO_2Me	DMA	Et_3N	24	reflux	68	119
6	**168** (6.8×10^{-7})	C_6H_5I	Ph	DMF	Et_3N	14	140	88	120
7	**158** (2.0×10^{-2})	C_6H_5Br	Ph	EGME/H_2O	K_2CO_3	20	120	98	121
8	**159** (2.0)	C_6H_5Br	Ph	EGME	Cs_2CO_3	2	120	68	121
9	**169** (2.0)	C_6H_5Br	Ph	EGME	Cs_2CO_3	2	120	71	121
10	**170** (5.0×10^{-2})	*p*-MeC_6H_4I	CO_2Me	DMF	K_3PO_4	15 minutes	140[a]	98	116
11	**170** (5.0×10^{-2})	*p*-MeC_6H_4I	CO_2Me	DMF	K_3PO_4	10 minutes	160[b]	94	116
12	**170** (5.0×10^{-2})	*p*-MeC_6H_4I	CO_2Me	DMF	K_3PO_4	1	140	97	116
13	**173** (0.1)	C_6H_5Br	Ph	DMF	K_3PO_4	12	75	95	117
14	**174** (1.0)	C_6H_5Br	Ph	NMP	K_2CO_3	6	140	85	122
15	**163** (0.5)	*p*-OHC_6H_4Br	Ph	NMP	K_2CO_3	24	150	64	94, 123
16[c]	**176** (7.5×10^{-2})	*p*-$MeOC_6H_4Cl$	Ph	NMP	$Ca(OH)_2$	6	160	82	124
17	**179** (1.0×10^{-2})	*m*-PyBr	Ph	DMF	K_2CO_3	24	150	93	125
18	**180** (1.0)	C_6H_5Br	CO_2*n*-Bu	DMA	Cs_2CO_3	3	145	71	126
19	**181** (1.0)	*p*-MeC_6H_4Br	CO_2*n*-Bu	DMA	Cs_2CO_3	3	145	61	126
20	**181** (1.0)	*p*-CNC_6H_4Cl	CO_2*n*-Bu	DMA	Cs_2CO_3	3	145	53	126

[a]Infrared irradiation was used.
[b]Microwave irradiation was used.
[c]LiBr and TBAB were added as additive.

In 2010, Broggini performed the intramolecular Heck reaction of *N*-allyl-2-halobenzylamine and stable σ-alkylpalladium Heck intermediates were isolated as seven-membered palladacycle in a fused ring structure or five-membered bridge-ring skeleton (Scheme 2.19) *(127)*. It is believed that the strong constraint from the bridged junction hapmers the *cisoid* conformation was essential for β-hydride elimination.

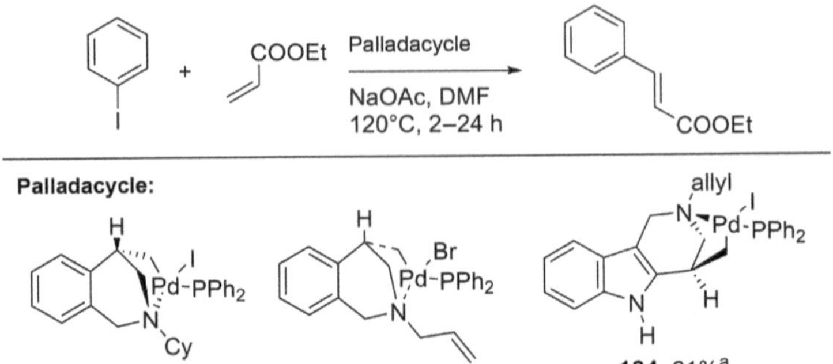

182 98% (0.001 mol% Pd)   **183** 76% (0.1 mol% Pd)   **184** 61%[a]

[a]0.1 mol% Pd, CH_3CN, reflux, 2 h

Scheme 2.19 σ-Alkylpalladacycle with a bridge junction in Heck reaction.

Sun and Ban reported a series of binuclear 4,5,9,10-tetra(arylimino)-pyrenylidenyldipalladium tetrachlorides complexes (**185a–d**) promoting the Heck reaction (Fig. 2.25) *(128)*. They exhibited higher activities than the mono-palladium analogues bearing 4,5-bis(arylimino)pyrenylidene derivatives (**186a–d**) *(129)*.

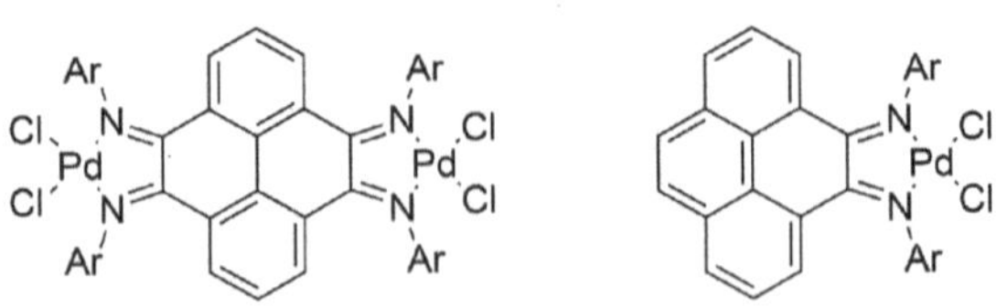

185   **186**

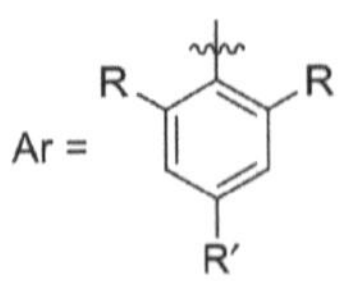

a R = Me, R′ = H
b R = Me, R′ = Me
c R = Et, R′ = H
d R = Et, R′ = Me

Fig. 2.25 Examples of mono/dinuclear arylimino-containing palladacycles.

Silica-supported imine palladacycle **187** was shown as an effective catalyst for Heck reaction of iodobenzene and acrylamide (Scheme 2.20) *(130)*. The catalyst could be recycled without deleterious effect on the reaction.

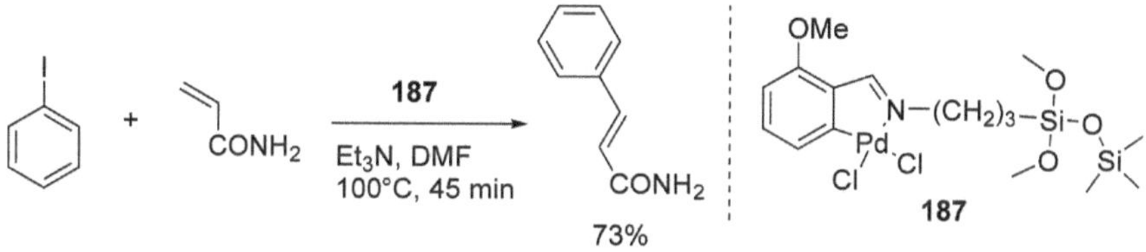

Scheme 2.20 Heck coupling catalyzed by silica-supported imine palladacycle.

In 2016, the Sabino group developed six different bis-imines and their corresponding amines-based tetradentate palladacycles **188–193** *(131)*. From the result of the Heck reaction between bromobenzene and styrene, both palladacycles derived from amines and imines performed well to afford the desirable product in moderate-to-good yield (Scheme 2.21).

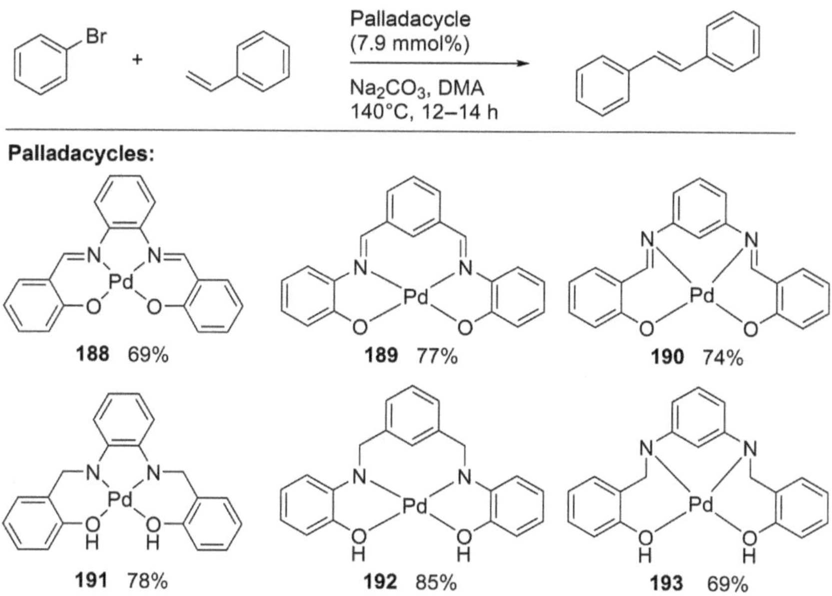

Scheme 2.21 Bis-imines and their corresponding amines-based palladacycles in Heck coupling reaction of bromobenzene and styrene.

4.1.3 Oxime-Based Palladayclces

Oxime-based five-membered palladacycles were prepared in 1970 by *ortho*-palladation of oximes with the alkali metal tetrachloropalldates *(132)*. They are excellent three- or four-atom Pd clusters for C—C bond formation. There are different scaffolds for the oxime-based palladacycles, such as the dimers derived from aromatic and aliphatic ketone oximes *(133)*.

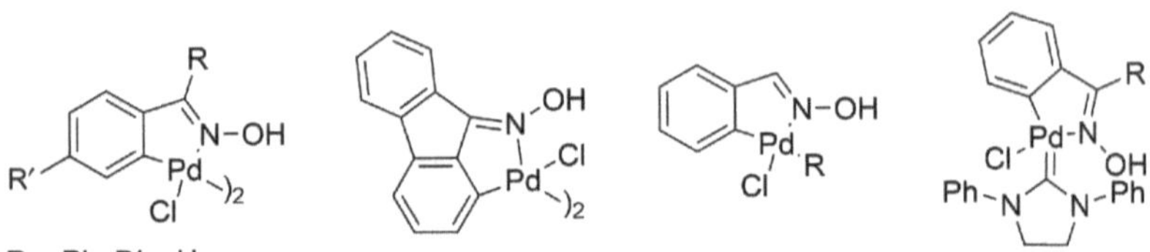

194 R = Ph, R′ = H
195 R = *p*-ClC_6H_4, R′ = Cl
196 R = *p*-$MeOC_6H_4$, R′ = OMe
197 R = Me, R′ = H
198 R = Me, R′ = OH

199

200 R = PPh_3
201 R = $P(OEt)_3$

202 R = Me
203 R = Ph

Fig. 2.26 Examples of oxime-based palladacycles.

In 2000, palladacycles **194–197** (Nájera's catalyst) were used for the Heck reaction of aryl iodides and acrylates with 0.01 mol% of Pd (Fig. 2.26) *(134)*. Since then, Nájera's catalyst derivatives were widely applied in the Heck reaction (Table 2.19) *(104,135)*. Aryl iodides and bromides were applicable coupling partners with the oxime-based palladacycles **194–197** and **200–203**. However, only activation aryl chlorides could be coupled under the general reaction conditions and usually high reaction temperature was required (130–160°C).

In 2005, the Kuck and Cao group demonstrated a multiple vinylation of tribenzotriquinacenes and fenestrindanes with styrenes or methyl acrylates with excellent product yields (Scheme 2.22) *(139)*. It is notable that this vinylation reaction is not well-performed when common Pd sources were applied, such as $Pd(OAc)_2$ and $Pd(PPh_3)_4$.

Electron-rich styrenes are always more challenging coupling partners. Nájera and co-workers further applied palladacycles **195** and **197** in the synthesis of methylated resveratrol, piceatannol, and pinosilvine by Heck reaction of electron-rich 3,5-dimethoxystyrenes (Fig. 2.26) *(140)*. It provided an alternative synthetic pathway for these useful compounds where they failed to be synthesized using the traditional $Pd(OAc)_2$ catalyst system.

Table 2.19 Oxime-based palladacycles in Heck reaction

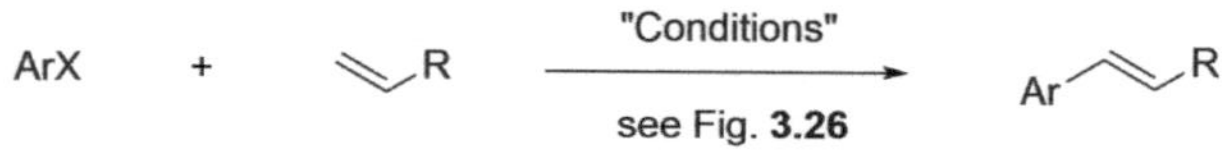

Entry	Palladacycle (mol% Pd)	ArX	R or alkene	Solvent	Base	Time (h)	Temp. (°C)	Yield (%)	Ref.
1	**194** (1.0×10^{-3})	C_6H_5I	CO_2Me	NMP	Et_3N	12	110	92	134
2	**195** (1.0×10^{-2})	C_6H_5I	CO_2Me	DMF	Et_3N	1.5	110	>99	134
3	**196** (1.0×10^{-2})	C_6H_5I	CO_2Me	DMF	Et_3N	6	110	85	134
4	**197** (0.1)	C_6H_5I	CO_2Et	DMA	*n*-Bu_3N	17	85	97	104
5	**200** (1.32×10^{-3})	*o*-naphBr	Ph	NMP	NaOAc	23	150	89	106
6	**201** (1.6×10^{-3})	*o*-naphBr	Ph	NMP	NaOAc	22	150	82	106
7[a]	**202** (1.8×10^{-2})	*p*-$MeOC_6H_4Cl$	Ph	NMP	NaOAc	24	150	54	106
8[a]	**203** (2.3×10^{-2})	*p*-$MeOC_6H_4Cl$	Ph	NMP	NaOAc	24	150	75	106
9[b]	**198** (1.0)	*p*-$CF_3C_6H_4Cl$	$CH(EtO)_2$	DMA	K_2CO_3	14	120	71	136
10[c]	**197** (3.0)	*m*-F-*p*-PhC_6H_3Br	ethylene	DMA	KOAc	18	105	100	137
11[a]	**195** (2.5×10^{-4})	*o*-$MeOC_6H_4Br$	Ph	water	K_2CO_3	6	100	78	138

[a]TBAB was added as an additive.
[b]TBAA was added as an additive.
[c]Phenothiazine was added as radical inhibitor.

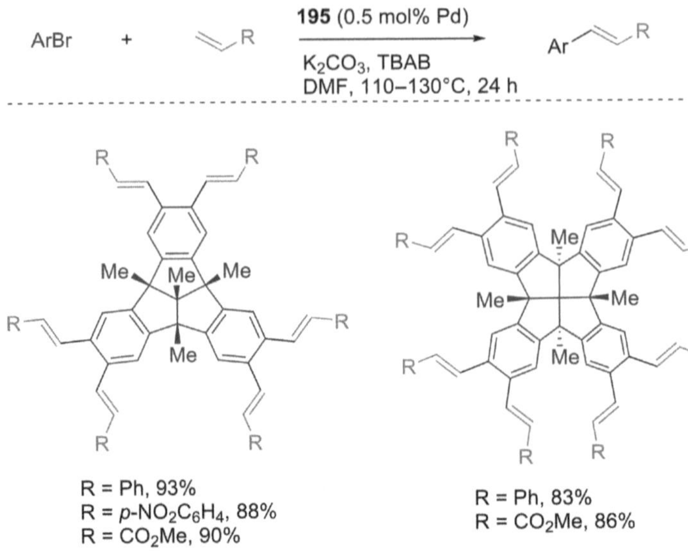

Scheme 2.22 Multi-Heck coupling of aryl bromides with oxime palladacycles.

By undergoing arylation reaction of Morita-Baylis-Hillman adducts with aryl iodides, α-benzyl-β-keto esters could be afforded and then transformed to spirocyclohexadienones using the palladacycle **195** catalyst system *(141)*. Hydroxylated benzylidene, benzyl-pyrrolizidinones, and pyrrolizidines could also be diastereoselectively synthesized by the Heck reaction of Morita–Baylis–Hillman adducts with **195** catalyst *(142)*. Palladacycle **197** was also a good catalyst for the functionalization of 4-pentenoic acid derivatives via Heck reaction and Rh-catalyzed cyclohydrocarbonylation to prepare 5-arylpipecolic acids and piperidine derivatives *(143)*.

Immobilized oxime-based palladacycles could be covalently anchored to various inorganic and organic supports. An imidazolium group was added to **197** to give the palladacycle **204** (Fig. 2.27) *(144)*. A polystyrene-supported palladacycle derived from Kaiser oxime resin (**205**) was further developed by Nájera in 2008 *(145)*. The catalytic activity was maintained for the Heck reaction even in the aqueous solvent environment with 0.01 mol% Pd loading. There are different PVP matrices for the palladacycle **197** to coordinate: (a) PVP as a powder form; (b) the same PVP matrix as a coating of the glass surface inside a megaporous glass rod which is part of a PASSflow microreactor; and (c) megaporous glass Raschig rings coated with the PVP patrix *(146)*. These supported oxime-based palladacycles were used in the Heck coupling reaction under continuous flow conditions. Good product yields were obtained for the arylation of acrylates and styrenes with aryl iodides and bromides in batch mode using the powder form of **206**. In 2009, an ionophilic phosphonium-appended carbopalladacycle **207** was developed for Heck cross-coupling reaction which showed a better result than **204** *(147)*.

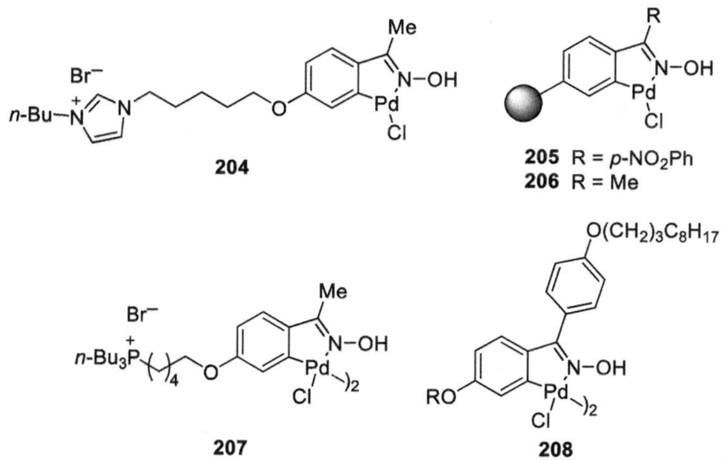

Fig. 2.27 Examples of supported oxime-based palladacycles.

Fluorous oxime palladacycles **208** was developed by Lam and Lo group in 2012 (Fig. 2.27) *(148)*. Under the microwave irradiation, *N*-vinylphthalimide was successfully coupled with various aryl halides and ethyl 3-arylpropionoates and cinnamaldehydes were synthesized through the Heck coupling reactions with moderate recovery of the catalyst. It provides a useful synthetic pathway for the important *N*-vinylphthalimide owing to their biological activity and also starting compounds for the synthesis of phenethylamines. **198** and **205** were also an efficient catalyst system for such a coupling reaction *(149)*. Under similar reaction conditions, Sonogashira coupling of various alkynes and aryl bromides was also demonstrated *(150)*.

Heck arylation of allyl alcohols gave the regio- and chemoselectivity problems that β- and γ-arylation can be occurred (Scheme 2.23). It was suggested that the β-arylation goes through a cationic pathway while γ-arylation takes place through a neutral mechanism *(151)*. Using supported-oxime palladacycle **205** with dicyclohexylmethylamine as the base, γ-arylation took place to give β-arylated aldehydes and ketones *(152)*. Interestingly, palladacycle **195** did not afford any coupling products for the Heck coupling of but-3-en-ol and iodobenzene while palladacycle **198** gave the best product yield.

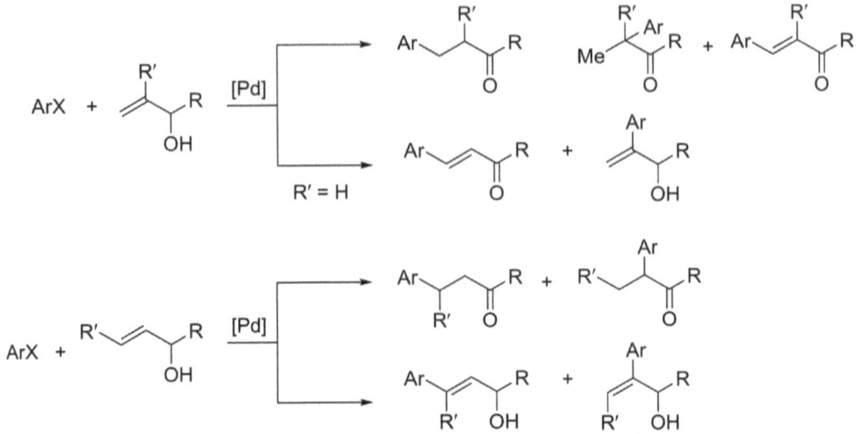

Scheme 2.23 Possible products of Heck reaction of allyl alcohol.

4.1.4 N-Heterocycles C,N or C,N,N-Donor Palladacycles

Since the preparation of 2,2′-bipyridine over 110 years ago, versatile *N*-heterocyclic compounds were used as chelating ligands in coordination and organometallics chemistry. Here, various common *N*-heterocycles palladacycles will be summarized for the Heck reaction. Pyridine is widely used as ligands in homogeneous catalysis and it was introduced to coordinate with the palladium species to form the palladacycles. Fig. 2.28 summarized most of the palladacycles with pyridine as the main scaffold (Table 2.20).

A three-component cascade involving aryl iodide/bromide, allene, and a secondary amine was carried out. In the presence of **222**, **223**, or **227**, corresponding 2-arylallylamines were obtained in 55%–94% yield over 24 h (Scheme 2.24) *(155)*.

The Kantam group developed a *N,N″,N‴,O*-tetrafunctional palladacycle **235** which comprised of multidonor amido/pyridyl/phenolic/amine ligands for the Heck coupling of deactivated aryl halides (Fig. 2.28). Various aryl chlorides were used as coupling partners although only moderate yields were obtained *(161)*. Later, the same research group further applied this palladacycle in the Heck coupling of vinyl heteroaromatic compounds with aryl halides *(162)*.

A polymer-supported di(2-pyridyl)methylamine-palladium dichloride complex covalently anchored to a styrene-maleic anhydride co-polymer supported-palladacycle **238** was developed (Fig. 2.29). It catalyzed the Heck reaction in neat water with conventional heating and microwave irradiation *(156)*. The phosphovanadotungstate polyanion $[P_2W_{15}V_3O_{62}]^{9-}$ is used to support and stabilize the palladacycle with the inorganic framework. In

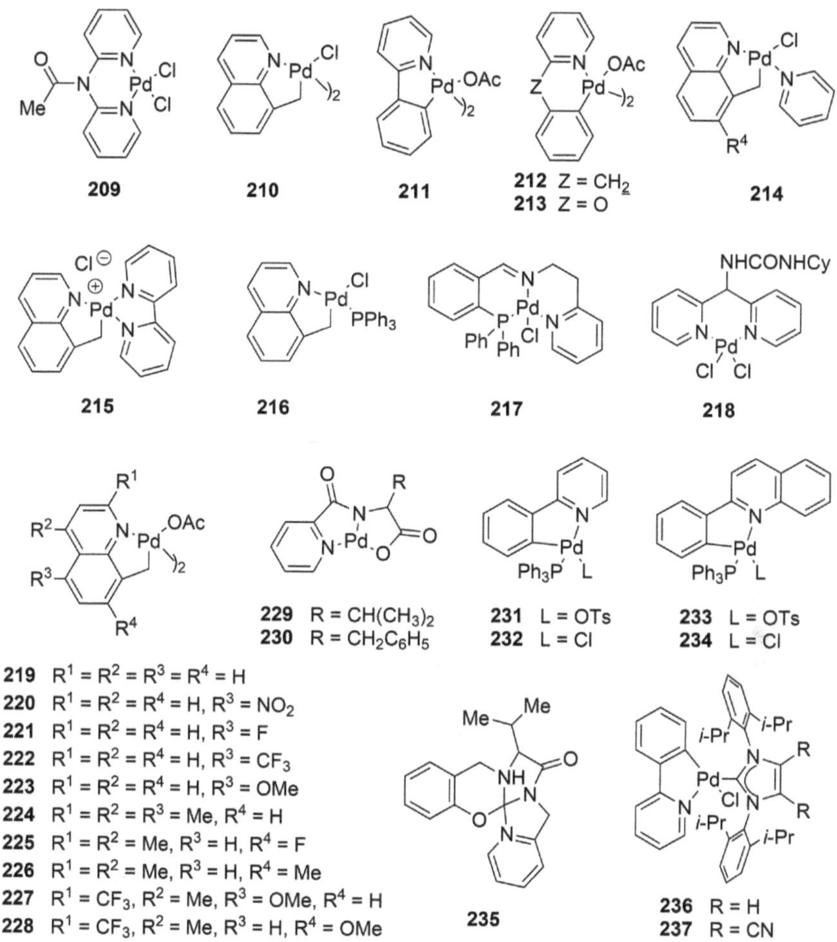

Fig. 2.28 Examples of palladacycles with pyridine moiety.

2013, a series of special design of diol-amide POM-Pd hybrid polyoxometalate palladacycles was developed and applied in the Heck coupling reaction *(163)*. By DFT modeling, it showed that the POM ligand provided the strong electron-withdrawing effect to beneficial to the C-H insertion process. Recently, Menéndez reported a graphene anchored palladacycle **239** which showed excellent cyclability and air stability, and was reused over eight cycles without any loss of activity in the Heck coupling *(159)*. Graphene and its related materials have their unique properties in the electron and thermal conductivity. They are suitable to use as a support of palladium catalyst due to the chemical inertness and intrinsic porosity.

Table 2.20 Pyridine-containing palladacycles in Heck reaction

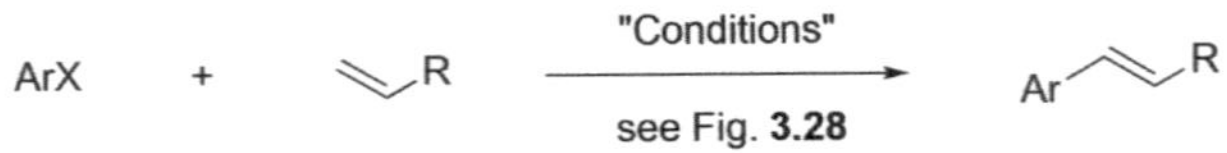

Entry	Palladacycle (mol% Pd)	ArX	R	Solvent	Base	Time (h)	Temp. (°C)	Yield (%)	Ref.
1[a]	**209** (3.3×10^{-3})	C_6H_5Cl	Ph	DMA	K_2CO_3/ Na_2CO_3	90	140	89	153
2	**210** (0.1)	C_6H_5I	Ph	DMF	n-Bu_3N	7.5	85	95	104
3	**211** (0.1)	C_6H_5I	CO_2Et	DMA	n-Bu_3N	5.5	85	89	104
4	**212** (0.1)	C_6H_5I	CO_2Et	DMA	n-Bu_3N	3	85	82	104
5	**213** (0.1)	C_6H_5I	CO_2Et	DMA	n-Bu_3N	6	85	98	104
6	**214** (0.1)	C_6H_5I	CO_2Et	DMA	n-Bu_3N	6.5	85	87	104
7	**216** (0.1)	C_6H_5I	CO_2Et	DMA	n-Bu_3N	22	85	88	104
8	**217** (1.0)	C_6H_5Br	Ph	DGME[b]	Cs_2CO_3	8	120	86	154
9	**223** (2.0×10^{-5})	C_6H_5I	CO_2n-Bu	DMF	CsOAc	96	140	98	155
10	**227** (2.0×10^{-5})	C_6H_5I	CO_2n-Bu	DMF	CsOAc	96	140	78	155
11[a]	**218** (0.1)	p-Me(CO)C_6H_4Br	Ph	H_2O	i-Pr_2NH	10 minutes	100[d]	96	156
12	**229** (1.0×10^{-2})	p-MeOC_6H_4Br	CO_2n-Bu	DMF	Na_2CO_3	20	145	92	157
13	**230** (1.0×10^{-2})	p-MeOC_6H_4Br	CO_2n-Bu	DMF	Na_2CO_3	20	145	89	157
14	**231** (1.3×10^{-2})	BMN[c]	ethylene	NMP	NaOAc	1.5	110	98	158
15	**233** (1.3×10^{-2})	BMN[c]	ethylene	NMP	NaOAc	1.5	110	93	158
16	**239** (0.3)	p-MeC_6H_4Br	Ph	DMF	Na_2CO_3	7.5	140	89	159
17[a]	**236** (0.1)	p-MeC(O)C_6H_4Cl	CO_2n-Bu	DMF	Na_2CO_3	16	140	10	160
18[a]	**237** (0.1)	p-MeC(O)C_6H_4Cl	CO_2n-Bu	DMF	Na_2CO_3	16	140	31	160
19[a]	**237** (0.1)	p-HOC_6H_4Br	CO_2n-Bu	DMF	Na_2CO_3	16	140	84	160

[a]TBAB was added as an additive.
[b]DGME = ethylene glycol monomethyl ether.
[c]BMN = 2-bromo-6-methoxynaphthalene.
[d]Microwave irradiation was used.

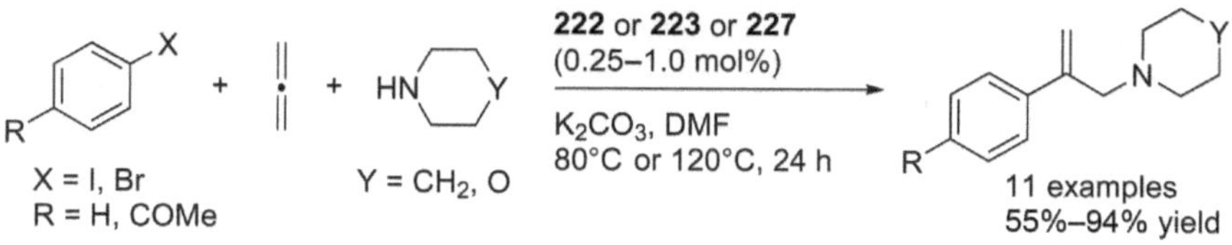

Scheme 2.24 A three-component reaction of ArI/Br, allene, and a secondary amine.

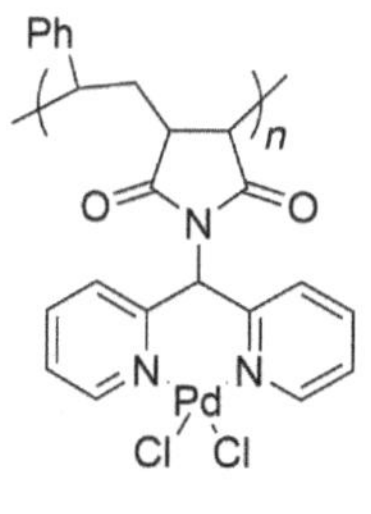

238

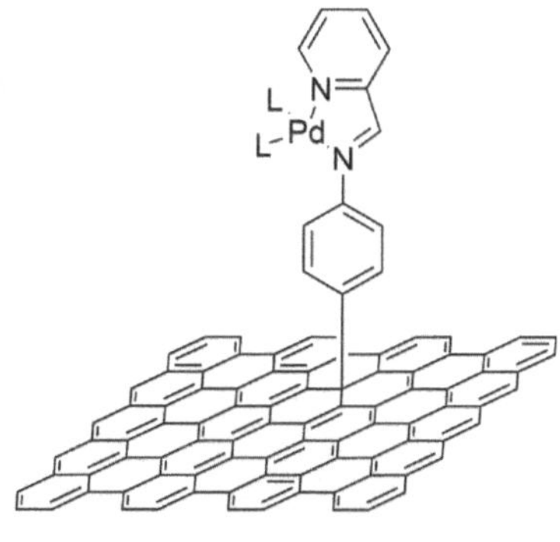

239 TRGO-NPy-Pd

Fig. 2.29 Examples of polymer/graphene supported-palladacycles with pyridine moiety.

Apart from the most common pyridine-containing palladacycles, other *N*-heterocycles can be part of the palladacycles, such as benzimidazole, benzoxazole, benzothiazole, quinolone, and pyrazole (Fig. 2.30). They were successfully applied in Heck coupling reaction and the results are summarized in Table 2.21.

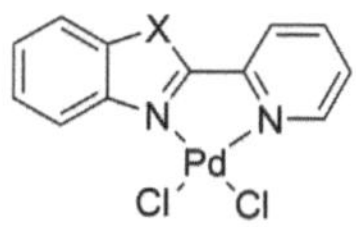

240 X = NH
241 X = O
242 X = S
243 X = NMe

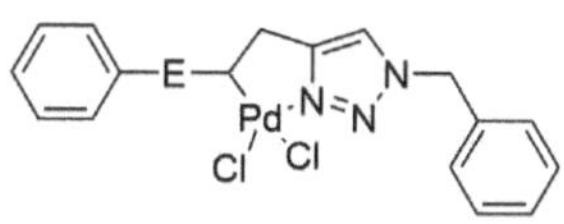

244 E = S
245 E = Se

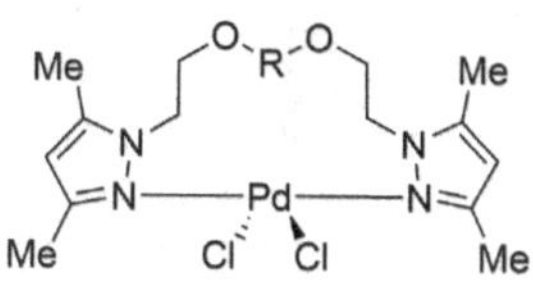

246 R = Et
247 R = *o*-xylyl
248 R = *m*-xylyl
249 R = *p*-xylyl

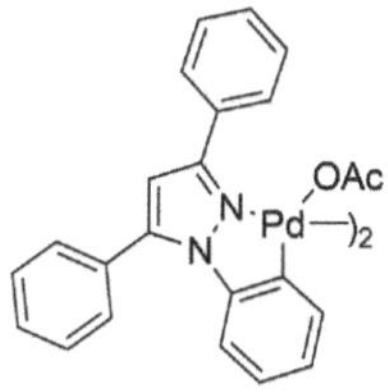

250

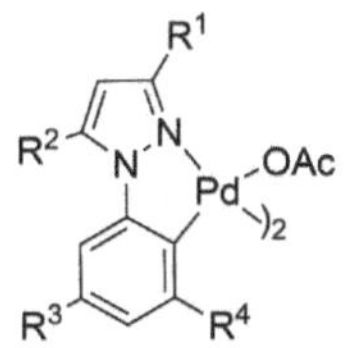

251 $R^1 = R^2 = Me, R^3 = F, R^4 = H$
252 $R^1 = CF_3, R^2 = Me, R^3 = F, R^4 = H$
253 $R^1 = Me, R^2 = CF_3, R^3 = F, R^4 = H$
254 $R^1 = R^2 = Me, R^3 = R^4 = F$
255 $R^1 = CF_3, R^2 = Me, R^3 = R^4 = F$
256 $R^1 = Me, R^2 = CF_3, R^3 = R^4 = F$

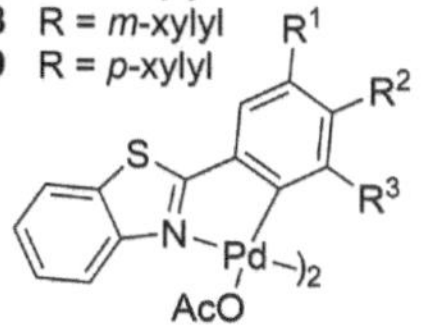

257 $R^1 = F, R^2 = R^3 = H$
258 $R^1 = R^3 = H, R^2 = OMe$
259 $R^1 = R^3 = F, R^2 = H$
260 $R^1 = CF_3, R^2 = R^3 = H$
261 $R^1 = CF_3, R^2 = H, R^3 = F$
262 $R^1 = OMe, R^2 = R^3 = H$

Fig. 2.30 Other *N*-heterocycles palladacycles.

Table 2.21 Other *N*-heterocycles palladacycles in Heck cross-coupling reaction

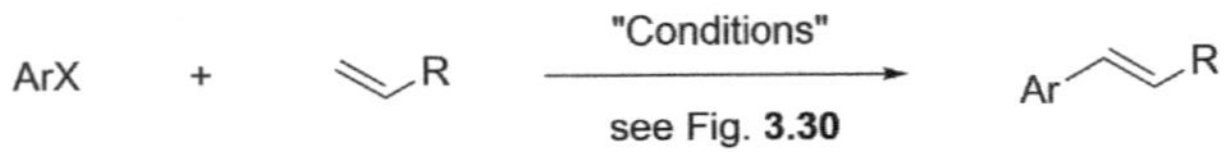

Entry	Palladacycle (mol% Pd)	ArX	R	Solvent	Base	Time (h)	Temp. (°C)	Yield (%)	Ref.
1	**240** (1.0)	*p*-MeC_6H_4Br	CO_2*t*-Bu	DMF	K_2CO_3	—	120	97	164
2	**241** (1.0)	*p*-MeC_6H_4Br	CO_2*t*-Bu	DMF	K_2CO_3	—	120	83	164
3	**242** (1.0)	*p*-MeC_6H_4Br	CO_2*t*-Bu	DMF	K_2CO_3	—	120	75	164
4	**243** (1.0)	*p*-MeC_6H_4Br	CO_2*t*-Bu	DMF	K_2CO_3	—	120	61	164
5	**244** (0.5)	C_6H_5Br	CO_2*n*-Bu	DMF	K_2CO_3	24	110	75	165
6	**245** (0.5)	C_6H_5Br	CO_2*n*-Bu	DMF	K_2CO_3	36	110	78	165
7[a]	**250** (0.4)	C_6H_5Br	Ph	NMP	K_3PO_4	30	160	98	166
8	**246** (0.1)	C_6H_5Cl	Ph	DMF	Et_3N	72	140	49	167
9	**247** (0.1)	C_6H_5Cl	Ph	DMF	Et_3N	33	140	28	167
10	**251** (5.0×10^{-7})	C_6H_5I	CO_2*n*-Bu	DMF	K_2CO_3	48	110	78	168

[a]TBAB was added as an additive.

In 2012, the Malinakova group applied different *N*-heterocycle-palladacycles in the three-component coupling of arylboronic acids, allenes, and imines to synthesize the homoallylic amines *(39)*. The pyridine-, benzo [*h*]quinoline-, benzothiazole-, quinoline-palladacycle were tested and **263** gave the best yield of the product (Scheme 2.25).

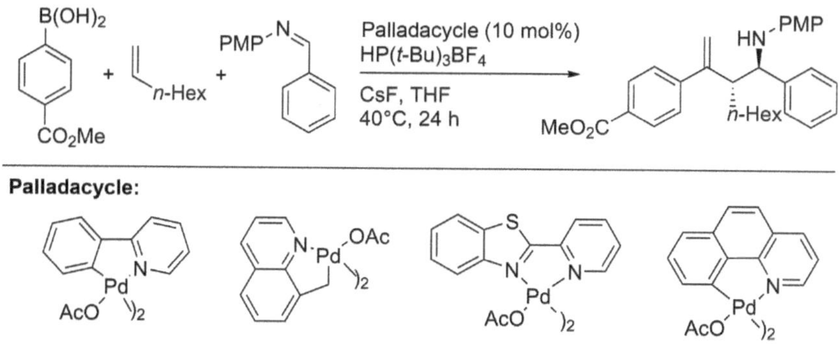

211 61% **219** 50% **263** 76% **264** 21%

Scheme 2.25 Three-component Heck reaction of boronic acids, allenes and imines.

Bimetallic complexes of porphyrinphenanthroline was a complicated catalyst system that the bimetallic complexes can be [Co/Pd], [Ni/Pd], [Zn/Pd], [Cu/Pd], and [Mg/Pd] showed in Fig. 2.31 *(169)*. They were employed in the Heck coupling of iodobenzene and butyl acrylate and palladacycle **267** performed the best among these five porphyrinphenanthroline palladacycles.

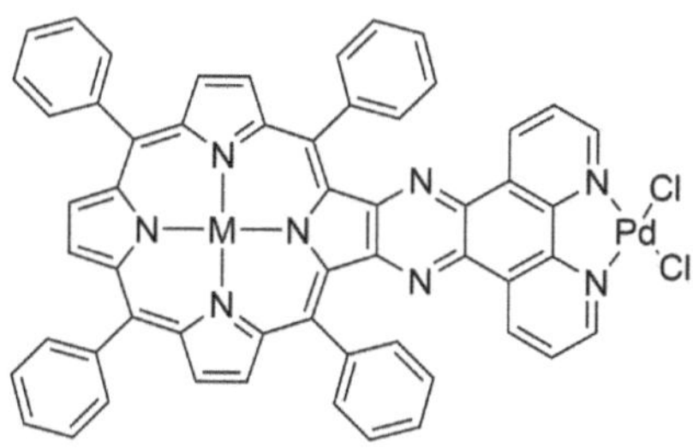

265 M = Co **267** M = Zn **269** M = Mg
266 M = Ni **268** M = Cu

Fig. 2.31 Examples of porphyrinphenanthroline palladacycles.

4.1.5 Other Nitrogen-Containing Palladacycles

In the past few decades, a large number of structurally different nitrogen palladacycles were tested in the Mizoroki-Heck reaction. Apart from the previously described common amines, imines, and oxime structure, other nitrogen-containing palladacycles were also described. Sulfilimine, *(170)* hydrazine, *(116)* benzodiazepine, *(171)* benzamidinate, *(172)* and diketiminatophosphine *(138, 173)* are some of the examples that are a part of the palladacycles for Heck coupling.

In 2015, a series of bis(*N*-substituted oxamate)palladate(II) complexes were reported where the substituent are electron-withdrawing/donating/bulky groups (Fig. 2.32, Table 2.22) *(174)*. The Heck coupling of iodobenzene/bromobenzene and alkyl acrylate/styrene were examined with this series of palladacycles. Competitive reaction between aryl halides and two different olefins (ethyl acrylate vs styrene) was carried out. The coupling of ethyl acrylate is faster than that of styrene by using oxamate-based palladacycles. This investigation result was just the opposite trend with respect to that reported by using phosphapalladacycle *(18)*.

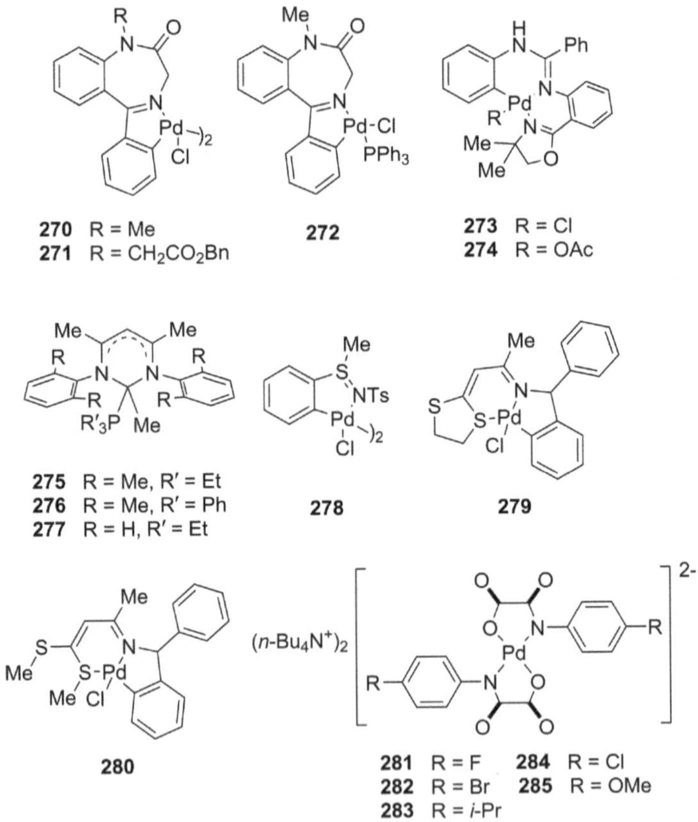

Fig. 2.32 Other nitrogen-containing palladacycles.

Table 2.22 Other nitrogen-containing palladacycles in Heck reaction

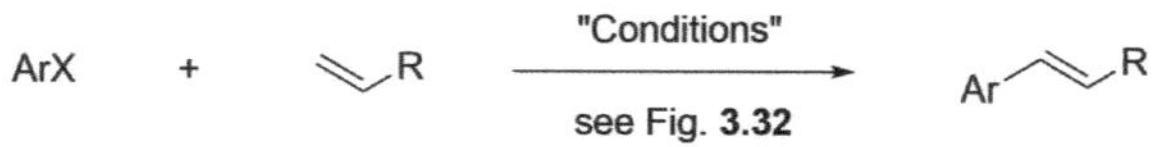

Entry	Palladacycle (mol% Pd)	ArX	R	Solvent	Base	Time (h)	Temp. (°C)	Yield (%)	Ref.
1	**278** (2.5×10^{-4})	C_6H_5I	Ph	NMP	Et_3N	12	140	94	170
2[a]	**270** (0.1)	C_6H_5Br	Ph	DMA	NaOAc	24	120	85	171
3[a]	**270** (0.1)	*p*-MeC(O)C_6H_4Cl	CO_2*n*-Bu	DMA	NaOAc	24	120	31	171
4	**272** (0.1)	*p*-MeC(O)C_6H_4Br	CO_2*n*-Bu	DMA	NaOAc	24	120	78	171
5	**274** (1.0)	*p*-MeC_6H_4Br	Ph	toluene	Cs_2CO_3	8	110	92	172
6	**274** (2.0)	*p*-MeC(O)C_6H_4Cl	Ph	DMA	Cs_2CO_3	24	135	80	172
7[a]	**275** (1.0)	*p*-MeC_6H_4Cl	CO_2Et	DMF-H_2O	K_3PO_4	8	85	90	173
8[a]	**276** (1.0)	*p*-MeC_6H_4Cl	CO_2Et	DMF-H_2O	K_3PO_4	8	85	52	173
9[a]	**277** (1.0)	*p*-MeC_6H_4Cl	CO_2Et	DMF-H_2O	K_3PO_4	8	85	80	173
10	**278** (5.0×10^{-2})	*p*-MeC_6H_4I	CO_2Me	DMF	K_3PO_4	2	140[b]	96	116
11	**280** (5.0×10^{-2})	*p*-MeC_6H_4I	CO_2Me	DMF	K_3PO_4	3	140[b]	87	116
12	**281** (0.5)	C_6H_5Br	Ph	TBAB	Et_3N	4	120	92	174
13	**282** (0.5)	C_6H_5Br	Ph	TBAB	Et_3N	4	120	0	174
14	**284** (0.5)	C_6H_5Br	Ph	TBAB	Et_3N	4	120	66	174
15	**285** (0.5)	C_6H_5Br	Ph	TBAB	Et_3N	4	120	36	174

[a]TBAB was added as an additive.
[b]Infrared irradiation was used.

Sudalai and co-workers further mentioned the synthesis of sulfonamide and hydrazine-based palladacycles in 2006 (Scheme 2.26) *(138)*. They are stable with air and moisture and even soluble in water which allowed the Heck reactions to be carried out in aqueous medium.

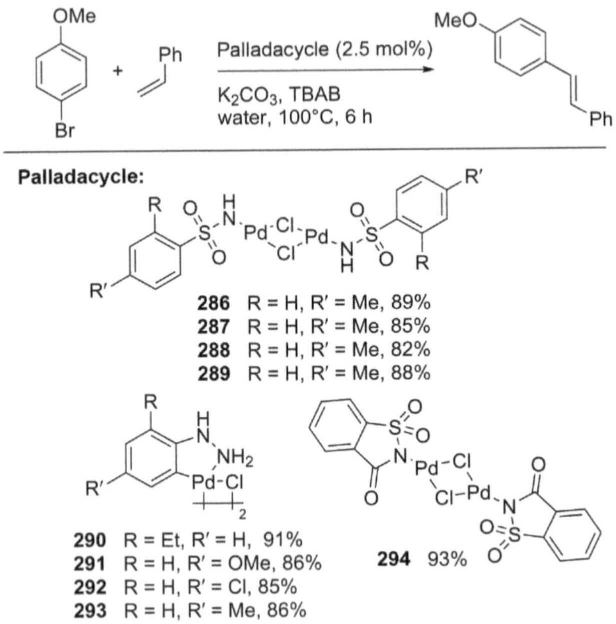

Scheme 2.26 Sulfonamide- and hydrazine-based palladacycles in Heck reaction.

ZnO nanoplated supported hydrazine-based palladacycles were recently explored in the Heck coupling by Rosatmizadeh *(175)*. They act as a highly efficient heterogeneous catalyst for Heck reaction between aryl iodides, bromides, and even chlorides with various olefins. They can be reused for eight sequential cycles without remarkable decrease in the catalytic activity (Scheme 2.27).

4.2 Sonogashira Cross-Coupling Reaction

The amine-derived palladacycles were presented that show good activity in the Sonogashira reaction under different reaction conditions. Jin and co-workers *(176)* reported a water-stable pyridylazetidine-derived Pd(II) complex **296** (Fig. 2.33) for the Sonogashira coupling reaction in aqueous mixture (DMA/water). Various aryl iodides, bromides, and chlorides were

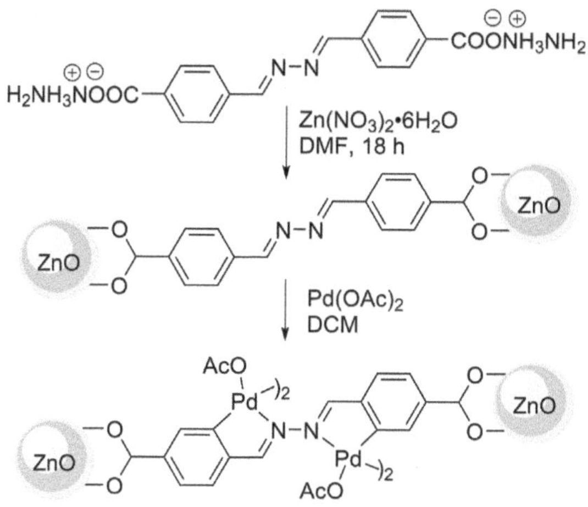

295

Scheme 2.27 Synthesis of ZnO nanoplates supported hydrazine-based palladacycles.

coupled with aryl/alkyl alkynes in good-to-excellent yields with low catalyst loading (0.01–0.5 mol%), Notably, a stoichiometric amount of CuI was necessary for the reaction.

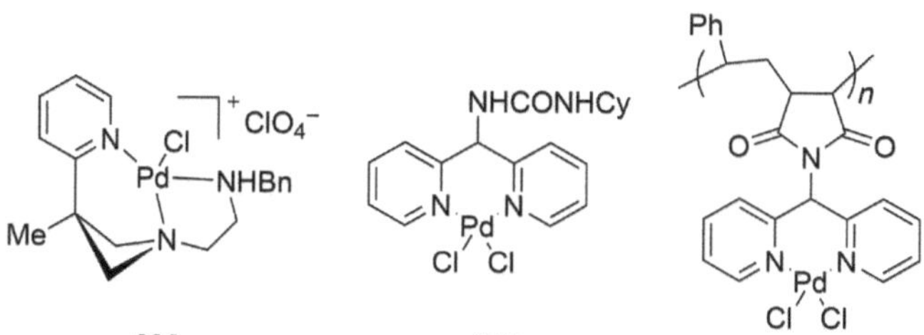

296 **218** **238**

Fig. 2.33 Examples of pyridylazetidine- and di-2-pyridylmethylamine-derived palladacycles.

Pyridine-derived palladacycles showed good catalytic activity in the Sonogashira reaction. Monomeric and polymeric di-2-pyridylmethylamine-based palladium complexes **218** and **238** (Fig. 2.33) were successfully used in the Sonogashira coupling of aryl iodides and bromides in water and in NMP under homogeneous and heterogeneous conditions *(156,177,178).* The (dipyridin-2-ylmethyl)amine-derived palladium(II) chloride complex **218** showed a higher catalyst efficacy in NMP than in water, and no deactivation was observed after several cycles.

Palladacycles **297–299** bearing fluorous-ponytails in the structure of 2,2′-bipyridine were employed in the Sonogashira reaction of aryl iodides or bromides with phenylacetylene or 1-hexyne at room temperature (Fig. 2.34). Only perfluorinated solvents could be used for such reaction protocol. Palladacycle **300** derived from benzoquinoline was found as the most effective palladacycle from a series of related pyridine-based palladium complexes in the Sonogashira reaction *(179)*. Recently, the first new 32-membered macrocyclic dinuclear palladium complex with two aza-crown macrocycles, bearing two pyridine arms, palladacycle **301** (Fig. 2.34), was prepared. It was employed as a moisture/air-stable catalyst for copper- and phosphine-free Sonogashira cross-coupling reaction of aryl halides with phenylacetylene in DMSO *(180)*.

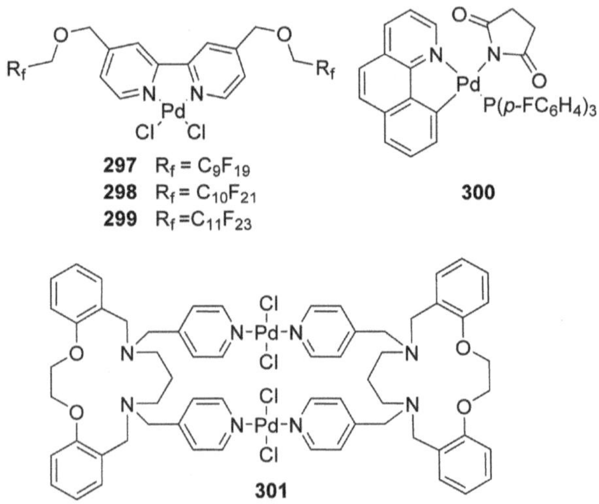

297 $R_f = C_9F_{19}$
298 $R_f = C_{10}F_{21}$
299 $R_f = C_{11}F_{23}$

300

301

Fig. 2.34 Examples of pyridine-based palladacycles.

Other nitrogen-derived palladacycles were developed and depicted in Fig. 2.35 *(181)*. Unsymmetrical palladacycles **302–304** with benzimidazolium-pyrazole were synthesized and employed in the coupling of aryl bromides with phenylacetylene *(182)*. The C,N-palladacycle **305**, N-S(Se)-palladacycles **306–309** (Fig. 2.35) were used as efficient homogeneous catalyst for Sonogashira reaction between various aryl halides and phenylacetylene, and the reaction could be performed without the need for copper-cocatalyst under aerobic conditions *(183)*. Furthermore, the catalytic system showed high catalytic activity (0.05–0.1 mol% catalyst loading) to give excellent yields for aryl iodides and bromides and even aryl chlorides. Saccharin-derived palladacycle **294** was employed in the Sonogashira reaction of various aryl halides with arylacetylenes in neat water at 100°C *(138)*.

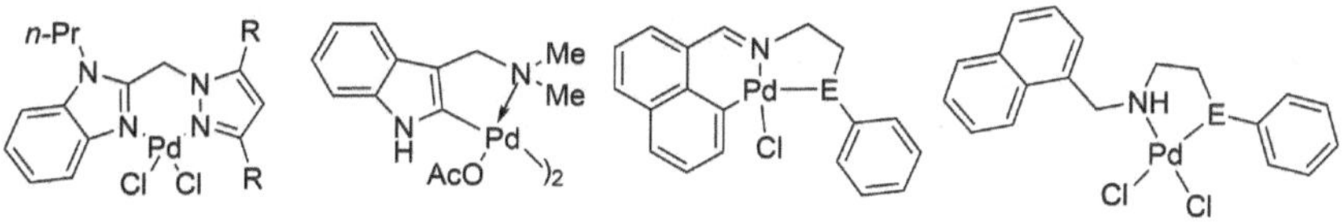

302 R = H **305** **306** E = S **308** E = S
303 R = Me **307** E = Se **309** E = Se
304 R = *t*-Bu

Fig. 2.35 Other nitrogen-derived palladacycles.

In 2011, β-diketiminatophosphine palladacycles **275–277** were employed in the Sonogashira coupling of aryl chlorides with alkynes *(184)*. 0.5 mol% of the catalyst was sufficient to catalyze the coupling. A one-pot double Sonogashira of aryl dichlorides was performed to afford dialkynylbenzene derivatives.

The addition of primary amines to the C=C bond of diphenylalkenyl iminophosphoranes yielded a new subtype of *N,N*-bidentate ligands, which reacted with $PdCl_2$(COD) to give the corresponding *σN,σN*-palladium complex. It contains secondary amino groups, bearing an intrinsically chiral nitrogen atom, and iminophosphorane units. The catalytic activity of palladacycle **310** (Fig. 2.36) was investigated in a copper- and amine-free Sonogashira cross-coupling reaction between aryl iodides and aryl bromides, and terminal alkynes using TBAF as an additive in DMF at 110°C *(181)*. Cyclopalladated allylamines **144**, and **311** (Fig. 2.36) were used as catalysts for the coupling of iodoarenes and activated bromoarenes with terminal alkynes at room temperature *(181)*. The poisoning tests indicated the involvement of soluble Pd(0) species as the catalytically active species generated from the palladacycle **144** (Table 2.23).

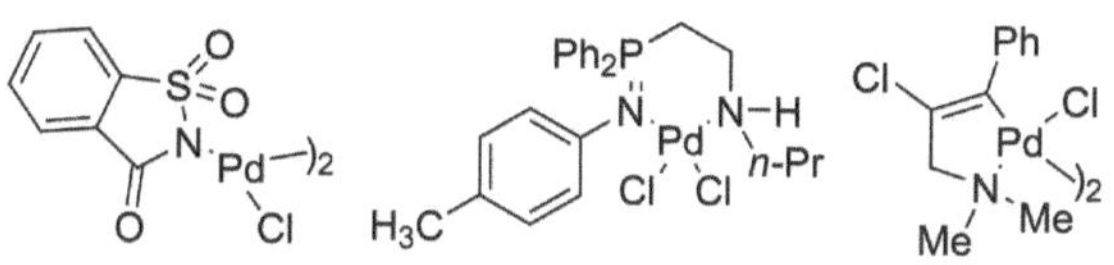

294 **310** **144**

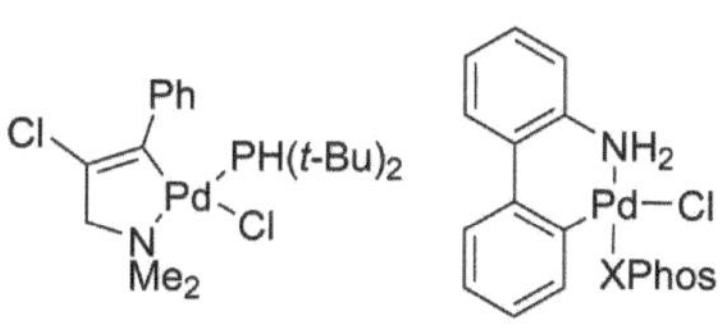

311 **312**

Fig. 2.36 Examples N,C- and N,N-type palladacycles.

Table 2.23 Other nitrogen-containing palladacycles in Sonogashira reaction

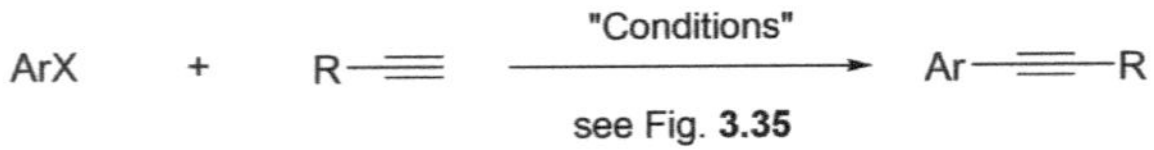

Entry	Palladacycle (mol% Pd)	ArX	R	Solvent	Base	Time (h)	Temp. (°C)	Yield (%)	Ref.
1	**310** (1.0)	*p*-$MeOC_6H_4I$	Ph	DMF	TBAF	1	110	98	180b
2	**310** (1.0)	*p*-MeC_6H_4Br	*o*-NH_2Ph	DMF	TBAF	8	110	90	180b
3	**310** (1.0)	*p*-MeC_6H_4Br	*o*-NH_2Ph	DMF	TBAF	36	110	85	180b
4	**310** (1.0)	*p*-$MeOC_6H_4I$	*n*-Hexyl	DMF	TBAF	38	110	89	180b
5	**144** (1.0)	C_6H_5I	Ph	DMA	TBAOAc	4	30	97	180a
6	**311** (1.0)	C_6H_5I	Ph	DMA	TBAOAc	4	30	43	180a
7[a]	**312** (1.0)	*p*-$SMeC_6H_4Br$	*o,o′*-$(CF_3)_2C_6H_3$	CH_3CN	Cs_2CO_3	4	90	91	185
8	**144** (1.0)	*p*-CNC_6H_4Br	Ph	DMA	TBAOAc	4	30	95	180a
9	**144** (0.1)	*p*-CNC_6H_4Br	Ph	DMA	TBAOAc	4	50	90	180a
10	**144** (1.0×10^{-2})	*p*-CNC_6H_4Br	Ph	DMA	TBAOAc	4	120	82	180a
11	**144** (1.0×10^{-3})	*p*-CNC_6H_4Br	Ph	DMA	TBAOAc	4	150	40	180a
12	**311** (0.1)	*p*-CNC_6H_4Br	Ph	DMA	TBAOAc	4	50	92	180a
13	**311** (1.0×10^{-3})	*p*-CNC_6H_4Br	Ph	DMA	TBAOAc	4	150	39	180a
14[b]	**275** (0.5)	*o*-MeC_6H_4Cl	*o*-MeC_6H_4	DMF-H_2O	piperidine	6	50	86	184
15[b]	**275** (1.0)	*o*-ClC_6H_4Cl	*o*-MeC_6H_4	DMF-H_2O	Piperidine	8	70	89	184

[a]XPhos was added as a co-catalyst.
[b]TBAB was added as an additive.

In 2011, Buchwald and co-workers reported the use of a precatalyst for the continuous flow Heck alkynylations *(185)*. The palladacycle **312** is named as a "second-generation XPhos precatalyst" that easily generated the catalytically active LPd(0) species upon treatment with a base. Different Buchwald's group ligands were used to synthesize the precatalyst and tested in the alkynylations reaction. XPhos gave the best result while DavePhos the second, and SPhos, RuPhos gave a moderate result. Notably that one example of aryl tosylate was demonstrated to couple with 1-hexyne to give 68% product yield at a higher catalyst loading.

The use of supported amine-derived palladacycles in the Sonogashira reaction was rare. The Bakherad group reported a polystyrene-supported palladium(II) ethylenediamine complex **313** used as a highly active catalyst for the heteroannulation of copper-cocatalyzed Sonogashira reaction of aryl iodides with alkynylated pyridinium salt. Further in situ heterocyclization was carried out to form the 2-benzylimidazol[2,1-*b*]pyridines *(186)*. In addition, the study showed that this catalyst could be recovered by filtration and reused up to five times although observing a certain loss of activity. A 0.5 mol% of the palladacycle was leached to the solution during the course of the reaction. Later, the same research group prepared two other different polystyrene-supported palladium(II) complexes, such as polystyrene-supported palladium(II) *N,N*-bis(naphthylideneimino)diethylenetriamine complex **314** *(187)* and polystyrene-supported palladium(II) 1-phenyl-1,2-propanedione-2-oxime thiosemi-carbazone complex **315** (Fig. 2.37) *(188)*. They were found to be highly active catalysts for the copper-free Sonogashira coupling reaction of aryl halides with aryl and alkyl acetylenes in an air atmosphere at room temperature. The catalyst could be reused up to ten times with a slightly low activity. A water soluble salen-palladacycle **316** was developed for the copper-free Sonogashira reaction of aryl iodides and terminal alkynes in neat water at 60°C (Fig. 2.37) *(189)*. Notably, sodium lauryl sulfate was added as a surfactant for this reaction.

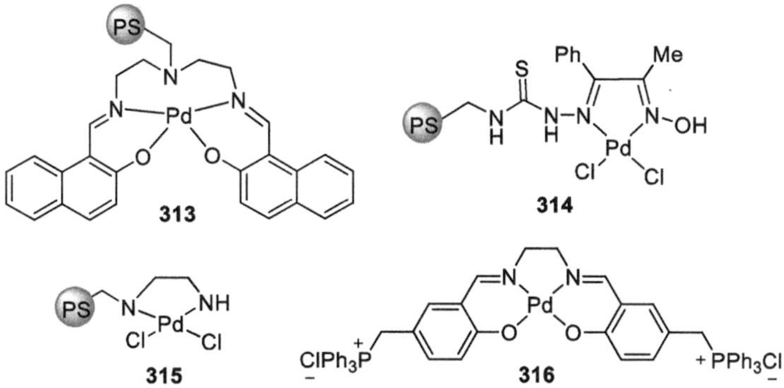

Fig. 2.37 Examples of the supported amine-derived palladacycles.

A new polystyrene-anchored Schiff base-derived palladacycle **317** (Fig. 2.38) was obtained and employed in the aqueous Sonogashira coupling by the Islam group *(190)*. High yields of the corresponding acetylenic compounds were obtained when aryl iodides were coupled with arylacetylene. However, only moderate and poor product yields were given when aryl bromides or chlorides were used as the coupling partners. The catalyst could be reused up to five times without significant loss of the catalytic activity or palladium leaching to the solution. In 2010, Ondruschka and co-workers mixed 2-pyridinealdoxime and polysaccharide chitosan with palladium acetate to generate palladium(II) complex **318** (Fig. 2.38) *(191)*. 1,10-Phenanthroline moiety was anchored to a porous crosslinked chloromethylated polystyrene monolith filling a capillary column for the preparation of a supported palladium(II) complex **319** (Fig. 2.38) *(192)*. The polystyrene-supported palladacycles were used in the coupling reaction of iodobenzene and phenylacetylene with microwave irradiation or in a microfluidic device. Poly-(3,6-dibenzaldimino-*N*-vinylcarbazole)-anchored palladacycle **320** was prepared and used for the copper-free cross-coupling of aryl halides with phenylacetylene as a reusable catalyst in water *(193)*. The coupling of aryl iodides and bromides with phenylacetylene proceeded well with high yield. However, only a very small amount of the coupling product was obtained when aryl chlorides were used as substrates. This immobilized catalyst could be easily separated and reused for further reactions for more than five times without noticeable loss in the catalytic activity.

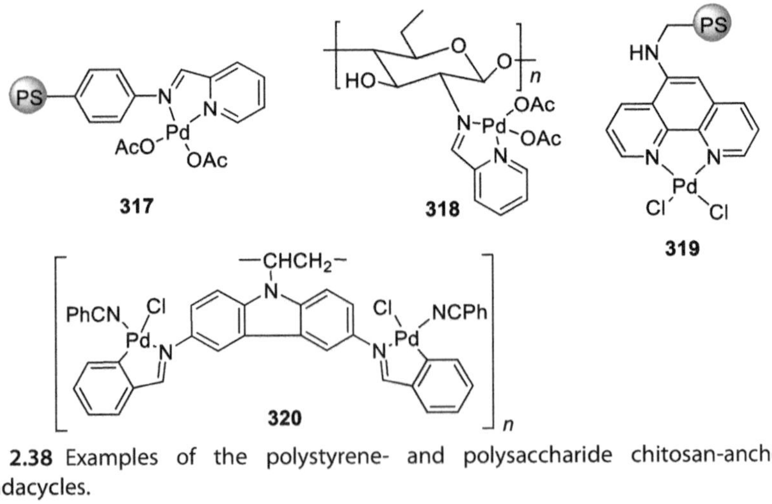

Fig. 2.38 Examples of the polystyrene- and polysaccharide chitosan-anchored palladacycles.

Superparamagnetic cobalt-iron nanoparticles with a $CoFe_2O_4$ structure were functionalized with Schiff-base groups on the surface to form immobilized bidentate palladium(II) complex **321** (Fig. 2.39). It was used as an efficient catalyst for the Sonogashira reaction without the addition of phosphine ligands *(194)*. The immobilized palladium complex could be reused several times without significant degradation in catalytic activity and no palladium leaching was detected.

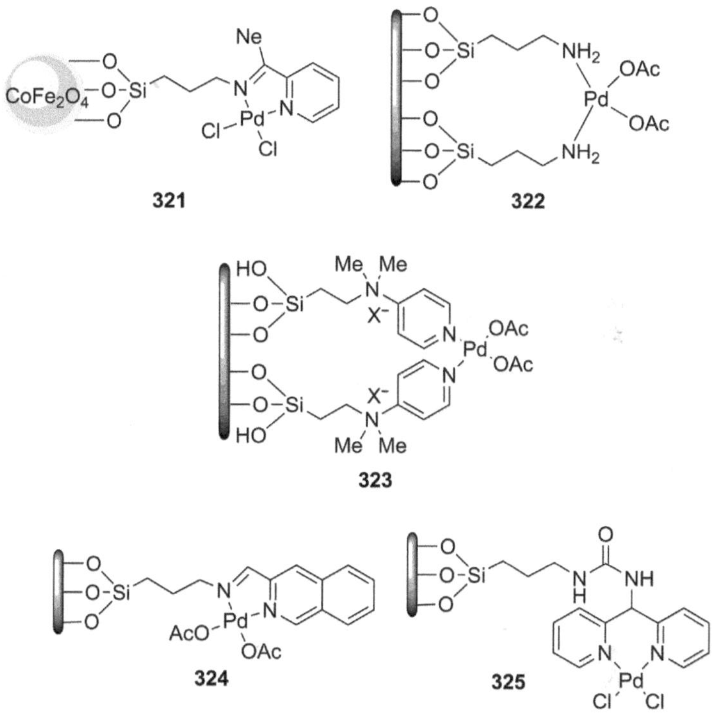

Fig. 2.39 Examples of the silica-supported palladacycles.

The 1,2-diaminocyclohexane-derived palladacycle **322** (Fig. 2.39) *(195)* anchored on amorphous silica gel and pyridine-derived palladacycle **323** *(196)* with functionalized nanostructured silica were prepared by Ahn, Moreau, and co-workers, respectively. They were employed in the copper-free Sonogashira coupling of aryl iodides with phenylacetylene. Quinoline-2-carboimine-derived palladacycle **324s** and di(2-pyridyl) methylamine-derived palladacycle **325** immobilized on the hybrid silica materials were used in the heterogeneous copper-free Sonogashira coupling of aryl iodides and bromides with arylacetylenes *(197,198)*.

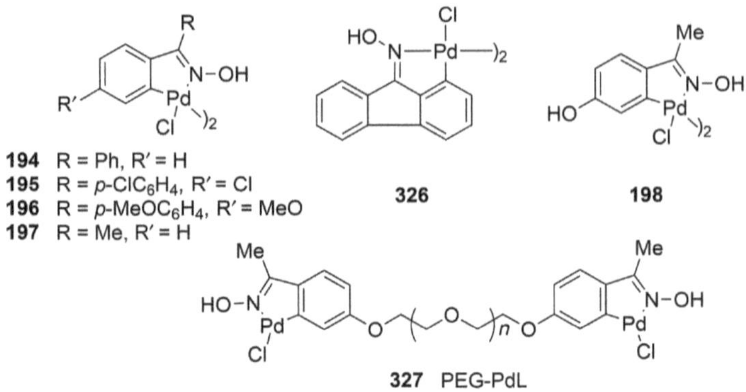

Fig. 2.40 Oxime-based palladacycles.

Oxime-derived palladacycles showed a good activity in the Sonogashira reaction. The chloro-bridge oxime-derived palladacycles **194–198, 326**, and **327** were prepared from different several benzo- and acetophenone-derived oximes by the Nájera group (Fig. 2.40) *(134)*. The preliminary studies demonstrated that palladacycle **196** was an efficient catalyst (0.5 mol%) for the conventional copper-cocatalyzed Sonogashira coupling of iodobenzene and phenylacetylene in pyrrolidine at 90°C (Scheme 2.28).

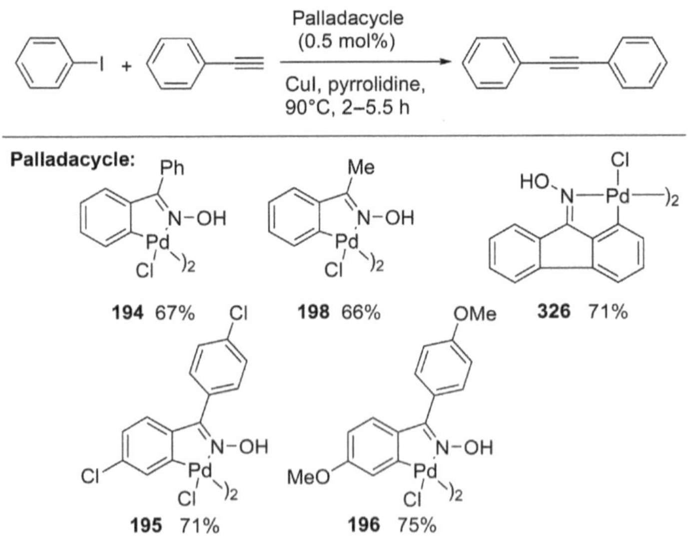

Scheme 2.28 Oxime-based palladacycles for the Sonogashira reaction.

The same research group continued their studies on the copper- and amine-free coupling of aryl iodides, and bromides, heterocyclic bromides, and vinyl bromides with terminal alkynes using 0.1–0.5 mol% of palladacycle **195** *(199)*. **195** is also an effective promoter of the sila-Sonogashira coupling reaction of aryl iodides and bromides with alkynyl silanes in the presence of CuI or TBAB as co-catalysts (Table 2.24).

Table 2.24 Sonogashira-type reactions catalyzed by oxime-derived palladacycle **195**[199]

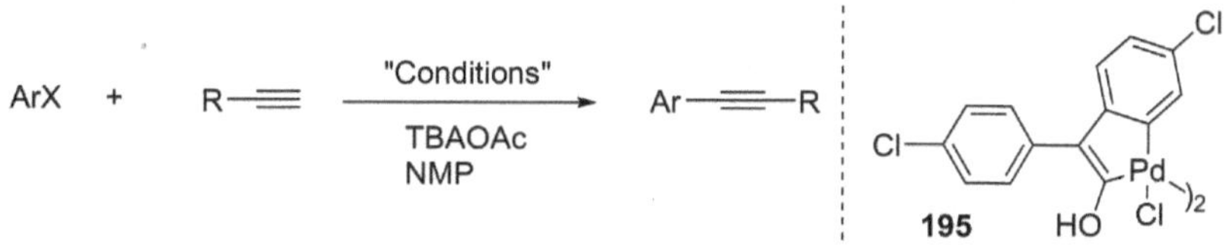

Entry	195 (mol% Pd)	ArX	R	Time (h)	Temp. (°C)	Yield (%)
1	0.1	*p*-ClC_6H_4I	Ph	1	110	73
2	0.1	*p*-$MeOC_6H_4I$	Ph	1	110	80
3	0.5	*p*-ClC_6H_4Br	Ph	7.5	110	90
4	0.1	*p*-ClC_6H_4Br	*n*-Hexyl	2	110	78
5	0.25	*p*-$MeOC_6H_4Br$	TIPS	1	130	85
6	0.1	*o*-ThienylBr	*n*-Hexyl	3	110	62
7	0.1	$PhCH{=}CH_2Br$	TIPS	1	110	63

Table 2.25 Sonogashira coupling reaction of acylation with terminal alkynes[200]

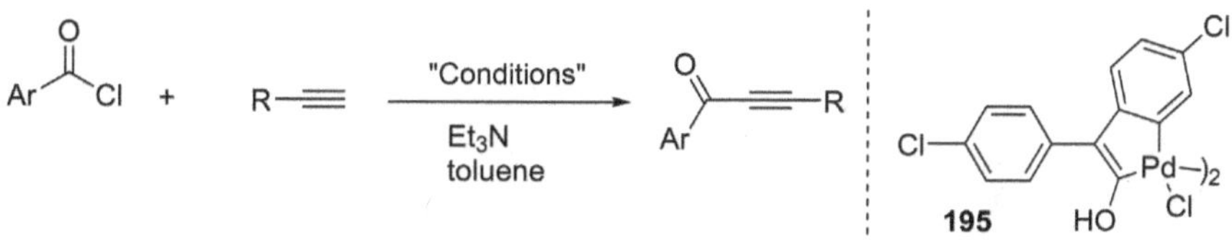

Entry	195 (mol% Pd)	Ar	R	Time (h)	Temp. (°C)	Yield (%)
1	0.2	Ph	Ph	2	110	75
2	0.5	Ph	Ph	23	25	96
3	0.2	Ph	TIPS	7	110	82
4	0.2	*p*-ClC_6H_4	Ph	2.5	110	72
5	0.2	*p*-$MeOC_6H_4$	Ph	4	110	70
6	0.2	*o*-Furanyl	Ph	4	110	50
7	0.2	$PhCH{=}CH_2$	Ph	3	110	77
8	0.2	Cyclohexyl	Ph	4	110	60
9	0.2	*t*-Bu	Ph	4	110	99

Dichlorobenzophenone oxime-derived palladacycle **195** showed a high catalytic activity for the copper-free acylation of terminal alkynes with different carboxylic acid chlorides to give the corresponding ynones in good yields *(200)*. The reaction could be run smoothly even under air at low catalyst loading. This new protocol allowed the reaction to be performed even at room temperature, or under microwave irradiation conditions (Table 2.25).

Table 2.26 Oxime-based palladacycle **195** used in carbonylative Sonogashira cross-coupling[201]

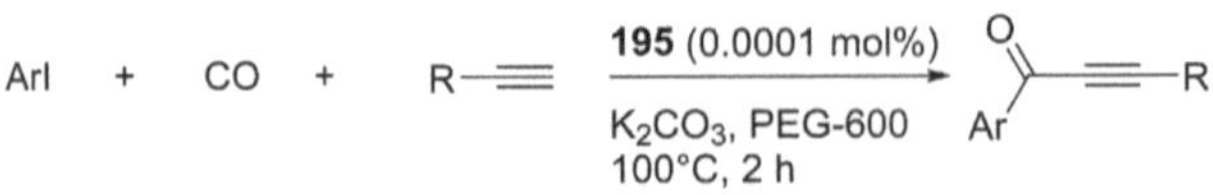

Entry	Ar	R	Yield (%)
1	*p*-$MeOC_6H_4$	Ph	89
2	*o*-$MeOC_6H_4$	Ph	80
3	*p*-MeC_6H_4	Ph	79
4	*p*-MeC_6H_4	*p*-$MeOC_6H_4$	90
5	Ph	*p*-$Me_2NC_6H_4$	89
6	*o*-Thienyl	Ph	83
7	*p*-$MeOC_6H_4$	*n*-Hexyl	69
8	*p*-$MeOC_6H_4$	Cyclopropyl	84

Recently, Bhanage and co-workers reported the carbonylative Sonogashira cross-coupling of aryl iodides with carbon monoxide and aryl/alkyl acetylenes with palladacycle **195** *(201)*. Low Pd loading down to 1×10^{-5} mol% was sufficient for the reaction to proceed. The palladacycle **195** could be recycled up to four cycles with only a marginal decrease in activity using PEG-600 as an environmentally benign solvent system (Table 2.26).

Nájera and co-workers employed the palladacycle **195** in the Sonogashira coupling of deactivated aryl chlorides with aryl- and alkyl-substituted terminal alkynes with water as solvent under microwave irradiation *(202)*. 2-Dicyclohexylphosphanyl-2′,4′,6′-triisopropylbiphenyl (**XPhos**) was added as ligand (0.2–2.0 mol%) and SDBS was added as surfactant (Scheme 2.29).

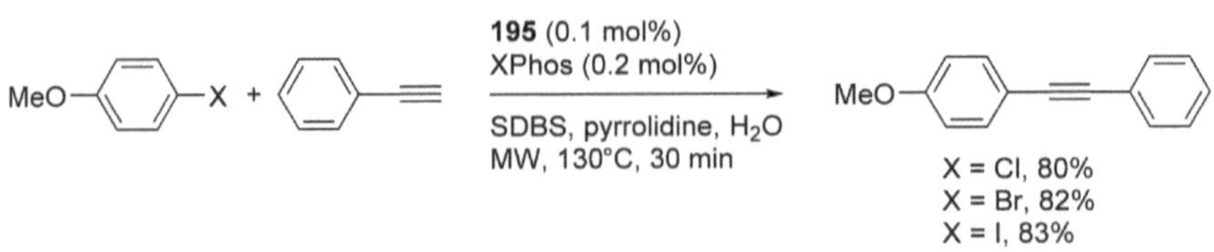

Scheme 2.29 Sonogashira coupling of deactivated aryl halides catalyzed by palladacycle **195**.

Oxime-derived palladacycle **195** was further used for the synthesis of dihydroisobenzofurans via heteroannulation by sequential Sonogashira cross-coupling/cyclization reactions between terminal alkynes and 2-(hydroxymethyl)bromobenzene and chlorobenzene under microwave irradiation. XPhos was added as ancillary ligand for the coupling reaction (Scheme 2.30) *(203)*. Under the optimized reaction conditions, functionalized 2-bromo- and 2-chlorobenzaldehydes were suitable coupling partners in the domino process affording phthalans in good yields.

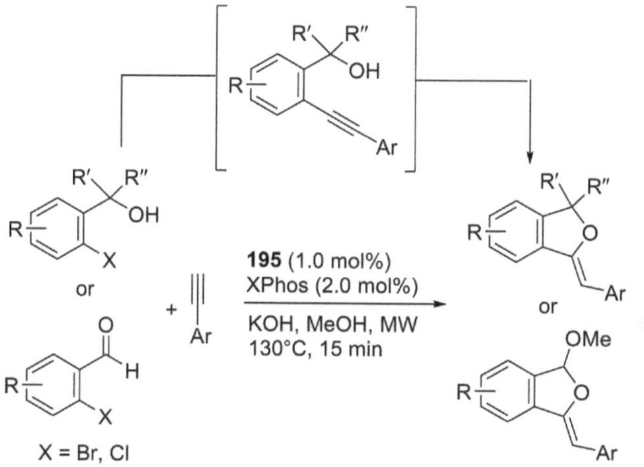

Scheme 2.30 Sonogashira/intramolecular hydroalkoxylation toward 1,3-dihydroisobenzofurans.

García and co-workers prepared an oxime-derived palladacycle **198** for the alkynylation reaction of aryl halides with phenylacetylene using ionic liquids or PEG as recyclable solvent (Scheme 2.28) *(204)*. The palladacycle **198** generally suffered extensive decomposition under the reaction conditions. However, it was found that the decomposition did not occur upon prolonged heating in PEG, giving way to PEG-stabilized active nanoparticles in a homogeneous recyclable system. The oxime-derived palladacycle

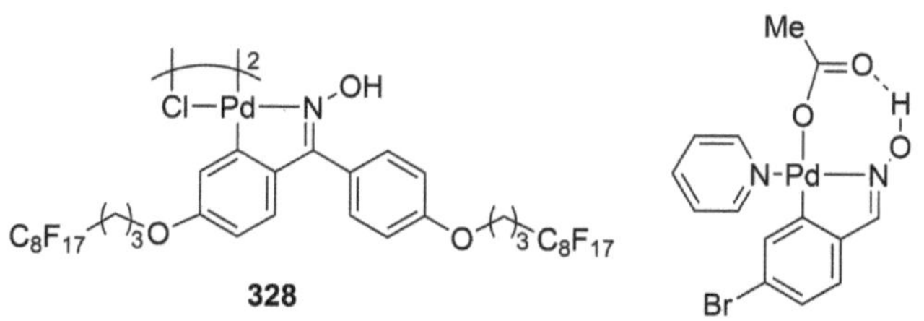

328

329

Fig. 2.41 Examples of oxime-based palladacycles.

198 anchored to soluble PEG to give the palladacycle **327** which was employed in Sonogashira coupling in PEG at 150°C (Fig. 2.40) *(205)*. The results showed that this PEG-anchored catalyst has high catalytic activity and could be reused for up to 10 cycles with no significant decrease in catalytic activity.

A fluorous oxime-based palladacycle **328** was prepared and used as precatalyst for the copper-free Sonogashira reaction of aryl iodides, bromides with phenylacetylene using pyrrolidine as the base in aqueous media under microwave irradiation at high temperature (140°C) with short reaction time. The catalyst could be reused for up to five cycles without observing loss of catalytic activity or palladium leaching to the solution (Fig. 2.41) *(150)*. Later, a highly active palladium nanoparticle oxime-derived palladacycle **329** was immobilized on Fe_3O_4/oleic acid. It allowed the copper-free Sonogashira cross-coupling reaction to be performed under air and mixed aqueous medium (Fig. 2.41) *(206)*. The coupling products were obtained in high yields with low catalyst loading (0.05 mol%) and the heterogeneous catalyst can be separated by an external magnet and reused for up to six runs without loss of activity.

5. N-Heterocyclic Carbene-Based Palladacycles

The metal *N*-Heterocyclic carbene (NHC) complexes were widely employed in the field of catalysis and developed rapidly over the decades. The NHCs are excellent σ-donor and poor π-acceptor ligands. By coordinating with poly(NHC) ligands, the stability of the metal complex improved by the chelating effect. Due to the high reactivity and stability of the metal complex, a wide variety of chelating metal NHC complexes was prepared.

The modification of NHC backbones would affect the electronic properties of the palladacycles and so affect the catalytic activity. It can be a change in the ring size (e.g., four-, five-, six-member ring); the nature of the heteroatoms attached to the carbine carbon (e.g., oxygen, nitrogen ligands); steric bulkiness (*i*-propyl, *t*-butyl group attachment). Regarding the bis-NHC-based palladacycle, different motifs were synthesized and extra coordination positions may also be added such as amine, pyridine. For brevity, those palladacycles having a bidentate ligand containing both NHC and phosphine are not summarized in the previous sections and the pincer-NHC-based palladacycle will be discussed in the next section (Fig. 2.42).

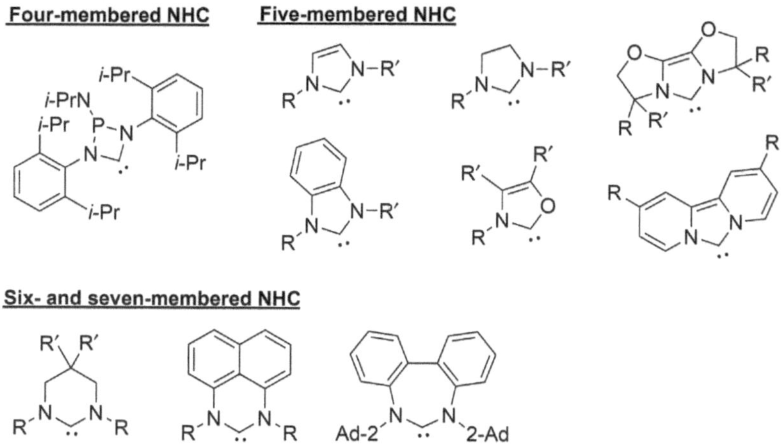

Fig. 2.42 Type of NHC coordinated in the palladacycles.

5.1 Heck Cross-Coupling Reaction

5.1.1 Imidazol-2-ylidene NHC-Based Palladacycles

The first example of NHC-based palladacycle in the Heck reaction was reported by Herrmann in 1995 *(207)*. Palladacycle **330** was easily prepared with high yields and it is extraordinarily stable to heat, oxygen and moisture. It composed of two imidazoles with a methylene bridge in between the two rings. The NHC derived from imidazole stabilized the Pd(II) and Pd(0) complexes due to the donor properties. Since then, several investigations in Heck reactions were released with the five-membered ring imidazol-2-ylidene NHC palladacycle. There are a number of imidazol-2-ylidene NHC-based palladacycles and they are easy to modify and thus different catalytic activities were shown (Fig. 2.43).

Lee and co-workers investigated the *N*-substituent on the catalytic efficacies of methylene bridged imidazol-2-ylidene NHC-based palladacycles with different alkyl chains and substituted benzyl groups *(208)*. Later, Strassner further extended the investigation on substituted aromatic rings and halide groups *(209)*. It is significant that the aryl substituted NHC-palladacycle is much more active than the alkyl substituted one but not the benzyl substituted one. Recently, a detailed mechanism and electronic effect DTF study was carried out by the same group *(210)*. Different mechanistic pathways were investigated and the most reasonable mechanism is a cationic catalytic cycle with Pd(0) and Pd(II). The aryl substitution with electron-donating groups at the *para*-position (**344** vs **342**; **335** vs **334**) reduced the reaction barrier of the rate-determining step.

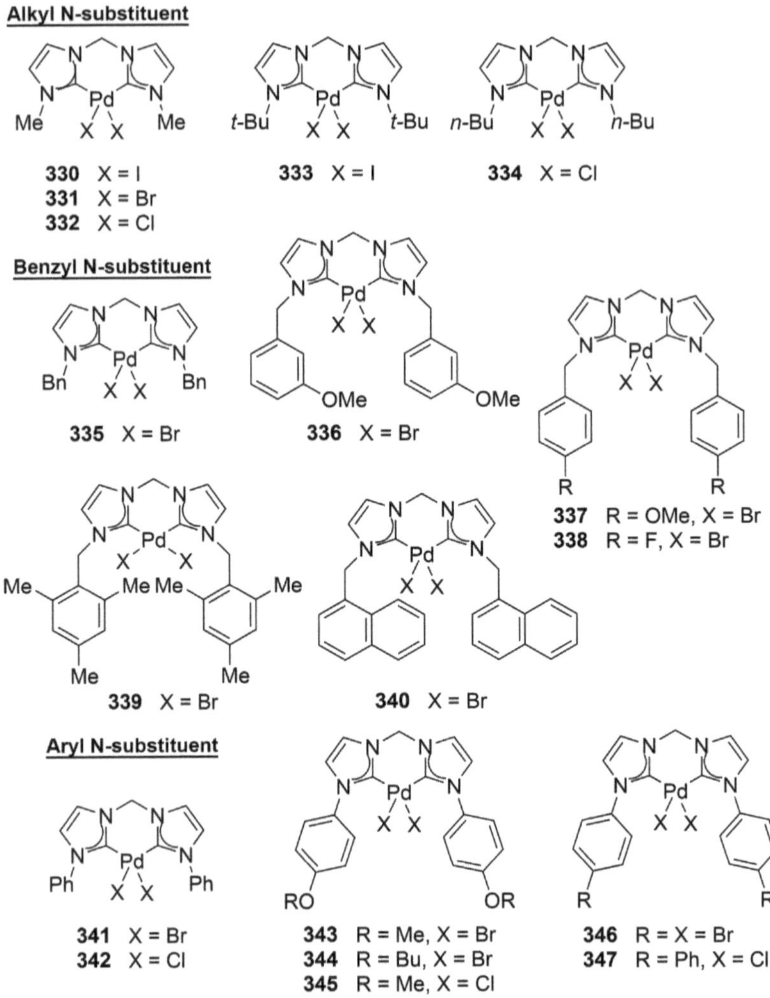

Fig. 2.43 Examples of *N*-substituted Imidazol-2-ylidene NHC-based palladacycles.

The *N*-position of the imidazole ring was substituted with different functional chain/structure with the *cis*-conformation. Also, replacing the methylene bridge by ethylene or another alkyl-chain (e.g., oxoether linkage) could have a significant effect on the Heck coupling reaction (Scheme 2.31; Table 2.27).

Upon replacing the methylene bridge by an ethylene linkage, the catalytic efficacy of palladacycle **350–354** was comparable to that of **335–340** *(208)*. Since then, different bridge-chains imidazol-2-ylidene NHC-based palladacycles have been developed and employed in the Heck coupling (Fig. 2.44).

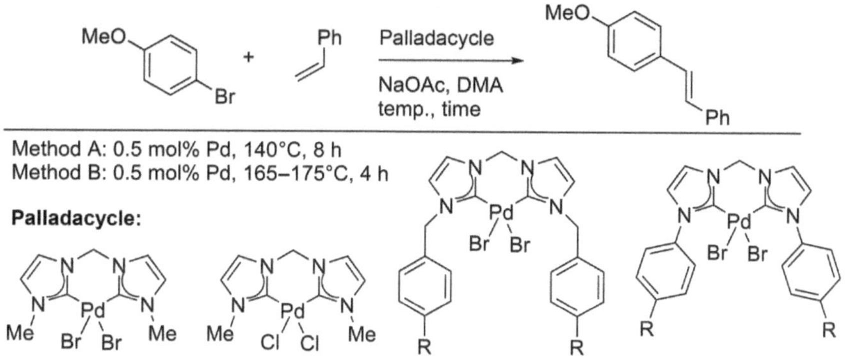

331 **A**: 32%, **B**: 43%

332 **A**: 30%

335 R = H, **B**: 91%
336 R = OMe, **B**: 89%

341 R = H, **A**: 34%
343 R = OMe, **A**: 52%

Scheme 2.31 Effect of *N*-substituents in NHC-based palladacycle in Heck coupling (method A,[209] method B[208]).

The impact of the chain length of the alkyl bridge $-(CH_2)_n$ ($n = 1$–4) between the two imidazolium units on catalytic efficacy was studied by Strassner (Scheme 2.32; Table 2.28). 4-Bromoacetophenone and styrene could be coupled successfully. However, they are not active enough to activate the 4-chloroacetophenone except **349** gave 10% of the product with 0.5 mol% Pd in 30 h *(215)*. Similarly, Huynh also demonstrated that the chain with two-carbon gave a better result *(216)*. A combination of imidazolium units and benzimidazolium unit joining together to form a multi-NHC-palladacycle with three carbon bridge was synthesized. They showed a higher catalytic efficacy with the palladacycles having both two units of imidazole or benzimidazole.

Apart from the simple alkyl chain bridge, an oxoether-bridged imidazol-2-ylidene NHC-based palladacycles with *cis*-chelating configuration was synthesized by Cavell *(218)*. Palladacycle **355** was highly active to catalyze the reaction of 4-bromoacetopheone with *n*-butyl acrylate with 8.4×10^{-4} mol% Pd loading. Phenyl ring bridged-imidazol-2-ylidene NHC-based palladacycle is also an interesting design which gave a rigid structure for coordination *(209)*. Later, anthracene and xanthene skeletons were applied as the connection bridge to give a more complex design to the imidazol-2-ylidene NHC-based palladacycles *(219)*. Superior performance was observed with xanthene-based palladacycle.

Table 2.27 Effect of *N*-substituent in the methylene-bridged imidazol-2-ylidene NHC-based palladacycles in Heck reaction

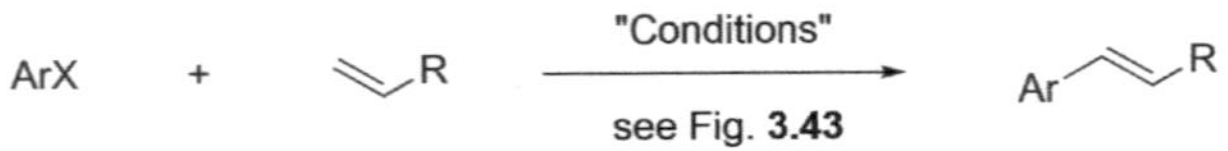

Entry	Palladacycle (mol% Pd)	ArX	R	Solvent	Base	Time (h)	Temp. (°C)	Yield (%)	Ref.
1	**330** (0.2)	C_6H_5I	Ph	DEA	NaOAc	4	reflux	84	211
2	**330** (0.67)	*p*-$MeOC_6H_4Br$	CO_2*n*-Bu	DMA	NaOAc	50	125	20	207
3[a]	**330** (1.0)	*p*-$CHOC_6H_4Cl$	CO_2*n*-Bu	DMA	NaOAc	24	140	>99	207
4	**330** (1.0×10^{-3})	*p*-$MeOC_6H_4Br$	CO_2*n*-Bu	DMA	NaOAc	24	120	13	212
5	**331** (0.5)	*p*-$MeC(O)C_6H_4Br$	Ph	DMA	NaOAc	8	140	56	209
6	**332** (0.5)	*p*-$MeC(O)C_6H_4Br$	Ph	DMA	NaOAc	8	140	80	209
7	**333** (1.0)	C_6H_5Br	CO_2*n*-Bu	IL[b]	NaOAc	12	120	76	213
8	**333** (1.0)	C_6H_5Cl	Ph	TBAB	NaOAc	17	150	45	16
9	**336** (0.5)	*p*-$MeOC_6H_4Br$	Ph	DMA	NaOAc	4	165-175	87	208
10	**338** (0.5)	*p*-$MeOC_6H_4Br$	Ph	DMA	NaOAc	4	165-175	86	208
11	**340** (0.5)	*p*-$MeOC_6H_4Br$	Ph	DMA	NaOAc	4	165-175	92	208
12	**339** (0.5)	*p*-$MeOC_6H_4Br$	Ph	DMA	NaOAc	4	165-175	86	208
13	**340** (3.0)	*p*-$MeC(O)C_6H_4Cl$	Ph	DMA	NaOAc	24	165-175	81	208
14	**340** (3.0)	*p*-MeC_6H_4Cl	Ph	DMA	NaOAc	24	165-175	68	208
15	**341** (1.0×10^{-3})	*p*-$MeOC_6H_4Br$	CO_2*n*-Bu	DMA	NaOAc	24	120	88	212
16	**344** (0.5)	*p*-$MeOC_6H_4Br$	Ph	DMA	NaOAc	8	140	88	209
17	**346** (0.5)	*p*-$MeOC_6H_4Br$	Ph	DMA	NaOAc	8	140	66	209
18	**347** (0.2)	C_6H_5I	Ph	DMA	Et_3N	24	110	80	214

[a] TBAB was added as an additive.
[b] IL = ionic liquid refer to reference [213].

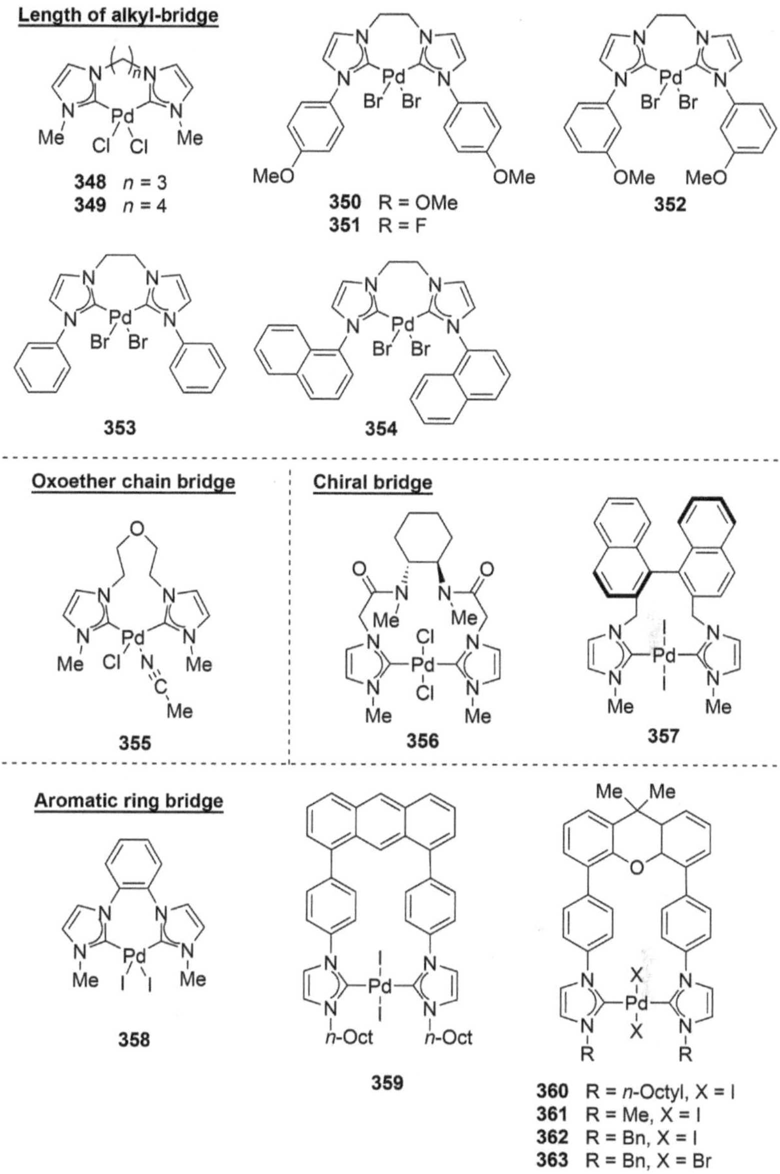

Fig. 2.44 Examples of bridged imidazol-2-ylidene NHC-based palladacycles.

The NHC-based palladium bearing different anionic co-ligands highly affect the catalytic efficacy. In general, the anionic co-ligands are halide anions, such as iodides, bromides or chlorides. In 2011, Huynh studied the co-ligand effect which showed the reactivity in the Heck reaction with the order of $SCN^- < I^- < CF_3COO^-$ (Scheme 2.33) *(220)*.

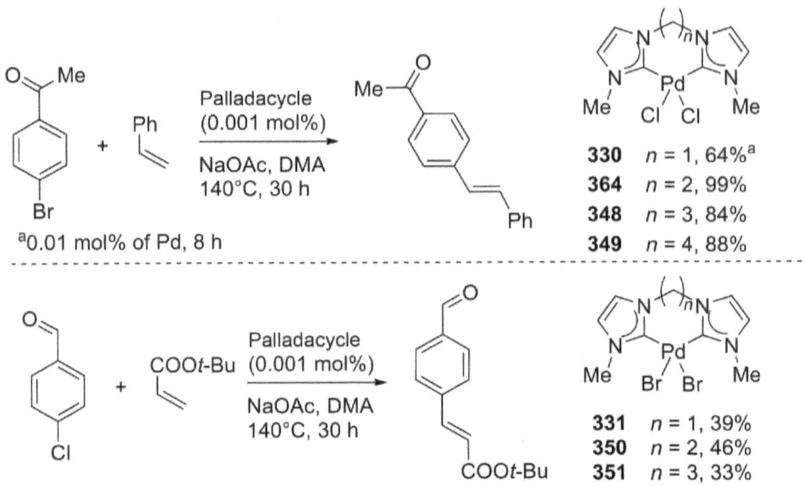

Scheme 2.32 Effect of alkyl chain length in the catalytic efficacy.

In 2000, the Baker group synthesized a few Pd-bidentate ligands containing two imidazolium units within a cyclophane skeleton (Fig. 2.45; Table 2.29). The cyclophane structure can be expanded by altering the alkyl chain in between the phenyl ring and the imidazolium unit. The rigidity of the cyclophane may confer added stability on the Pd-carbene structure. Baker and co-workers also compared the cyclophane-type Pd-complexes **368** with the noncyclophane-type Pd-complexes **371** with the Heck coupling reaction *(221,222)*. It is obvious that the cyclophane-type Pd-complexes displayed lower activity than the noncyclophane-type.

In 2013, Hwang developed homoleptic Pd(II) complexes with one and two chelating NHC-ligands with the pyridinium cation *(223)*. In the presence of 1.2 mol% Pd, bromobenzene was coupled with styrene using Cs_2CO_3 as the base and water as solvent at 140°C in 8 h. It was found that palladacycle **376** performed better than **375** in the Heck reaction (85% vs 60%) (Fig. 2.46). It was suggested that palladacycle **376** provided a lower steric hindrance and the corresponding monopalladium led to shorter induction periods in the catalytic process *(224)*. It is notable that the palladacycle with pyridinium auxiliary showed a higher activity when compared to that without pyridinium group but just the *n*-butyl carbon chain *(225)*.

In 2006, Biffis compared the above modified NHC-based palladacycle to conclude the effect of *N*-substituent, the type of NHC, and the connection

Table 2.28 Effect of bridged structure of imidazol-2-ylidene NHC-based palladacycles in Heck reaction

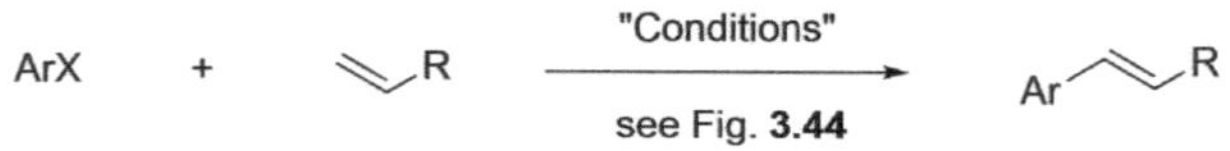

Entry	Palladacycle (mol% Pd)	ArX	R	Solvent	Base	Time (h)	Temp. (°C)	Yield (%)	Ref.
1	**350** (0.5)	*p*-MeOC$_6$H$_4$Br	Ph	DMA	NaOAc	4	165–175	88	208
2	**352** (0.5)	*p*-MeOC$_6$H$_4$Br	Ph	DMA	NaOAc	2	165–175	62	208
3	**351** (0.5)	*p*-MeOC$_6$H$_4$Br	Ph	DMA	NaOAc	2	165–175	79	208
4	**353** (0.5)	*p*-MeOC$_6$H$_4$Br	Ph	DMA	NaOAc	4	165–175	90	208
5	**354** (3.0)	*p*-MeC(O)C$_6$H$_4$Cl	Ph	DMA	NaOAc	24	165–175	96	208
6	**354** (3.0)	*p*-MeC$_6$H$_4$Cl	Ph	DMA	NaOAc	24	165–175	72	208
7	**355** (8.4×10^{-4})	*p*-MeC(O)C$_6$H$_4$Br	CO$_2$*n*-Bu	DMA	NaOAc	20	120	88	218
8[a]	**358** (1×10^{-3})	*p*-MeOC$_6$H$_4$Br	CO$_2$*n*-Bu	DMA	NaOAc	24	120	77	212
9	**359** (0.5)	*p*-NO$_2$C$_6$H$_4$I	CO$_2$*n*-Bu	DMF	NaOAc	26	120	92	219
10	**360** (0.5)	*p*-NO$_2$C$_6$H$_4$I	CO$_2$*n*-Bu	DMF	NaOAc	16	120	80	219
11	**361** (0.5)	*p*-NO$_2$C$_6$H$_4$Br	CO$_2$*n*-Bu	DMF	NaOAc	17	140	70	219
12	**363** (0.5)	*p*-NO$_2$C$_6$H$_4$I	CO$_2$*n*-Bu	DMF	NaOAc	16	120	92	219
13	**357** (1.0–2.0)	C$_6$H$_5$I	CO$_2$Et	DMA	NaOAc	38	140	95	217
14	**357** (1.0)	C$_6$H$_5$Br	CO$_2$Et	NMP	Cs$_2$CO$_3$	24	130	79	217

[a]TBAB was added as an additive.

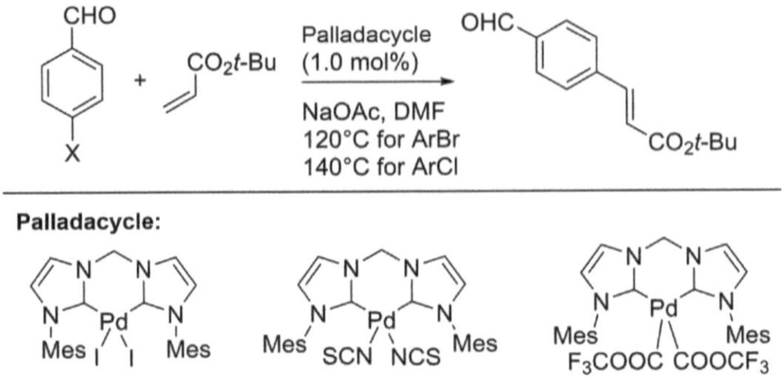

365
X = Br, 89%
X = Cl, 8%; 74%[a]

366
X = Br, 85%

367
X = Br, 97%
X = Cl, 15%; 75%[a]

[a]TBAB (1.5 equiv.) was added

Scheme 2.33 Effect of coordinating group (co-ligand) of palladacycles.

bridge to the catalytic activity of the Heck coupling *(212)*. Recently, Hazari synthesized the first families of bidentate trans-chelating NHC ligands **378–381** with the flexible linker and they are catalytically active for the Heck coupling of 4-bromoacetopheone and *n*-butyl acrylates (Scheme 2.34) *(226)*.

5.1.2 Functionalized Imidazol-2-ylidene NHC-Based Palladacycles

Functionalized imidazol-2-ylidene NHC-based palladacycles are developed for Heck coupling and the most common skeleton is the *N*-donor substituent, such as pyridyl-, pyrazolyl-, pyrimidyl-, pyrazyl-, oxazolyl-structure. (Diphosphino)alkyl-functionalized imidazol-2-ylidene NHC-based palladacycles are another choice for Heck coupling.

5.1.2.1 Pyridyl-Imidazol-2-ylidene NHC-Based Palladacycles

Pyridine is a common ligand to coordinate with metal for catalysis. Thus, it is attractive to merge the feature of pyridyl-group with the imidazol-2-ylidene NHC-based palladacycles to catalyze the Heck coupling reaction. In 2000, Danopoulos and Hursthouse reported a pyridine functionalized NHC-palladium-complex **382** which gave excellent catalytic efficacy in the Heck coupling of aryl iodides and alkyl acrylate *(227)*. However, it performed poorly

Table 2.29 Imidazolium-linked *ortho*-cyclophanes palladacycles in Heck reaction

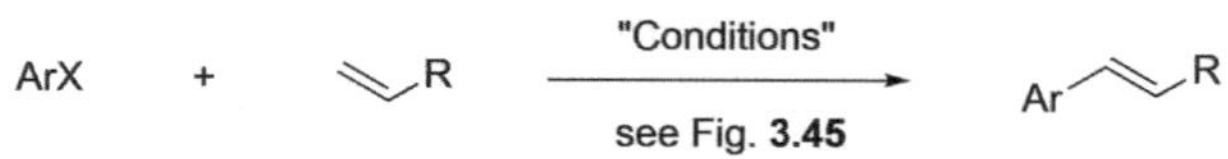

Entry	Palladacycle (mol% Pd)	ArX	R	Solvent	Base	Time (h)	Temp. (°C)	Yield (%)	Ref.
1	**368** (1.4×10^{-4})	C_6H_5I	CO_2Me	DMF	Et_3N	24	140	20	221
2	**372** (6.3×10^{-4})	C_6H_5I	CO_2Me	DMF	Et_3N	24	140	52	221
3	**373** (3.6×10^{-4})	C_6H_5I	CO_2Me	DMF	Et_3N	24	140	46	221
4	**369** (2.0×10^{-4})	C_6H_5I	CO_2*n*-Bu	DMF	Et_3N	24	140	63	222
5	**371** (2.0×10^{-4})	C_6H_5I	CO_2*n*-Bu	DMF	Et_3N	24	140	60	222
6	**373** (2.0×10^{-4})	C_6H_5I	CO_2*n*-Bu	DMF	Et_3N	24	140	79	222
7	**374** (2.0×10^{-4})	C_6H_5I	CO_2*n*-Bu	DMF	Et_3N	24	140	84	222

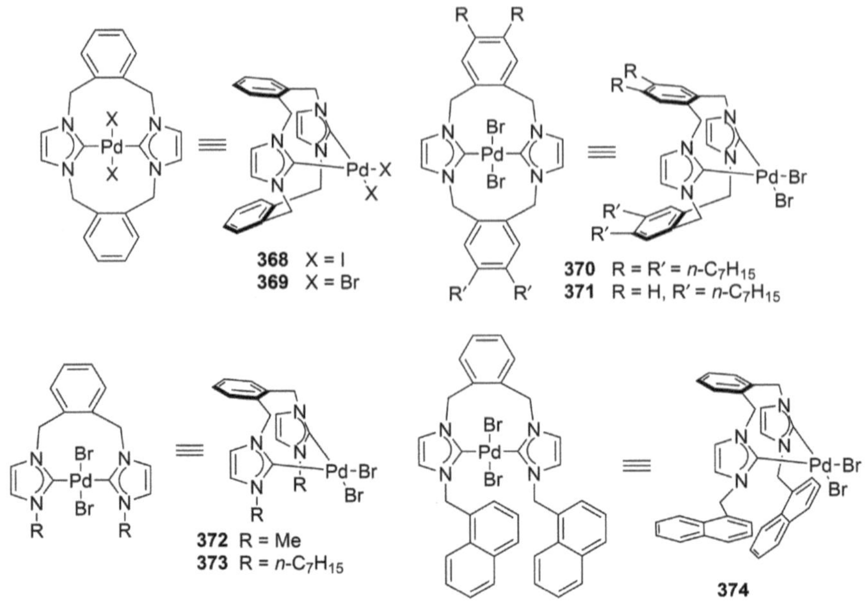

Fig. 2.45 Imidazolium-linked *ortho*-cyclophanes palladacycles.

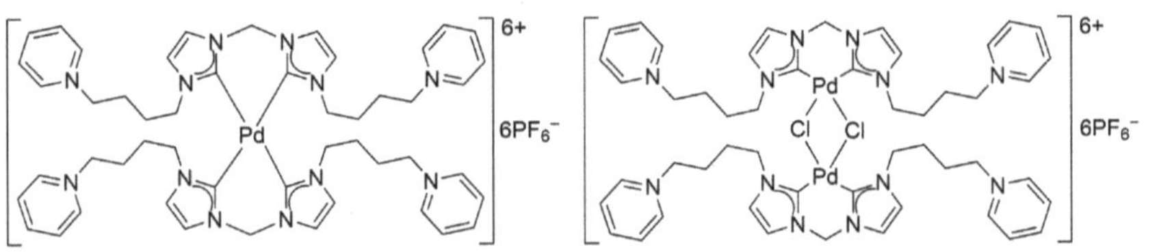

375 **376**

Fig. 2.46 Pd-NHC-palladacycle with mono- and bis-chelate carbene ligands.

for the coupling of bromobenzene. Sterically bulky catalysts usually improve the catalytic efficacy; Liu reported pyridyl-imidazol-2-ylidene NHC-based palladacycles by adding the sterically hindered mesitylene substituted at the pyridyl unit and the *N*-position of the imidazolium unit (**387**). In general, 0.5 mol % of the palladacycle could catalyze the Heck coupling of (hetero)aryl bromides. Sterically hindered 2-bromomesitylene could also be coupled to give 83% yield *(228)*. By adding a different substituent in the *N*-position of the imidazolium units (methyl group for **386** and *N*-heterocycles for **390**, **391**), the catalytic efficacy of those pyridyl-imidazol-2-ylidene NHC-based palladacycles were tested (Table 2.30, entries 4 vs 6 vs 9).

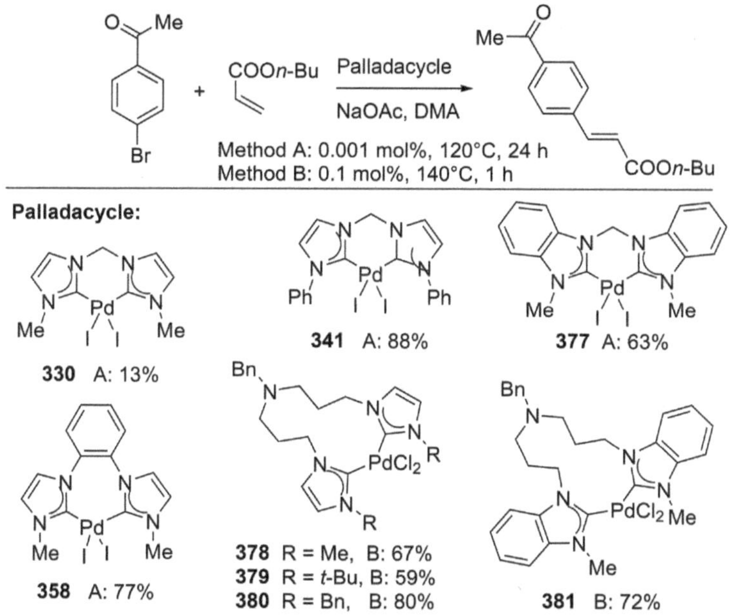

Scheme 2.34 Comparison of NHC palladacycle reactivity on Heck reaction.

A series of palladacycles with bidentate pyridyl-, picolyl-NHC were reported to catalyze the Heck coupling of aryl bromides and methyl acrylate *(232)*. A mechanism study was carried out based on in situ EXAFS study. Palladacycles **383** and **389** were compared with methylene and ethylene bridge respectively (Fig. 2.47). It was suggested that they may have shown different activities due to the different formation rate of dissociation of the cyclic coordination. Palladacycle **383** with a six-membered chelate ring may undergo pyridine dissociation/oxidative addition faster than that of **334** which is more robust and stable.

In 2008, Chen reported a one-pot sequence of Heck/Heck and Heck/Sonogashira coupling reactions of aryl dihalides with palladacycle **390** to afford unsymmetrically substituted arenes in excellent yields *(233)*. A single catalyst could offer the regioselective formation of two or more C—C bond which generates less waste and obviates the tedious purification procedures which would be very attractive for industrial use.

Table 2.30 Pyridyl-imidazol-2-ylidene NHC-based palladacycles in Heck reaction

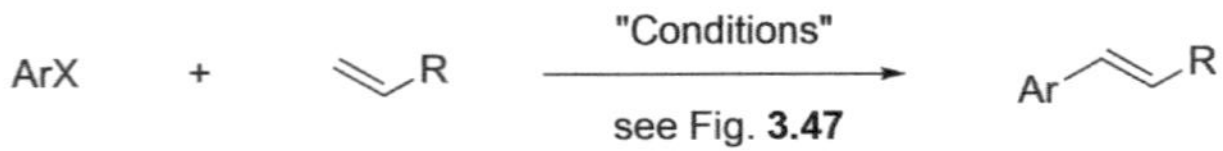

Entry	Palladacycle (mol% Pd)	ArX	R	Solvent	Base	Time (h)	Temp. (°C)	Yield (%)	Ref.
1	**382** (7.0×10^{-5})	C_6H_5I	CO_2Me	NMP	Et_3N	18	140	98	227
2	**382** (7.0×10^{-3})	C_6H_5Br	CO_2Me	NMP	Et_3N	75	130	10	227
3	**385** (2.0×10^{-2})	C_6H_5I	CO_2n-Bu	DMA	NaOAc	5	120	92	229
4	**386** (1.6×10^{-4})	*p*-MeC(O)C_6H_4Br	CO_2n-Bu	DMA	NaOAc	48	120	53	229
5[a]	**387** (5.0×10^{-2})	*p*-$Me_2NC_6H_4Br$	CO_2n-Bu	NMP	*i*-Pr_2NH	12	140	83	228
6	**390** (5.0×10^{-4})	*p*-MeC(O)C_6H_4Br	CO_2n-Bu	DMA	NaOAc	0.5	140	57	230
7[b]	**390** (0.5)	*p*-MeC_6H_4Br	CO_2n-Bu	DMA	NaOAc	6	140	89	230
8[b]	**390** (0.5)	*o*-MeC_6H_4Br	CO_2n-Bu	DMA	NaOAc	6	140	13	230
9	**391** (1.0×10^{-2})	*p*-MeC(O)C_6H_4Br	CO_2n-Bu	DMA	NaOAc	2	120	100	231
10[b]	**391** (1.0)	*p*-MeC(O)C_6H_4Cl	CO_2n-Bu	DMA	NaOAc	20	120	87	231

[a]PPh_3 was added as an additive.
[b]TBAB was added as an additive.

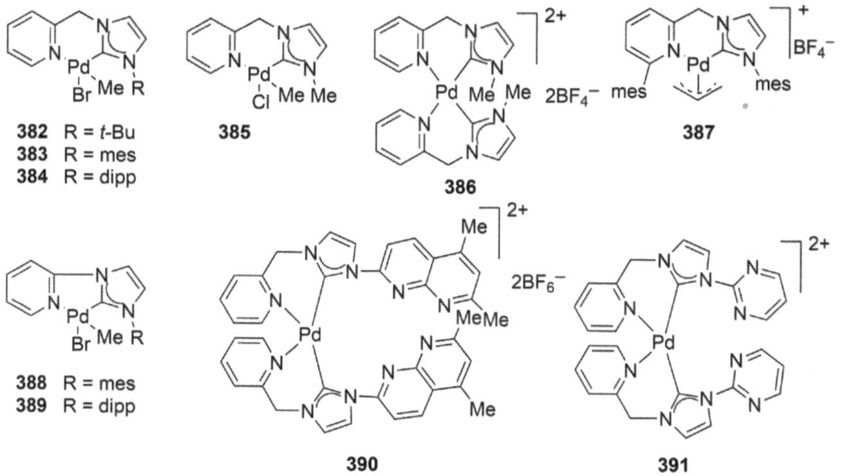

Fig. 2.47 Examples of pyridyl-Imidazol-2-ylidene NHC-based palladacycles.

5.1.2.2 Other Imidazol-2-ylidene NHC-based palladacycles

A hemilabile pyrazolyl-functionalized imidazol-2-ylidene NHC-based palladacycle was synthesized by Shreeve and successfully employed in the Heck coupling with ionic liquid (ILs) as solvent *(234)*. Activated aryl iodides and bromides are well-coupled with *n*-butyl acrylates and they are able to recycle the catalytic system for at least three cycles without loss in performance. It was suggested that the excellent recyclability in ILs arises from its strong Pd (II)-carbene and the weak Pd(II)-nitrogen bond. When replacing the mesitylene group with a *n*-butyl carbon chain, a comparable result was obtained *(235)*.

Palladacycle **394** is an unusual pentacoordinated palladacycle in which the Pd is surrounded by two imidazolylidene, two pyrimidines, and an acetonitrile molecule in a square-pyramidal geometry (Fig. 2.48) *(231)*. It showed excellent catalytic efficacy in Heck coupling with aryl bromides (Table 2.31). When 4-bromotoluene was used, TBAB was added to improve the product yield from 58% to 90%. Even steric hindered 2-bromotoluene could be successfully coupled to give 78% yield under the same reaction condition. It is believed that the addition of TBAB shortens the Pd(II)/Pd(0) activation induction period and improves the catalytic cycle. In 2009, Strassner and co-workers further developed a series of novel pyrimidyl-functionalized imidazol-2-ylidene NHC-based palladacycles with different substituents with aryl ring or alkyl chain *(236)*. Among the six palladacycles shown in

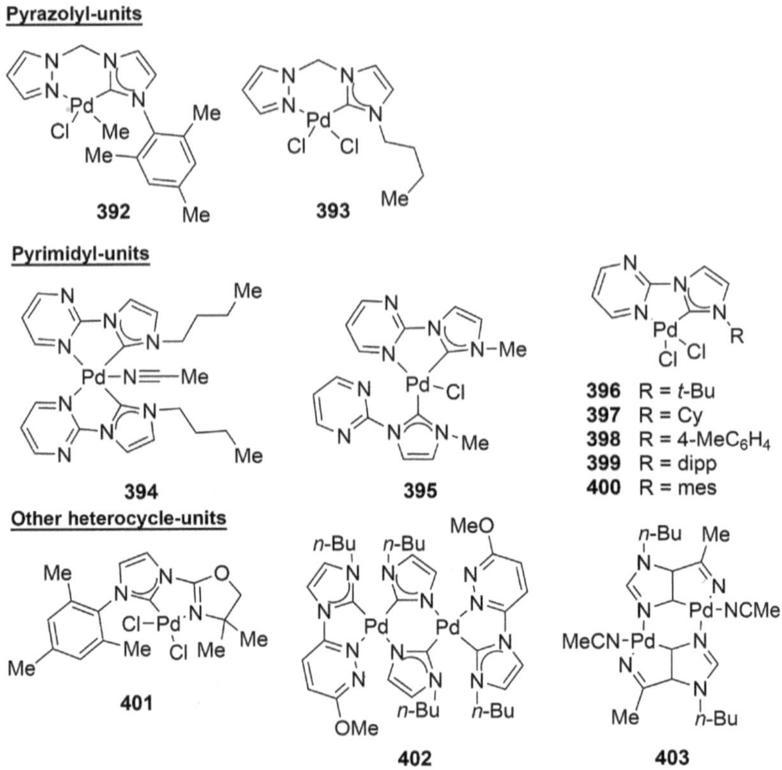

Fig. 2.48 Examples of imidazol-2-ylidene NHC-based palladacycles.

the publication, **395** showed the best performance in which 0.0005 mol% of Pd was sufficient to catalyse the reaction to give 84% while others gave comparable results with 0.01 mol% of Pd.

Other heterocycles imidazol-2-ylidene NHC-based palladacycles were also developed (Fig. 2.48). The oxzaolyl-based was reported in 2002 when a distorted square planar configuration was observed using X-ray diffraction with the imidazolyl and the oxazolinyl ring lying in the molecular plane *(237)*.

In 2010, Lee developed a special imidazol-2-ylidene NHC-based palladacycle **404** consist of multicomponent with pyrazole and pyridine units. It showed excellent catalytic performance that sterically congested 2,4,6-diisopropylbromobenzene could couple with styrene and a wide range of aryl chlorides including challenging 2-chlorotoluene could be coupled to give excellent product yields *(239)*. Binaphthyl-2,2′-diamine tridentate-imidazol-2-ylidene NHC-based palladacycle **405** is also a good choice for

Table 2.31 Heterocycle-imidazol-2-ylidene NHC-based palladacycles in Heck reaction

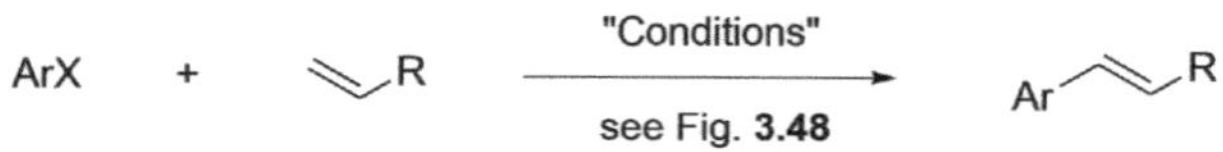

Entry	Palladacycle (mol% Pd)	ArX	R	Solvent	Base	Time (h)	Temp. (°C)	Yield (%)	Ref.
1	**392** (2.0)	*p*-MeC(O)C_6H_4Br	CO_2*n*-Bu	[BMIm][PF_6]	Et_3N	8	120	77, 73, 81[a]	234
2	**392** (2.0)	*p*-MeC_6H_4I	CO_2*n*-Bu	[BMIm][PF_6]	Et_3N	8	120	92. 89. 94[a]	234
3	**393** (2.0)	*p*-MeC_6H_4I	CO_2*n*-Bu	[BMIm][PF_6]	Na_2CO_3	12	120	93	235
4[b]	**394** (1.0)	*p*-MeC_6H_4Br	CO_2*n*-Bu	DMA	NaOAc	20	140	90	231
5[b]	**394** (1.0)	*o*-MeC_6H_4Br	CO_2*n*-Bu	DMA	NaOAc	20	140	78	231
6	**395** (5.0×10^{-4})	*p*-MeC(O)C_6H_4Br	Ph	DMA	NaOAc	26	140	84	236
7	**396** (1.0×10^{-2})	*p*-MeC(O)C_6H_4Br	Ph	DMA	NaOAc	8	140	83	236
8	**398** (1.0×10^{-2})	*p*-MeC(O)C_6H_4Br	Ph	DMA	NaOAc	8	140	88	236
9[b]	**401** (0.2)	*p*-MeC(O)C_6H_4Cl	Ph	DMA	K_3PO_4	18	135	100	237
10[b]	**402** (0.5)	*p*-MeC_6H_4Br	CO_2*n*-Bu	DMA	NaOAc	24	130	69	238
11[b]	**403** (0.5)	*p*-MeC_6H_4Br	CO_2*n*-Bu	DMA	NaOAc	24	130	75	238

[a]Isolated yield for each recycle three times.
[b]TBAB was added as an additive.

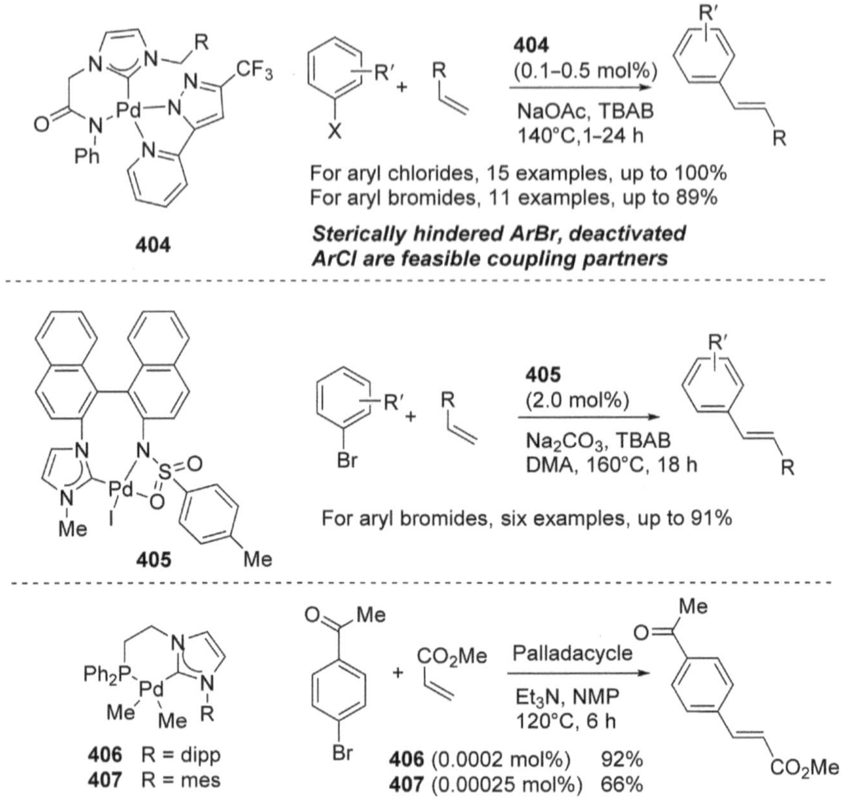

Scheme 2.35 Other Imidazol-2-ylidene NHC-based palladacycles in Heck coupling.

the Heck coupling *(240)*. (Diphenylphosphino)alkyl-functionalized imidazol-2-ylidene NHC-based palladacycles **406–407** with sterically hindered *N*-substituent were developed to show moderate catalytic activity for the Heck coupling of acrylates with aryl bromides but not chlorides (Scheme 2.35).

5.1.3 Benzimidazol-2-ylidene NHC-Based Palladacycles

Similar to that of imidazol-2-ylidene NHC-based palladacycles, the connecting bridge between two benzimidazolium units could affect the catalytic activity in Heck coupling reaction (Fig. 2.49; Table 2.32). Kunz reported three palladacycles **408–410** to investigate the effect of methylene and ethylene bridged benzimidazol-2-ylidene NHC-based palladacycle in the catalysis *(241)*.

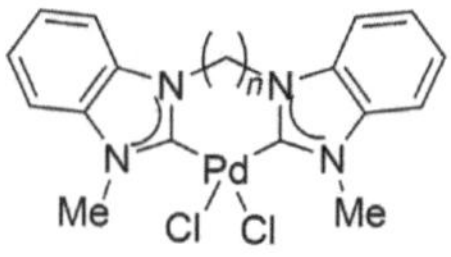

377 n = 1, X = I
408 n = 1, X = Br
409 n = 2, X = Br

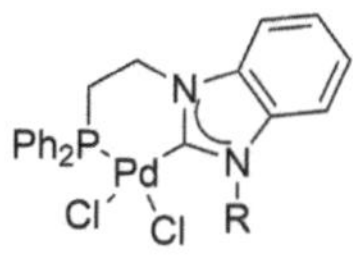

410 R = Et
411 R = *n*-Pr
412 R = *n*-Bu

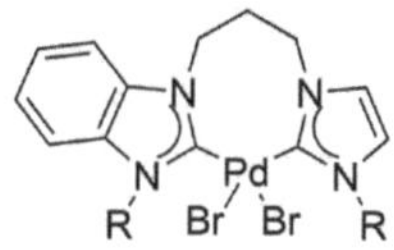

413 R = Me
414 R = Bn

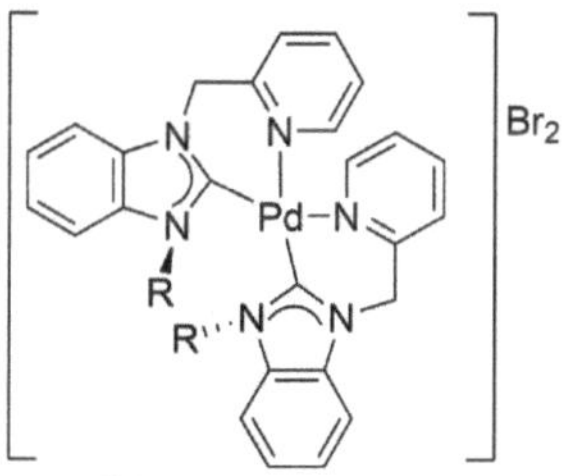

415 R = Me
416 R = Et
417 R = *n*-Pr
418 R = *n*-Bu

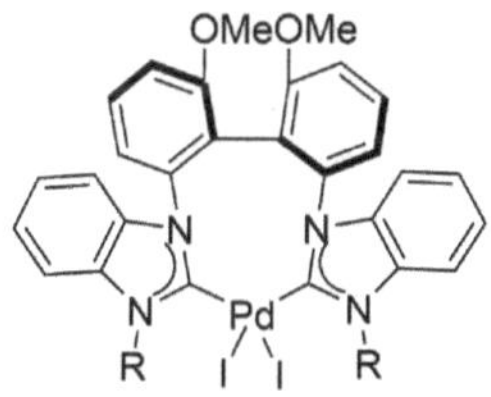

419 R = Et
420 R = Me

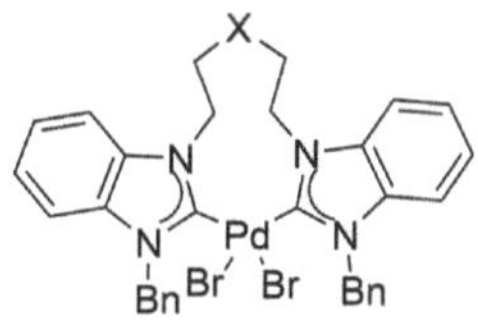

421 X = S
422 X = S=O

Fig. 2.49 Examples of benzimidazol-2-ylidene NHC-based palladacycles.

In 2008, Jung developed a chiral tridentate NHC-amidate alkoxide with oxygen bridging dimer palladacycles **423** and **424** (Scheme 2.36) *(246)*. They facilitated an asymmetric oxidative Heck reaction of arylboronic acids and acyclic alkenes offering high enantioselectivities.

As mentioned in the previous subsection, cyclophane-type NHC-based palladacycles were synthesized with the imidazolium units by the Baker group in 2000. Later, the same group further extended the cyclophane skeleton to the palladacycles with the benzimidazolium units for the Heck coupling reaction *(247)*. It is notable that the palladacycle is in *cis*-chelating configurations. The electronic effect of the substituent at the benzimidazol-2-ylidene NHC-aryl ring was also studied. Again, the noncyclophane-type palladacycle gave a higher product yield than the cyclophane one. Moreover, the slightly electron-donating *n*-butoxy substituent gave a better activity to facilitate the coupling reaction. It is important that the **426** existed in both *cis*- vs *trans*- and *syn*- and *anti*-configuration. The *n*-butoxy at 4-,7-positions in the *cis* form provided the steric bulkiness that resulted in an equilibrium between *syn*- and *anti*-configuration in solutions (Fig. 2.50; Table 2.33).

Table 2.32 Benzimidazol-2-ylidene NHC-based palladacycles in Heck reaction

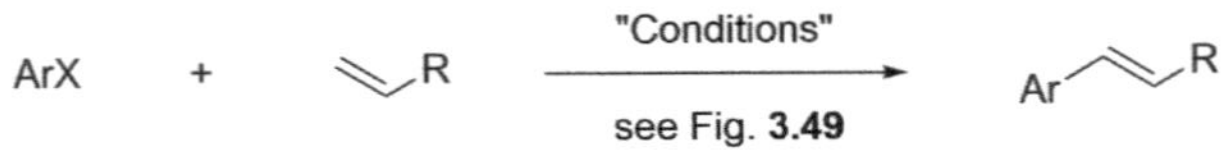

Entry	Palladacycle (mol% Pd)	ArX	R	Solvent	Base	Time (h)	Temp. (°C)	Yield (%)	Ref.
1	**377** (1.0)	C_6H_5Br	CO_2*n*-Bu	dioxane	Cs_2CO_3	16	150	67	241
2	**408** (1.0)	C_6H_5Br	CO_2*n*-Bu	dioxane	Cs_2CO_3	16	150	44	241
3	**409** (1.0)	C_6H_5Br	CO_2*n*-Bu	dioxane	Cs_2CO_3	16	150	83	241
4[a]	**420** (0.1)	C_6H_5Br	CO_2*n*-Bu	DMA	Na_2CO_3	18	140	88	242
5	**410** (1.0)	*p*-MeC(O)C_6H_4Br	Ph	DMA	NaOAc	2	110	84	243
6	**411** (1.0)	*p*-MeC(O)C_6H_4Br	Ph	DMA	NaOAc	2	110	78	243
7	**412** (1.0)	*p*-MeC(O)C_6H_4Br	Ph	DMA	NaOAc	2	110	79	243
8	**415** (1.0)	*p*-CHOC_6H_4Br	Ph	DMA	NaOAc	2	110	30	244
9	**415** (1.0)	*p*-CHOC_6H_4Br	Ph	DMA	NaOAc	24	110	98	244
10	**421** (1.0)	*p*-MeC_6H_4Br	CO_2*t*-Bu	DMF	NaOAc	24	140	53	245
11	**422** (1.0)	*p*-MeC_6H_4Br	CO_2*t*-Bu	DMF	NaOAc	24	140	52	245

[a]TBAB was added as an additive.

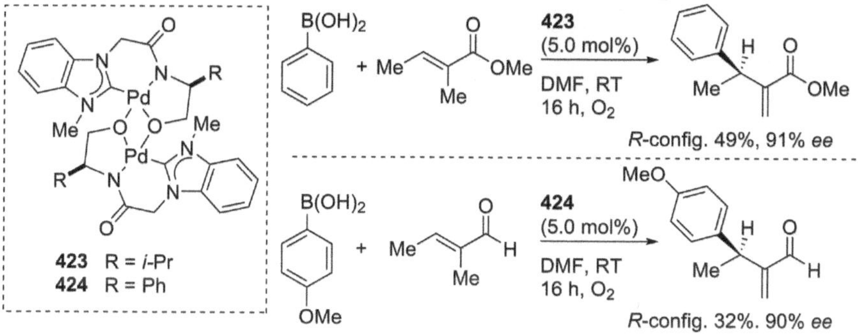

Scheme 2.36 Oxidative Heck reaction of arylboronic acids with acyclic alkenes.

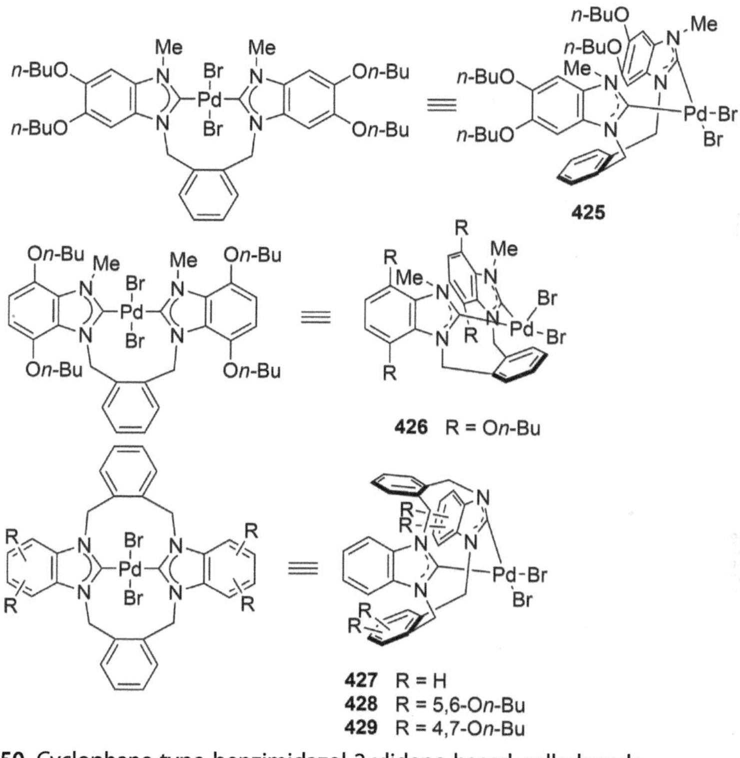

Fig. 2.50 Cyclophane-type benzimidazol-2-ylidene-based palladacycle.

5.1.4 Other NHC-Based Palladacycles

Cyclometalated-NHC palladacycles **430–431** were synthesized and tested in the Heck coupling of *n*-butyl acrylate with a number of aryl bromides up to 99% product yield *(248)*. Activated 4-chloroacetophenone also reacted smoothly (Fig. 2.51; Table 2.34).

Table 2.33 Benzimidazol-2-ylidene NHC-based palladacycles in Heck reaction *(247)*

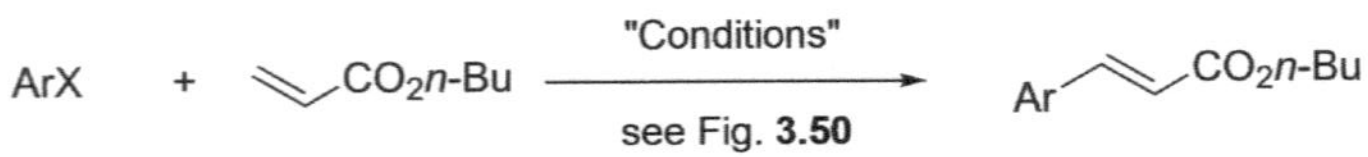

Entry	Palladacycle (mol% Pd)	ArX	Solvent	Base	Time (h)	Temp. (°C)	Yield (%)
1	**427** (5.0×10^{-5})	C_6H_5I	DMF	K_2CO_3	24	120	49
2	**428** (5.0×10^{-5})	C_6H_5I	DMF	K_2CO_3	24	120	48
3	**429** (5.0×10^{-5})	C_6H_5I	DMF	K_2CO_3	24	120	72
4	**425** (5.0×10^{-5})	C_6H_5I	DMF	K_2CO_3	24	120	92
5	**426** (5.0×10^{-5})	C_6H_5I	DMF	K_2CO_3	24	120	74
6	**427** (1.0)	C_6H_5Br	DMF	K_2CO_3	24	120	13
7	**428** (1.0)	C_6H_5Br	DMF	K_2CO_3	24	120	22
8	**429** (1.0)	C_6H_5Br	DMF	K_2CO_3	24	120	48
9	**425** (1.0)	C_6H_5Br	DMF	K_2CO_3	24	120	42
10	**426** (1.0)	C_6H_5Br	DMF	K_2CO_3	24	120	21
11[a]	**425** (1.0)	*p*-$CHOC_6H_4Cl$	DMA	NaOAc	24	165	18
12[a]	**426** (1.0)	*p*-$CHOC_6H_4Cl$	DMA	NaOAc	24	120	14

[a]TBAB was added as an additive.

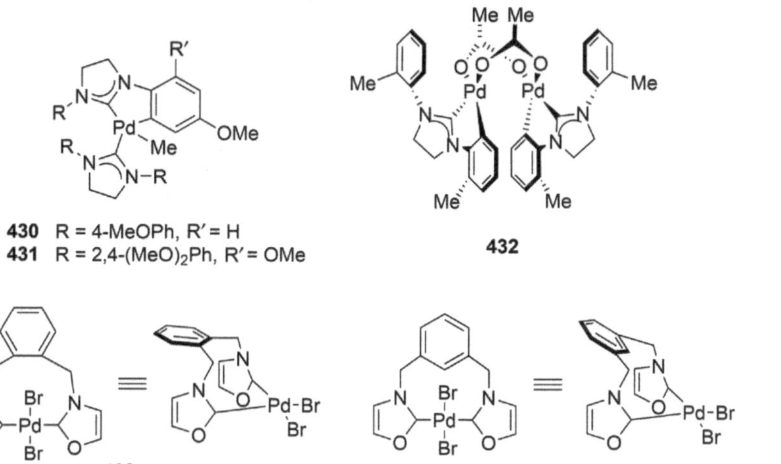

430 R = 4-MeOPh, R′ = H
431 R = 2,4-(MeO)$_2$Ph, R′ = OMe
432
433
435

Fig. 2.51 Other NHC-based palladacycles.

A hexacarbene palladacycle **435** was constructed and employed in the Heck coupling reaction. It has a structure related to **436**, and it showed a better catalytic activity than the monomeric catalyst (Fig. 2.52) *(250)*. Later, a work of developing a bridge connecting two or four **436** together was carried out. A dinuclear tetracarbene **440** and a dendritic octacarbene were prepared and applied in Heck coupling. The former one is less active than the latter one but both of them are more active than the nondendritic mononuclear palladacycle *(251)*.

Imidazo[1,2-*a*]pyridine and C2-phenyl substituted imidazole-based NHC palladacycles were developed and synthesized in high yields (Scheme 2.37) *(252)*. A dimeric form of palladacycle with bridging bromides was also obtained. They are air-stable and applied in the Heck coupling of 4-chloroacetopheone and styrene to give excellent catalytic efficacy.

5.2 Sonogashira Cross-Coupling Reaction

The use of *N*-heterocyclic carbenes (NHCs) as alternative ligands in palladium-catalyzed Sonogashira reactions is rapidly gaining in popularity. Among the different metal-NHC complexes, the Pd-NHC complexes have attracted much attention in recent years. Fu *(253)* and Glorius *(254)* reported the application of carbene ligands (Table 2.35) in the Sonogashira-type

Table 2.34 Other NHC-based palladacycles in Heck reaction

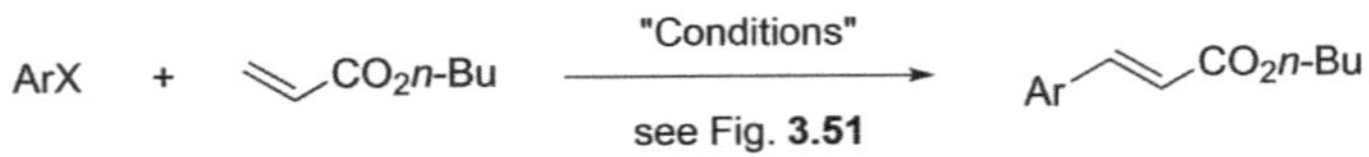

Entry	Palladacycle (mol% Pd)	ArX	Solvent	Base	Time (h)	Temp. (°C)	Yield (%)	Ref.
1	**431** (1.5)	C_6H_5Br	Dioxane	Et_3N	18	110	99	248
2	**432** (1.5)	C_6H_5Br	Dioxane	Et_3N	18	110	90	248
3	**432** (1.5)	*p*-$MeC(O)C_6H_4Cl$	TBAB	NaOAc	18	140	70	248
4	**433** (5.0×10^{-2})	*p*-$MeOC_6H_4Br$	NMP	NaOAc	5	135	85	249
5	**434** (5.0×10^{-2})	*p*-$MeOC_6H_4Br$	NMP	NaOAc	5	135	100	249
6	**435** (0.48)	*p*-$CHOC_6H_4Br$	DMA	—	5	125	74	250
7	**436** (0.5)	*p*-$CHOC_6H_4Br$	DMA	——	5	125	50	250
8[a]	**437** (0.5)	*p*-$CHOC_6H_4Br$	DMA	NaOAc	2	125	77	251
9[a]	**438** (0.5)	*p*-$CHOC_6H_4Br$	DMA	NaOAc	20	125	69	251
10	**439** (0.5)	*p*-$CHOC_6H_4Br$	DMA	NaOAc	1	125	99	251
11	**440** (6.3×10^{-2})	*p*-$CHOC_6H_4Br$	DMA	NaOAc	1.5	125	96	251

[a]TBAB was added as an additive.

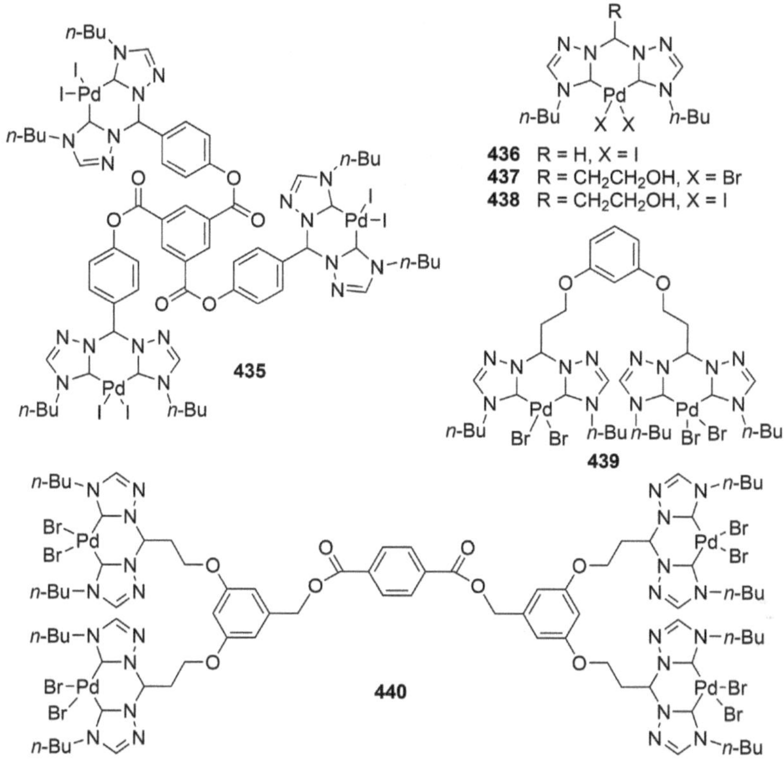

Fig. 2.52 Mono-, di-, tri- and dendritic multicarbene palladacycles.

reaction of unactivated primary and secondary alkyl bromides in the presence of CuI as cocatalyst with high catalytic efficiency under mild reaction conditions.

Previously, the chelating NHC-palladacycle **330** was applied in the Sonogashira coupling of activated aryl bromides with phenylacetylene by Hermann and co-workers. Unfortunately, an extremely low conversion was given for the coupling between bromoanisole and phenylacetylene (Fig. 2.53) *(255)*. Liu and co-workers prepared the NHC-palladacycle **447** and used it as catalyst (1.0 mol%) in the coupling reaction of iodobenzene with phenylacetylene in 1,4-dioxane/H_2O as solvent *(256)*. This palladacycle displayed moderate-to-high efficacy in the coupling of various aryl halides (Fig. 2.53).

Novel *N*-phosphanyl-*N*-heterocyclic carbenes (NHCPs) featuring a saturated tetrahydropyrimid-2-ylidenering were synthesized and used as chelating ligands with palladium(II), forming in high yield of palladacycles

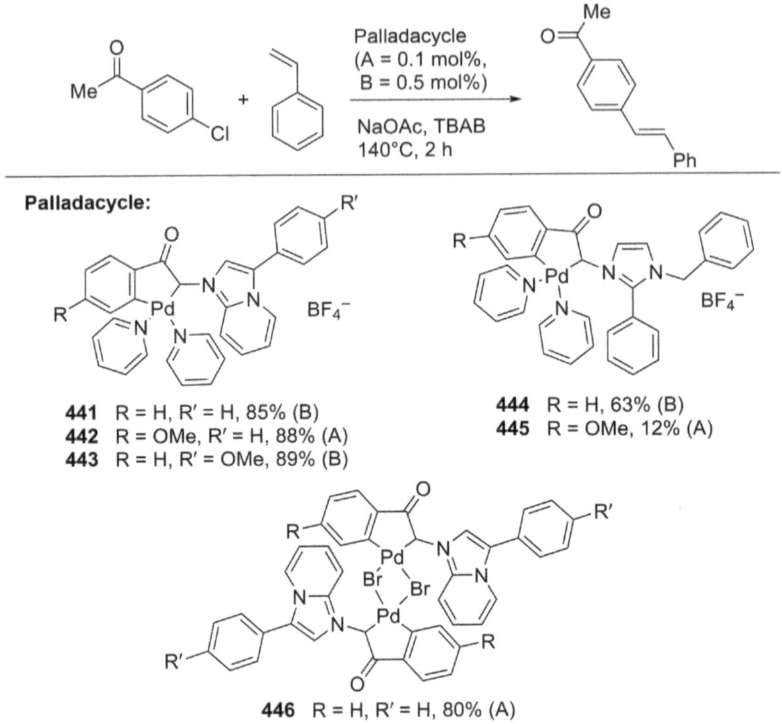

Scheme 2.37 Imidazo[1,2-*a*]pyridine and C2-phenyl substituted imidazole-based NHC palladacycles.

448–450 that are indefinitely stable in the solid state even without strict exclusion of air and moisture *(257)*. The resulting palladacycles were employed as precatalysts in copper- and amine-free Sonogashira couplings of activated aryl bromides with phenylacetylene in DMF using K_2CO_3 as the base. When deactivated aryl bromides, such as 4-bromoanisole were used as the coupling partners, low product yields were obtained (15%–54% yields); whereas with aryl chloride, such as 4-chloroacetophenone, used as substrate, the catalysts were completely inactive (Scheme 2.38).

Besides unsupported NHC-derived palladacycles, a couple of polymer-supported NHC-palladium complexes, which is applicable to the Sonogashira reaction, have been developed. For instance, the crosslinked polystyrene-supported NHC-palladium complex **451** was synthesized and employed in the copper-free Sonogashira reaction of aryl iodides with terminal alkynes, using piperidine or cesium carbonate as the base and

Table 2.35 Sonogashira coupling of primary and secondary alkyl bromides with terminal alkynes

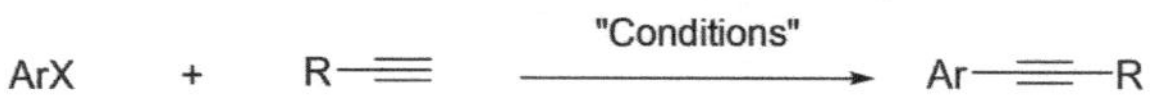

Entry	Catalyst	ArX	R	Solvent	Base	Time (h)	Temp (°C)	Yield (%)
1[a]	$[Pd(\pi\text{-allyl})Cl]_2$ 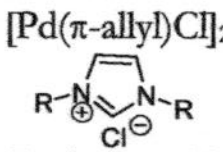 R = 1-adamantyl	$NC(CH_2)_3Br$	*n*-Hex-C≡CH	DMF/Et_2O	Cs_2CO_3	16	45	79
2[a]	$[Pd(\pi\text{-allyl})Cl]_2$ 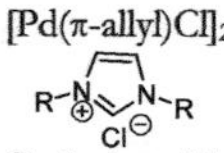 R = 1-adamantyl	$AcO(CH_2)_4Br$	$Cl(CH_2)_4C{\equiv}CH$	DMF/Et_2O	Cs_2CO_3	16	45	73
3[a]	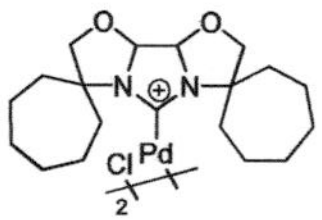	CyclopentylBr	*n*-Oct-C≡CH	DMF/DME	Cs_2CO_3	16	60	65
4[a]	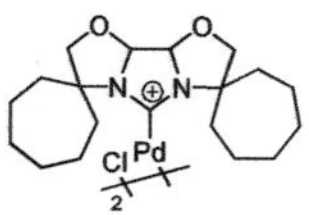	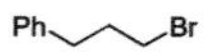	*n*-Oct-C≡CH	DMF/DME	Cs_2CO_3	16	60	77
5[a]	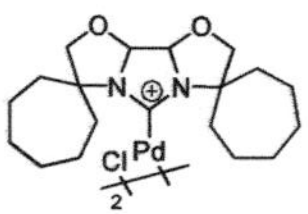	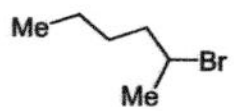	*n*-Oct-C≡CH	DMF/DME	Cs_2CO_3	16	60	66

[a]CuI was added as a co-catalyst.

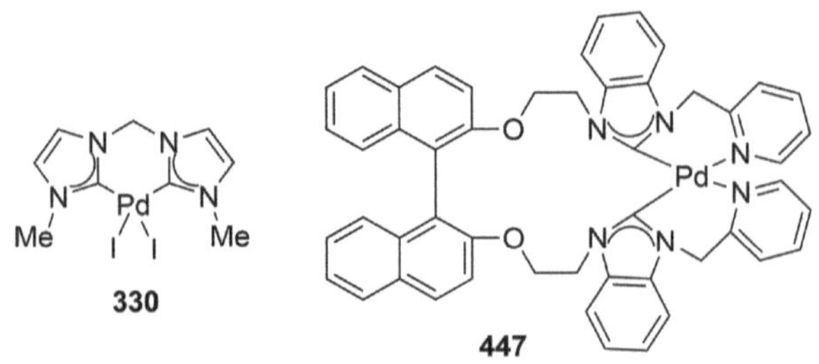

Fig. 2.53 Examples of chelating bis(NHC)-palladacycles.

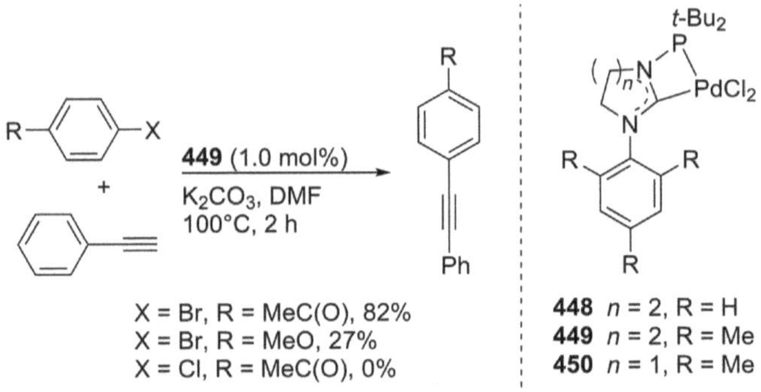

Scheme 2.38 Sonogashira reaction catalyzed by NHC-palladium complexes.

DMF/water (1:1) as solvent at 60–100°C (Fig. 2.54) *(258)*. This polymer-supported NHC-palladium catalyst has consistent activity after four runs. The recovered catalyst was analyzed by inductively coupled plasma-atomic emission spectrometer (ICP-AES) and no significant decrease in palladium content on polymer beads was observed.

Very recently, a new poly(NHC-palladium complex) immobilized on nano silica **452** was developed by Moghadam and Khajehzadeh (Fig. 2.54) *(259)*. This (Pd-NHCs)$_n$@nSiO$_2$ was applied as a highly efficient heterogeneous catalyst for copper-free Sonogashira reaction of aryl halides with phenylacetylene in water at 90°C (Table 2.36). Furthermore, the catalyst is stable and can be recycled and reused eight times without a significant loss of its catalytic activity. Notably, palladacycle **452** also catalyzed the Heck cross-coupling of aryl chlorides and styrene. Compared to the reaction conditions of Sonogashira reaction, DMF/H_2O (2:1) was used as solvent and K_2CO_3 was added as the base in the Heck coupling.

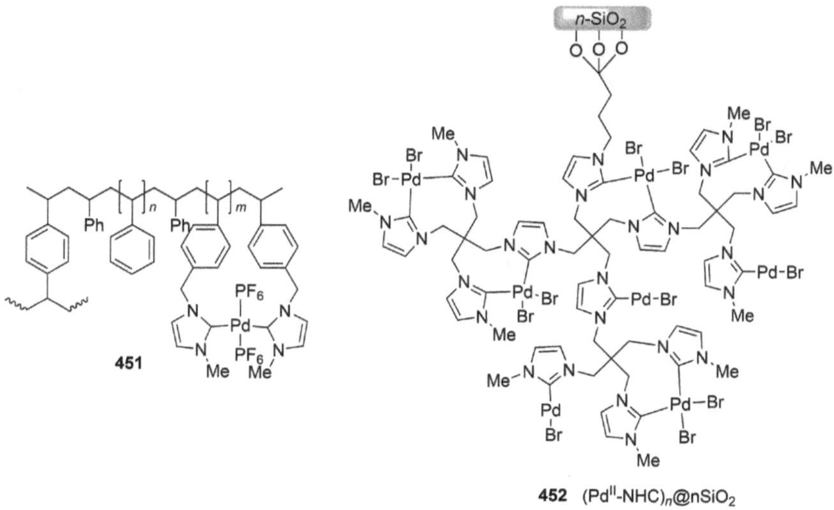

Fig. 2.54 Examples of supported NHC-palladacycles.

Table 2.36 Sonogashira cross-coupling reaction of aryl halides with phenylacetylene catalyzed by $(Pd^{II}\text{-NHCs})_n@n\text{SiO}_2$

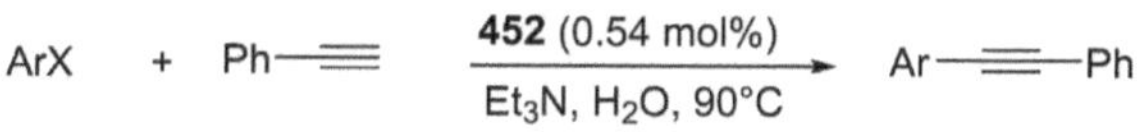

Entry	ArX	Time (min)	Yield (%)	Ref.
1	C_6H_5I	9	92	259
2	C_6H_5Br	13	90	259
3	C_6H_5Cl	15	87	259
4	*p*-MeC_6H_4I	14	85	259
5	*p*-$MeOC_6H_4I$	12	86	259
6	*m*-ClC_6H_4Cl	12	90	259
7	*p*-$NO_2C_6H_4Cl$	17	83	259
8	*m*-$NO_2C_6H_4Cl$	22	84	259

The *N*-heterocyclic carbene palladium complexes can be supported on SiO_2-pA-Cyan (Fig. 2.55). The palladacycle **453** showed high efficiency in the copper-free Sonogashira coupling of aryl halides with phenylacetylene in H_2O/DMF at 80°C *(260)*. Aryl chlorides were not suitable coupling substrates while aryl iodides and bromides could couple with phenylacetylene smoothly to give good yields under the standard conditions. The recovery and reusability of the catalyst was demonstrated five times without detriment to its catalytic activity. Palladacycle **455** was stabilized by the carbene-phosphine ligand (Fig. 2.55); low catalyst loading (0.04 mol%) was sufficient to catalyze the Sonogashira coupling in water medium *(261)*. This coupling was demonstrated to be amenable for the successful conversion of a range of aryl bromides and aryl iodides to give the corresponding coupled products in good-to-excellent yield in a short reaction time; whereas the aryl chlorides implied longer reaction time and lower yields were obtained. PdNPs immobilized in polymeric NHC grafted-silica **454** exhibited an excellent activity in Sonogashira coupling reaction of various aryl halides, as well as heterocyclic halides with phenylacetylene. The addition of TBAB (0.01 mol%) was necessary when aryl chlorides were used as coupling substrates (Fig. 2.55) *(262)*.

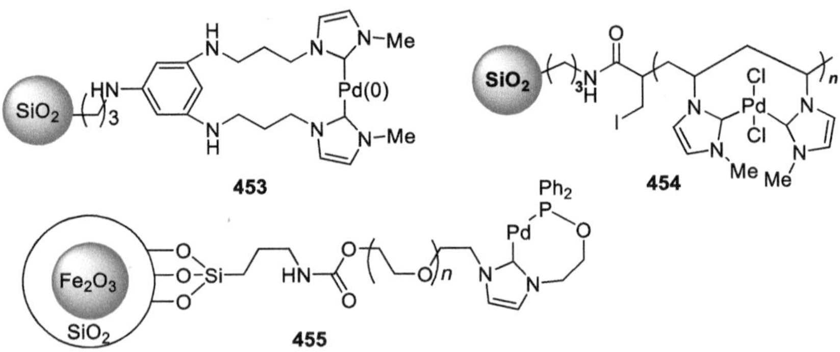

Fig. 2.55 Examples of SiO_2-immobilized NHC-palladacycles.

6. Pincer-Based Palladacycles

Pincer ligands are chelating rings binding to three adjacent coplanar sites incorporating two fused palladacycles. The palladium-pincer complex can be represented by two side arms with a metal center. In some cases, one arm is with a hemilabile feature. In the beginning of the development, the most common class of the pincer ligands consists of two phosphorus donors and a so-called PCP palladacycle. Variations have been developed in which the phosphorus donors can be replaced by sulfur, and nitrogen donors. The general term to describe the pincer palladacycle is therefore LZL′ palladacycle where L and L′ represent different neutral donors at the side arms (such as SR, SeR or TeR) while Z is carbon or nitrogen (Fig. 2.56).

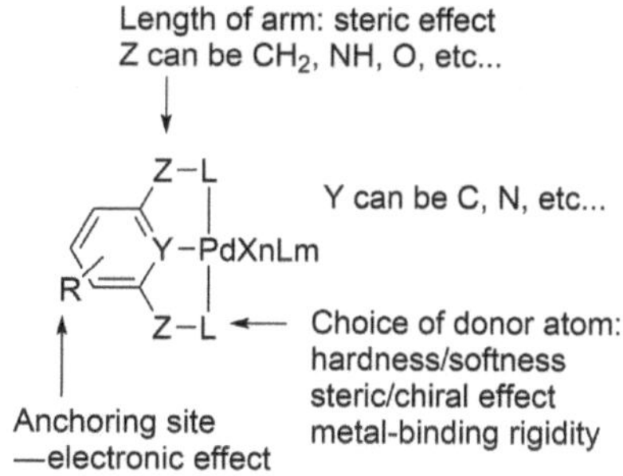

Fig. 2.56 General structure of pincer palladacycle.

In general, most of the pincer-based palladacycles show high thermal stability due to their firm tridentate coordination. Under ambient conditions, the oxidative state of Pd is 2+. In the d^8 square planar pincer-based palladacycle, only one free coordination site is available for the catalysis. This avoids the formation of unwanted side products from the ligand exchange process *(263)*. Pincer-based palladacycle is attractive to chemists as they are highly tunable with steric and electronic properties by varying the side arms of the pincer ligands, and thus directly affect the reactivity of the catalyst system. They can even be used in much asymmetric catalysis by adding a chiral group on the side arms.

6.1 Heck Cross-coupling Reaction

6.1.1 PCP- and PNP-Pincer Palladacycles

Phosphorus has been a popular donor atom in organometallic chemistry, owing to its ability to stabilize the metal centers in high and low oxidation states. Phosphines or phosphites are common ligands that are used as a part of the palladacycles.

The first PCP-pincer palladacycles were studied in 1997 by Milstein and co-workers. Palladacycles **456** and **458** were developed as efficient catalysts for the Heck coupling reaction of aryl bromides/iodides and styrene/acrylates (Fig. 2.57) *(264)*. Despite the electron-rich bulky side arms as in phosphine pincers, dialkyl, or trialkylphosphines side arms with aliphatic (sp^3) carbon backbones, pincer palladacycles have also received much attention.

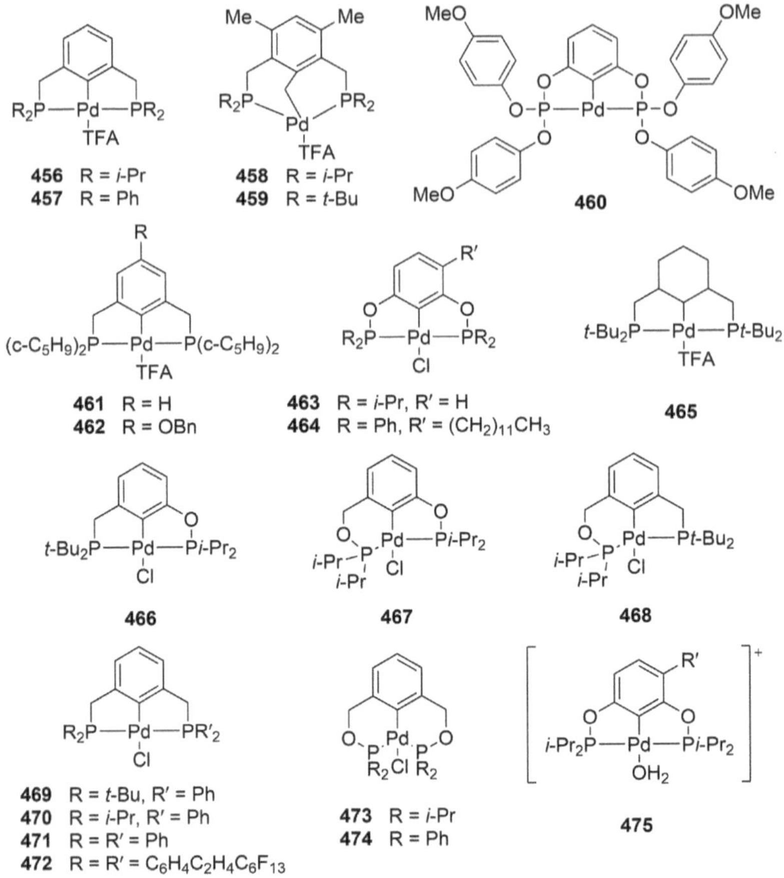

Fig. 2.57 Examples of PCP-pincer palladacycle in Heck coupling reaction.

Palladacycle **461** was examined in an intramolecular Heck coupling reaction *(265)*. When a 1,4-diene was added to the reaction mixture, no coupling product can be obtained so that it effectively poisoned the PCP palladacycle. Among the PCP pincer ligands stated in Fig. 2.57, **461** showed the best reactivity in the Heck coupling reaction with a broad scope of aryl chloride, even the sterically hindered one *(266)*. However, it should be noted that extreme high reaction temperature (180°C) was required, and how the additive of hydroquinone improved the product yield remains unknown.

In 2004, Eberhard studied the effect of ligand structure and electronics on the catalytic activity by comparing the Heck reaction of phenyl halides and styrene with a series of PCP-pincer ligands (**463** vs **466–468**) *(267)*.

Palladacycles **463** and **467** gave 72% and 61% product yields correspondingly at 100°C in 6 h while **466** and **468** only gave 8% and 5% (Table 2.37). Increasing the temperature to 180°C, **463** and **467** gave the desired products in 65% and 43% within 30 min. For **466**, it showed an induction period in the first 15 min and gradually increased to 58% after 6 h. However, palladacycle **468** kept in low activity with only 5% yield. Other Hg drop tests, quantitative poisoning experiment, and NMR studies were carried out. Preliminary studies showed that the introduction of P*t*-Bu_2 groups leads to more stable palladacycles but lower catalytic activity to the Heck coupling. At the same year, the Jensen group also synthesized the PCP-pincer palladacycles with different phosphine side arms *(268)*. Surprisingly, **469** did not catalyze the Heck coupling even when iodobenzene was used. Palladacycle **470** only catalyzed the reaction of iodobenzene and styrene with 5.0 mol% catalyst loading but not bromo- or chlorobenzene.

There are few reports about the pincer palladacycles incorporating perfluoroalkyl groups. In 2011, Stuart developed a recyclable perlfuoroalkylated PCP pincer palladacycle **472**, which promoted the Heck coupling of aryl iodides and bromides *(269)*. It can be easily recovered by fluorous solid-phase extraction and reused four times without loss in catalytic activity (Table 2.37).

An aminophosphine-based PCP-pincer palladacycle was easily prepared by sequential addition of 1,1′,1″-phosphinetriyltripiperidine and 1,3-diaminobenzene or resorcinol to the solutions of [Pd(cod)$(Cl)_2$] in toluene under N_2 *(274)*. A number of challenging substrate, deactivated, and sterically hindered aryl bromides could be coupled with various olefins with 20 ppm of Pd. It was highly attractive that they could also catalyze deactivated and sterically hindered aryl chlorides by increasing reaction time and reaction temperature with only 0.01 mol% of catalyst (Scheme 2.39).

One of the design ideas was to develop the linear and branched conjugated PCP-pincer palladacycle by palladation of the special ligands with $Pd(BF_4)_2$(MeCN) in boiling acetonitrile *(275)*. **478–481** were synthesized and they exhibited considerable catalytic activity in the Heck coupling of iodobenzene and ethyl acrylate (Fig. 2.58).

From the above PCP-pincer palladacycle, the center-region is all phenyl rings with the carbon center atom. Other heterocycles, such as triazoles as the center part of the PCP-pincer palladacycle, was synthesized by Gandelman in 2008 *(276)*. They gave the name "Pincer Click Ligands" for the idea to "click" two monomeric groups with coordinating atoms and functionalized with azidomethyl and propargyl units to provide the pincer framework. It is

Table 2.37 PCP-pincer palladacycles in Heck reaction

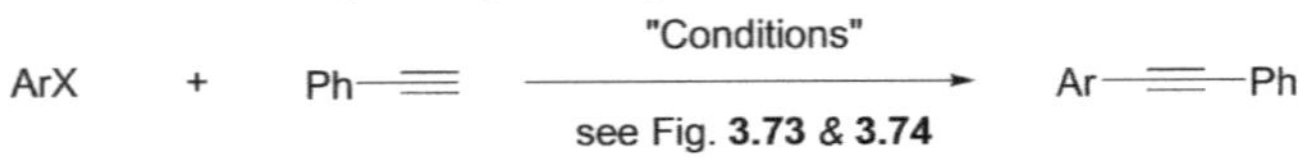

Entry	Palladacycle (mol% Pd)	ArX	R	Solvent	Base	Time (h)	Temp. (°C)	Yield (%)	Ref.
1	**456** (7.0×10^{-4})	C_6H_5I	CO_2Me	NMP	Na_2CO_3	60	40	100	264
2	**458** (7.0×10^{-4})	C_6H_5Br	CO_2Me	NMP	Na_2CO_3	63	140	93	264
3[a]	**460** (1.0×10^{-5})	C_6H_5I	CO_2*n*-Bu	NMP	Na_2CO_3	22	180	86	270
4	**461** (0.67)	*o*-MeC_6H_4Cl	Ph	Dioxane	CsOAc	24	180	85	266
5	**463** (5.8×10^{-4})	C_6H_5Br	CO_2Me	NMP	Na_2CO_3	90	160	82	271
6	**471** (1.0)	C_6H_5Br	CO_2Me	NMP	$KHCO_3$	24	120	82	269
7	**472** (1.0)	C_6H_5Br	CO_2Me	NMP	$KHCO_3$	24	120	79	269
8	**473** (1.0×10^{-2})	C_6H_5Cl	Ph	NMP	Na_2CO_3	—	180	94	272
9	**475** (2.0×10^{-2})	*m*-$CO_2HC_6H_4I$	*p*-$CO_2HC_6H_4$	Water	$NaHCO_3$/NaOH	0.4	100	16	49
10	**464** (0.1)	*p*-MeC_6H_4Br	Ph	DMF	Na_2CO_3	6	140	70	273

[a]Hydroquinone was added as an additive.

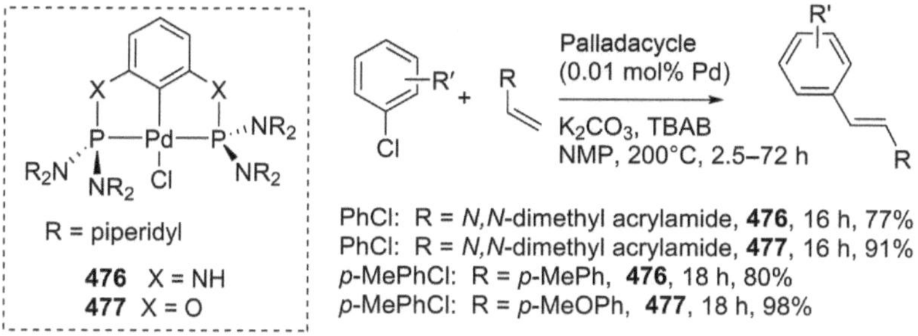

Scheme 2.39 Aminophosphine-based PCP-pincer palladacycle in Heck coupling of aryl chloride.

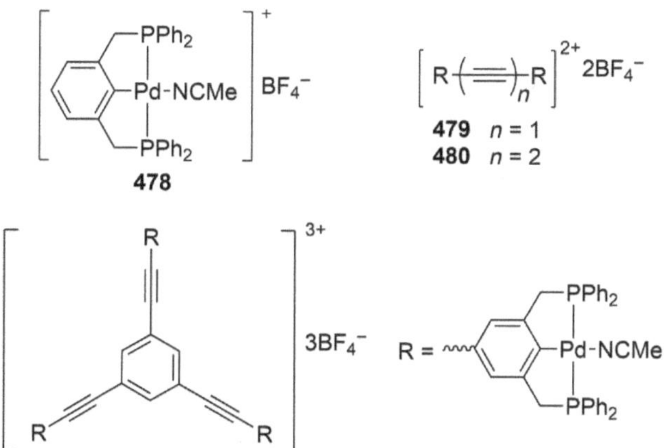

Fig. 2.58 Mono-, di-, and tri-nucleus PCP-pincer palladacycle.

believed that the two side arms in the 1,4-position and the relatively acidic C—H bond between them had a positive effect on the direct insertion of a metal atom. Palladacycle **482** (94%) gave a slightly better product yield than **483** (88%) in the Heck coupling of bromobenzene and methyl acrylate with 7×10^{-4} mol% of Pd in the presence of Na_2CO_3 in DMF at 140°C for 48 h (Fig. 2.59).

The central carbon atom in the PCP-pincer palladacycle can come from the *N*-heterocyclic carbene which is named as $PC^{NHC}P$ complexes *(277)*. A new imidazolium salt, 1,3-bis(2-diphenylphosphanylethyl)-3*H*-imidazol-1-ium chlorides, can be combined with the phosphine and palladium complexes (Fig. 2.60). Two different kinds of $PC^{NHC}P$-pincer palladacycle **484–486** were synthesized. The X-ray structure showed the chiral twisting of the

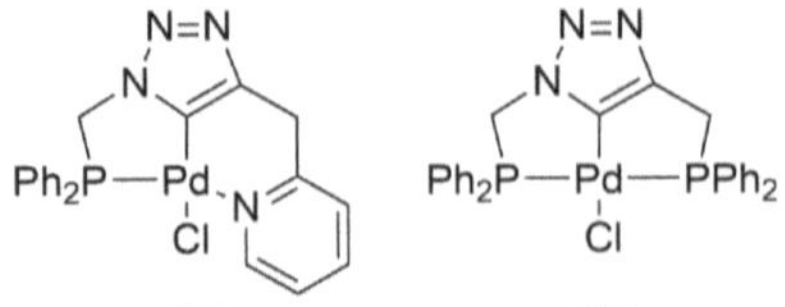

482 483

Fig. 2.59 Examples of PCP-pincer palladacycle.

central imidazole ring from the metal coordination plane. They showed excellent activity in the Heck coupling of bromobenzene and styrene or *n*-butyl acrylate with 0.5 mol% Pd in the presence of NaOAc and DMA at 165°C for 2–4 h. In 2006, Hahn further developed the $PC^{NHC}P$ palladacycle **487** by replacing the imidazole to benzimidazole and employed in the Heck reaction of *p*-bromobenzaldehyde and styrene *(243)*. 1.0 mol% of Pd could catalyze the coupling to give 72% product yield within 2 h. Further lowering the Pd loading (0.01 mol%) and lengthening the reaction time (24 h) could still catalyze the coupling to give 100% yield.

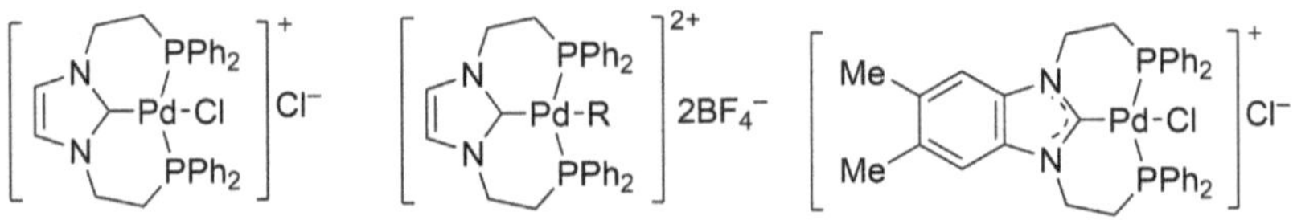

484 485 R = NCMe 486 R = Py 487

Fig. 2.60 Examples of $PC^{NHC}P$-pincer palladacycle.

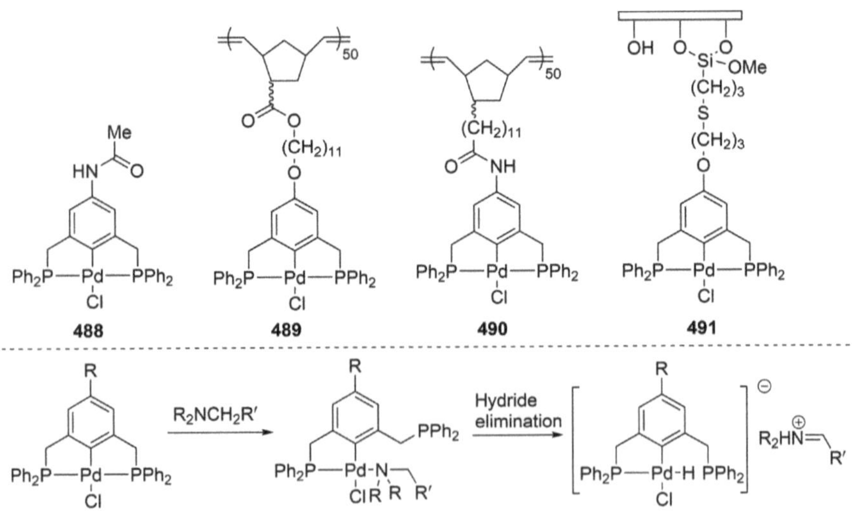

Fig. 2.61 Immobilized PCP-pincer palladacycle and proposed decomposition step.

Immobilized supported PCP-pincer palladacycles **488–491** were synthesized by Sherrill, Jones, and Weck in 2005 *(278)*. The PCP-pincer palladacycles that are covalently tethered onto polymer or silica supports with amide/ether linkages were evaluated in the Heck reaction to investigate their stability during the catalysis. The proposed initial decomposition of the PCP-pincer palladacycles with organic base was depicted in Fig. 2.61.

Later, the Tamami group developed a PCP-pincer palladacycle supported on modified Merrifield resin **492** as a heterogeneous catalyst for the Heck coupling reactions of aryl iodides and *n*-butyl acrylate and Sonogashira coupling of aryl iodides and phenylacetylene (Scheme 2.40).[492]

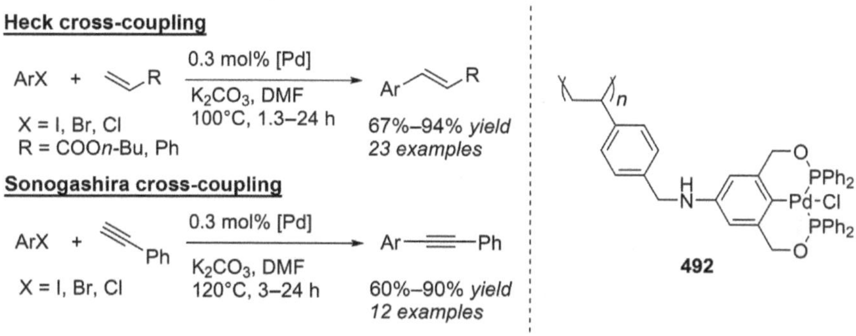

Scheme 2.40 Modified Merrifield resin-supported PCP pincer palladium nanoparticles in Heck and Sonogashira coupling reactions.

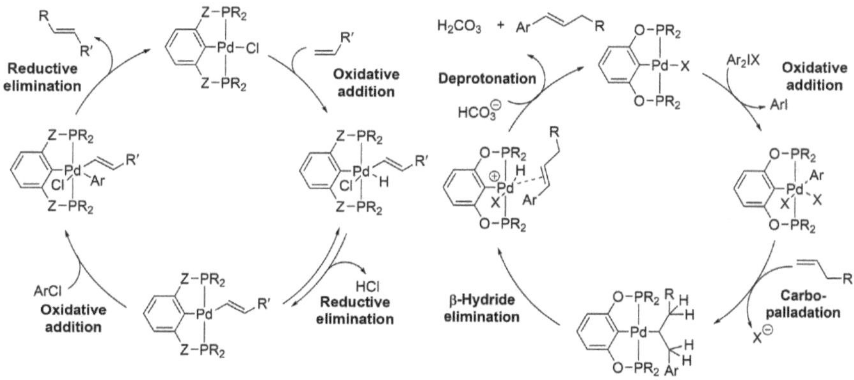

Fig. 2.62 Alternative proposed mechanisms for the olefination via Pd(II)/Pd(IV) cycle proposed by Jensen (left) and Szabó (right).

In general, the classical catalytic cycle of the Heck reaction is based on the Pd(0)/Pd(II) cycle in which the first step is the oxidative addition of aryl halide to the Pd(0) catalyst to give the Pd(II) complex. After that, alkene exchanges with the ligand to give the olefin Pd-complex which then undergoes insertion of the olefin and finally undergoes β-hydride elimination, providing the coupling product (refer to Fig. 2.1). However, the two alternative mechanisms Pd(II)/Pd(IV) cycle was suggested for the PCP palladacycle in the Heck coupling from Jensen, *(266)* Whitcombe, *(280)* Szabó, *(281)* and Frech *(282)*. Oxidative addition of the alkenes to the Pd(II) complex initially occurred and reductive elimination of HCl then took place from the Pd(IV) species and it is suggested to be the rate determining step to generate back the Pd(II) species. One more oxidative addition of ArCl occurred to generate the Pd(IV) species again and further generate the coupling product through a second reductive elimination and restore the pincer catalyst. The formation of the intermediate after the first oxidative addition was supported by NMR determination (Fig. 2.62). Apart from the PCP-pincer palladacycles, the Liang group developed a PNP-pincer palladacycle for the Heck reaction (Scheme 2.41) *(283)*. The amido pincer complexes of palladium is highly effective for coupling of styrene and aryl iodides with turnover numbers of up to 4.5×10^7.

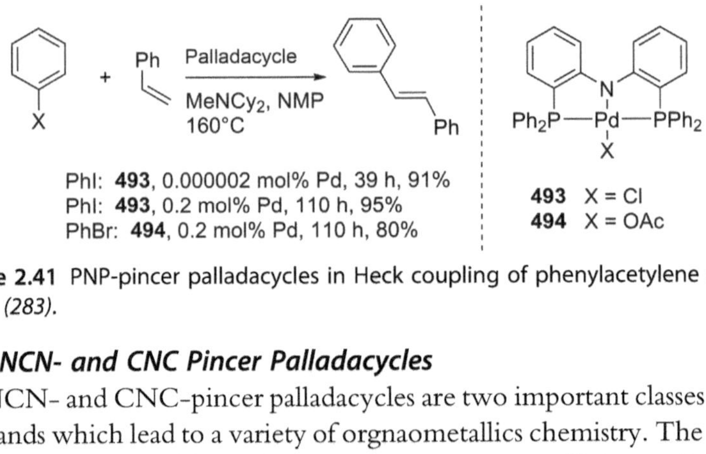

Scheme 2.41 PNP-pincer palladacycles in Heck coupling of phenylacetylene and aryl halides *(283)*.

6.1.2 NCN- and CNC Pincer Palladacycles

The NCN- and CNC-pincer palladacycles are two important classes of pincer ligands which lead to a variety of orgnaometallics chemistry. The NCN-pincer bearing two additional nitrogen atoms was an efficient palladacycle for Heck coupling reaction (Fig. 2.63).

The CNC-pincer offers two metal-carbon σ bonds in the terdentate mode when the NCN-pincer only has one metal-carbon σ bond. The metal-carbon σ bonds enhanced the stability of the palladacycles which minimized the metal leaching during the catalysis. Compared with PCP-pincer palladacycle, NCN- and CNC-pincers gain the benefit of diverse functional groups availability and its endurance against oxygen under aerobic

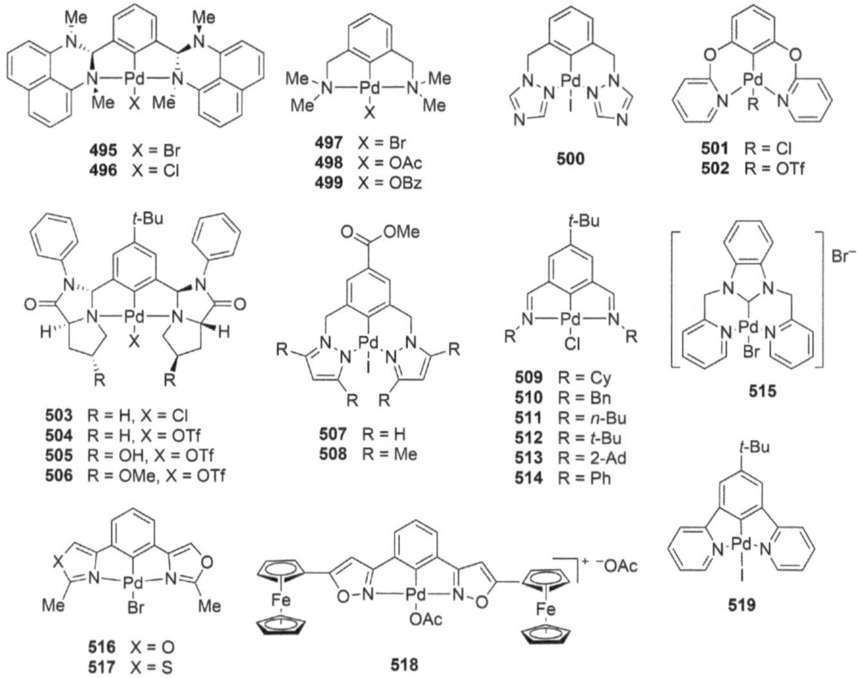

Fig. 2.63 Examples of NCN-pincer palladacycles.

conditions. Much effort has been made to modify the side arms of the NCN- and CNC-pincer palladacycle such as introducing a chirality group to make it a chiral catalyst. Modification of the central benzene ring was also widely investigated. A summary of the NCN-pincer palladacycles used in the Heck coupling is listed in Table 2.38 for a brief comparison.

In general, the central ring of a NCN-pincer palladacycle is a phenyl ring with center carbon coordinated with the Pd. In 2001, two carbene-NCN-pincer complexes were synthesized and employed in the Heck coupling of 4-bromoacetophenone and *n*-butyl acrylate to give moderate yield *(292)*. Interestingly, there was 5%–10% of brominated side-product after the coupling (Fig. 2.64). It is suggested that the pyridyl side-arms of **520** hindered the β-hydride elimination by blocking the coordination site of the Pd and thus it underwent reductive elimination from the intermediates instead. The Shang group reported a bis(ferrocene-isoxazole/isoxazoline) ligand (NNN- and NCN-pincer ligands) for the preparation of pincer palladacycles **518**, which have been used as catalyst in the coupling of olefin and aryl halides in a mixture of DMF/water as solvent and TBAB as additive *(293)*.

Table 2.38 NCN-pincer palladacycles in Heck reaction

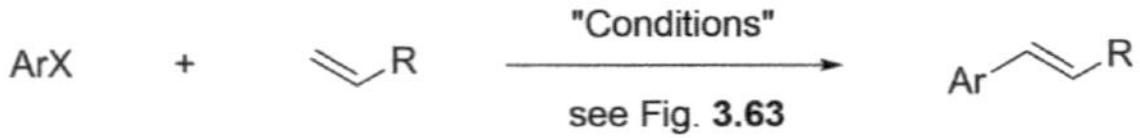

Entry	Palladacycle (mol% Pd)	ArX	R	Solvent	Base	Time (h)	Temp. (°C)	Yield (%)	Ref.
1	**495** (1.0×10^{-4})	*p*-MeC_6H_4I	CO_2Me	DMF	Et_3N	12	110	96	284
2	**495** (0.1)	*p*-$CHOC_6H_4Br$	CO_2Me	DMF	Et_3N	12	110	35	284
3	**496** (1.0×10^{-3})	*p*-$NO_2C_6H_4Br$	CO_2Me	DMF	Et_3N	12	110	52	284
4	**497** (1.0×10^{-4})	*p*-MeC_6H_4I	CO_2Me	DMF	Et_3N	12	110	25	284
5[a]	**500** (1.0)	*p*-$NO_2C_6H_4Cl$	CO_2*n*-Bu	DMA	NaOAc	15	140	89	285
6	**503** (1.0×10^{-4})	C_6H_5I	CO_2Me	NMP	$NaHCO_3$	22	140	87	286
7	**508** (0.1)	*p*-$MeC(O)C_6H_4Br$	CO_2Me	DMF	K_2CO_3	18	140	86	287
8	**509** (1.0)	C_6H_5I	CO_2Me	NMP	*n*-Bu_3N	2	100	90	288
9	**511** (1.0×10^{-4})	*o*-MeC_6H_4I	CO_2*n*-Bu	NMP	*n*-Bu_3N	21	140	83	288
10	**513** (1.0)	C_6H_5I	CO_2Me	NMP	*n*-Bu_3N	2	100	86	288
11	**514** (1.0)	C_6H_5I	CO_2Me	NMP	*n*-Bu_3N	2	100	89	288
12	**501** (1.0×10^{-3})	C_6H_5I	CO_2Me	DMF	Et_3N	14	110	87	289
13	**502** (1.0×10^{-7})	C_6H_5I	CO_2Me	DMF	Et_3N	6 days	110	84	289
14	**502** (1.0×10^{-2})	*p*-MeC_6H_4Br	CO_2Me	DMF	Et_3N	14	110	51	289
15	**498** (5.0)	Ar_2IX[b]	*p*-BrC_6H_4	THF	$NaHCO_3$	17	50	95	281
16	**515** (1×10^{-2})	*p*-$MeC(O)C_6H_4Br$	Ph	DMAC	NaOAc	2	110	57	244
17	**516** (1.0)	*p*-$MeC(O)C_6H_4Br$	Ph	NMP	KOAc	7	120	84	290
18	**517** (1.0)	*p*-$MeC(O)C_6H_4Br$	Ph	NMP	KOAc	7	120	75	290
19	**519** (4.0×10^{-2})	C_6H_5I	Ph	DMF	Na_2CO_3	25	250	82	291

[a]TBAB was added as an additive.
[b]Aryl iodonium salts were used.

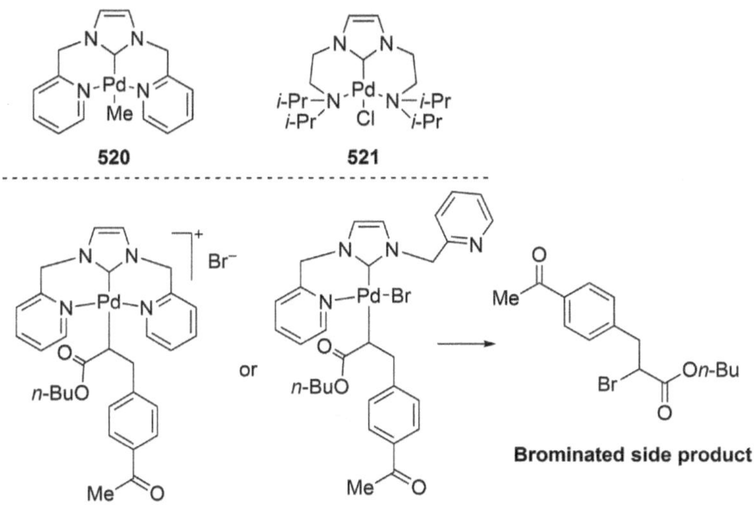

Fig. 2.64 Carbene-NCN-pincer palladacycle in Heck coupling and its side product formation.

In 2015, the Yang and Mao group synthesized a pincer palladacycle containing ring expanded six-membered NHC based on the 1,3-dipicolyl-3, 4,5,6-tetrahydropyrimidin-2-ylidenes **522–523** *(294)*. It showed a high reactivity in the Heck reaction of bromobenzene or *para*-substituted phenyl bromides with olefin. Moderate to excellent yields were obtained so that **522** and **523** gave comparable results at the stated reaction conditions (Scheme 2.42).

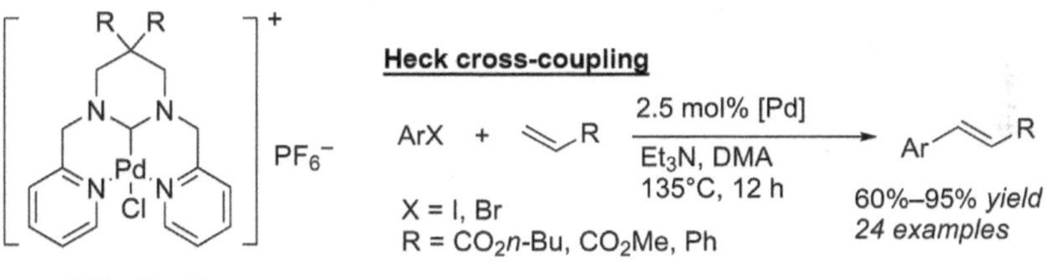

522 R = H
523 R = Me

Scheme 2.42 1,3-Dipicolyl-3,4,5,6-tetrahydropyrimidin-2-ylidenes NCN-pincer palladacycle in Heck reaction.

Tridentate pincer palladacycles of 2,6-lutidinyl-biscarbene were synthesized which exhibit helical C_2 symmetrical structure (Fig. 2.65) *(295)*. The racemic mixtures of chiral **524** can exhibit the Heck coupling reaction. **524–527** could also activate the aryl chlorides substrates. In 2000, Peris and Crabtree developed a robust CNC-pincer palladacycle **537** and compared the catalytic activity with **331** and **1** in the Heck coupling of iodobenzene and styrene (Table 2.39) *(211)*. It was found that **331** was easily to decompose in DMA above 70°C and palladacycle **1** also deposited Pd black after 17 h *(207)*. **537** was more stable under heat that remain unchanged after 24 h.

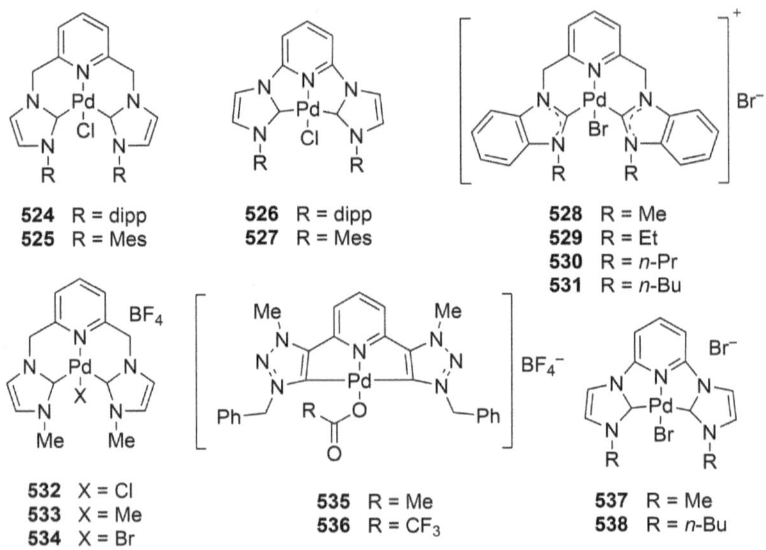

Fig. 2.65 Examples of CNC-pincer palladacycles.

The Cavell group further investigated the pyridine-bridged NHC-pincer palladacycles with the *N*-substituent effect on the catalytic performance and stability *(300)*. *N*-Position of the NHC was substituted with methyl, mesityl, *tert*-butyl, and di*iso*propyl groups that correlated with the proximity of the aryl rings. The palladacycles with bulkier substituent at the *N*-position of the NHC generally havebetter catalytic performance in the Heck coupling reaction (Scheme 2.43).

Tridentate di-NHC-pyridine-pincer palladacycles were employed in the Heck couplings. The NHC or the pyridine would dissociate from the palladacycle to release coordination site for the coupling catalysis. In 2007, Douthwaite synthesized a di-NHC-amine-palladaycle replacing the pyridine with amine chain to study the relationship between palladacycle

Table 2.39 CNC-pincer palladacycles in Heck cross-coupling reaction

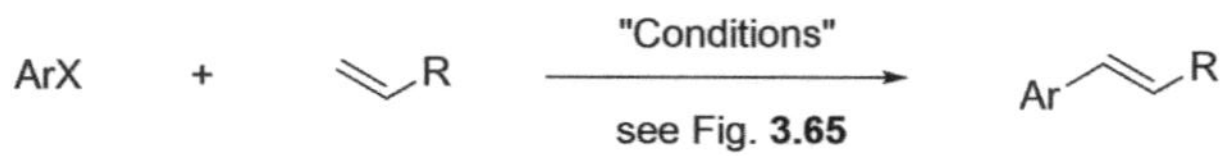

Entry	Palladacycle (mol% Pd)	ArX	R	Solvent	Base	Time (h)	Temp. (°C)	Yield (%)	Ref.
1	**537** (0.2)	C_6H_5I	Ph	DEA	NaOAc	4	reflux	85	211
2[a]	**538** (0.2)	*p*-MeC(O)C_6H_4Cl	Ph	DMA	NaOAc	0.5	145	90	296
3	**532** (1.03×10^{-3})	*p*-Me$CO_2C_6H_4Br$	CO_2*n*-Bu	DMA	Pr_4NBr	66	120	72	297
4	**533** (1.01×10^{-3})	*p*-Me$CO_2C_6H_4Br$	CO_2*n*-Bu	DMA	Pr_4NBr	111	120	71	297
5	**535** (1.0)	*o*-MeC_6H_4Br	CO_2*t*-Bu	DMF	NaOAc	24	120	73	298
6	**536** (1.0)	*p*-CHOC_6H_4Br	CO_2*t*-Bu	DMF	NaOAc	24	120	93	298
7[a]	**535** (1.0)	*p*-CHOC_6H_4Cl	CO_2*t*-Bu	DMF	NaOAc	24	120	74	298
8[a]	**534** (0.2)	*p*-MeC(O)C_6H_4Cl	Ph	DMA	NaOAc	0.25	165	91	299

[a]TBAB was added as an additive.

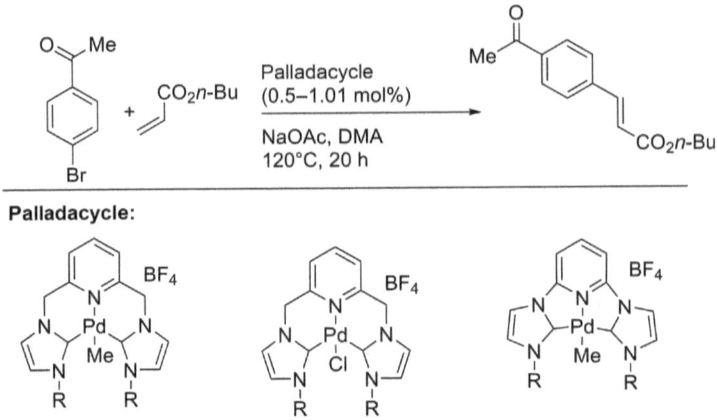

539 R = Me, 17%
540 R = mes, 16%
541 R = dipp, 25%
542 R = *t*-Bu, 73%
543 R = mes, 25%
544 R = *t*-Bu, 66%
545 R = Me, 42%
546 R = mes, 62%
547 R = dipp, 35%

Scheme 2.43 Effect on *N*-substituent to the CNC-pincer palladacycle catalytic performance.

structure and catalytic properties (Scheme 2.44) *(301)*. The result again showed the phenomenon of bulkier mesityl group gave high catalytic activity than the *tert*-butyl-substituent, hence preventing the associative substitution process.

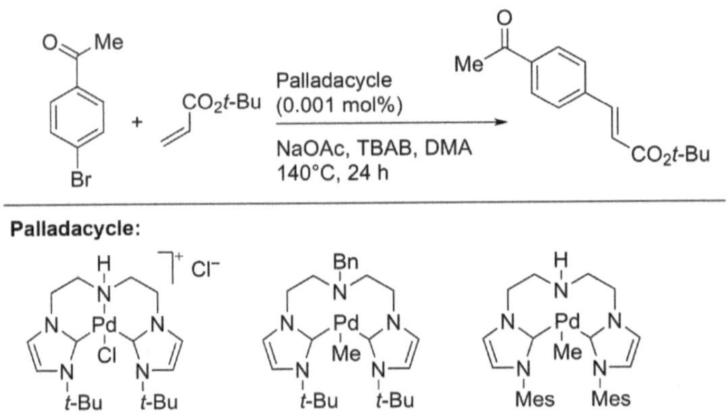

548 56% **549** 40% **550** 93%

Scheme 2.44 Di-NHC-amine- palladaycle in Heck reaction.

6.1.3 SCS-, SNS-, SeCSe- and SeNSe-Pincer Palladacycles

Although phosphorus and nitrogen are common donor atoms in the palladacycles, sulfur and selenium give another choice as the donor atoms. The sulfur donor atom can act as σ-donor, π-donor, and even π-acceptor ligands.

It can be used to tune the electronic properties of the metal center so that they can accommodate both hard and soft ancillary ligands and metal center *(302)*. Selenium shares similar properties to sulfur; however, fewer investigations were reported in this area.

The SCS-pincer palladacycle started to appear in 1980, but its actual application for Heck coupling reaction was known in 1999. In comparison to the analogous PCP-pincer palladacycle, the SCS-pincer palladacycle avoided the use of air sensitive phosphine ligands which are more attractive to apply. Unfortunately, they were found to be only activating aryl iodides but not aryl bromides in the Heck coupling reactions. **553** and **554**, water-soluble palladacycles, were synthesized by Ogo and Fukuzumi and employed in the Heck coupling of 3-I-C_6H_4COOH) with 4-CH_2=CH-(C_6H_4)COOH in water *(49)*. Although these palladacycles performed with high reactivity of Suzuki–Miyaura coupling of aryl iodides, no reaction product can be obtained when they employed in the Heck coupling (Fig. 2.66; Table 2.40).

A series of hetero-multimetallic tetrakis(SCS-pincer palladium)-(metallo)porphyrin hybrids was designed (Fig. 2.66) *(303)*. Manganese, nickel, and magnesium were introduced into the porphyrin macrocycle and they were employed in the Heck reaction of iodobenzene and styrene. The catalytic activity increased for the metalloporphyrin series MnCl < 2H < Ni < Mg which is related to the electron richness of the porphyrin ring.

The SCS-pincer palladacycles were attached to various polymeric supports for the Heck coupling reactions (Fig. 2.67). In 1999, a poly(ethylene glycol)-supported 5-oxo-SCS-pincer palladacycle was synthesized by Bergbreiter and employed in the Heck coupling of iodobenzene and styrene without catalyst deactivation even recycled three times *(304)*. This polymer-support made the palladacycle to be more stable so that the 5-oxo-counterparts decompose slowly and hence improve the reactivity of the catalyst. Later, the same group reported another polymer-bound SCS- and SeCSe-pincer palladacycle with the poly(*N*-isopropylacrylamide) (PNIPAM) support and PNODAM-support in the Heck coupling under thermomorphic conditions *(309)*. A homogeneous SCS-pincer palladacycle was covalently immobilized onto mesoporous silica support *(310)*. 0.13 mol% of **528** was used to couple the iodobenzene with *n*-butyl acrylate to give <99% product yield.

Pincer type organochalcogen palladacycle is rare due to their difficult preparation and modification process. Indeed, the SeCSe- and SeNSe-pincer

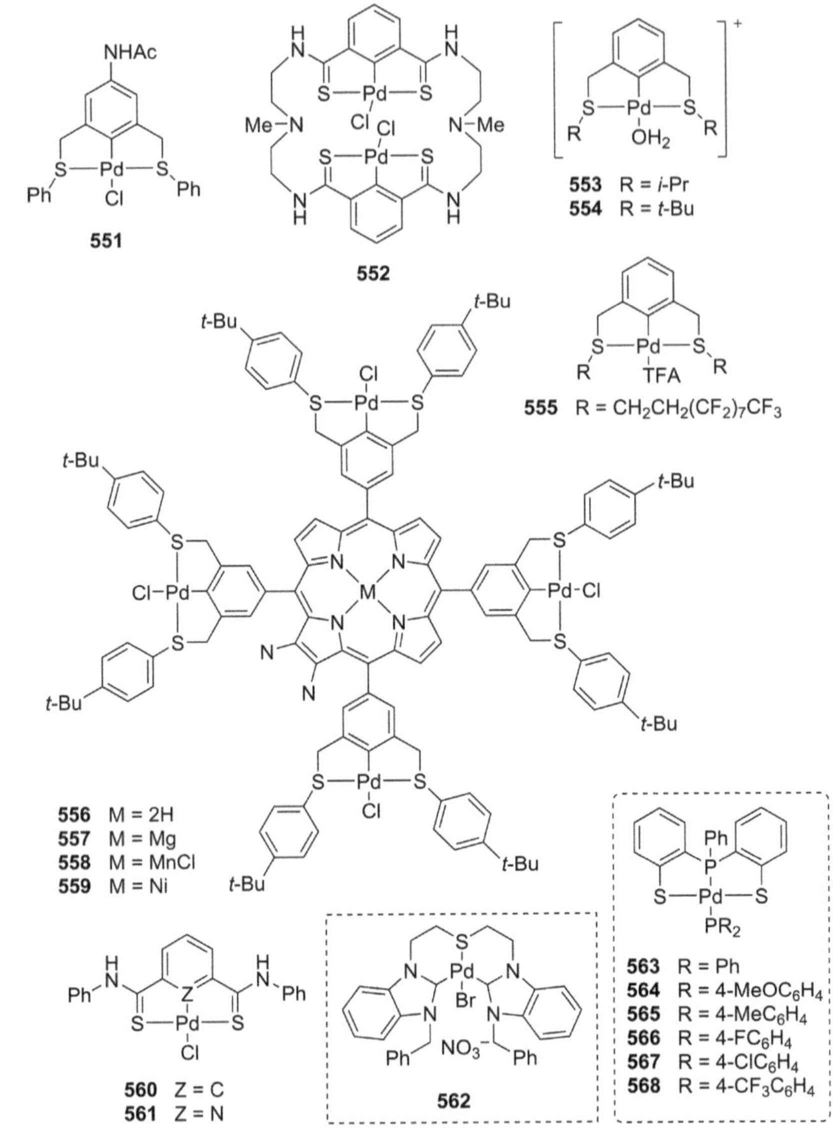

Fig. 2.66 SCS- and SNS- and other S-containing pincer palladacycle.

palladacycles are easier to prepare. However, only a few examples of these selenium-containing pincer palladacycles were employed in Heck coupling (Fig. 2.68; Table 2.41).

In order to prove that pincer-chelation is highly related to the catalytic activity, Beletskaya synthesized **578** and **584** for the comparison *(311)*. They are

Table 2.40 SCS- and SNS-pincer palladacycles in Heck reaction

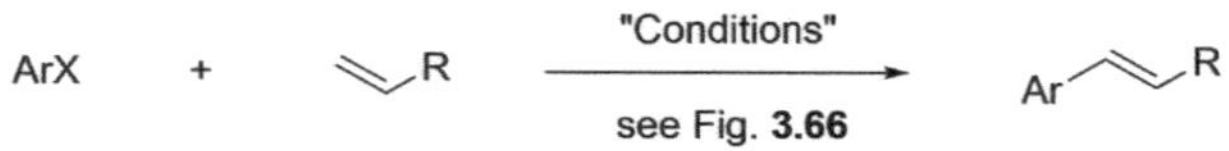

Entry	Palladacycle (mol% Pd)	ArX	R	Solvent	Base	Time (h)	Temp. (°C)	Yield (%)	Ref.
1	**551** (0.1)	C_6H_5I	Ph	DMF	Et_3N	6.5	115	92	304
2[a]	**552** (6.0×10^{-2})	*p*-MeC_6H_4I	Ph	DMA	NaOAc	2	165	82	305
3	**560** (0.2)	*p*-MeC_6H_4I	Ph	DMA	NaOAc	0.25	165	89	306
4	**561** (0.2)	*p*-MeC_6H_4I	Ph	DMA	NaOAc	16	165	53	306
5	**555** (0.23)	C_6H_5I	CO_2Me	DMF	*i*-Pr_2NEt	4	125	97	307
6[a]	**562** (1.0)	*p*-MeC_6H_4Br	CO_2*t*-Bu	DMF	NaOAc	24	140	47	245
7	**563** (0.17)	C_6H_5Br	Ph	DMF	Cs_2CO_3	4	160	65	308
8	**564** (0.17)	C_6H_5Br	Ph	DMF	Cs_2CO_3	4	160	64	308
9	**566** (0.17)	C_6H_5Br	Ph	DMF	Cs_2CO_3	4	160	57	308
10	**567** (0.17)	C_6H_5Br	Ph	DMF	Cs_2CO_3	4	160	61	308

[a]TBAB was added as an additive.

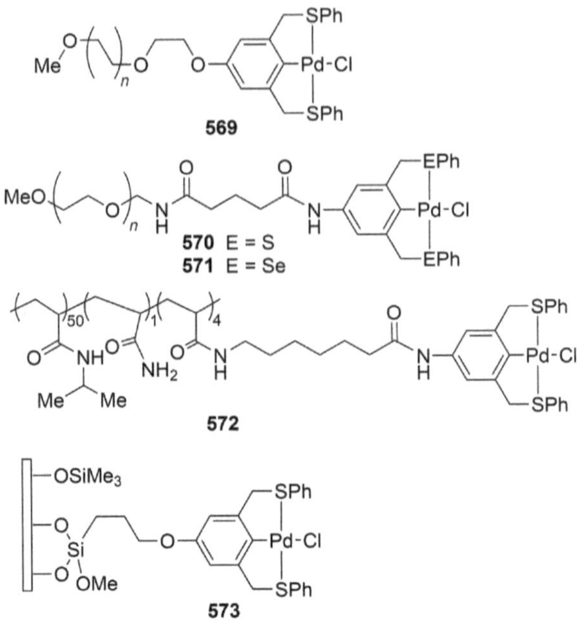

Fig. 2.67 Examples of polymer-/silica-supported SCS- and SeCSe-pincer palladacycles.

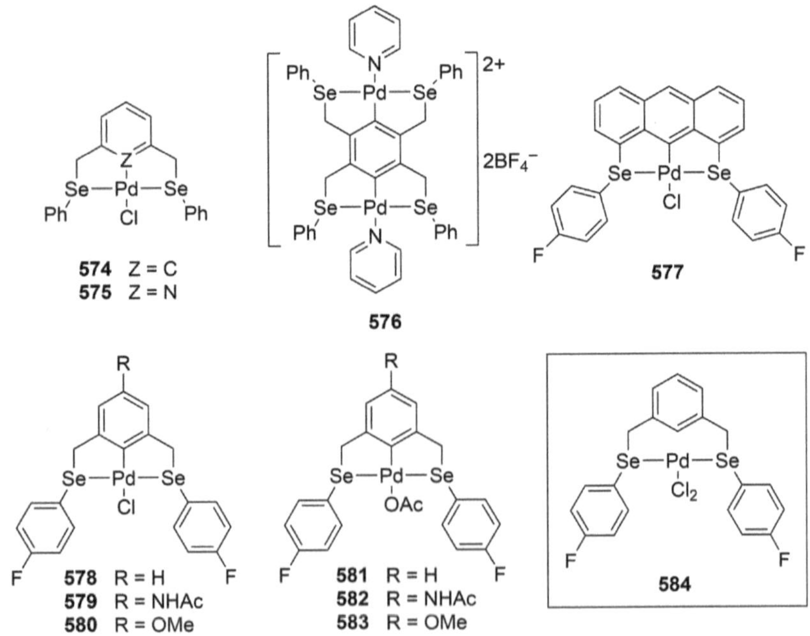

Fig. 2.68 Examples of SeNSe-, SeCSe-pincer palladacycle.

Table 2.41 SeCSe-/SeNSe-pincer palladacycles in Heck reaction

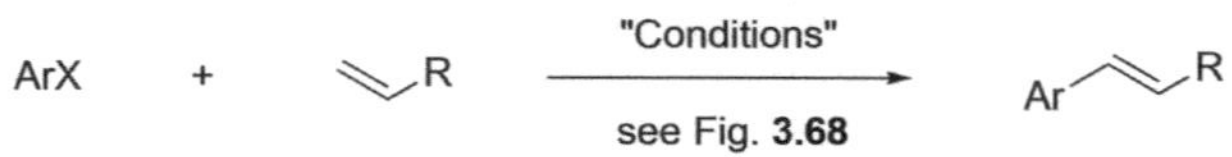

Entry	Palladacycle (mol% Pd)	ArX	R	Solvent	Base	Time (h)	Temp. (°C)	Yield (%)	Ref.
1	**574** (3.5×10^{-4})	C_6H_5Br	CO_2*n*-Bu	DMA	Na_2CO_3	48	140	61	312
2[a]	**574** (2.5×10^{-4})	*o*-$MeOC_6H_4Br$	Ph	water	K_2CO_3	6	100	51	138
3	**575** (1.0×10^{-3})	*p*-$MeOC_6H_4Br$	CO_2*n*-Bu	*p*-xylene	*n*-$BuNH_2$	24	100	43	313
4	**576** (1.0×10^{-3})	*p*-$MeOC_6H_4Br$	Ph	DMA	*n*-$BuNH_2$	18	100	92	314
5	**577** (1.0×10^{-3})	*p*-FC_6H_4I	CO_2Me	DMA	Et_3N	10	120	100	311
6	**578** (1.0×10^{-2})	*p*-FC_6H_4I	CO_2Me	DMA	Et_3N	12	100	100	311
7	**581** (1.0×10^{-3})	*p*-FC_6H_4I	CO_2Me	DMA	Et_3N	20	100	98	311
8	**584** (1.0×10^{-2})	*p*-FC_6H_4I	CO_2Me	DMA	Et_3N	3	100	100	311

[a]TBAB was added as an additive.

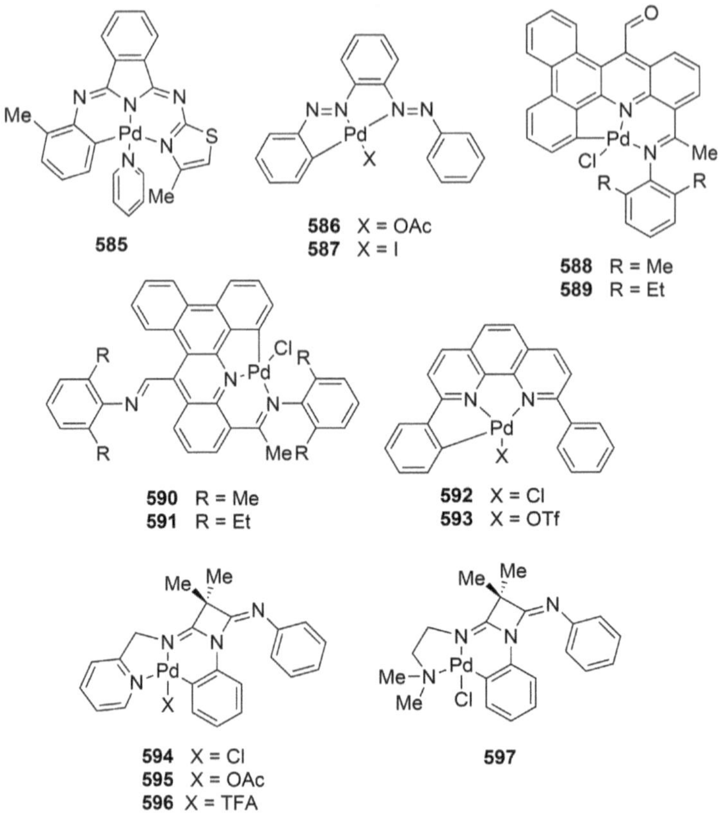

Fig. 2.69 Examples of CNN-pincer palladacycle.

structurally the same ligands but a different mode of chelation and thus the ability to release Pd(0) species, the former being the SeCSe-pincer palladacycles while the latter being simple SeSe chelates palladacycles. From the Heck coupling reaction, the pincer chelated palladacycle **578** took longer reaction to give the desired products. More robust pincer palladacycles supports a much slower but quantitative reaction.

6.1.4 CNN-Pincer Palladacycles

Unsymmetrical CNN-pincer palladacycle gives rise to the *trans* influence of the carbon donor of the complex. It may gain additional support by a well-positioned steric constraint and give the hemilabile feature to the palladacycle. The hemilabile features give two reactive sites at the palladium ion in the *cis* position and thus affect the catalytic activity of the palladacycle.

Table 2.42 CNN-pincer palladacycles in Heck reaction

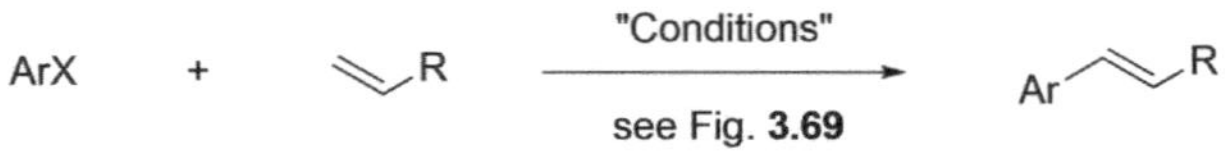

Entry	Palladacycle (mol% Pd)	ArX	R	Solvent	Base	Time (h)	Temp. (°C)	Yield (%)	Ref.
1	**585** (0.25)	2,6-$Me_2C_6H_4I$	CO_2Me	DMF	Na_2CO_3	18	140	78	315
2	**585** (0.25)	C_6H_5Br	CO_2Me	DMF	Na_2CO_3	18	140	33	315
3	**586** (1.0×10^{-2})	*p*-$CHOC_6H_4Br$	Ph	DMF	K_2CO_3	18	135	84	316
4	**587** (1.0×10^{-3})	*p*-$CHOC_6H_4Br$	Ph	DMF	K_2CO_3	18	135	83	317
5	**588** (2.0×10^{-3})	C_6H_5Br	Ph	DMF	Na_2CO_3	8	140	70	317
6	**589** (2.0×10^{-3})	C_6H_5Br	Ph	DMF	Na_2CO_3	8	140	66	317
7	**590** (2.0×10^{-3})	C_6H_5Br	Ph	DMF	Na_2CO_3	8	140	67	317
8	**591** (2.0×10^{-3})	C_6H_5Br	Ph	DMF	Na_2CO_3	8	140	62	317
9	**592** (1.0×10^{-4})	C_6H_5I	CO_2*n*-Bu	NMP	*n*-Bu_3N	15	140	99	318
10	**593** (1.0×10^{-4})	C_6H_5I	CO_2*n*-Bu	NMP	*n*-Bu_3N	15	140	33	318
11	**592** (1.0×10^{-2})	*p*-MeC(O) C_6H_4Cl	Ph	NMP	NaOAc	48	160	25	318
12	**594** (1.0)	*p*-MeC(O) C_6H_4Br	Ph	DMA	KF	3	135	75	319
13	**595** (1.0)	*p*-MeC(O) C_6H_4Br	Ph	DMA	KF	3	135	99	319
14	**596** (1.0)	*p*-MeC(O) C_6H_4Br	Ph	DMA	KF	3	135	95	319
15	**597** (1.0×10^{-2})	*p*-MeC(O) C_6H_4Br	Ph	DMA	KF	24	135	98	319

A few CNN-pincer palladacycles have recently been synthesized and employed in the Heck coupling reaction (Fig. 2.69; Table 2.42).

The CNN-pincer palladacycle **592** was applied in a ten-gram-scale synthesis of the UV-B sunscreen agent octinoxate (2-ethylhyexyl 4-methoxycinnamate) for which only 0.0001 mol% of Pd loading was required. This provided an attractive reaction protocol in the industrial synthesis (Scheme 2.45).

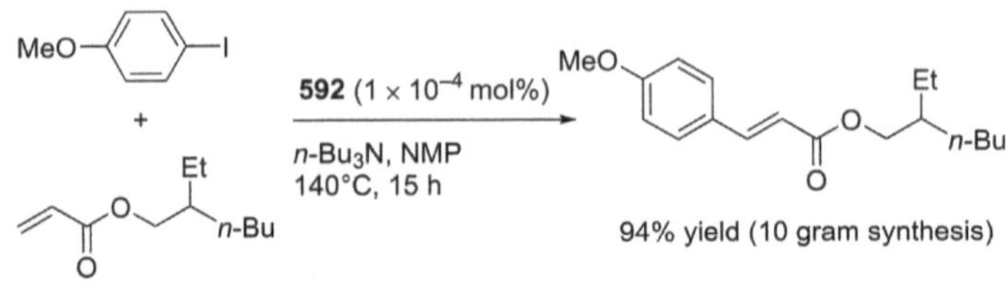

Scheme 2.45 Ten-gram-scale synthesis of the UV-B sunscreen agent octinoxate with palladacycle **592**.

6.1.5 LCL′- and LNL′-Pincer Palladacycles (L = C, N, or O; L′ = C, N, O, S, Se, P)

There are a number of other pincer palladacycles bearing noncontiguous donor atoms that bind to a metal center. These donor atoms can be C, N, O, S, Se, or P as discussed before. However, most of them are in a symmetric manner and the unsymmetric pincer palladacycle and unusual and rare pincer palladacycle will be discussed *(320)*. The steric and electronic properties at each coordinating site are different and the synergistic or antagonistic effect can be studied through binding different ligands to the metal center (Figs. 2.70–2.72; Tables 2.43–2.45).

6.2 Sonogashira Cross-Coupling Reaction

There are examples of the use of bis-chelated palladacycles, usually referred to as "pincer" complexes, in Sonogashira cross-coupling reactions. The Eberhard group reported a PCP-pincer complex **463** (Fig. 2.73), which is reactive enough to couple a wide range of activated and nonactivated aryl chlorides with phenylacetylene using Cs_2CO_3 as the base in the presence $ZnCl_2$ as co-catalyst. However, the reaction was performed under harsh conditions (DMSO, 160°C) (Table 2.46) *(332)*. Another PCP-pincer palladacycle **628** (Fig. 2.73) was used in the coupling of aryl bromides and phenylacetylene *(333)*. A higher catalyst efficiency in Sonogashira coupling was shown for PCP-pincer complex **629** (Fig. 2.73). 0.005 mol% of

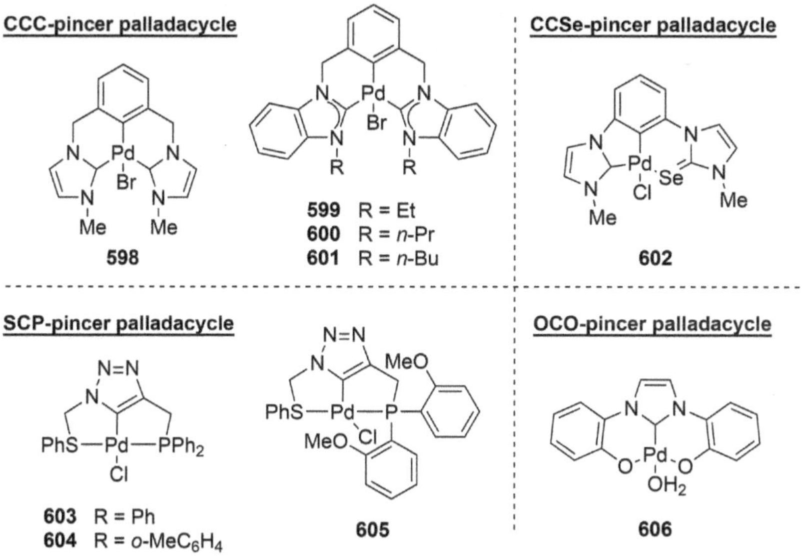

Fig. 2.70 Examples of other LCL′-pincer palladacycles.

palladacycles was sufficient for the coupling of aryl or heteroaryl iodides or bromides with aryl or heteroaryl alkynes (TON up to 20,000) at 140°C *(334)*.

The *N*-heterocyclic NCN-pincer palladium complexes **630–631** (Fig. 2.74) were employed recently in the coupling between aryl iodides and terminal alkynes (0.1 mol% catalyst loading) in pyrrolidine at 100°C *(335)*. The PCN-pincer palladacycle **635** was employed as catalyst (2.0 mol %) in the copper-free coupling reaction of aryl iodides and bromides with phenylacetylene or 1-hexyne at room temperature or 70°C *(336)*.

Very recently, the Singh group *(337)* and the Movassagh group *(338)* used the ONS/Se- and SeCSe-pincer palladacycles **632–634** (Fig. 2.74) as efficient and robust catalysts in the Sonogashira reaction (Table 2.46). *β*-Bromostyrene was coupled successfully with aromatic and aliphatic terminal alkynes in the presence of **634** to afford substituted 1,3-enynes with high *E*:*Z* ratios. Especially, the catalyst **634** maintained its catalytic activity by simple filtration and reused up to six consecutive runs for this copper- and amine-free Sonogashira reaction.

Considering water or aqueous solution is a nonflammable, safe, and cheap solvent. Domínguezhas reported a copper-free Sonogashira alkynylation of aryl halides with PCN-pincer palladacycle **636** in neat water (Fig. 2.75)

Table 2.43 LCL′-pincer palladacycles in Heck reaction

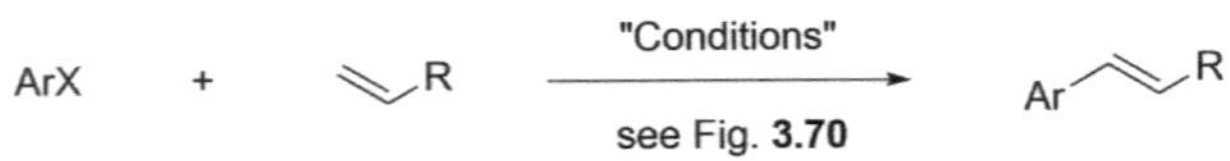

Entry	Palladacycle (mol% Pd)	ArX	R	Solvent	Base	Time (h)	Temp. (°C)	Yield (%)	Ref.
1[a]	**598** (0.2)	*p*-$CHOC_6H_4Cl$	Ph	DMA	NaOAc	0.25	165	91	299
2	**599** (1.0)	C_6H_5Br	Ph	DMA	NaOAc	24	110	31	321
3	**600** (1.0)	C_6H_5Br	Ph	DMA	NaOAc	24	110	40	321
4	**601** (1.0)	C_6H_5Br	Ph	DMA	NaOAc	24	110	38	321
5	**600** (1.0)	*p*-$MeOC_6H_4Br$	Ph	DMA	NaOAc	24	110	46	321
6	**601** (1.0)	*p*-$MeOC_6H_4Br$	Ph	DMA	NaOAc	24	110	50	321
7	**603** (7.0×10^{-4})	C_6H_5Br	CO_2Me	DMF	Na_2CO_3	48	140	29	276
8	**605** (7.0×10^{-4})	C_6H_5Br	CO_2Me	DMF	Na_2CO_3	48	140	5	276
9	**604** (1.0)	C_6H_5I	Ph	DMA	K_2CO_3	20	140	94	322
10	**606** (0.2)	C_6H_5I	Ph	DMA	Cs_2CO_3	6	150	73	323

[a]TBAB was added as an additive.

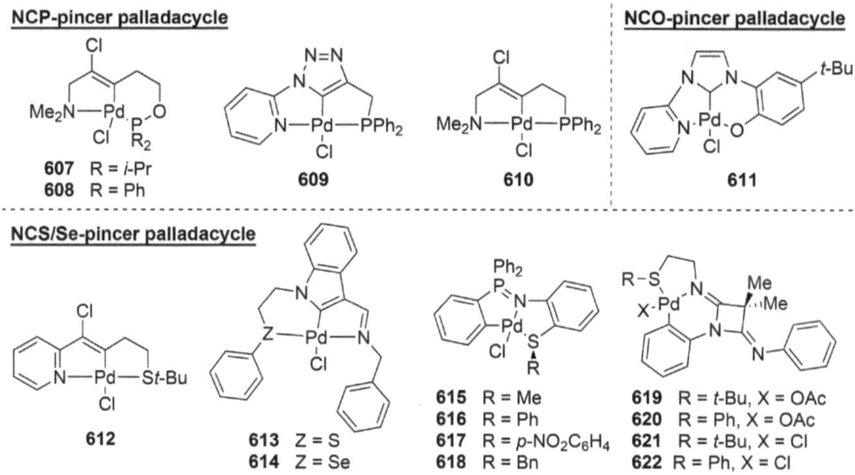

Fig. 2.71 Examples of NCL-pincer palladacycles.

(339). In 2015, Rosa and co-workers further employed palladacycle in various cross-coupling reactions including Heck and Sonogashira coupling. The effects of base, solvent, and temperature were studied and it showed that temperature was statistically significant to the coupling yield *(340)*.

The Shang group reported a bis(ferrocene-isoxazole/isoxazoline) ligand (NNN- and NCN-pincer ligands) for the preparation of pincer palladacycles **637** and **518** (Fig. 2.75) *(341)*. An iron NNC-pincer palladacycle **638** bearing a 3-butyl-1-(1,10-phenanthrolin-2-yl)imidazolydinene ligand was reported by Chen and Gu (Fig. 2.75) *(342)*. This pincer palladacycle (1.0 mol%) was evaluated in the copper-free Sonogashira reaction of aryl iodides and bromides with phenylacetylene or 1-hexeyne in the presence of triphenylphosphine (1.0 mol%). Recently, Wang reported three classes of NCN-diimine palladacycles **639–643** with different substituents at the center phenyl ring. The palladacycle with the nitro group (**643**) showed the best catalytic efficacy in which the electron-withdrawing properties would enhance the activity in Sonogashira coupling *(343)*.

A tridentate CNC-dicarbene pincer palladacycle **538** was evaluated in the reaction of iodobenzene and 4-bromoacetophenone with phenylacetylene by Crabtree and co-workers (Fig. 2.76). The coupled product from iodobenzene could be obtained in 92% yield, whereas, only 10% yield was obtained after 19 h from 4-bromoacetophenone *(296)*. Subsequently,

Table 2.44 NCL-pincer palladacycle in Heck cross-coupling reaction

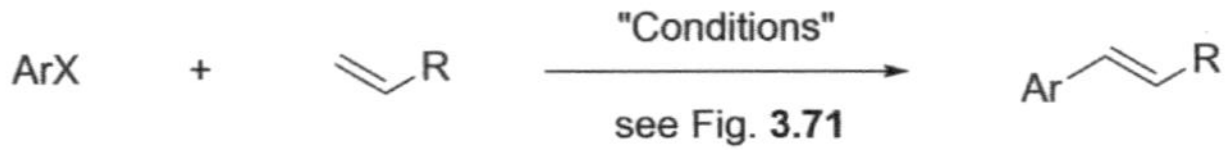

Entry	Palladacycle (mol % Pd)	ArX	R	Solvent	Base	Time (h)	Temp (°C)	Yield (%)	Ref.
1[a]	**607** (0.5)	C_6H_5Br	CO_2*n*-Bu	DMF	Na_2CO_3	17	130	99	324
2[a]	**607** (1.0)	*p*-CNC_6H_4Cl	CO_2*n*-Bu	DMF	Na_2CO_3	23	130	84	324
3	**609** (7.0×10^{-4})	C_6H_5Br	CO_2Me	DMF	Na_2CO_3	48	140	88	276
4[a]	**610** (1.0×10^{-2})	*p*-MeC_6H_4Br	CO_2*n*-Bu	DMA	NaOAc	24	150	85	325
5[a]	**608** (1.0×10^{-2})	*p*-MeC_6H_4Br	CO_2*n*-Bu	DMA	NaOAc	24	150	50	325
6[a]	**612** (1.0×10^{-2})	*p*-MeC_6H_4Br	CO_2*n*-Bu	DMA	NaOAc	24	150	20	325
7	**613** (0.1)	*p*-MeC_6H_4Br	CO_2*n*-Bu	DMA	K_2CO_3	24	140	33	326
8	**614** (0.1)	*p*-MeC_6H_4Br	CO_2*n*-Bu	DMA	K_2CO_3	24	140	60	326
9	**615** (1.0)	C_6H_5Br	Ph	DMF	Na_3PO_4	0.25	110[b]	0	327
10	**616** (1.0)	C_6H_5Br	Ph	DMF	Na_3PO_4	0.25	110[b]	46	327
11	**617** (1.0)	C_6H_5Br	Ph	DMF	Na_3PO_4	0.25	110[b]	45	327
12	**618** (1.0)	C_6H_5Br	Ph	DMF	Na_3PO_4	0.25	110[b]	38	327
13	**619** (0.1)	*p*-Me(CO) C_6H_4Br	Ph	DMF	Cs_2CO_3	6	135	75	328
14	**611** (0.2)	C_6H_5I	Ph	DMA	Cs_2CO_3	6	150	99	323

[a]TBAB was added as an additive.
[b]Microwave irradiation (100 W) was used.

Table 2.45 LNL′-pincer palladacycles in Heck reaction

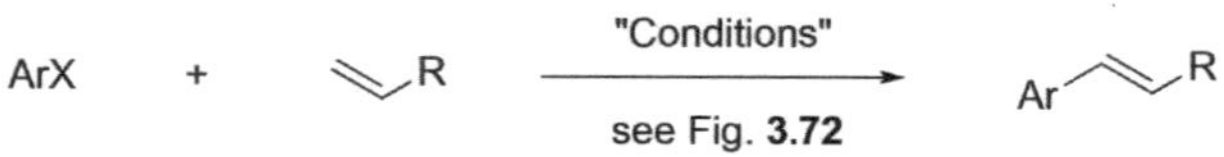

Entry	Palladacycle (mol% Pd)	ArX	R	Solvent	Base	Time (h)	Temp (°C)	Yield (%)	Ref.
1	**623** (3.0)	C_6H_5I	CO_2*n*-Bu	Toluene	KOAc	0.25	reflux	100	329
2[a]	**623** (10.0)	*o*-$CO_2HC_6H_4I$	CO_2Me	D_6-acetone	—	6	RT	83	330
3	**623** (10.0)	*o*-$CO_2HC_6H_4I$	CO_2Me	D_6-DMSO	—	1	140	25	330
4	**627** (0.1)	*o*-$MeOC_6H_4Br$	Ph	DMF	NaOAc	24	150	46	331

[a] $AgClO_4$ was added as an additive.

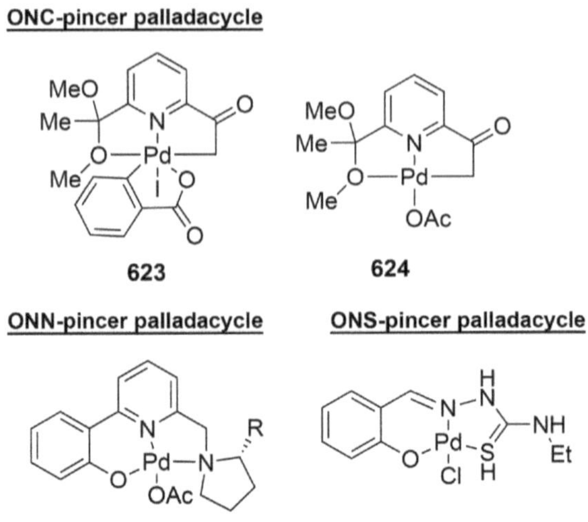

625 R = H
626 R = CONH*t*-Bu

Fig. 2.72 Examples of LNL′-pincer palladacycles.

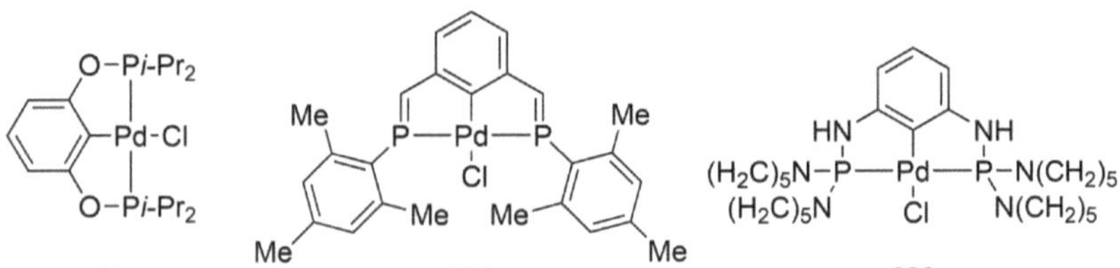

463 **628** **629**

Fig. 2.73 Examples of PCP-pincer palladacycles.

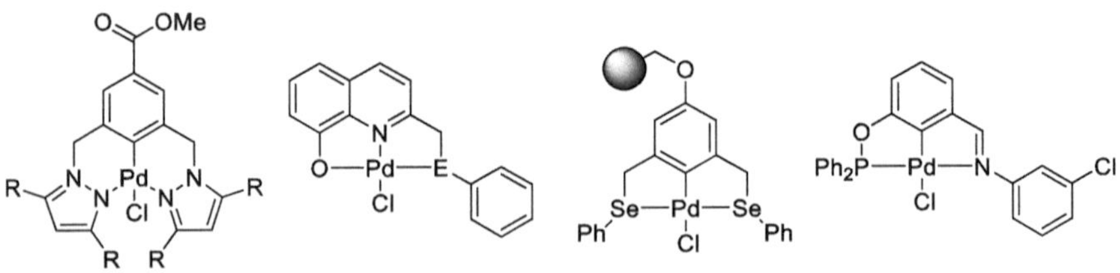

630 R = H
631 R = Me
632 E = S
633 E = Se
634 (PS-[PdCl-(SeCSe)])
635

Fig. 2.74 Examples of other pincer palladacycles.

this palladacycle was immobilized onto clays, and successfully applied on the coupling of phenylacetylene with aryl iodides and three activated bromides in the presence of 5.0 mol% CuI as cocatalyst *(344)*. The CNC-dicarbene pincer palladacycle **644** displayed a high activity in a copper- and amine-free

Table 2.46 Examples of copper and amine-free Sonogashira reaction using pincer palladacycles

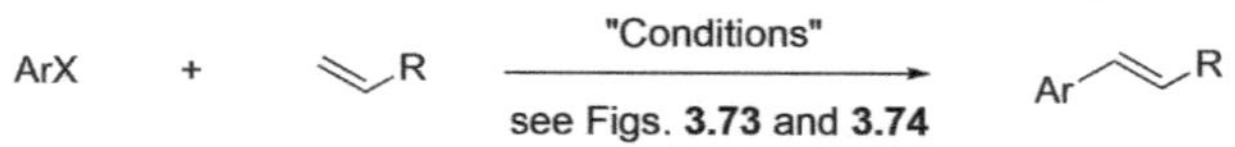

Entry	Palladacycle (mol% Pd)	ArX	R	Solvent	Base	Time (h)	Temp. (°C)	Yield (%)	Ref.
1[a]	**463** (5.0)	C_6H_5Cl	Ph	Dioxane	Cs_2CO_3	24	160	61	332
2[a]	**463** (5.0)	*p*-MeC(O)C_6H_4Cl	Ph	Dioxane	Cs_2CO_3	24	160	91	332
3	**630** (0.1)	C_6H_5I	Ph	Pyrrolidine	–	6	100	92	335
4	**631** (0.1)	C_6H_5I	Ph	Pyrrolidine	–	6	100	96	335
5	**632** (0.5)	*p*-CHOC_6H_4Br	Ph	DMF	K_2CO_3	8	90	84	337
6	**633** (0.5)	*p*-CHOC_6H_4Br	Ph	DMF	K_2CO_3	10	90	89	337
7	**634** (0.3)	C_6H_5I	Ph	NMP	TBAF	0.5	70	91	338
8	**634** (0.3)	C_6H_5Br	Ph	NMP	TBAF	4	70	88	338
9	**634** (0.3)	C_6H_5Cl	Ph	NMP	TBAF	24	70	41	338
10	**634** (0.3)	C_6H_5I	*n*-Bu	NMP	TBAF	3	70	66	338
11	**634** (0.3)	β-StyrylBr	*n*-Bu	NMP	TBAF	2	70	86	338

[a] $ZnCl_2$ was added.

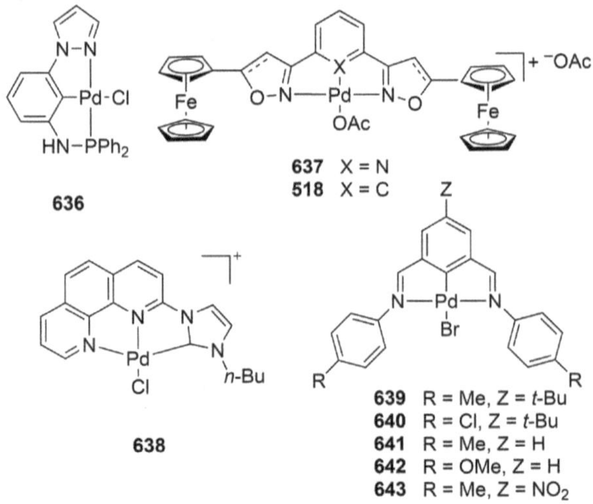

Fig. 2.75 Examples of NCN-pincer palladacycles.

Sonogashira reaction of 4-bromobenzaldehyde and phenylacetylene *(345)*. The experiment result showed that this catalyst also exhibited good activity for the coupling of activated, deactivated, and steric bulky aryl bromides, as well as heteroaryl bromides. In addition, the catalyst was reused for up to six runs without significant loss of its activity and an overall TON of 56000 was reached.

Meldal reported a dipeptide-containing supported monocarbene palladacycle **645**, which served as the cross-coupling NHC-palladated catalyst of activated aryl iodides, such as 1-iodo-2-nitrobenzene, with terminal alkynes, such as TMSA (Fig. 2.76) *(346)*. The reaction was performed in the presence of CuI as co-catalyst at 50°C. It showed no loss of activity after eight runs.

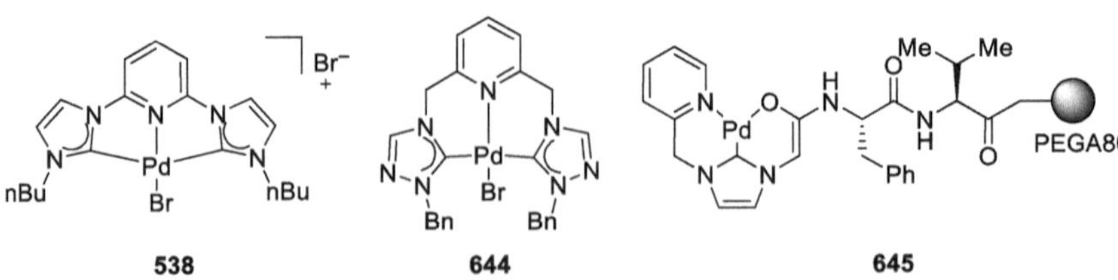

Fig. 2.76 The CNC-pincer and supported pincer palladacycles.

7. Other Palladacycles

7.1 Heck Cross-Coupling Reaction

In 2002, Gladysz reported a thermomorphic fluorous imine **168** for the Heck coupling *(120)*. After this initial publication, another generation of the catalyst was developed in which the nitrogen-atom was replaced by sulfur-atom to give a new thioether palladacycle **646** (Fig. 2.77; Table 2.47) *(347)*. The catalytic activities of these two palladacycles were compared and the imine-palladacycle was found to be more active than the corresponding thioether-palladacycle. Later, Dupont reported other sulfur-containing palladacycles **647–649**, which mainly derived from the *ortho*-palladation of benzylic thioethers (Fig. 2.77). They were employed in the Heck reaction of aryl halides and styrene, acrylic ester. Unfortunately, only highly activated *p*-nitrochlorobenzene was able to react with low product yields *(348)*.

Thiourea is very useful in palladium-catalysis. Apart from the thiourea-based palladacycle, a series of arylcarbothioamide-derived palladacycle was synthesized (**597–604**) (Figs. 2.69 and 2.70) *(349)*. It takes advantage of easy tunability of the electronic and steric effect of the palladacycle by altering the substituents on the aromatic ring and the *N*-position. Interestingly, the eight furancarbothioamide-derived palladacycles gave similar reactivity in the Heck reaction of iodobenzene and methyl acrylate.

Sulfilimine and sulfoxide-based palladacycles are other attractive catalysts for Heck coupling. They are superior to phosphine palladacycles due to their high stability toward moisture, air, and high reaction temperature. Sudalai reported the use of these two sulfur-containing palladacycles in the Suzuki coupling initially and revealed that **610** gave the best catalytic performance *(170)*. This **610** was later applied in both Heck coupling and Sonogashira coupling.

The supramolecular dithiolate palladacycles were developed for Heck coupling (Fig. 2.78). They are highly stable and robust. The tetranuclear palladacycle showed a slightly higher catalytic activity than the octanuclear analog in the coupling of aryl bromides with styrene *(350)*.

A series of organoselenium palladacycles was synthesized with different bridging units (Fig. 2.79) *(311)*. The alkyl-bridged, alkene-bridged, pyridyl-bridged, phenyl-bridged, naphthyl-bridged, and anthracene-bridged diselenium palladacycles were tested in the benchmarked coupling substrates, i.e., 4-fluoroiodobenzene and methyl acrylate. The catalytic activity decreased when the number of carbon in the alkyl bridge was increased

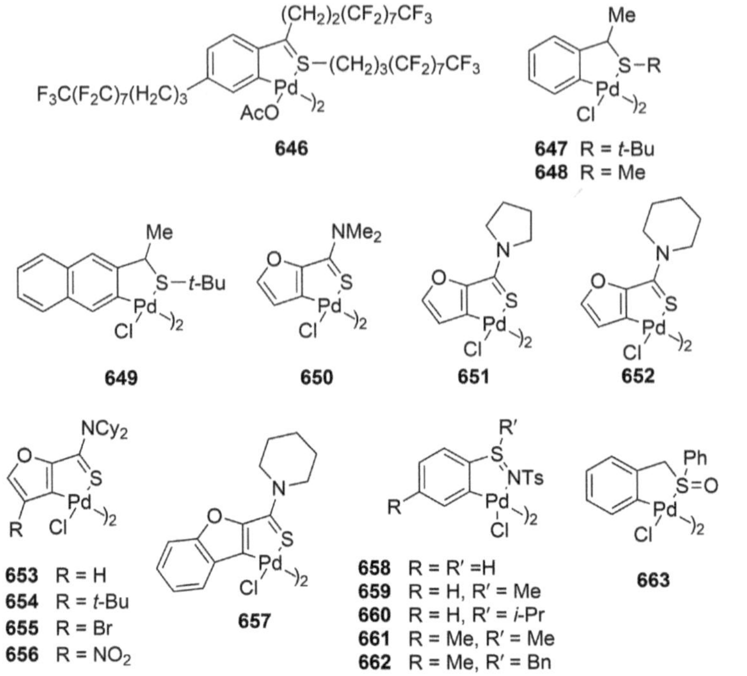

Fig. 2.77 Examples of sulfur-containing palladacycles.

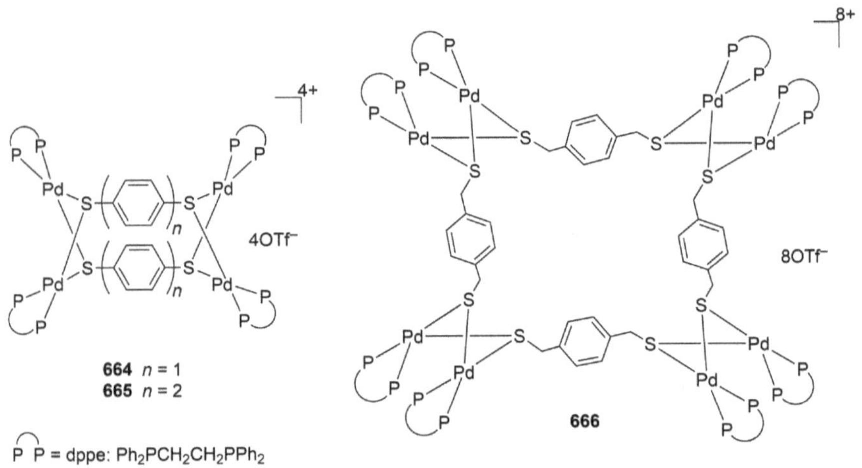

Fig. 2.78 Examples of supramolecular dithiolate palladacycles.

(Palladacycles **669–673**). They showed comparable results for the other bridged-organoselenium palladacycles at the catalyst loading of 0.01 mol% (Table 2.47). A series of NO-palladacycles with a Se-Ph tail was also prepared where the Se-Ph tail connects with the main skeleton with a tunable carbon chain length *(351)*.

Table 2.47 Sulfur- or selenium-containing palladacycles in Heck reaction

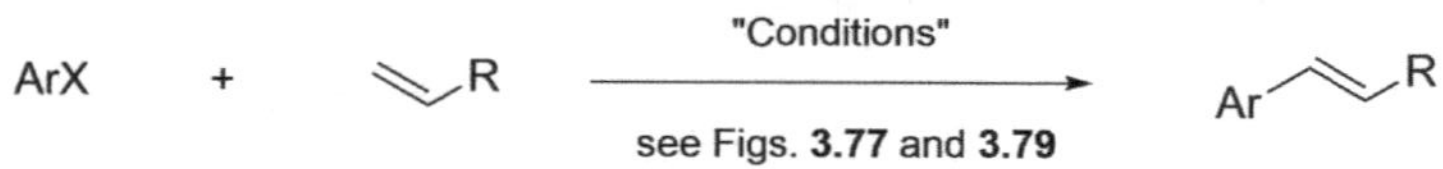

Entry	Palladacycle (mol% Pd)	ArX	R	Solvent	Base	Time (h)	Temp, °C	Yield, %	Ref.
1	**646** (1.6×10^{-3})	*p*-MeC(O)C_6H_4Br	CO_2Me	DMF	Et_3N	48	140	50	347
2	**653** (1.0×10^{-3})	C_6H_5I	CO_2Me	DMA	Et_3N	6	130	97	349
3	**663** (2.5×10^{-4})	*p*-OHC_6H_4I	CO_2*n*-Bu	NMP	Et_3N	12	140	65	170
4	**670** (1.0×10^{-3})	*p*-$MeOC_6H_4Br$	Ph	DMA	*n*-$BuNH_2$	18	100	26	314
5	**671** (9.0×10^{-4})	*p*-CNC_6H_4Br	CO_2*n*-Bu	DMA	Na_2CO_3	24	140	88	312
6	**671** (9.0×10^{-4})	*m*-PyBr	Ph	DMA	NaOAc	43	140	40	312
7	**672** (1.0×10^{-2})	*p*-FC_6H_4I	CO_2Me	DMF	Et_3N	2	100	98	311
8	**675** (0.1)	*p*-FC_6H_4I	CO_2Me	DMF	Et_3N	3	100	78	311
9	**676** (0.1)	*p*-FC_6H_4I	CO_2Me	DMF	Et_3N	7	100	32	311
10	**684** (1.0×10^{-2})	*p*-FC_6H_4I	CO_2Me	DMF	Et_3N	4	100	100	311

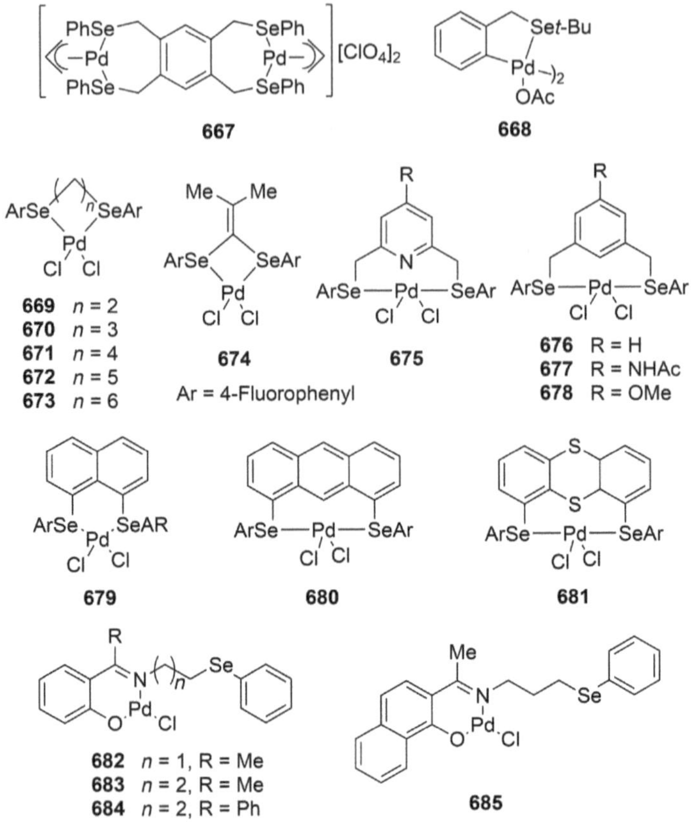

Fig. 2.79 Examples of organoselenium palladacycles.

Carbocyclic carbenes derived from aromatic nitrones act as ancillary ligands coordinated with palladium to form an unusual palladacycle for Heck coupling reaction *(352)*. This skeleton contains a bidentate ligand, featuring an anionic carbene attached to a nitrone group. The iminium nitrogen potentially stabilizes the charge on the phenyl ring and the anoionic oxygen serves as a directing group. It was found to be highly active to promote Heck coupling of aryl bromides and styrene (Scheme 2.46).

A 16-electron borapalladacycle **688** formed by coordinating 8-quinolyldimesitylborane with bis(benzonitrile))dichloropalladium(II) was employed in catalyzing the Heck reactions of aryl halides and styrene/ *n*-butyl acrylate (Scheme 2.46) *(353)*.

In 2005, Gladysz developed a new palladacycle that contains half-sandwich metal cyclopentadienyl moieties in the backbone *(354)*. The palladacycle contains a chiral Rhenium fragment and the enantiomeric palladacycle (*S,S*)-**689** and (*R*)-**690** was evaluated as the catalyst for Heck coupling of methyl acrylate and aryl bromide and iodides (Scheme 2.46).

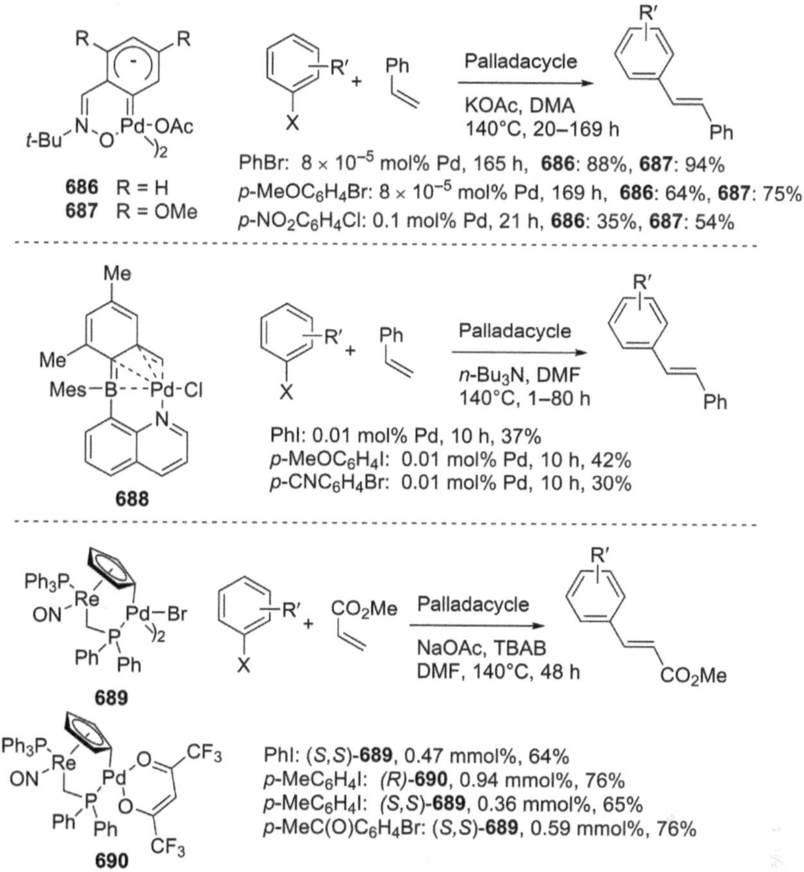

Scheme 2.46 Other palladacycle in Heck coupling.

7.2 Sonogashira Cross-Coupling Reaction

Jin and co-workers reported the palladium(II) β-oxoiminatophosphane complexes (Fig. 2.80), which have high catalytic performance in the copper-free Sonogashira coupling reaction. The palladacycle **691** was examined in the coupling of aryl iodides, bromides, and even chlorides with terminal alkynes *(355)*. This system gave high product yields even under low catalyst loadings (0.01–0.1 mol%) at 40–50°C. Remarkably, the amount of catalyst was permitted to be reduced to 0.0002 mol% in the coupling between iodobenzene and phenylacetylene, affording 91% yield of the desired diphenylacetylene. In addition to the homogeneous system, an anchored palladacycle **692** was prepared via the immobilization of the palladium(II) *β*-oxoiminatophosphane complex to a magnetic nanoparticle formed by coating of commercial iron(III) oxide with silica (Fig. 2.80). The catalyst **692** was then tested in different cross-coupling reactions. Chloroarenes were able to cross-couple with various terminal alkynes in high yields (74%–96% yield) without copper-cocatalyst in the presence of piperidine and TBAB in water at 60°C *(356)*. A dinuclear tetraphenyl oxalic amidinate-derived palladacycle **693** (Fig. 2.80) was examined as a precatalyst in the copper-free Sonogashira reaction between 4-bromoacetophenone and phenylacetylene *(357)*. A moderate TON (up to 490) was observed. Nandurkar and Bhanage employed an inexpensive, air-stable palladacycle **694** bearing bis(2,2,6,6-tetramethyl-3,5-heptanedionate) group for the copper-free Sonogashira coupling reactions of aryl iodides with aliphatic and aromatic alkynes in water at room temperature *(358)*. Nevertheless, this catalyst system showed no activity toward bromobenzene.

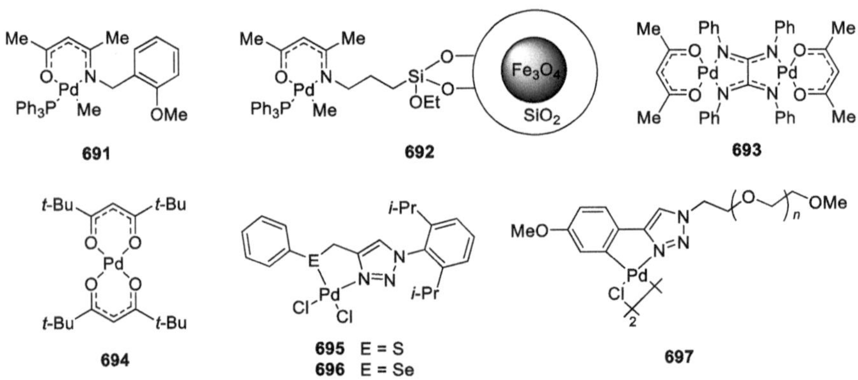

Fig. 2.80 Examples of other nitrogen-containing palladacycles.

Triazoles can coordinate through the donor atoms in conjunction with two nitrogen atoms and one carbon atom of the heterocyclic ring. Several triazole-based palladacycles were reported recently and used as a precatalyst in the Sonogashira reaction. Triazole-based organosulfur/selenium palladacycles **695** and **696** were synthesized from the complexation of 1-(2,6-diisopropylphenyl)-4-(phenylthio/selenomethyl)-1*H*-1,2,3-triazole ligand with [Pd(CH_3CN)$_2$$Cl_2$]. These catalysts were explored for the coupling of various aryl bromides with phenylacetylene in the presence of CuI as cocatalyst using potassium carbonate as the base in DMF (Table 2.48)

Table 2.48 Triazole-based palladacycles in Sonogashira reaction of aryl halides with phenylacetylene

ArX + Ph—≡ —"Conditions" / see Fig. **3.80**→ Ar—≡—Ph

Entry	Palladacycle (mol% Pd)	ArX	Additives	Solvent	Base	Time (h)	Temp, °C	Yield, %	Ref.
1	**695** (1.0)	*p*-$CHOC_6H_4Br$	CuI	DMF	K_2CO_3	8	120	87	359
2	**695** (2.0)	*p*-CNC_6H_4Br	CuI	DMF	K_2CO_3	12	120	91	359
3	**695** (0.1)	*p*-$MeOC_6H_4Br$	CuI	DMF	K_2CO_3	12	120	43	359
4	**696** (1.0)	*p*-$CHOC_6H_4Br$	CuI	DMF	K_2CO_3	12	120	52	359
5	**696** (2.0)	*p*-CNC_6H_4Br	CuI	DMF	K_2CO_3	12	120	88	359
6	**697** (5.0×10^{-2})	C_6H_5I	—	water	Cs_2CO_3	12	reflux	90	360
7	**697** (5.0×10^{-2})	C_6H_5Br	—	water	Cs_2CO_3	12	reflux	91	360
8	**697** (0.1)	*p*-MeC_6H_4Br	—	water	Cs_2CO_3	12	reflux	87	360
9	**697** (0.1)	*p*-PhC_6H_4Br	—	water	Cs_2CO_3	12	reflux	90	360
10	**697** (0.1)	*p*-ClC_6H_4Br	—	water	Cs_2CO_3	12	reflux	83	360

(359). The corresponding PEG-containing triazole-palladacycle **697** was also probed in the copper-free aqueous Sonogashira cross-coupling of aryl iodides and aryl bromides with phenylacetylene (Table 2.48) *(360)*. Catalyst loading ranging from 0.05 to 0.1 mol% was found to be enough to promote this reaction and fortunately no homocoupling by-product was detected. After removing the products by simple extraction, the catalyst could be reused for three times without significant loss of activity.

A new family of sulfilimine-base palladacycle **659** was synthesized and assayed in a copper-free Sonogashira coupling of aryl halides with phenylacetylene using triethylamine as both solvent and base at 80°C (Scheme 2.47). Good yield was obtained in the arylation of phenylacetylene with iodobenzene, whereas low yields were obtained when bromobenzene and chlorobenzene were used as substrates *(170)*. On the other hand, the polystyrene-anchored azo-based palladacycle **698** was shown to behave as a highly efficient heterogeneous catalyst in the copper-free Sonogashira reaction involving various halides under water medium in aerobic conditions (Scheme 2.47). The coupling reaction gave the diarylacetylene products in good-to-excellent yields using aryl iodides and aryl bromides as the arene sources, whereas aryl bromides required higher reaction temperature and extended reaction time. Unfortunately, the Sonogashira coupling of chlorobenzene resulted in a very low yield (6%). The catalyst could be recovered by simple filtration and reused for more than six runs without significant loss of its activity *(361)*.

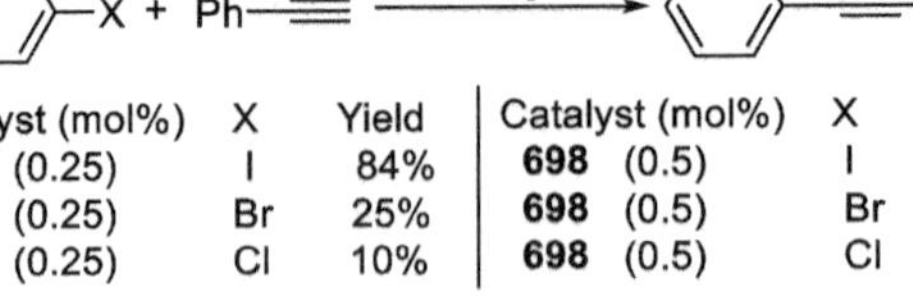

659 **698**

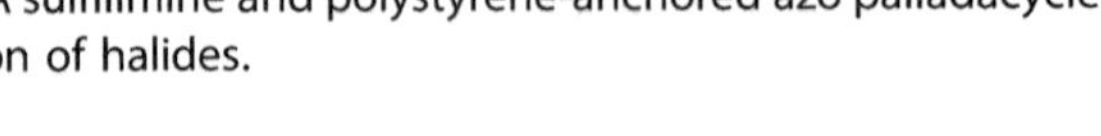

Catalyst (mol%)	X	Yield	Catalyst (mol%)	X	Yield
659 (0.25)	I	84%	**698** (0.5)	I	88%
659 (0.25)	Br	25%	**698** (0.5)	Br	84%
659 (0.25)	Cl	10%	**698** (0.5)	Cl	6%

Scheme 2.47 A sulfilimine and polystyrene-anchored azo palladacycles for the copper-free alkynylation of halides.

The polymer-supported thiopseudourea palladacycle **699** as a heterogeneous catalyst for a copper- and solvent-free Sonogashira reaction was developed by Keesara (Scheme 2.48). The desired cross-coupling products from aryl iodides and arylacetylenes were obtained in high yields in the presence

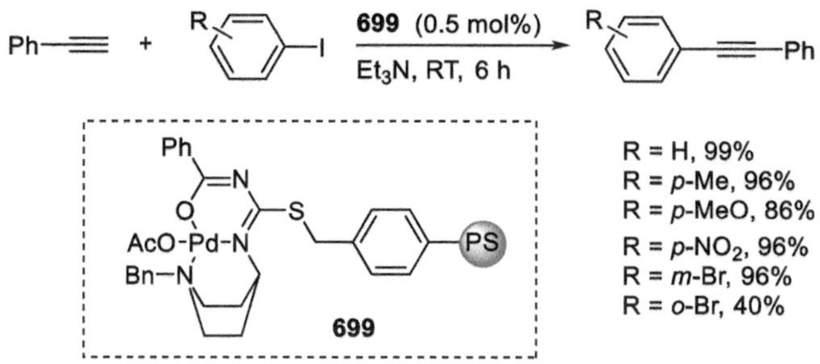

Scheme 2.48 A solvent- and copper-free alkynylation catalyzed by the polymer supported thiopseudourea palladacycle.

of triethylamine as the base at room temperature conditions *(362)*. In addition, it was found even applicable for gram scale synthesis with excellent coupling product yields.

8. Conclusions and Prospects

Heck and Sonogashira couplings were first disclosed more than 30 years ago. Most of the initial investigations were focused on aryl iodides, aryl bromides, and aryl triflates. Nevertheless, aryl chlorides are still challenging substrates, yet were rarely employed at that time. Harsh and unique reaction conditions or high catalyst loading were always necessary. With the palladacycles developed in the past decade, milder reaction conditions (they can even perform at room temperature), lower catalyst loadings, and more environmentally friendly catalytic systems were established. Since the first report of Herrmann-Beller's catalyst in 1995, palladacycles have become a new direction for exploring new cross-coupling reactions. In fact, they often show excellent catalytic activity and in particular with the high turnover numbers. They feature high tunability resulting from modulating their electronic and steric properties, in which they can be easily achieved by altering the type of donor group, its substituents, the size of the palladacyclic ring, and the nature of X ligands. Nowadays, palladacycles are widely applied in Heck and Sonogashira cross-coupling chemistry. They have attractive characteristics that are easy to handle, stable in moisture and air, and can potentially be utilized in large-scale synthesis. We anticipate that the applied research to utilize the palladacycles in catalysis will continue to grow considerably in tackling existing problems in fine chemical and drug syntheses. More robust palladacyclic catalysts systems with superior performance will be attainable.

References

1. (a) Heck, R. F. *J. Am. Chem. Soc.* **1968**, *90*, 5518; (b) Mizoroki, T.; Mori, K.; Ozaki, A. *Bull. Chem. Soc. Jpn.* **1971**, *44*, 581; (c) Heck, R. F.; Nolly, J. P. *J. Org. Chem.* **1972**, *37*, 2320.
2. Drahl, C. *Chem. Eng. News* **2010**, *88*, 31.
3. (a) Bader, R. R.; Baumeister, P.; Blaser, H. U. *Chimia* **1996**, *50*, 99; (b) de Vries, J. G. *Can. J. Chem.* **2001**, *79*, 1086; (c) Kobetić, R.; Biliškov, N. *Chem. Ind.* **2007**, *56*, 391.
4. For selected reviews, see: (a) I.P. Beletskaya and A.V. Cheprakov, Chem. Rev. 100, 2000, 3009; (b) Jagtap, S. *Catalysts* **2017**, 7, 267
5. Sonogashira, K.; Tohda, Y.; Hagihara, N. *Tetrahedron Lett.* **1975**, *16*, 4467.
6. For Highlight and reviews, see: (a) H. Plenio, Angew. Chem. Int. Ed. 47, 2008, 6954; (b) Doucet, H.; Hierso, J.-C. *Angew. Chem. Int. Ed.* **2007**, *46*, 834; (c) Chinchilla, R.; Najera, C. Chem. Rev. **2007**, *107*, 874; (d) Viciu, M. S.; Nolan, S. P. Arylation reactions of alkynes: the Sonogashira reaction. In: *Modern Arylation Methods;* Ackermann, L., Ed.; Wiley-VCH: Weinheim, 2009; pp 183–220; (e) Chinchilla, R.; Nájera, C. *Chem. Soc. Rev.* **2011**, *40*, 5084; (f) Bakherad, M. *Appl. Organometal. Chem.* **2013**, *27*, 125
7. Diederich, F.; Stang, P.; Tykwinski, R. R. *Acetylene Chemistry: Chemistry, Biology and Material Science.* Wiley-VCH: Weinheim, 2005.
8. For recent selected book chapters and reviews: (a) A. de Meijere, S. Bräseand M. Oestreich, Metal-Catalyzed Cross-Coupling Reactions and More, 1, 2 and 3, Wiley-VCH: Weinheim, 2013; (b) Colacot, T. J. *New Trends in Cross-Coupling: Theory and Applications.* RSC: Cambridge, 2015; (c) Oestreich, M. *The Mizoroki-Heck Reaction.* John Wiley & Sons: Germany, 2009; (d) Molnár, Á. *Palladium-Catalyzed Coupling Reactions: Practical Aspects and Future Developments.* Wiley-VCH: Weinheim, 2013; (e) Li, H.; Seechurn, C. C. C. J.; Colacot, T. J. *ACS Catal.* **2012**, *2*, 1147; (f) Evano, G.; Gaumont, A.-C.; Alayrac, C.; Wrona, I. E.; Giguere, J. R.; Delacroix, O.; Bayle, A.; Jouvin, K.; Theunissen, C.; Gatignol, J.; Silvanus, A. C. *Tetrahedron* **2014**, *70*, 1529; (g) Gildner, P. G.; Colacot, T. J. *Organometallics* **2015**, *34*, 5497; (h) Roy, D.; Uozumi, Y. *Adv. Synth. Catal.* **2018**, *360*, 602
9. (a) Cope, A. C.; Siekman, R. W. *J. Am. Chem. Soc.* **1965**, *87*, 3272; (b) Cope, A. C.; Friedrich, E. C. *J. Am. Chem. Soc.* **1968**, *90*, 909.
10. For selected reviews and book chapters: (a) J. Dupont, M. Pfefferand J. Spencer, Eur. J. Inorg. Chem. 2001, 1917; (b) Muñoz, M. P.; Martín-Matute, B.; Fernández-R, C.; Cárdenas, D. J.; Echavarren, A. E. *Adv. Synth. Catal.* **2001**, *343*, 338; (c) Bedford, R. B. *Chem. Commun.* **2003**, 1787; (d) Dunina, V. V.; Gorunova, O. N. *Russ. Chem. Rev.* **2004**, *73*, 309; (e) Beletskaya, I. P.; Cheprakov, A. V. *J. Org. Chem.* **2004**, *689*, 4055; (f) Dupont, J.; Consorti, C. S.; Spencer, J. *Chem. Rev.* **2005**, *105*, 2527; (g) Djukic, J.-P.; Hijazi, A.; Flack, H. D.; Bernardinelli, G. *Chem. Soc. Rev.* **2008**, *37*, 406; (h) Dupont, J.; Pfeffer, M. Palladacycles. In: *Synthesis, Characterization and Applications;* Wiley-VCH: Weinheim, 2008;; (i) Mo, D.-L.; Zhang, T.-K.; Ge, G.-C.; Huang, X.-J.; Ding, C.-H.; Dai, L.-X.; Hou, X.-L. *Synlett* **2014**, *25*, 2686; (j) Ratti, R. *Can. Chem. Trans.* **2014**, *2*, 467; (k) Bruneau, A.; Roche, M.; Alami, M.; Messaoudi, S. *ACS Catal.* **2015**, *5*, 1386
11. Herrmann, W. A.; Brossmer, C.; Öfele, K.; Reisinger, C.-P.; Priermeier, T.; Beller, M.; Fischer, H. *Angew. Chem. Int. Ed.* **1995**, *34*, 1844.
12. Herrmann, W. A.; Böhn, V. P. W.; Reisinger, C.-P. *J. Org. Chem.* **1999**, *576*, 23.
13. (a) Dunina, V. V.; Gorunova, O. N. *Russ. Chem. Rev.* **2004**, *73*, 309; (b) Dunina, V. V.; Gorunova, O. N. *Russ. Chem. Rev.* **2005**, *74*, 871.
14. Herrmann, W. A.; Brossmer, C.; Reisinger, C.-P.; Riermeier, T. H.; Öfele, K.; Beller, M. *Chem. Eur. J.* **1997**, *3*, 1357.

15. Herrmann, W. A.; Böhm, V. P. W. *J. Org. Chem.* **1999**, *572*, 141.
16. Herrmann, W. A.; Böhm, V. P. W. *Chem. Eur. J.* **2000**, *6*, 1017.
17. Zapf, A.; Beller, M. *Chem. Eur. J.* **2001**, *7*, 2908.
18. Böhm, V. P.; Herrmann, W. A. *Chem. Eur. J.* **2001**, *7*, 4191.
19. (a) Rosner, T.; Bars, J. L.; Pfaltz, A.; Blackmond, D. G. *J. Am. Chem. Soc.* **2001**, *123*, 1848; (b) Rosner, T.; Pfaltz, A.; Blackmond, D. G. *J. Am. Chem. Soc.* **2001**, *123*, 4621; (c) Mathew, J. S.; Klussmann, M.; Iwamura, H.; Valera, f.; Futran, A.; Emanuelsson, E. A. C.; Blackmond, D. G. *J. Org. Chem.* **2006**, *71*, 4711; (d) Blackmond, D. G.; Schultz, T.; Mathew, J. S.; Loew, C.; Rosner, T.; Pfaltz, A. *Synlett* **2006**, 3135.
20. (a) de Vries, A. H. M.; Mulders, J. M. C.; Mommers, J. H. M.; Henderickx, H. J. W.; de Vries, J. G. *Org. Lett.* **2003**, *5*, 3285;; (b)J.G. de Vries, Dalton Trans. 2006, 421.
21. (a) Sud, A.; Deshpande, R. M.; Chaudhari, R. V. *J. Mol. Catal. A Chem.* **2007**, *270*, 144; (b) Svennebring, A.; Sjöberg, P. J. R.; Larhed, M.; Nilsson, P. *Tetrahedron* **2008**, *64*, 1808.
22. (a) Beller, M.; Riermeier, T. H. *Tetrahedron Lett.* **1996**, *37*, 6535; (b) Beller, M.; Riermeier, T. H. *Eur. J. Inorg. Chem.* **1998**, 29.
23. Beller, M.; Köhler, K.; Hartung, C. G. *Org. Lett.* **1999**, *1*, 709.
24. Prashad, M.; Liu, Y.; Mak, X. Y.; Har, D.; Repič, O.; Blacklock, T. J. *Tetrahedron Lett.* **2002**, *43*, 8559.
25. (a) Nejjar, A.; Pinel, C.; Djakovitch, L. *Adv. Synth. Catal.* **2003**, *345*, 612; (b) Noël, S.; Djakovitch, L.; Pinel, C. *Tetrahedron Lett.* **2006**, *47*, 3839; (c) Tarabay, J.; Al-Maksoud, W.; Jaber, F.; Pinel, C.; Prakash, S.; Djakovitch, L. *Appl. Catal., A* **2010**, *388*, 124.
26. (a) Pan, K.; Noël, S.; Pinel, C.; Djakovitch, L. *J. Org. Chem.* **2008**, *693*, 2863; (b) Al-Maksoud, W.; Menuel, S.; Jahjah, M.; Monflier, E.; Pinel, C.; Djakovitch, L. *Appl. Catal., A* **2014**, *469*, 250.
27. Konda, V.; Rydfjord, J.; Sävmarker, J.; Larhed, M. *Org. Process Res. Dev.* **2014**, *18*, 1413.
28. (a) Tietze, L. F.; Schirok, H.; Wöhrmann, M.; Schrader, K. *Eur. J. Org. Chem.* **2000**, 2433; (b) Tietze, L. F.; Petersen, S. *Eur. J. Org. Chem.* **2001**, 1619; (c) Tietze, L. F.; Kahle, K.; Raschke, T. *Chem. Eur. J.* **2002**, *8*, 401; (d) Sünnemann, H. W.; Hofmeister, A.; Magull, J.; de Meijere, A. *Chem. Eur. J.* **2006**, *12*, 8336; (e) Tietze, - L. F.; Brasche, G.; Grube, A.; Böhnke, N.; Stadler, C. *Chem. Eur. J.* **2007**, *13*, 8543.
29. (a) Chen, W.; Xu, L.; Xiao, J. *Tetrahedron Lett.* **2001**, *42*, 4275; (b) Chen, W.; Xu, L.; Hu, Y.; Osuna, A. M. B.; Xiao, J. *Tetrahedron* **2002**, *58*, 3889; (c) Wende, M.; Seidel, F.; Gladysz, J. A. *J. Fluor. Chem.* **2003**, *124*, 45; (d) Xu, L.; Mo, J.; Baillie, C.; Xiao, J. *J. Org. Chem.* **2003**, *687*, 301.
30. Eriksson, L.; Winberg, K. J.; Claro, R. T.; Sjöberg, S. *J. Org. Chem.* **2003**, *68*, 3569.
31. Lagisetty, P.; Zhang, L.; Lakshman, M. K. *Adv. Synth. Catal.* **2008**, *350*, 602.
32. Donaldson, L. R.; Haigh, D.; Hulme, A. N. *Tetrahedron* **2008**, *64*, 4468.
33. Tietze, L. F.; Krimmelbein, I. K. *Chem. Eur. J.* **2008**, *14*, 1541.
34. Rixson, J. E.; Chaloner, T.; Heath, C. H.; Tietze, L. F.; Stewart, S. G. *Eur. J. Org. Chem.* **2012**, 544.
35. Gibson, S.; Foster, D. F.; Eastham, G. R.; Tooze, R. P.; Cole-Hamilton, D. J. *Chem. Commun.* **2001**, 779.
36. Wang, C.; Fu, Y.; Li, Z.; Guo, Q.-X. *Chin. J. Chem.* **2008**, *26*, 358.
37. (a) Luo, F.-T.; Lin, C.-A. *Tetrahedron Lett.* **2003**, *44*, 7565; (b) Luo, F.-T.; Xue, C.; Ko, S.-L.; Shao, Y.-D.; Wu, C.-J.; Kuo, Y.-M. *Tetrahedron* **2005**, *61*, 6040.
38. Djakovitch, L.; Heise, H.; Köhler, K. *J. Org. Chem.* **1999**, *584*, 16.
39. Shiota, A.; Malinakova, H. C. *J. Org. Chem.* **2012**, *704*, 9.
40. Shaw, B. L.; Perera, S. D.; Staley, E. A. *Chem. Commun.* **1998**, 1361.

41. Frey, G. D.; Reisinger, C.-P.; Herdtweck, E.; Herrmann, W. A. *J. Org. Chem.* **2005**, *690*, 3193.
42. Frey, G. D.; Schütz, J.; Herdtweck, E.; Herrmann, W. A. *Organometallics* **2005**, *24*, 4416.
43. Albisson, D. A.; Bedford, R. B.; Scully, P. N. *Tetrahedron Lett.* **1998**, *39*, 9793.
44. Matano, Y.; Matsumoto, K.; Shibano, T.; Imahori, H. *J. Porphyr. Phthalocya.* **2011**, *15*, 1172.
45. Karami, K.; Ghasemi, M.; Naeini, N. H. *Tetrahedron Lett.* **2013**, *54*, 1352.
46. Nadri, S.; Joshaghani, M.; Rafiee, E. *Organometallics* **2009**, *28*, 6281.
47. Yan, X.; Liu, Y.; Xi, C. *Appl. Organomet. Chem.* **2008**, *22*, 341.
48. Deleted in Review.
49. Ogo, S.; Takebe, Y.; Uehara, K.; Yamazaki, T.; Nakai, H.; Watanabe, Y.; Fukuzumi, S. *Organometallics* **2006**, *25*, 331.
50. (a) So, C. M.; Zhou, Z.; Lau, C. P.; Kwong, F. Y. *Angew. Chem. Int. Ed.* **2008**, *47*, 6402; (b) Choy, P. Y.; Chow, W. K.; So, C. M.; Lau, C. P.; Kwong, F. Y. *Chem. Eur. J.* **2010**, *16*, 9982; (c) Wong, S. M.; Yuen, O. Y.; Choy, P. Y.; Kwong, F. Y. *Coord. Chem. Rev.* **2015**, *293–294*, 158.
51. Adhikary, A.; Guan, H. *ACS Catal.* **2015**, *5*, 6858.
52. Shaw, B. L.; Perera, S. D. *Chem. Commun.* **1998**, 1863.
53. Sabounchei, S. J.; Ahmadi, M. *Inorg. Chim. Acta* **2013**, *405*, 15.
54. Sabounchei, S. J.; Ahmadi, M.; Panahimehr, M.; Bagherjeri, F. A.; Nasri, Z. *J. Mol. Catal. A Chem.* **2014**, *383–384*, 249.
55. Gümgüm, B.; Biricik, N.; Durap, F.; Özdemir, I.; Gürbüz, N.; Ang, W. H.; Dyson, P. J. *Appl. Organomet. Chem.* **2007**, *21*, 711.
56. Sabounchei, S. J.; Ahmadi, M.; Azizi, T.; Panahimehr, M. *Synlett* **2014**, *25*, 336.
57. Sabounchei, S. J.; Hashemi, A.; Yousefi, A.; Derakhshandeh, P. G.; Karamian, R.; Asadbegy, M.; Hecke, K. V. *Polyhedron* **2017**, *135*, 1.
58. Chinchilla, R.; Nájera, C. *Chem. Rev.* **2007**, *107*, 874.
59. Fulmer, D. A.; Shearouse, W. C.; Medonza, S. T.; Mack, J. *Green Chem.* **2009**, *11*, 1821.
60. Farina, V. *Adv. Synth. Catal.* **2004**, *346*, 1553.
61. Herrmann, W. A.; Reisinger, C.-P.; Öfele, K.; Broβmer, C.; Beller, M.; Fischer, H. *J. Mol. Cata. A Chem.* **1996**, *108*, 51.
62. Herrmann, W. A.; Öfele, K.; Preysing, D.v.; Schneider, S. K. *J. Org. Chem.* **2003**, *687*, 229.
63. Bakherad, M.; Keivanloo, A.; Bahramian, B.; Mihanparast, S. *Tetrahedron Lett.* **2009**, *50*, 6418.
64. Tamami, B.; Ghasemi, F. *Collect. Czech. Chem. Commun.* **2011**, *76*, 1967.
65. Arques, A.; Auñon, D.; Molina, P. *Tetrahedron Lett.* **2004**, *45*, 4337.
66. Ruiz, J.; Cutillas, N.; López, F.; López, G.; Bautista, D. *Organometallics* **2006**, *25*, 5768.
67. Nishide, K.; Liang, H.; Ito, S.; Yoshifuji, M. *J. Org. Chem.* **2005**, *690*, 4809.
68. Finke, A. D.; Elleby, E. C.; Boyd, M. J.; Weissman, H.; Moore, J. S. *J. Org. Chem.* **2009**, *74*, 8897.
69. Sabounchei, S. J.; Ahmadi, M.; Nasri, Z.; Shams, E.; Panahimehr, M. *Tetrahedron Lett.* **2013**, *54*, 4656.
70. Sabounchei, S. J.; Ahmadi, M. *Catal. Commun.* **2013**, *37*, 114.
71. (a) Marziale, A. N.; Schlüter, J.; Eppinger, J. *Tetrahedron Lett.* **2011**, *52*, 6355; (b) Błaszczyk, I.; Trzeciak, A. M. Ziółkowski. *Catal. Lett.* **2009**, *133*, 262.
72. Gonthier, E.; Breinbauer, R. *Synlett* **2003**, 1049.
73. Suzuka, T.; Okada, Y.; Osshiro, K.; Uozumi, Y. *Tetrahedron Lett.* **2010**, *66*, 1064.
74. Méry, D.; Heuzé, K.; Astruc, D. *Chem. Commun.* **2003**, 1934.

75. (a) Heuzé, K.; Méry, D.; Gauss, D.; Astruc, D. *Chem. Commun.* **2003**, 2274; (b) Heuzé, K.; Méry, D.; Gauss, D.; Blais, J.-C.; Astruc, D. *Chem. Eur. J.* **2004**, *10*, 3936.
76. Rosario-Amorin, D.; Gaboyard, M.; Clérac, R.; Nlate, S.; Heuzé, K. *Dalton Trans.* **2011**, *40*, 44.
77. (a) Wang, Y.; Huang, B.; Sheng, S.; Cai, M. *J. Chem. Res.* **2007**, 728; (b) Cai, M.; Sha, J.; Xu, Q. *Tetrahedron* **2007**, *63*, 4642.
78. Hao, W.; Sha, J.; Sheng, S.; Cai, M. *J. Mol. Catal. A Chem.* **2009**, *298*, 94.
79. Zhang, L.-Y.; Wang, L. *Chin. J. Chem.* **2008**, *26*, 1601.
80. Lucas, N. T.; Hook, J. M.; McDonagh, A. M.; Colbran, S. B. *Eur. J. Inorg. Chem.* **2005**, 496.
81. Sokolov, V. I. C. *Chin. J. Org. Chem.* **2018**, *38*, 75.
82. Zhang, Y.-M.; Yang, F.; Zheng, R.; Tang, J. C. *Chin. J. Org. Chem.* **2001**, *21*, 533.
83. Yang, F.; Zhang, Y.l.; Zheng, R.; Tang, J.; He, M. *J. Org. Chem.* **2002**, *651*, 146.
84. Sokolov, V. I.; Bulygina, L. A.; Khrushcheva, N. S.; Ikonnikov, N. S. *Russ. Chem. Bull.* **2010**, *59*, 1400.
85. Wang, H.-X.; Wu, H.-F.; Yang, X.-L.; Ma, N.; Wan, L. *Polyhedron* **2007**, *26*, 3857.
86. Wu, Y.; Hou, J.; Yun, H.; Cui, X.; Yuan, R. *J. Org. Chem.* **2001**, *793*, 637–639.
87. Hou, J.-J.; Yang, L.-R.; Cui, X.-L.; Wu, Y.-J. *Chin. J. Chem.* **2003**, *21*, 717.
88. Mu, B.; Li, T.; Xu, W.; Zeng, G.; Liu, P.; Wu, Y. *Tetrahedron* **2007**, *63*, 11475.
89. (a) Wu, Y.; Yang, F.; Zhang, J.; Cui, X.; Gong, J.; Song, M.; Li, T. *Chin. Sci. Bull.* **2010**, *55*, 2784; (b) Wang, J.; Mu, B.; Fu, Z.; Wang, L.; Li, T.; Wu, Y. C. *J. Catal.* **2014**, *35*, 1059.
90. Nagy, T. Z.; Csámpai, A.; Kotschy, A. *Tetrahedron* **2005**, *61*, 9767.
91. Xu, C.; Gong, J.-F.; Yue, S.-F.; Zhu, Y.; Wu, Y.-J. *Dalton Trans.* **2006**, 4730.
92. Ren, G.; Cui, X.; Yang, E.; Yang, F.; Wu, Y. *Tetrahedron* **2010**, *66*, 4022.
93. Leng, Y.; Yang, F.; Wei, K.; Wu, Y. *Tetrahedron* **2010**, *66*, 1244.
94. Iyer, S.; Jayanthi, A. *Tetrahedron Lett.* **2001**, *42*, 7877.
95. Iyer, S.; Kulkarni, G. M.; Ramesh, C. *Tetrahedron* **2004**, *60*, 2163.
96. Alonso, D. A.; Nájera, C.; Pacheco, C. M. *Adv. Synth. Catal.* **2002**, *344*, 172.
97. For some selected references: (a) A. Kasahara, T. Izumi and M. Maemura, Bull. Chem. Soc. Jpn. 50, 1977, 1878; (b) Utepova, I. A.; Chupakhin, O. N.; Charushin, V. N. *Heterocycles* **2008**, *76*, 39
98. Chupakhin, O. N.; Utepova, I. A.; Kovalev, I. S.; Rusinov, V. L.; Starikova, Z. A. Eur. J. Org. Chem. **2007**, *5*, 857.
99. Balam-Villarreal, J. A.; Sandoval-Chávez, C. I.; Ortega-Jiménez, F.; Toscano, R. A.; Carreón-Castro, M. P.; López-Cortés, J. U. G.; Ortega-Alfaro, M. C. *J. Org. Chem.* **2016**, *818*, 7.
100. Yang, F.; Cui, X.; Li, Y.-N.; Zhang, J.; Ren, G.-R.; Wu, Y. *Tetrahedron* **2007**, *63*, 1963.
101. Huang, M.; Feng, Y.; Wu, Y. *Tetrahedron* **2012**, *68*, 376.
102. Hu, H.; Yang, F.; Wu, Y. *J. Org. Chem.* **2013**, *78*, 10506.
103. Bai, S.-Z.; Xu, C.; Li, H.-M.; Wang, Z.-Q.; Fu, W.-J. *Molecules* **2012**, *17*, 5532.
104. Beletskaya, I. P.; Kashin, A. N.; Karlstedt, N. B.; Mitin, A. V.; Cheprakov, A. V.; Kazankov, G. M. *J. Org. Chem.* **2001**, *622*, 89.
105. Consorti, C. S.; Zanini, M. L.; Leal, S.; Ebeling, G.; Dupont, J. *Org. Lett.* **2003**, *5*, 983.
106. Iyer, S.; Jayanthi, A. *Synlett* **2003**, *8*, 1125.
107. Kantchev, E. A. B.; Peh, G.-R.; Zhang, C.; Ying, J. Y. *Org. Lett.* **2008**, *10*, 3949.
108. Peh, G.-R.; Kantchev, E. A. B.; Zhang, C.; Ying, J. Y. *Org. Biomol. Chem.* **2009**, 7, 2110.
109. Consorti, C. S.; Flores, F. R.; Dupont, J. *J. Am. Chem. Soc.* **2005**, *127*, 12054.
110. Chen, K.; Pullarkat, S. A. *Org. Biomol. Chem.* **2012**, *10*, 6600.

111. Schnyder, A.; Indolese, A. F.; Studer, M.; Blaser, H.-U. *Angew. Chem. Int. Ed.* **2002**, *41*, 3668.
112. Karami, K.; Rahimi, N.; Shehni, M. B. *Tetrahedron Lett.* **2012**, *53*, 2428.
113. Karami, K.; Moghadam, Z. K.; Hosseini-Kharat, M. *Catal. Commun.* **2014**, *43*, 25.
114. Xu, C.; Afewerki, S.; Tai, C.-W.; Córdova, A.; Hedin, N. *ChemistrySelect* **2016**, *1*, 5801.
115. Blackmond, D. G.; Rosner, T.; Pfaltz, A. *Org. Process Res. Dev.* **1999**, *3*, 275.
116. Ortega-Jiménez, F.; Domínguez-Villa, F. X.; Rosas-Sánchez, A.; Penieres-Carrillo, G.; López-Cortés, J. G.; Ortega-Alfaro, M. C. *Appl. Organomet. Chem.* **2015**, *29*, 556.
117. Zhao, Y.; Zhao, N.; Zou, M.; Li, T.; Li, G.; Li, J.; Wu, Y. *Appl. Organomet. Chem.* **2016**, *30*, 540.
118. Ohff, M.; Ohff, A.; Milstein, D. *Chem. Commun.* **1999**, 357.
119. Bravo, J.; Cativiela, C.; Navarro, R.; Urriolabeitia, E. P. *J. Org. Chem.* **2002**, *650*, 157.
120. C. Rocaboyand J.A. Gladysz, Org. Lett. 4, 2002, 1993.
121. Zotto, A. D.; Prat, F. I.; Baratta, W.; Zangrando, E.; Rigo, P. *Inorg. Chim. Acta* **2009**, *362*, 97.
122. Yilmaz, M. K.; Keleş, M. *J. Turkish chem.* **2018**, *5*, 133.
123. Iyer, S.; Kulkarni, G. M.; Ramesh, C.; Sattar, A. K. *Ind. J. Chem.* **2005**, *244B*, 1894.
124. Röhlich, C.; Köhler, K. *Chem. Eur. J.* **2010**, *16*, 2363.
125. Tang, Y.; Zeng, Y.; Hu, Q.; Huang, F.; Jin, L.; Mo, W.; Sun, N.; Hu, B.; Shen, Z.; Hu, X.; Sun, W.-H. *Adv. Synth. Catal.* **2016**, *358*, 2642.
126. Chitanda, J. M.; Prokopchuk, D. E.; Quail, J. W.; Foley, S. R. *Dalton Trans.* **2008**, 6023.
127. Beccalli, E. M.; Borsini, E.; Brenna, S.; Gallil, S.; Rigamonti, M.; Broggini, G. *Chem. Eur. J.* **2010**, *16*, 1670.
128. Zhao, T.; Xing, Q.; Song, K.; Ban, Q.; Liang, T.; Liu, Q.; Sun, W.-H. *RSC Adv.* **2015**, *5*, 14228.
129. Song, K.; Kong, S.; Liu, Q.; Sun, W.-H.; Redshaw, C. *J. Org. Chem.* **2014**, *751*, 453.
130. Yang, X.; Zhang, C.; Dai, W.; Liu, J.; Wei, M. *Chin. J. Catal.* **2012**, *33*, 878.
131. Andrade, A. P. S.; Arantes, L. M.; Kadooca, J. Y.; Carvalho, R. L.; de Fátima, Â.; Sabino, A. A. *ChemistrySelect* **2016**, *5*, 886.
132. Onoue, H.; Minami, K.; Nakagawa, K. *Bull. Chem. Soc. Jpn.* **1970**, *43*, 3480.
133. (a) Alacid, E.; Alonso, D. A.; Botella, L.; Nájera, C.; Pacheco, M. C. *Chem. Rec.* **2006**, *6*, 117; (b) Alonso, D. A.; Nájera, C. *Chem. Soc. Rev.* **2010**, *39*, 2891.
134. Alonso, D. A.; Nájera, C.; Pacheco, M. C. *Org. Lett.* **2000**, *2*, 1823.
135. (a) Iyer, S.; Ramesh, C. *Tetrahedron Lett.* **2000**, *41*, 8981; (b) Iyer, S.; Jayanthi, A. *Synlett* **2003**, 1125.
136. (a) Nájera, C.; Botella, L. *Tetrahedron* **2005**, *61*, 9688; (b) Botella, L.; Nájera, C. *J. Org. Chem.* **2005**, *70*, 4360.
137. Smith, C. R.; Rajan Babu, T. V. *Tetrahedron* **2010**, *66*, 1102.
138. Kumar, N. S. C. R.; Raj, I. V. P.; Sudalai, A. *J. Mol. Catal. A Chem.* **2006**, *269*, 218.
139. Cao, X.-P.; Barth, D.; Kuck, D. *Eur. J. Org. Chem.* **2005**, 3482.
140. Botella, L.; Nájera, C. *Tetrahedron* **2004**, *60*, 5563.
141. (a) Pirovani, R. V.; Ferreira, B. R. V.; Coelho, F. *Synlett* **2009**, 2333; (b) Ferreira, B. R. V.;Pirovani, R. V.; Souza-Filho, L. G.; Coelho, F. *Tetrahedron* **2009**, *65*, 7712.
142. de Luna Freire, K. R.; Tormena, C. F.; Coelho, F. *Synlett* **2011**, *14*, 2059.
143. Arena, G.; Cini, E.; Petricci, E.; Randino, R.; Taddei, M. *Org. Chem. Front.* **2015**, *2*, 526.
144. Corma, A.; García, H.; Leyva, A. *Tetrahedron* **2004**, *60*, 8553.
145. (a) Alacid, E.; Nájera, C. *ARKIVOC* **2008**, *viii*, 50; (b) Alacid, E.; Nájera, C. *Eur. J. Org. Chem.* **2008**, 3102; (c) Alacid, E.; Nájera, C. *Synfacts* **2008**, 1008.

146. Mennecke, K.; Solodenko, W.; Kirschning, A. *Synthesis* **2008**, *10*, 1589.
147. Tindale, J. J.; Ragogna, P. J. *Can. J. Chem.* **2010**, *88*, 27.
148. Susanto, W.; Chu, C.-Y.; Ang, W. J.; Chou, T.-C.; Lo, L.-C.; Lam, Y. *J. Org. Chem.* **2012**, *77*, 2729.
149. Alacid, E.; Nájera, C. *Adv. Synth. Catal.* **2008**, *350*, 1316.
150. Susanto, W.; Chu, C.-Y.; Ang, W. J.; Chou, T.-C.; Lo, L.-C.; Lam, Y. *Green Chem.* **2012**, *14*, 77.
151. Nájera, C. *ChemCatChem* **2016**, *8*, 1865.
152. Alacid, E.; Nájera, C. *Adv. Synth. Catal.* **2007**, *349*, 2572.
153. Buchmeiser, M. R.; Wurst, K. *J. Am. Chem. Soc.* **1999**, *121*, 11101.
154. Zotto, A. D.; Zangrando, E.; Baratta, W.; Felluga, A.; Martinuzzi, P.; Rigo, P. *Eur. J. Org. Chem.* **2005**, 4707.
155. Evans, P.; Hogg, P.; Grigg, R.; Nurnabi, M.; Hinsley, J.; Sridharan, V.; Suganthan, S.; Korn, S.; Collard, S.; Muir, J. E. *Tetrahedron* **2005**, *61*, 9696.
156. Gil-Moltó, J.; Karlström, S.; Nájera, C. *Tetrahedron* **2005**, *61*, 12168.
157. Kantam, M. L.; Srinivas, P.; Yadav, J.; Likhar, P. R.; Bhargava, S. *J. Org. Chem.* **2009**, *74*, 4882.
158. Atla, S. B.; Kelkar, A. A.; Puranik, V. G.; Bensch, W.; Chaudhari, R. V. *J. Org. Chem.* **2009**, *694*, 683.
159. Fernández-García, L.; Blanco, M.; Blanco, C.; Álvarez, P.; Granda, M.; Santamaría, R.; Menéndez, R. *J. Mol. Catal. A Chem.* **2016**, *416*, 140.
160. Baier, H.; Kelling, A.; Schilde, U.; Holdt, H.-J. *Z. Anorg. Allg. Chem.* **2016**, *642*, 140.
161. Kantam, M.; Annapurna, M.; Likhar, P. R.; Srinivas, P.; Mirzadeh, N.; Bhargava, S. K. *J. Org. Chem.* **2013**, *723*, 129.
162. Annapurna, M.; Reddy, P. V.; Singh, S. P.; Kantam, M. L. *Tetrahedron* **2013**, *69*, 10940.
163. Riflade, B.; OBle, J.; Chenneberg, L.; Derat, E.; Hasenknopf, B.; Lacôte, E.; Thorimbert, S. *Tetrahedron* **2013**, *69*, 5772.
164. Haneda, S.; Gan, Z.; Eda, K.; Hayashi, M. *Organometallics* **2007**, *26*, 6551.
165. Saleem, F.; Rao, G. K.; Kumar, A.; Kumar, S.; Singh, M. P.; Singh, A. K. *RSC Adv.* **2014**, *4*, 56102.
166. Mamidala, R.; Mukundam, V.; Dhanunjayarao, K.; Venkatasubbaiah, K. *Dalton Trans.* **2015**, *44*, 5805.
167. M. Guerrero, J. Pons and J. Ros, J. Org. Chem. 695, 2010, 1957.
168. Gai, X.; Grigg, R.; Ramzan, M. I.; Sridharan, V.; Collard, S.; Muir, J. E. *Chem. Commun.* **2000**, 2053.
169. (a) Lin, R.-S.; Li, M.-R.; Liu, Y.-H.; Peng, S.-M.; Liu, S.-T. *Inorg. Chim. Acta* **2010**, *363*, 3523; (b) Fujimoto, K.; Yoneda, T.; Yorimitsu, H.; Osuka, A. *Angew. Chem. Int. Ed.* **2014**, *53*, 1127.
170. Thakur, V. V.; Kumar, N. S. C. R.; Sudalai, A. *Tetrahedron Lett.* **2004**, *45*, 2915.
171. Spencer, J.; Sharratt, D. P.; Dupont, J.; Monteiro, A. L.; Reis, V. I.; Stracke, M. P.; Rominger, F.; McDonald, I. M. *Organometallics* **2005**, *24*, 5665.
172. Wu, K.-M.; Huang, C.-A.; Peng, K.-F.; Chen, C. T. *Tetrahedron* **2005**, *61*, 9679.
173. Lee, D.-H.; Taher, A.; Hossain, S.; Jin, M.-J. *Org. Lett.* **2011**, *13*, 5540.
174. Fortea-Pérez, F. R.; Rothenpieler, B. L.; Marino, N.; Armentano, D.; Munno, G. D.; Julve, M.; Stiriba, S.-E. *Inorg. Chem. Front.* **2015**, *2*, 1029.
175. Nouri, F.; Rostamizadeh, S.; Azad, M. *Inorg. Chim. Acta* **2018**, *471*, 664.
176. Lee, D.-H.; Lee, Y. H.; Harrowfield, J. M.; Lee, I.-M.; Lee, H. I.; Lim, W. T.; Kim, Y.; Jin, M.-J. *Tetrahedron* **2009**, *65*, 1630.
177. (a) Gil-Moltó, J.; Nájera, C. *Adv. Synth. Catal.* **2006**, *348*, 1874; (b) Nájera, C.; Gil-Moltó, J.; Karlström, S.; Falvello, L. R. *Org. Lett.* **2003**, *5*, 1451.
178. Gil-Moltó, J.; Nájera, C. *Eur. J. Org. Chem.* **2005**, 4073.

179. Fairlamb, I. J. S.; Kapdi, A. R.; Lee, A. F.; Sánchez, G.; López, G.; Serrano, J. L.; García, L.; Pérez, J.; Pérez, E. *Dalton Trans.* **2004**, 3970.
180. Ghanbari, B.; Shahhoseini, L.; Hosseini, H.; Bagherzadeh, M.; Owczarzak, A.; Kubicki, M. *J. Org. Chem.* **2018**, *866*, 72.
181. (a) Consorti, C. S.; Flores, F. R.; Rominger, F.; Dupont, J. *Adv. Synth. Catal.* **2006**, *348*, 133; (b) Alajarin, M.; Lopez-Leonardo, C.; Llamas-Lorente, P.; Raja, R.; Bautista, D.; Orenes, R.-A. *Dalton Trans.* **2012**, *41*, 12259; (c) Artem'ev, A. V.; Kuimov, V. A.; Matveeva, E. A.; Bagryanskaya, I. Y.; Govdi, A. I.; Vasilevsky, S. F.; Rakhmanova, M. I.; Samultsev, D. O.; Gusarova, N. K.; Trofimov, B. A. *Inorg. Chem. Commun.* **2017**, *86*, 94.
182. Li, F.; Hor, T. S. A. *Adv. Synth. Catal.* **2008**, *350*, 2391.
183. (a) Karami, K.; Naeini, N. H. *Turk. J. Chem.* **2015**, *39*, 1199; (b) Bhaskar, R.; Sharma, A. K.; Yadav, M. K.; Singh, A. K. *Dalton Trans.* **2017**, *46*, 15235.
184. Lee, D.-H.; Kwon, Y.-J.; Jin, M.-J. *Adv. Synth. Catal.* **2011**, *353*, 3090.
185. Shu, W.; Buchwald, S. L. *Chem. Sci.* **2011**, *2*, 2321.
186. (a) Bakherad, M.; Bahramian, B.; Nasr-Isfahani, H.; Keivanloo, A.; Sang, G. *Chin. J. Chem.* **2009**, *27*, 353; (b) Bakherad, M.; Bahramian, B.; Nasr-Isfahani, H.; Keivanloo, A.; Doostmohammadi, N. *J. Heterocycl. Chem.* **2009**, *46*, 100.
187. Bakherad, M.; Amin, A. H.; Keivanloo, A.; Bahramian, B.; Raeissi, M. *Tetrahedron Lett.* **2010**, *51*, 5653.
188. Bakherad, M.; Keivanloo, A.; Bahramian, B.; Jajarmi, S. *Appl. Catal. A* **2010**, *390*, 135.
189. Bakherad, M.; Keivanloo, A.; Bahramian, B.; Hashemi, M. *Tetrahedron Lett.* **2009**, *50*, 1557.
190. Islam, S. M.; Mondal, S.; Roy, A. S.; Mondal, P.; Mobarak, M.; Hossain, D.; Pandit, P. *Transit. Met. Chem.* **2010**, *35*, 305.
191. Leonhardt, S. E. S.; Stolle, A.; Ondruschka, B.; Cravotto, G.; De Leo, C.; Jandt, K. D.; Keller, T. F. *Appl. Catal. A* **2010**, *379*, 30.
192. Gömann, A.; Deverell, J. A.; Munting, K. F.; Jones, R. C.; Rodemann, T.; Canty, A. J.; Smith, J. A.; Guijt, R. M. *Tetrahedron* **2009**, *65*, 1450.
193. Islam, M.; Mondal, P.; Roy, A. S.; Tuhina, K. *Synthesis* **2010**, 2399.
194. Phan, N. T. S.; Le, H. V. *J. Mol. Catal. A Chem.* **2011**, *334*, 130.
195. Lee, G.; Youn, H.-K.; Jin, M.-J.; Cheong, W.-J.; Ahn, W.-S. *Microporous Mesoporous Mater.* **2010**, *132*, 232.
196. Polshettiwar, V.; Hesemann, P.; Moreau, J. J. E. *Tetrahedron* **2007**, *63*, 6784.
197. Komura, K.; Nakamura, H.; Sugi, Y. *J. Mol. Catal. A Chem.* **2008**, *293*, 72.
198. Trilla, M.; Pleixats, R.; Man, M. W. C.; Bied, C.; Moreau, J. J. E. *Adv. Synth. Catal.* **2008**, *350*, 577.
199. (a) Alonso, D. A.; Nájera, C.; Pacheco, M. C. *Tetrahedron Lett.* **2012**, *43*, 9365; (b) Alonso, D. A.; Nájera, C.; Pacheco, M. C. *Adv. Synth. Catal.* **2003**, *345*, 1146.
200. Alonso, D. A.; Nájera, C.; Pacheco, M. C. *J. Org. Chem.* **2004**, *69*, 1615.
201. Gautam, P.; Bhanage, B. M. *ChemistrySelect* **2016**, *1*, 5463.
202. Buxaderas, E.; Alonso, D. A.; Nájera, C. *Eur. J. Org. Chem.* **2013**, 5864.
203. Buxaderas, E.; Alonso, D. A.; Nájera, C. *Adv. Synth. Catal.* **2014**, *356*, 3415.
204. Corma, A.; García, H.; Leyva, A. *Tetrahedron* **2005**, *61*, 9848.
205. Corma, A.; García, H.; Leyva, A. *J. Catal.* **2006**, *240*, 87.
206. Karami, K.; Najvani, S. D.; Naeini, N. H.; Hervés, P. *Chin. J. Catal.* **2015**, *36*, 1047.
207. Herrmann, W. A.; Elison, M.; Fischer, J.; Köcher, C.; Artus, G. R. J. *Angew. Chem. Int. Ed.* **1995**, *34*, 2371.
208. Lee, H. M.; Lu, C. Y.; Chen, C. Y.; Chen, W. L.; Lin, H. C.; Chiu, P. L.; Cheng, P. Y. *Tetrahedron* **2004**, *60*, 5807.
209. Taige, M. A.; Zeller, A.; Ahrens, S.; Goutal, S.; Herdtweck, E.; Strassner, T. *J. Org. Chem.* **2007**, *692*, 1519.

210. Allolio, C.; Strassner, T. *J. Org. Chem.* **2014**, *79*, 12096.
211. Peris, E.; Loch, J. A.; Mata, J.; Crabtree, R. H. *Chem. Commun.* **2001**, 201.
212. Tubaro, C.; Biffis, A.; Gonzato, C.; Zecca, M.; Basato, M. *J. Mol. Catal. A Chem.* **2006**, *248*, 93.
213. Jin, C.-M.; Twamley, B.; Shreeve, J. M. *Organometallics* **2005**, *24*, 3020.
214. Yagyu, T.; Tsubouchi, A.; Nishimura, Y. *J. Org. Chem.* **2014**, *767*, 1.
215. Ahrens, S.; Zeller, A.; Taige, M.; Strassner, T. *Organometallics* **2006**, *25*, 5409.
216. Huynh, H. V.; Jothibasu, R. *J. Org. Chem.* **2011**, *696*, 3369.
217. Clyne, D. S.; Jin, J.; Genest, E.; Gallucci, J. C.; Rajan Babu, T. V. *Org. Lett* **2000**, *2*, 1125.
218. Nielsen, D. J.; Cavell, K. J.; Skelton, B. W.; White, A. H. *Organometallics* **2006**, *25*, 4850.
219. Saito, S.; Yamaguchi, H.; Muto, H.; Makino, T. *Tetrahedron Lett.* **2007**, *48*, 7498.
220. Yeung, A. D.; Ng, P. S.; Huynh, H. V. *J. Org. Chem.* **2011**, *696*, 112.
221. Baker, M. V.; Skelton, B. W.; White, A. H.; Williams, C. C. *J. Chem. Soc. Dalton Trans.* **2001**, 111.
222. Baker, M. V.; Brown, D. H.; Simpson, P. V.; Skelton, B. W.; White, A. H.; Williams, C. C. *J. Org. Chem.* **2006**, *691*, 5845.
223. Lee, C.-S.; Lai, Y.-B.; Lin, W.-J.; Zhuang, R. R.; Hwang, W.-S. *J. Org. Chem.* **2013**, *724*, 235.
224. Liu, J.-Q.; Gou, X.-X.; Han, Y.-F. *Chem. Asian J.* **2018**, *13*, 2257.
225. Lee, C.-S.; Pal, S.; Yang, W.-S.; Hwang, W.-S.; Lin, I. J. B. *J. Mol. Catal. A Chem.* **2008**, *280*, 115.
226. Blakemore, J. D.; Chalkley, M. J.; Farnaby, J. H.; Guard, L. M.; Hazari, N.; Incarvito, C. D.; Luzik, E. D.; Suh, H. W. *Organometallics* **2011**, *30*, 1818.
227. Tulloch, A.; Danopoulos, A. A.; Tooze, R. P.; Cafferkey, S. M.; Kleinhenz, S.; Hursthouse, M. B. *Chem. Commun.* **2000**, 1247.
228. Wang, C.-Y.; Liu, Y.-H.; Peng, S.-M.; Chen, J.-T.; Liu, S. T. *J. Org. Chem.* **2007**, *692*, 3976.
229. McGuinness, D. S.; Cavell, K. J. *Organometallics* **2000**, *19*, 741.
230. Zhang, X.; Xi, Z.; Liu, A.; Chen, W. *Organometallics* **2008**, *27*, 4401.
231. Ye, J.; Chen, W.; Wang, D. *Dalton Trans.* **2008**, 4015.
232. Fiddy, S. G.; Evans, J.; Neisius, T.; Newton, M. A.; Tsoureas, N.; Tulloch, A. A. D.; Danbopoulos, A. A. *Chem. Eur. J.* **2007**, *13*, 3652.
233. Zhang, X.; Liu, A.; Chen, W. *Org. Lett.* **2008**, *10*, 3849.
234. Wang, R.; Twamley, B.; Shreeve, J. M. *J. Org. Chem.* **2006**, *71*, 426.
235. Wang, R.; Zeng, Z.; Twamley, B.; Piekarski, M. M.; Shreeve, J. M. *Eur. J. Org. Chem.* **2007**, 655.
236. Meyer, D.; Taige, M. A.; Zeller, A.; Hohlfeld, K.; Ahrens, S.; Strassner, T. *Organometallics* **2009**, *28*, 2142.
237. César, V.; Bellemin-Laponnaz, S.; Gade, L. H. *Organometallics* **2002**, *21*, 5204.
238. Ye, J.; Zhang, X.; Chen, W.; Shimada, S. *Organometallics* **2008**, *27*, 4166.
239. Sie, M.-H.; Hsieh, Y.-H.; Tsai, Y.-H.; Wu, J.-R.; Chen, S.-J.; Kumar, P. V.; Lii, J.-H.; Lee, H. M. *Organometallics* **2010**, *29*, 6473.
240. Chen, T.; Gao, J.; Shi, M. *Tetrahedron* **2006**, *62*, 6289.
241. Nonnenmacher, M.; Kunz, D.; Rominger, F.; Oeser, T. *J. Org Chem.* **2007**, *692*, 2554.
242. Liu, L.-j.; Wang, F.; Shi, M. *Eur. J. Inorg. Chem.* **2009**, 1723.
243. Hahn, F. E.; Jahnke, M. C.; Pape, T. *Organometallics* **2006**, *25*, 5927.
244. Jahnke, M. C.; Pape, T.; Hahn, F. E. *Eur. J. Inorg. Chem.* **2009**, 1960.
245. Huynh, H. V.; Yuan, D.; Han, Y. *Dalton Trans.* **2009**, 7262.
246. Sakaguchi, S.; Yoo, K. S.; O'Neill, J.; Lee, J. H.; Stewart, T.; Jung, K. W. *Angew. Chem. Int. Ed.* **2008**, *47*, 9326.

247. Baker, M. V.; Brown, D. H.; Simpson, P. V.; Skelton, B. W.; White, A. H. *Dalton Trans.* **2009**, 7294.
248. Stylianides, N.; danopoulos, A. A.; Pugh, D.; Hancock, F.; Zanotti-Gerosa, A. *Organometallics* **2007**, *26*, 5627.
249. Tubaro, C.; Biffis, A.; Basato, M.; Benetollo, F.; Cavell, K. J.; Ooi, L.-l. *Organometallics* **2005**, *24*, 4153.
250. Díez-Barra, E.; Guerra, J.; Rodríguez-Curiel, R. I.; Merino, S.; Tejeda, J. *J. Org. Chem.* **2002**, *660*, 50.
251. Hornillos, V.; Guerra, J.; de Cózar, A.; Prieto, P.; Merino, S.; Maestro, M. A.; Díez-Barra, E.; Tejeda, J. *Dalton Trans.* **2011**, *40*, 4095.
252. Lo, C. H.; Lee, H. M. *Organometallics* **2018**, *37*, 1150.
253. Eckhardt, M.; Fu, G. C. *J. Am. Chem. Soc.* **2003**, *125*, 13642.
254. Altenhoff, G.; Würtz, S.; Glorius, F. *Tetrahedron Lett.* **2006**, *47*, 2925.
255. Herrmann, W. A.; Reisinger, C.-P.; Spiegler, M. *J. Org. Chem.* **1998**, *557*, 93.
256. Liu, Q.-X.; Cai, K.-Q.; Zhao, Z.-X. *RSC Adv.* **2015**, *5*, 85568.
257. Marchenko, A.; Koidan, G.; Hurieva, A. N.; Vlasenko, Y.; Kostyuk, A.; Biffis, A. *Organometallics* **2016**, *35*, 762.
258. Kim, J.-H.; Lee, D.-H.; Jun, B.-H.; Lee, Y.-S. *Tetrahedron Lett.* **2007**, *48*, 7079.
259. Khajehzadeh, M.; Moghadam, M. *J. Org. Chem.* **2018**, *863*, 60.
260. Ghiaci, M.; Zarghani, M.; Khojastehnezhad, A.; Moeinpour, F. *RSC Adv.* **2014**, *4*, 15496.
261. Bahadorikhalili, S.; Ma'mani, L.; Mahdavi, H.; Shafiee, A. *RSC Adv.* **2015**, *5*, 71297.
262. B. Tamami, F. Farjadian, S. Ghasemiand H. Allahyari, New J. Chem. 37, 2013, 2011.
263. Selander, N.; Szabó, K. J. *Chem. Rev.* **2011**, *111*, 2048.
264. Ohff, M.; Ohff, A.; van der Boom, M. E.; Milstein, D. *J. Am. Chem. Soc.* **1997**, *119*, 11687.
265. Kiewel, K.; Liu, Y.; Bergbreiter, D. E.; Sulikowski, G. A. *Tetrahedron Lett.* **1999**, *40*, 8945.
266. (a) Morales-Morales, D.; Redón, R.; Yung, C.; Jensen, C. M. *Chem. Commun.* **2000**, 1619; (b) Morales-Morales, D.; Grause, C.; Kasaoka, K.; Redón, R.; Cramer, R. E.; Jensen, C. M. *Inorg. Chim. Acta* **2000**, *300–302*, 958.
267. Eberhard, M. R. *Org. Lett.* **2004**, *6*, 2125.
268. Naghipour, A.; Sabounchei, S. J.; Morales-Morales, D.; Hernández-Ortega, S.; Jensen, C. M. *J. Org. Chem.* **2004**, *689*, 2494.
269. Duncan, D.; Hope, E. G.; Singh, K.; Stuart, A. M. *Dalton Trans.* **2011**, *40*, 1998.
270. Miyazaki, F.; Yamaguchi, K.; Shibasaki, M. *Tetrahedron Lett.* **1999**, *40*, 7379.
271. Sjövall, S.; Wendt, O. F.; Andersson, C. *J. Chem. Soc. Dalton Trans.* **2002**, 1396.
272. Naghipour, A.; Sabounchei, S. J.; Morales-Morales, D.; Canseco-González, D.; Jensen, C. M. *Polyhedron* **2007**, *26*, 1445.
273. Solano-Prado, M. A.; Estudiante-Negrete, F.; Morales-Morales, D. *Polyhedron* **2010**, *29*, 592.
274. Bolliger, J. L.; Blacque, O.; Frech, C. M. *Chem. Eur. J.* **2008**, *14*, 7969.
275. (a) Beletskaya, I. P.; Chuchurjukin, A. V.; Dijkstra, H. P.; van Klink, G. P. M.; van Koten, G. *Tetrahedron Lett.* **2000**, *41*, 1075; (b) Beletskaya, I. P.; Chuchuryukin, A. V.; van Koten, G.; Dijkstra, H. P.; van Klink, G. P. M.; Kashin, A. N.; Nefedov, S. E.; Eremenko, I. L. *Russ. J. Org. Chem.* **2003**, *39*, 1268.
276. Schuster, E. M.; Botoshansky, M.; Gandelman, M. *Angew. Chem. Int. Ed.* **2008**, *47*, 4555.
277. Lee, H. M.; Zeng, J. Y.; Hu, C.-H.; Lee, M.-T. *Inorg. Chem.* **2004**, *43*, 6822.
278. Sommer, W. J.; Yu, K.; Sears, J. S.; Ji, Y.; Zheng, X.; Davis, R. J.; Sherrill, C. D.; Jones, C. W.; Weck, M. *Organometallics* **2005**, *24*, 4351.
279. Tamami, B.; Nezhad, M. M.; Ghasemi, S.; Farjadian, F. *J. Org. Chem.* **2013**, *743*, 10.

280. Whitcombe, N. J.; Hii, K. K. M.; Gibson, S. E. *Tetrahedron* **2001**, *57*, 7449.
281. Aydin, J.; Larsson, J. M.; Selander, N.; Szabó, K. J. *Org. Lett.* **2009**, *11*, 2852.
282. Blacque, O.; Frech, C. M. *Chem. Eur. J.* **2010**, *16*, 1521.
283. Huang, M.-H.; Liang, L.-C. *Organometallics* **2004**, *23*, 2813.
284. Jung, I. G.; Son, S. U.; Park, K. H.; Chung, K.-C.; Lee, J. W.; Chung, Y. K. *Organometallics* **2003**, *22*, 4715.
285. Díez-Barra, E.; Guerra, J.; Hornillos, V.; Merino, S.; Tejeda, J. *Organometallics* **2003**, *22*, 4610.
286. Takenaka, K.; Uozumi, Y. *Adv. Synth. Catal.* **2004**, *346*, 1693.
287. Churruca, F.; San Martin, R.; Tellitu, I.; Dominguez, E. *Synlett* **2005**, 3116.
288. Takenaka, K.; Minakawa, M.; Uozumi, Y. *J. Am. Chem. Soc.* **2005**, *127*, 12273.
289. Yoon, M. S.; Ryu, D.; Kim, J.; Ahn, K. H. *Organometallics* **2006**, *25*, 2409.
290. Luo, Q.-L.; Tan, J.-P.; Li, Z.-F.; Qin, Y.'.; Ma, L.; Xiao, D.-R. *Dalton Trans.* **2011**, *40*, 3601.
291. Hirano, Y.; Saiki, Y.; Taji, H.; Matsukawa, S.; Yamamoto, Y. *Heterocycles* **2008**, *76*, 1585.
292. Magill, A. M.; McGuinness, D. S.; Cavell, K. J.; Britovsek, G. J. P.; Gibson, V. C.; White, A. J. P.; Williams, D. J.; White, A. H.; Skelton, B. W. *J. Org. Chem.* **2001**, *617-618*, 546.
293. (a) Shang, Y.-J.; Wu, J.-W.; Fan, C.-L.; Hu, J.-S.; Lu, B.-Y. *J. Org. Chem.* **2008**, *693*, 2963; (b) Yu, S.-Y.; Zhang, Z.-Q.; Yu, Z.-Y.; Shang, Y.-J. *Appl. Organomet. Chem.* **2014**, *28*, 657.
294. Yang, L.; Zhang, X.; Mao, P.; Xiao, Y.; Bian, H.; Yuan, J.; Mai, W.; Qu, L. *RSC Adv.* **2015**, *5*, 25723.
295. Tulloch, A. A. D.; Danopoulos, A. A.; Tizzard, G. J.; Coles, S. J.; Hursthouse, M. B.; Hay-Motherwell, R. S.; Motherwell, W. B. *Chem. Commun.* **2001**, 1270.
296. Loch, J. A.; Albrecht, M.; Peris, E.; Mata, J.; Faller, J. W.; Crabtree, R. H. *Organometallics* **2002**, *21*, 700.
297. Nielsen, D. J.; Cavell, K. J.; Skelton, B. W.; White, A. H. *Inorg. Chim. Acta* **2002**, *327*, 116.
298. Wang, H.; Zhang, B.; Yan, X.; Guo, S. *Dalton Trans.* **2018**, *47*, 528.
299. Gründemann, S.; Albrecht, M.; Loch, J. A.; Faller, J. W.; Crabtree, R. H. *Organometallics* **2001**, *20*, 5485.
300. Nielsen, D. J.; Cavell, K. J.; Skelton, B. W.; White, A. H. *Inorg. Chim. Acta* **2006**, *359*, 1855.
301. Houghton, J.; Dyson, G.; Douthwaite, R. E.; Whitwood, A. C.; Kariuki, B. M. *Dalton Trans.* **2007**, 3065.
302. Artero, V.; Fontecave, M. C. *Chem. Rev.* **2005**, *249*, 1518.
303. Suijkerbuijk, B. M. J. M.; Martínez, S. D. H.; van Koten, G.; Gebbink, R. J. M. K. *Organometallics* **2008**, *27*, 534.
304. Bergbreiter, D. E.; Osburn, P. L.; Liu, Y.-S. *J. Am. Chem. Soc.* **1999**, *121*, 9531.
305. Hossain, M. A.; Lucarini, S.; Powell, D.; Bowman-James, K. *Inorg. Chem.* **2004**, *43*, 7275.
306. Begum, R. A.; Powell, D.; Bowman-James, K. *Inorg. Chem.* **2006**, *45*, 964.
307. da Costa, R. C.; Jurisch, M.; Gladysz, J. A. *Inorg. Chim. Acta* **2008**, *361*, 3205.
308. Gómez-Benítez, V.; Hernández-Ortega, S.; Toscano, R. A.; Morales-Morales, D. *Inorg. Chim. Acta* **2007**, *360*, 2128.
309. (a) Bergbreiter, D. E.; Osburn, P. L.; Frels, J. D. *Adv. Synth. Catal.* **2005**, *347*, 172; (b) Bergbreiter, D. E.; Osburn, P. L.; Wilson, A.; Sink, E. M. *J. Am. Chem. Soc.* **2000**, *122*, 9058.
310. Yu, K.; Sommer, W.; Weck, M.; Jones, C. W. *J. Catal.* **2004**, *226*, 101.

311. Sigeev, A. S.; Peregudov, A. S.; Cheprakov, A. V.; Beletskaya, I. P. *Adv. Synth. Catal.* **2015**, *357*, 417.
312. Yao, Q.; Kinney, E. P.; Zheng, C. *Org. Lett.* **2004**, *6*, 2997.
313. Das, D.; Rao, G. K.; Singh, A. K. *Organometallics* **2009**, *28*, 6054.
314. Das, D.; Singh, P.; Singh, M.; Singh, A. K. *Dalton Trans.* **2010**, *39*, 10876.
315. Bröring, M.; Kleeberg, C.; Köhler, S. *Inorg. Chem.* **2008**, *47*, 6404.
316. Roy, S.; Pramanik, S.; Ghorui, T.; Pramanik, K. *RSC Adv.* **2015**, *5*, 22544.
317. Mahmood, Q.; Yue, E.; Zhang, W.; Solan, G. A.; Liang, T.; Sun, W.-H. *Org. Chem. Front.* **2016**, *3*, 1668.
318. Hamasaka, G.; Ichii, S.; Uozumi, Y. *Adv. Synth. Catal.* **2018**, *360*, 1833.
319. Chen, C.-T.; Chan, Y.-S.; Tzeng, Y.-R.; Chen, M.-T. *Dalton Trans.* **2004**, 2691.
320. For a review for pincer ligands, (a) A. Kumar, G.K. Rao and A.K. Singh, RSC Adv. 2, 2012, 12552; (b) Balakrishna, M. S. *Polyhedron* **2018**, *143*, 2
321. Hahn, F. E.; Jahnke, M. C.; Pape, T. *Organometallics* **2007**, *26*, 150.
322. Ghavale, N.; Manjare, S. T.; Singh, H. B.; Butcher, R. J. *Dalton Trans.* **2015**, *44*, 11893.
323. Yagyu, T.; Oya, S.; Maeda, M.; Jitsukawa, K. *Chem. Lett.* **2006**, *35*, 154.
324. Rosa, G. R.; Rosa, D. S. *RSC Adv.* **2012**, *2*, 5080.
325. Consorti, C. S.; Ebeling, G.; Flores, F. R.; Rominger, f.; Dupont, J. *Adv. Synth. Catal.* **2004**, *346*, 617.
326. Singh, M. P.; Saleem, F.; Rao, G. K.; Kumar, R. S.; Joshi, H.; Singh, A. K. *Dalton Trans.* **2016**, *45*, 6718.
327. Ramírez-Rave, S.; Morales-Morales, D.; Grévy, J.-M. *Inorg. Chim. Acta* **2017**, *462*, 249.
328. Chen, M.-T.; Huang, C.-A.; chen, C.-T. *Eur. J. Inorg. Chem.* **2006**, 4642.
329. Debono, N.; Iglesias, M.; Sánchez, F. *Adv. Synth. Catal.* **2007**, *349*, 2470.
330. Vicente, J.; Arcas, A.; Juliández, F.; Bautista, D. *Angew. Chem. Int. Ed.* **2011**, *50*, 6896.
331. Kovala-Demertzi, D.; Yadav, P. N.; Demertzis, M. A.; Jasiski, J. P.; Andreadaki, F. J.; Kostas, I. D. *Tetrahedron Lett.* **2004**, *45*, 2923.
332. Eberhard, M. R.; Wang, Z.; Jensen, C. M. *Chem. Commun.* **2002**, 818.
333. Deschamps, B.; Le Goff, X.; Ricard, L.; Le Floch, P. *Heteroat. Chem.* **2007**, *18*, 363.
334. Bolliger, J. L.; French, C. M. *Adv. Synth. Catal.* **2009**, *351*, 891.
335. Churruca, F.; San Martin, R.; Tellitu, I.; Domínguez, E. *Synlett* **2005**, 3116.
336. Zhang, B.-S.; Wang, C.; Gong, J.-F.; Song, M.-P. *J. Org. Chem.* **2009**, *694*, 2555.
337. Kumar, S.; Saleem, F.; Mishra, M. K.; Singh, A. K. *New J. Chem.* **2017**, *41*, 2745.
338. Mohammadi, E.; Movassagh, B. *New J. Chem.* **2018**, *42*, 11471.
339. Inés, B.; San Martin, R.; Churruca, F.; Domínguez, E.; Urtiaga, M. K.; Arriortua, M. I. *Organometallics* **2008**, *27*, 2833.
340. Rosa, D. S.; Antelo, F.; Lopes, T. J.; Quim, N. F.de Mouraand G.R.Rosa *Quim. Nova* **2015**, *38*, 605.
341. (a) Feng, Z.; Yu, S.; Shang, Y. *Appl. Organomet. Chem.* **2008**, *22*, 577; (b) Yu, S.; Wu, J.; He, X.; Shang, Y. *Appl. Organometal. Chem.* **2018**, *32*, e4156.
342. Gu, S.; Chen, W. *Organometallics* **2009**, *28*, 909.
343. Zhang, J.-H.; Li, P.; Hu, W.-P.; Wang, H.-X. *Polyhedron* **2015**, *96*, 107.
344. Mas-Marzá, E.; Segarra, A. M.; Claver, C.; Perisb, E.; Fernandez, E. *Tetrahedron Lett.* **2003**, *44*, 6595.
345. Huynh, H. V.; Lee, C.-S. *Dalton Trans.* **2013**, *42*, 6803.
346. Worm-Leonhard, K.; Meldal, M. *Eur. J. Org. Chem.* **2008**, 5244.
347. Rocaboy, C.; Gladysz, J. A. *New J. Chem.* **2003**, *27*, 39.
348. Gruber, A. S.; Zim, d.; Ebeling, G.; Monteiro, A. L.; Dupont, J. *Org. Lett.* **2000**, *2*, 1287.
349. Xiong, Z.; Wang, N.; Dai, M.; Li, A.; Chen, J.; Yang, Z. *Org. Lett.* **2004**, *6*, 3337.

350. Mane, P. A.; Dey, S.; Vivekananda, K. V. *Tetrahedron Lett.* **2017**, *58*, 25.
351. (a) Kumar, A.; Agarwal, M.; Singh, A. K. *Polyhedron* **2008**, *27*, 485; (b) Kumar, A.; Agarwal, M.; Singh, A. K. *J. Org. Chem.* **2008**, *693*, 3533; (c) Kumar, A.; Agarwal, M.; Singh, A. K.; Butcher, R. J. *Inorg. Chim. Acta* **2009**, *362*, 3208.
352. Yao, Q.; Zabawa, M.; Woo, J.; Zheng, C. *J. Am. Chem. Soc.* **2007**, *129*, 3088.
353. Tamang, S. R.; Hoefelmeyer, J. D. *Molecules* **2015**, *20*, 12979.
354. (a) Friedlein, F. K.; Hampel, F.; Gladysz, J. A. *Organometallics* **2005**, *24*, 4103; (b) Friedlein, F. K.; Kromm, K.; Hampel, F.; Gladysz, J. A. *Chem. Eur. J.* **2006**, *12*, 5267.
355. Lee, D.-H.; Qiu, H.; Cho, M.-H.; Lee, I.-M.; Jin, M.-J. *Synlett* **2008**, 1657.
356. Jin, M.-J.; Lee, D.-H. *Angew. Chem. Int. Ed.* **2010**, *49*, 1119.
357. Rau, S.; Lamm, K.; Görls, H.; Schöffel, J.; Walther, D. *J. Org. Chem.* **2004**, *689*, 3582.
358. Nandurkar, N. S.; Bhanage, B. M. *Tetrahedron* **2008**, *64*, 3655.
359. Kumar, S.; Saleem, F.; Singh, A. K. *Dalton Trans.* **2016**, *45*, 11445.
360. Zhang, G.; Zhang, W.; Luan, Y.; Han, X.; Ding, C. *Chin. J. Chem.* **2015**, *33*, 705.
361. Islam, S. M.; Mondal, P.; Roy, A. S.; Mondal, S.; Hossain, D. *Tetrahedron Lett.* **2010**, *51*, 2067.
362. Thogiti, S.; Parvathaneni, S. P.; Keesara, S. *J. Org. Chem.* **2016**, *822*, 165.

CHAPTER 3

Palladacycles as Efficient Precatalysts for Negishi and Buchwald-Hartwig Amination Reactions

José L. Serrano*, Tejpalsingh R. Girase†
*Department of Mining, Geological and Cartographic Engineering, Inorganic Chemistry Section, Technical University of Cartagena, Regional Campus of International Excellence "Campus Mare Nostrum", Cartagena, Spain
†Department of Chemistry, Institute of Chemical Technology, Mumbai, India

Contents

1. INTRODUCTION

The 2010 Nobel Prize in Chemistry recognized the individual work of professors Negishi, Heck, and Suzuki, but also the tremendous impact of cross-coupling reactions in both sound industrial applications and daily academic research *(1)*. Thus, metal-catalyzed cross-coupling reactions have played a pivotal role in synthetic chemistry over the last 30 years, changing the way synthetic organic and organometallic chemists approach the formation

Palladacycles
https://doi.org/10.1016/B978-0-12-815505-9.00004-4
© 2019 Elsevier Inc.
All rights reserved.

of carbon-carbon and carbon-heteroatom bonds. In this chapter we will discuss the contribution of palladacycles to the development of two of these specific reactions named after their discoverers: Negishi coupling and Buchwald-Hartwig amination. Since the main impact and applications of palladacycles in Negishi coupling correspond to the four generations (palladacycles G1–G4) developed by Buchwald and coworkers, we will first introduce these palladacycles in the frame of Buchwald-Hartwig amination. For this reason, we will start the chapter addressing this reaction, and next we will present the reports on palladacycles used as precatalysts in the Negishi coupling.

1.1 Buchwald-Hartwig Amination: A Brief Historic Perspective

The chemistry of nitrogen compounds has particularly benefited from this development of cross-coupling methods, with the advent of new mechanistically related routes to create C—N bonds in arylamines. This is so that the early comprehensive monograph edited by Diederich and Stang in 1998 *(2a)*, widely recognized now as a key contribution to the field, did not include palladium catalyzed aromatic C—N bond forming reactions but dedicated to this subject a reference chapter by Jiang and Buchwald in its second enlarged and updated 2004 edition *(2b)*. Nowadays this growing interest is maintained, as reflected by the number of recent reviews *(3)* and articles dedicated to Buchwald-Hartwig aryl amination (Scheme 3.1) that will be collected here.

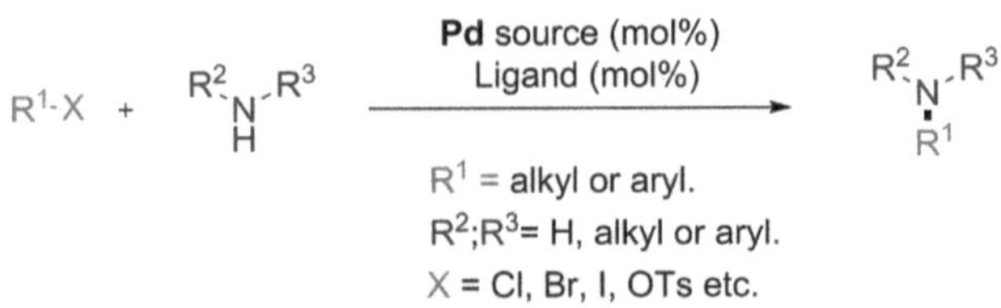

Scheme 3.1 General scheme for Buchwald-Hartwig amination.

It has been argued that the unstoppable growth in the field of C—N cross-coupling has been fostered by the absence of straightforward, flexible ways to create new C—N bonds among the previous synthetic methodologies, and a compelling demand of nitrogen-containing biologically active compounds for the pharmaceutical or agrochemical industries, among others *(4)*, thus combining again the interests of industry and academia. Some relevant processes for the preparation of arylamines were available prior to the breakthrough in the field of Buchwald and Hartwig *(5)* including acidic nitration followed by a reduction that was not always easy *(6)*, nucleophilic aromatic substitution methods introduced by Bunnett *(7)*, aryne chemistry *(8)*, and Cu-assisted Ullman methods *(9)*, although a step forward toward a more general method was desirable. The first report of a Pd-catalyzed route for the formation of anilines is acknowledged to Migita et al. in 1983, which used an stoichiometric amount of dialkyltin-amides and palladium tris(*o*-tolyl)phosphine complexes, in the course of an investigation on the application of organotin compound in organic chemistry *(10)*. This method had limited application but started a 12-year lapse of exciting developments before a tin-free palladium-catalyzed amination was reported simultaneously by the groups of Hartwig and Buchwald in 1995 *(5c, 5d)*. As early as 1996, Louie and Hartwig *(11)* first reported Herrmann-Beller palladacycle **1**, trans-di(μ-acetato)-bis[*o*-(di-*o*-tolylphosphino)benzyl]dipalladium(II), that contains C∘P orthometallated P(*o*-tolyl)$_3$, as useful preformed precatalyst in the coupling of 4′-bromobenzophenone and N-methylaniline, discussing its mechanistic relation with the Pd(0) species [Pd{P(*o*-tolyl$_3$)}$_2$] in the amination of aryl halides. Since then, several families of palladacycles, which will be reviewed in this chapter, have occasionally shown relevant roles in C—N bond forming reactions.

The early stages in the evolution of Pd-catalyzed C—N bond forming reactions have been previously reviewed *(2b, 12)* and will not be addressed in detail here. A complete and comparative overview, including the current improved Cu-catalyzed approach to C—N bond forming protocols, can be found in the review by Beletskaya and Cheprakov *(4a)*, while a more pharmaceutical perspective of Cu/Pd as "complementary competitors" is presented in the dedicated book chapter by Hesp and Genovino *(13)*. Both reviews conclude that Pd- and Cu-catalyzed approaches differ among other aspects in the role played by ligands, since Pd-catalyzed C—N cross-coupling is a clear example of metal-catalyzed process controlled by ligand design. In fact, the employment of several generations of ancillary ligands **L** and the understanding of their key function in the catalytic cycle (Fig. 3.1) defining catalytic activity, selectivity, and scope have set the pace of the advances in Pd-Catalyzed C—N cross-coupling *(3b, 4a, 6b, 13)*.

Fig. 3.2 displays a selection of key ligands relevant in the development of this field. The chronological advent of successive generations of ligands has been commonly used as a unifying thread to display the developments in Buchwald-Hartwig reactions. This will also be a guide to organize this chapter, emphasizing the incidence and contribution that palladacycles have had

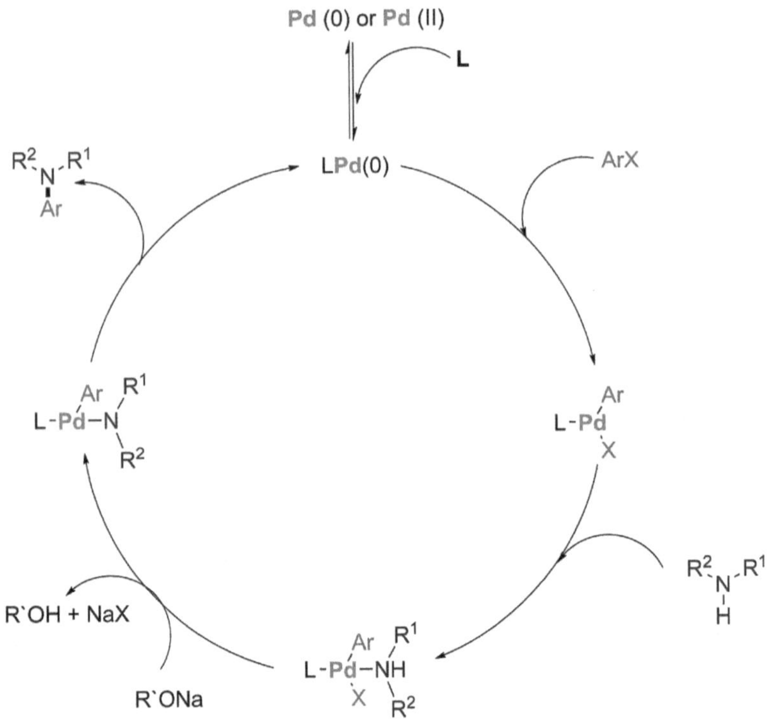

Fig. 3.1 General catalytic cycle for Buchwald Hartwig amination reaction.

in each stage. Before that descriptive section, a brief summary with general features of the Buchwald-Hartwig reaction, together with a discussion on the palladium sources available and some general considerations on in situ *vs* preformed catalysts, also applicable to Negishi coupling, will be presented.

1.2 General Features

1.2.1 Mechanism

Considerable knowledge has been gained about the mechanism of the Buchwald-Hartwig reaction *(2b, 14)* as a result of decades of both theoretical *(14b, 15)* and experimental studies *(16)* that still promote advances in this field today *(4b, 17)*. A good understanding of the four-step simplified catalytic cycle displayed in Fig. 3.1 (aryl halide oxidative addition, amine binding, dehydrohalogenation, and reductive elimination of product), has fostered the broad synthetic utility of the reaction and allowed a mechanism-inspired development of engineered ligands that should possess several crucial properties *(2b, 4a, 13)*. Thus, (i) bulky ligands are desirable to

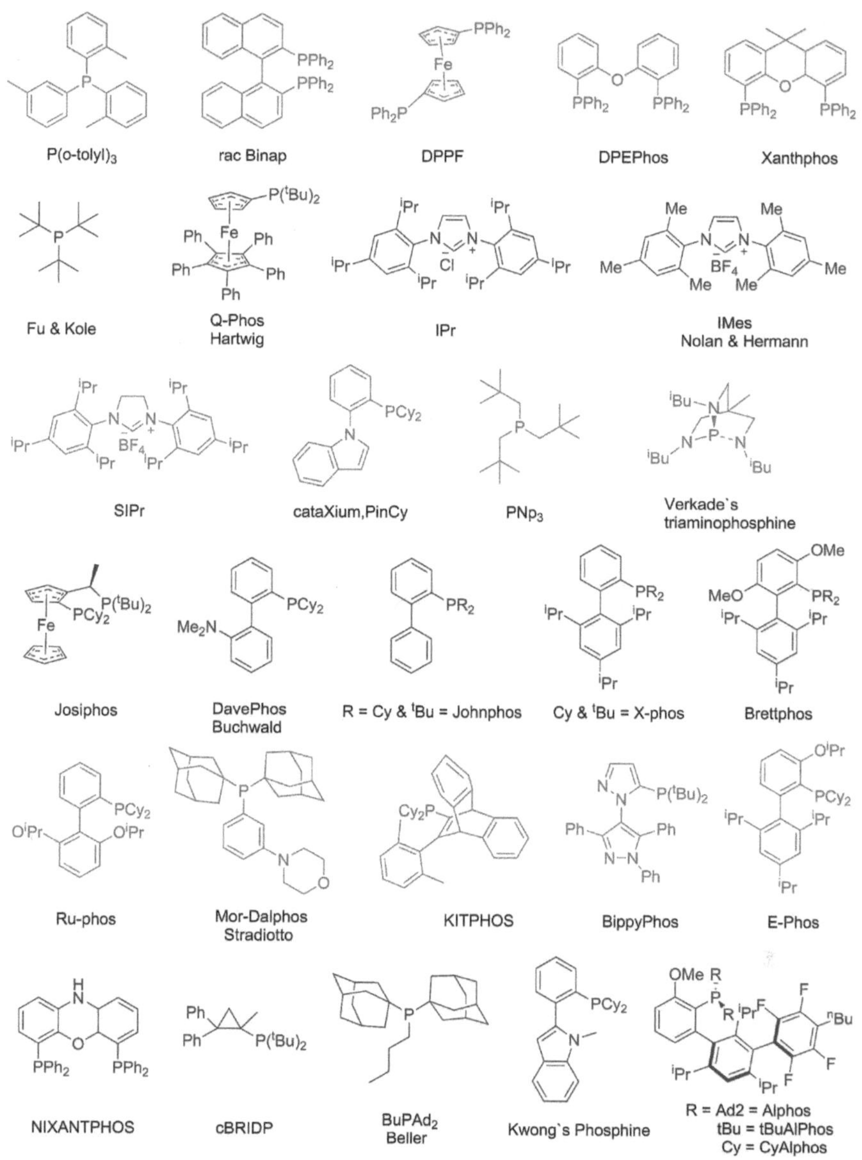

Fig. 3.2 Key ligands in Pd-catalyzed Buchwald-Hartwig amination-reaction.

stabilize monoligated Pd(0) species LPd against L_2Pd; (ii) electron-rich ligands are needed to activate the Pd center toward oxidative addition of the (hetero)aryl (pseudo)halide; (iii) either hemilability in bidentated ligands or a balance between donor/steric ability in monodentated ligands would be preferred to control the coordination environment; and (iv) steric strain around the coordination sphere of the Pd center would help to promote the product-forming step and the regeneration of mono-ligated Pd(0)-catalyst.

1.2.2 Role of Ancillary Ligands

These four requirements have directed and compassed the evolution of successive generations of ligands displayed in Fig. 3.2. From pioneer reports using the slightly bulky phosphine P(*o*-tolyl)$_3$ *(5)* through the second-generation of symmetrical chelating bisphosphines, such as BINAP *(18)*, DPPF *(19)* and later those van Leeuwen's ligands *(20)* with very-large bite angle DPEPPhos *(21)* and XantPhos *(22)*. They then gave pace to a third generation, with the sterically demanding and electron-rich phosphine trialkylphosphine P(t-Bu)$_3$ *(23)*, and several N-heterocyclic carbenes (NHCs) as prototype ligands. Notable contributions exploiting this design principle are the ligands QPhos *(24)*, IPr/SIPr *(25)*, CataCXium *(26)*, neopentylphosphines PNp$_3$ *(17c)*, and Verkade's triaminophosphines *(27)*. The most recent and important advance came with the advent of the four generation that includes potentially bidentated ligands with at least one strongly bonded phosphine and a second on/off position switched by an hemilabile ligand and/or steric effects. JosiPhos bidentate ligands display two different and strong phosphine binding sites and large bite angles *(28)*, but most examples to date contain besides the phosphine a more labile oxygen (MorDalPhos), a nitrogen (DavePhos) or an electron-rich benzene ring serving as a labile η2 ligand (JohnPhos) (Fig. 3.3) *(4a, 13)*.

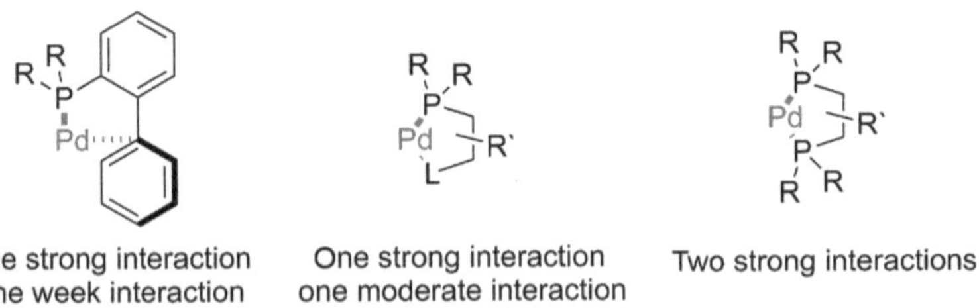

Fig. 3.3 Most common ligand design used in Pd-catalyzed C-N cross-coupling.

Among them, the best known and developed are the air/moisture stable (bulky)dialky-biarylphosphines, which perform very well with mild base and low temperature, allowing a broad substrate scope. First reports used DavePhos *(29)* and JohnPhos *(30)*, soon improved by adding bulky groups in the second phenyl ring such as in XPhos *(31)*, BrettPhos *(32)* or RuPhos

(33), among others. The presence of such groups in ortho-position effectively avoids the undesirable intramolecular ortho palladation of the ligands, which reduces the catalyst's activity *(34)*. A catalytically inactive palladacycle has also been reported using the $Pd_2(dba)_3/PNp_3$ catalyst system *(17c)*.

Inspired by Buchwald's biarylphosphine design concept but built in different scaffolds are the newer ligands shown in Fig. 3.2, such as MorDalPhos *(35)*, caged structures as KITPHOS *(36)*, or BippyPhos *(37)*. Also, the recently described EPhos in which the methoxy substituent at the C3-position in BrettPhos is changed by an $-O^{i}Pr$ group, *(17b)* and NIXANTPHOS, which in addition to a similar large bite angle to XantPhos, *(6b)* contains a relatively acidic phenoxazine N—H which overall confers extraordinary performance with inactivated aryl chlorides *(38)*. The search of efficient protocols for the synthesis of biphenyl based phosphine ligands is therefore a subject of current interest *(39)*.

Although the choice of ligand is a key factor in C—N cross-coupling reactions, the importance of other parameters, such as base or solvent, cannot be overestimated and are the object of continuous study *(2b, 3d, 4b, 14b, 17a, 40)*.

1.2.3 Palladium Source

With respect to the palladium source in C—N bond forming processes, which ideally should produce in situ a highly active 12-electron LPd(0) species under mild reaction conditions once mixed with the ligand of choice, most of the catalyst systems described use $Pd_2(dba)_3$ or $Pd(OAc)_2$. Both have their own advantages and drawbacks. Reduction to Pd(0) is needed when using $Pd(OAc)_2$, an unnecessary step when using $Pd_2(dba)_3$, which however contains noninnocent dibenzylideneacetone that has been reported to compete with **L** ligands then diminishing the activity of the catalyst *(2b, 40, 41)*. Differences have also been reported in obtaining pure reproducible samples of these Pd precursors depending on the synthetic method and the supplier *(42)*.

In this sense, a first approach to the role of palladacycles in the development of Buchwald-Hartwig amination reactions would place them among the most employed "one component" precatalysts, which share with other preformed complexes the advantages to yield the active Pd(0) species in a predetermined L:Pd ratio, to reduce the metal loading and to be appropriate for derivatizing via a high-throughput screening approach *(2b, 40, 41e)*. It has also sometimes been suggested that palladacycles did not meet their early promise and merely act as a reservoir of Pd(II), which requires reduction to Pd(0) to enter into the catalytic cycle *(3d, 43)*.

From a wider perspective which includes other cross-coupling reactions and not only palladacycles, the relevance of preformed compared to in situ catalysts has been highlighted in terms of improved selectivity and activity, simplified workup and general safety, and economy advantages in an industrial setting *(1b, 41a, 44)*. Some specific advantages of palladacycles *vs* in situ catalysts in amination reactions, such as avoiding the use of excess ligand to generate active LPd(0) species, have also been reported *(3a, 40, 41e, 45)* and will be pointed out in corresponding examples.

2. TYPES OF PALLADACYCLES USED IN BUCHWALD-HARTWIG C—N BOND FORMING REACTIONS

The relevant role of palladacycles in this field as Pd(II) sources of active LPd (0) species and the important amount and quality of the results reported since 1996, which have accompanied the aforementioned development of ancillary ligands, have motivated interesting classifications, which provide a clarifying global picture *(3a, 44a, 44b, 46)* of the palladacycles available for the preparation of aryl amines. Thus, a first rough classification divides palladacycles into two categories: (i) those that incorporate through carbopalladation as C∘P backbone the **L** ligand that later in the cycle will be found in **L**Pd (0); and (ii) those in which the palladacycle is formed via cyclopalladation of a sacrificial C∘N backbone that should dissociate without interference during the catalytic cycle. All palladacycles in this category incorporate an appropriate **L** ancillary ligand. Fig. 3.4 displays examples of these two

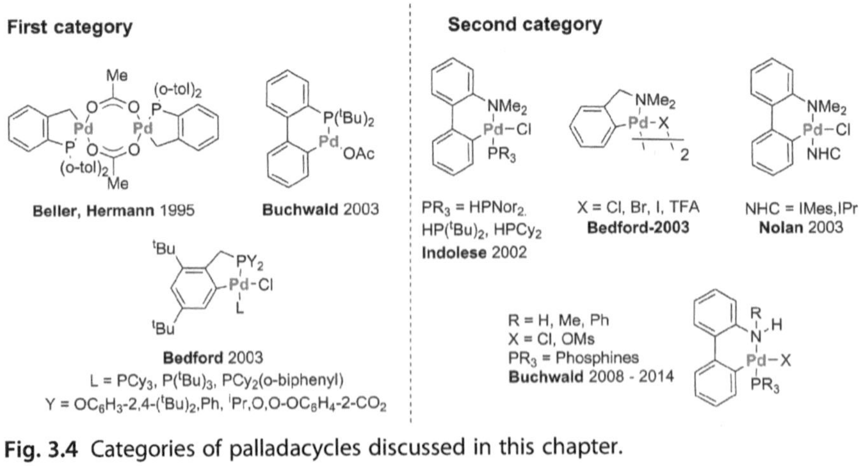

Fig. 3.4 Categories of palladacycles discussed in this chapter.

categories following Calacot's classification. Mixed phosphite-phosphine and phosphinite-phosphine palladacyclic complexes by Bedford and Blake would probably fit both categories, with a suggested particular mechanism for the formation of Pd(0) *(44d, 45c)*. General modes of activation of palladacycle precatalysts to yield palladium(0) species have been described since the initial studies by Louie and Hartwig on Herrmann-Beller palladacycle *(11, 44d)* to the modern Buchwald G1–G4 family *(46)*.

The archetypal examples of the first category are the Herrmann-Beller precatalyst *(47)* or Buchwald precatalyst with orthopalladated JohnPhos *(45d)*. Activation/reduction to LPd(0) is believed to happen by the action of an amine and a base *(11, 32, 44a)*. Buchwald *(46)*, Indolese *(48)*, Nolan *(49)*, and Bedford precatalysts *(50)* shown in Fig. 3.4 would be classified into the second category.

In order to address an exhaustive review of palladacycles' contribution to Buchwald-Hartwig amination reactions, taking into account the timeline of the different families that emerged and their relation with the development of the four generations of **L** ancillary ligands that have been explained earlier, we will discuss the following three groups of palladacycles: (i) Herrmann-Beller palladacycle and related that incorporate an ortho-metallated phosphine/P-donor atom; (ii) several C∘N Buchwald generations of catalysts commonly labeled as G1–G4, in some way developed to host the successive generations of ancillary **L** ligands, and all those other incorporating a C∘N backbone; (iii) Nolan palladacycles and related that incorporate NHC ligands instead of phosphines. Representative palladacyles of each of the three groups are drawn in Fig. 3.5.

2.1 Group 1: Herrmann-Beller Palladacycle and Others Incorporating a C^P Backbone

Nowadays it is widely recognized that the introduction of the Herrmann-Beller palladacycle in 1995 *(47a, 51)*, as catalyst in the Heck reaction, dramatically changed the perception of this family of compounds. Herrmann and Beller first synthesized his yellow palladacycle with 90%–95% yield by cyclometallation of $P(o\text{-tolyl})_3$ with $Pd(OAc)_2$ in toluene at ambient temperature for 16 h. This palladacycle proved to be an effective catalyst for the Heck reaction, with high TONs for several examples, and more importantly, it turned out to be more effective than the in situ generated catalyst. Indeed, trans-di(μ-acetato)-bis[*o*-(di-*o*-tolylphosphino)benzyl]dipalladium (II) soon became commercially available as a robust and effective catalyst for the olefination of aryl chlorides, and this unprecedented activity

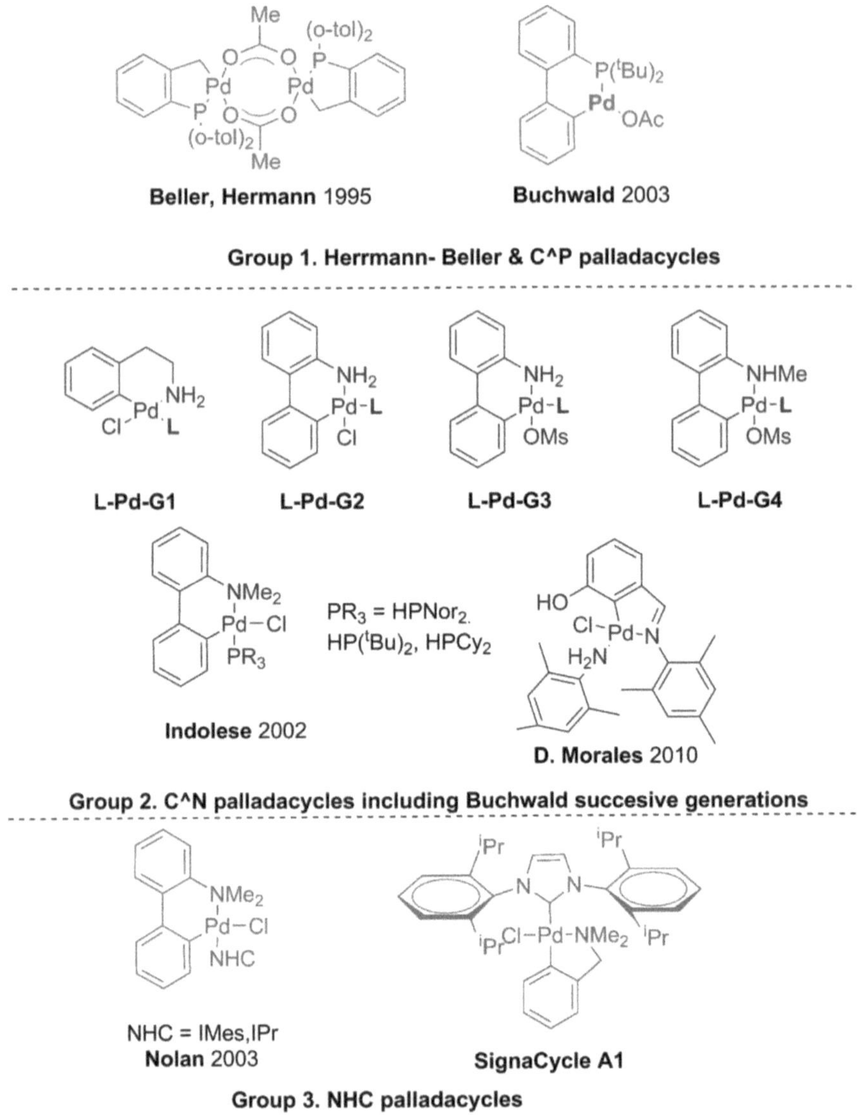

Fig. 3.5 Classification of palladacycles used in this chapter.

motivated a continued interest in cyclopalladated compounds as potential catalysts for cross-coupling chemistry *(43, 44a, 44d, 46)*.

Regarding the Buchwald-Hartwig amination reaction, again the Herrmann-Beller palladacycle produced pioneering studies and serves us as representative of this group that incorporates a CoP backbone. As mentioned in the introduction, the C—N bond forming reaction was initially shown to be catalyzed by Pd complexes with simple triarylphosphines, the usual ligands employed in Pd catalysis at that moment. Modestly bulky tris(-*o*-tolyl)phosphine had a crucial role as accompanying ligand in those pioneer studies, and its inclusion as cyclometallated ligand resulted in a natural move. It has indeed been described as a carbopalladation of the ligand itself, which will be involved later in the LPd(0) active species *(44a)*. Activation of these precatalysts required exogenous amine/base additives to produce the catalytically active species LPd(0).

Louie and Hartwig *(11)* explored the Herrmann-Beller palladacycle for the first time in 1996 as catalyst in the C—N cross–coupling reaction of 4-bromobenzophenone with *N*-methylaniline. This article is also well known because it advanced the mechanism of the formation of Pd(0) species from palladacycle **1** in Stille cross-coupling and amination reactions *(44d)*. A common Pd(0) catalyst was suggested since 1, 2, and 3 species displayed in Scheme 3.2 showed identical activities in this reaction.

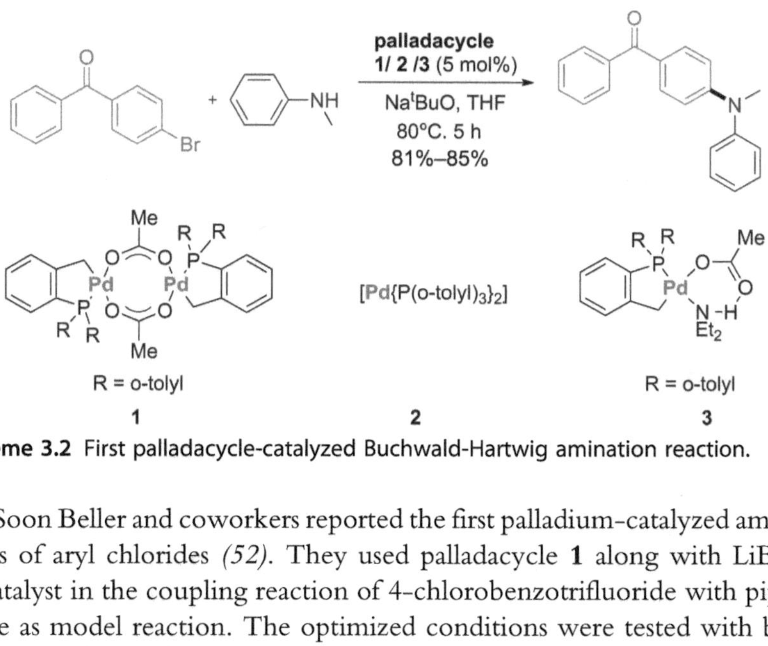

Scheme 3.2 First palladacycle-catalyzed Buchwald-Hartwig amination reaction.

Soon Beller and coworkers reported the first palladium-catalyzed aminations of aryl chlorides *(52)*. They used palladacycle **1** along with LiBr as cocatalyst in the coupling reaction of 4-chlorobenzotrifluoride with piperidine as model reaction. The optimized conditions were tested with both primary and secondary amines and aryl chlorides with electron-withdrawing substituents.

In 2003, Buchwald and Zim *(45d)* developed the synthesis of a one-component palladacyclic precatalyst, incorporating the biaryl phosphine ligand JohnPhos as CoP-backbone. Stirring in toluene JohnPhos with palladium acetate at RT for 16 h they obtained in good yield the corresponding palladacycle shown in Scheme 3.3. This precatalyst proved to be air, moisture, and thermally stable, so it avoided the use of a glovebox, and was an early example of the advantages of preformed *vs* in situ catalysts in terms of reducing time, Pd loading, and the excess of ligand. It showed an excellent reactivity in the amination of aryl chlorides at low catalyst loading, with moderate to excellent yields working from room temperature to 120° C. Several examples of coupling primary and secondary amines, and the use of different protocols with three different bases were reported.

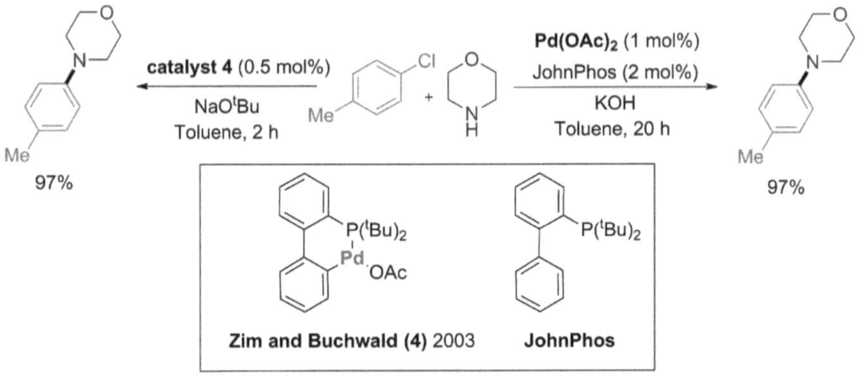

Scheme 3.3 Comparative study of Buchwald precatalyst *vs* palladium acetate in amination reaction of *p*-chlorotoluene and morpholine.

In 2003, Bedford and coworkers *(45c)* published their results on highly active mixed phosphite-phosphine and phosphinite-phosphine palladacyclic complexes for amination reaction of aryl chlorides. The best results were obtained with tri-tert-butylphosphine mononuclear derivatives of phosphite-based palladacycles, which worked well in the amination of a wide range of aryl chlorides and amine substrates at low catalyst loading (even 0.1 mol%). The catalysts formed in situ from dinuclear complexes **5** and tricyclohexylphosphine (PCy_3) or tri-tert-butylphosphine, as displayed in Scheme 3.4.

Palladacycles forming a strained four-membered ring described by Vilar and coworkers *(53)* have sometimes been reported as precatalyst in the Buchwald-Hartwig amination reaction *(44a, 44b)*, although these palladacycles were originally reported as "decomposition" products from the biaryl-bridged Pd(I) dimers, the active catalysts shown in Fig. 3.6.

Besides their unusual structures, these air, moisture, and thermally stable complexes generate catalytically active species for the amination of aryl chlorides and bromides at room temperature. Dinuclear Pd(I) complexes have

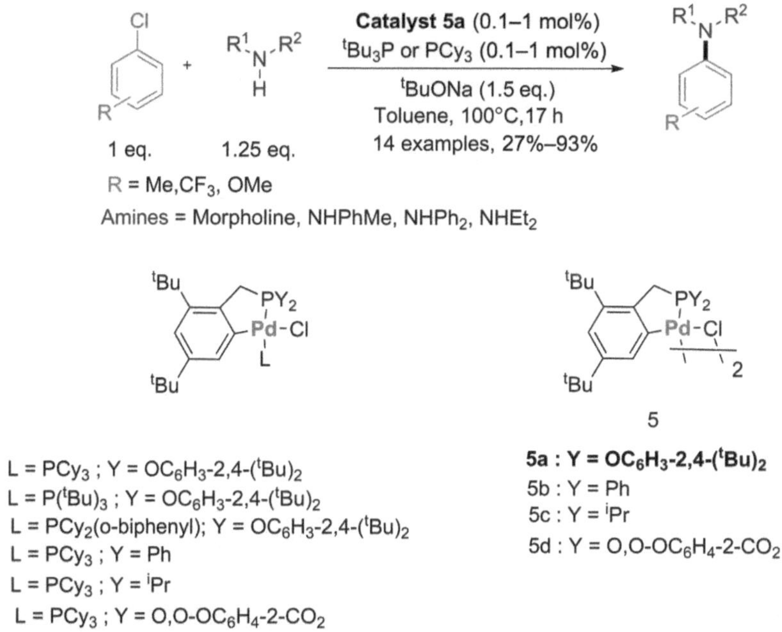

Scheme 3.4 Amination of aryl chlorides with Bedford's phosphite-phosphine palladacycles.

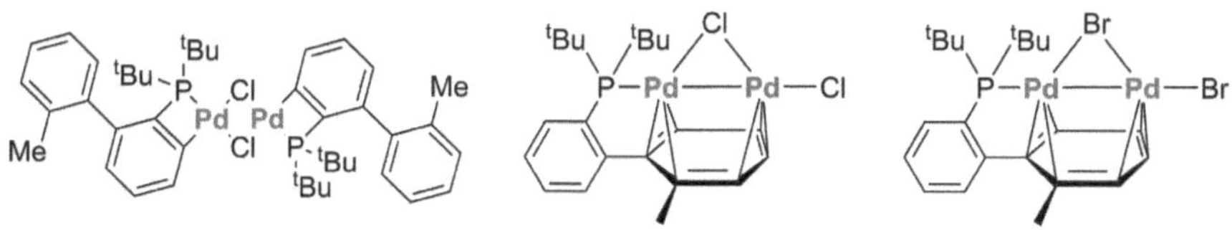

Vilar 2006 **6A** **6B**

Fig. 3.6 Vilar precatalysts in Buchwald-Hartwig amination reactions.

been shown to be an active class of precatalysts for cross-coupling reactions since the initial report of Hartwig *et al. (23f)* to very recent examples *(54)*. In contrast to base-activated Pd(II) precatalysts, these dimers either disproportionate to both Pd(0) and Pd(II) species upon thermal activation, or form Pd(0) after reduction.

2.2 Group 2: Buchwald Palladacycles G1–G4 and Others Containing a C^N Backbone

It was mentioned in the introduction that Colacot defined a second category of palladacycles as precatalysts to active LPd(0). Here would be included all those palladacycles that usually present a CoN backbone prone to dissociate during the catalytic cycle, and which also contain a monodentate ligand (typically phosphine or N-heterocyclic carbenes NHC) which would stay coordinated to the Pd center throughout the catalytic sequence. For this category of precatalysts, it is proposed that reductive elimination leads to the active LPd(0) species after transmetallation (NMe_2 palladacycles) or deprotonation (Buchwald NH palladacycles). From a wide perspective, all the examples reported hereafter as groups 2 and 3 would be included in this unique category, but for the sake of clarity we will present as group 2 the early reports by Indolese and Bedford and examples with ferrocenyl-containing backbones, together with Buchwald G1–G4 palladacycles. Then those including NHC ancillary ligands instead of phosphines will be presented as group 3.

2.2.1 Miscellaneous Palladacycles With a C^N Backbone

After the Herrmann-Beller palladacycle breakthrough, considerable attention was paid to other palladacycles containing for example 2-(dimethyl amino)biphenyl or N,N-dimethyl benzyl amine. These palladacycles, specifically the mononuclear adducts ligated with phosphines, exhibited good activity in cross-coupling reactions. As mentioned in the introduction, at that time the benefit from electron-rich and bulky monodentate ligands was well known, as tris(tertbutyl)phosphine or monodentate heterocyclic carbenes in cross-coupling reactions, so this ligand-trend was incorporated into palladacycles chemistry. In 2002, Indolese and Studer reported a first wide screening study on air stable, highly active palladacyclic complexes for C—C and C—N coupling reactions with aryl chlorides *(48)*. Again, some advantages of the one-component precatalysts were highlighted, and a simple procedure to prepare the phosphine adducts mentioned above was developed, as depicted in Scheme 3.5. The best results were obtained when a bis(2-norbornyl)phosphine ligand was incorporated, and this palladacyclic catalyst was soon commercialized as SK-CC01-A.

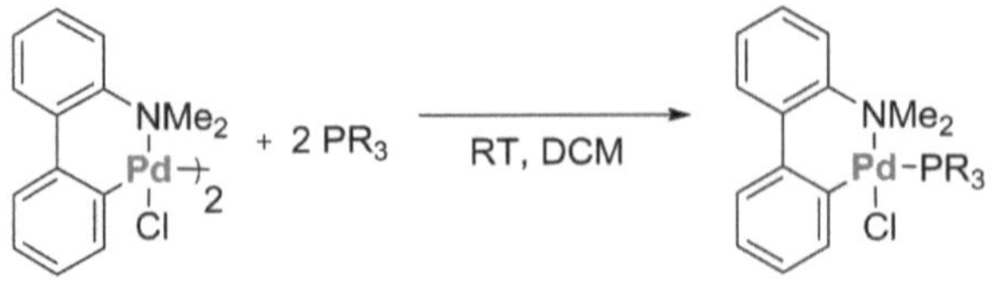

PR_3	Yields %
$HPNor_2$	100
$HP(^tBu)_2$	68
$HPCy_2$	91
PCy_3	88

Scheme 3.5 Synthesis of one-component palladacycles.

Encouraged by these results, the authors designed a parallel screening series of 140 experiments with seven palladium precatalysts (of which four are palladacycles), five secondary phosphines and four test reactions including nonconventional substrates such as aliphatic amines and deactivated anilines *(55)*. As a main outcome, the combination of a dimethyl aminomethyl ferrocene palladacycle and bis(2-norbornyl) phosphine produced the highest activity, thus disclosing a new family of palladacycles for Buchwald-Hartwig reactions (Scheme 3.6). Solvias commercialized this new catalyst under the name of SK-CC02-A.

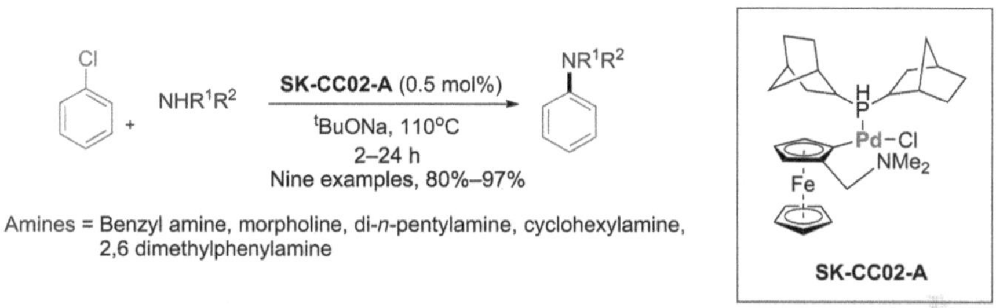

Scheme 3.6 Application of SK-CC02-A Solvias palladacycle.

Bedford and coworkers pointed out the importance of the palladium source employed, as a main conclusion of their work on amination of 4-chloroanisole with morpholine *(45b)*. They developed a series of ortho-palladated imine and N,N-dimethyl benzyl amine palladacycles, testing the latter in Buchwald-Hartwig amination. They specifically compared the performance of these palladacycles (the phosphine-free dinuclear TFA-bridged and the PCy_3 and PCy_2(*o*-biphenyl) adducts displayed in Fig. 3.7) against conventional palladium sources such as $Pd(OAc)_2$/phosphine. When palladium acetate was replaced by an orthometallated palladium source, important increases in TONs were observed. During these studies, the authors identified the formation of 4,6-bis(aryl)-3,4-dihydro-2*H*- *(1,4)*oxazines from aryl halides and morpholine.

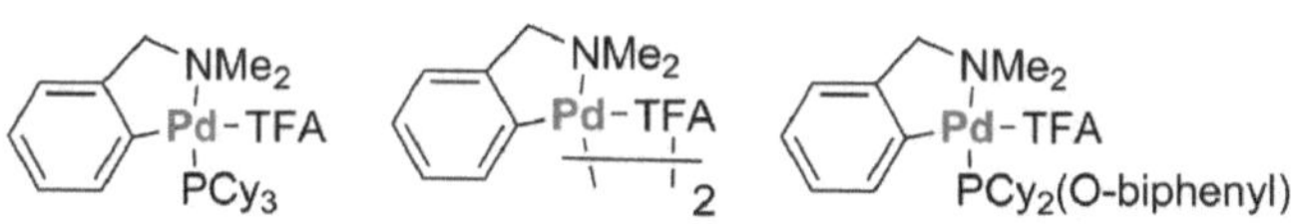

7A 7B 7C

Fig. 3.7 Bedford's N,N-dimethyl benzyl amine palladacycles.

In 2009, a new ferrocene-based ligand, 4,6-dimethyl-2-pyrimidinylferrocene, was synthesized and its monophosphine-palladacycle complexes were easily obtained from the subsequent cyclopalladation and bridge-splitting reactions. Their catalytic activity was evaluated in the Buchwald-Hartwig amination of a range of sterically hindered aryl chlorides, providing a reusable system in PEG-400 (Scheme 3.7) *(56)*.

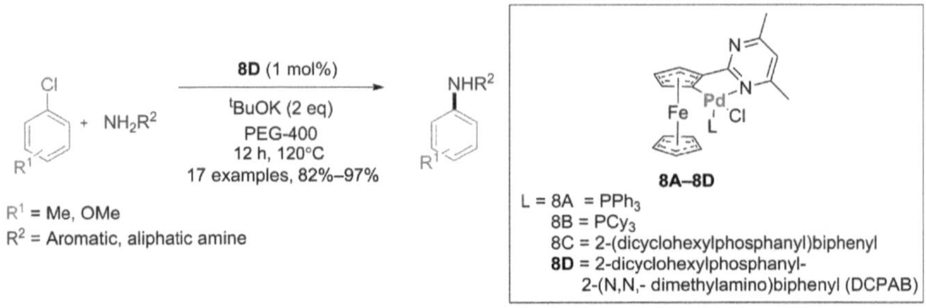

Scheme 3.7 Reusable ferrocene-based palladacycle in amination of aryl chlorides.

Similar cyclopalladated complexes bearing ferrocenylimines have also been reported as highly efficient palladium catalyst precursors in a wide variety of catalytic reactions, including Buchwald-Hartwig coupling *(57)*. Morales and coworkers reported imino-pseudopincer C∘N palladacycles in 2010 (Fig. 3.8) and examined their catalytic activity in Buchwald-Hartwig C—N cross-coupling of morpholine with a series of p-substituted bromobenzenes *(58)*.

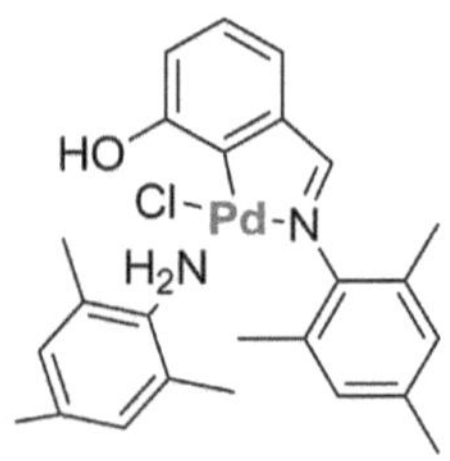

9

Fig. 3.8 Morales pseudopincer palladacycle.

Xu and coworkers reported an efficient method for the synthesis of palladacycles with polydentate ligands by a stepwise coupling route without additional palladium catalysis. The halide-containing ferrocenylpyridine palladacycles were both efficient catalysts and substrates in the amination reactions *(59)*.

2.2.2 Successive Generations of Buchwald Palladacycles G1–G4

We mentioned in the introduction that a crucial step in the development of Buchwald-Hartwig reaction was the advent of a wide series of ligands displaying a precise control of the coordination shell of palladium, and we referred to them as fourth generation ligands. These ligands combined steric bulk plus high donicity, and were prone to form only monoligated LPd(0) species. Thus, particularly biarylphosphine ligands acting as potentially bidentated, with at least one strongly bonded phosphine and a second on/off position switched by a hemilabile ligand and/or steric effects, have set the pace of modern advances in Buchwald-Hartwig reactions *(4a, 13, 40, 46)*.

In parallel, four generations of Buchwald palladacycles were developed to incorporate/accommodate such ligands in defined complexes, increasing catalytic efficiency, shortening reaction times, and minimizing excess of ligand through fast generation of the active LPd(0) species. Nowadays the amazing functionality and performance of these four generations of Buchwald palladacycles is clear from their commercial availability. Both Strem with its updated Product Booklet 02/18 "Buchwald ligands and precatalysts" https://www.strem.com/resource/2/product_booklets and Aldrich with its "Buchwald portfolio" https://www.sigmaaldrich.com/chemistry/chemical-synthesis/technology-spotlights/buchwald-guide.html offer a wide variety of complexes that provide easy access to Buchwald chemistry with palladacycles. Some compared features of Buchwald precatalysts G1–G4 are shown in Table 3.1. Specific details are given in the mentioned commercial guides and key references below, which report how Buchwald's group addressed the issues of catalyst activation through successive generations of precatalysts. While G1-palladacycles, including a phenethylamine-based backbone, exhibited a wide range of reactivity in diverse cross-coupling reactions, they required stronger conditions for activation, presented a short lifetime in solution and initially required a multistep synthesis which limited their scalability *(44b, 60)*. Second-generation biphenyl-based palladacycles G2 can be activated using a weak base at room temperature although cannot incorporate larger ligands such as BrettPhos and tBuXPhos *(61)*. This drawback was overcome with G3-palladacycles that replaced the chloride by a noncoordinating and more electron-poor methanesulfonate counterion, allowing for the incorporation of larger biaryl phosphine ligands *(41)*. In those uncommon cases when N-Substituted 2-aminobiphenyls are required (if the carbazole leaving group can inhibit the catalyst in some specific reaction), G4 precatalyst are used by prior methylation or arylation of the amino group on the biphenyl backbone *(62)*. Palladacycles with R = Ph have been referred to sometimes as L-Pd-G5.

Table 3.1 Basic comparison of four generations of Buchwald precatalysts

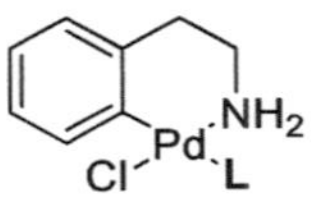 L-Pd-G1 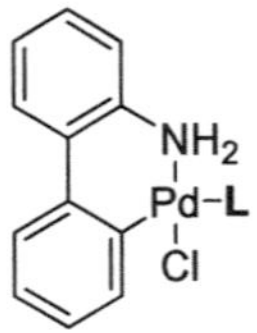 L-Pd-G2 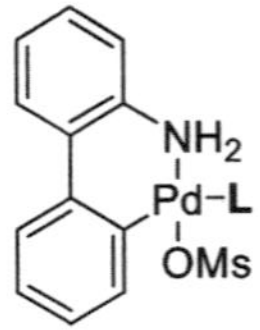 L-Pd-G3 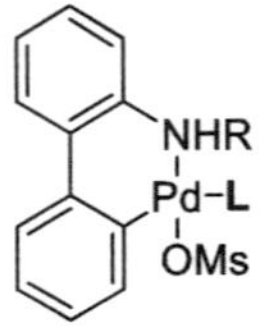 R = Me,Ph L-Pd-G4	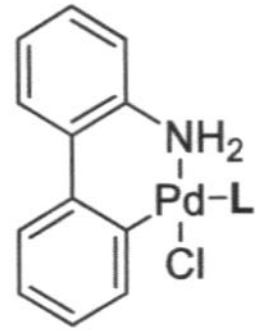 L-Pd-G2	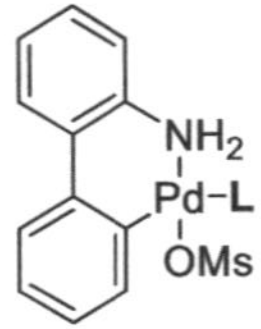 L-Pd-G3	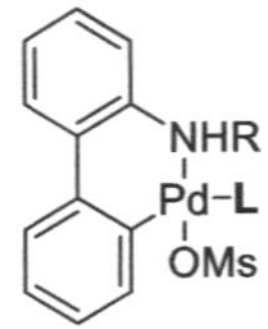 R = Me,Ph L-Pd-G4
1. 3 step preparation	**1.** Single one pot preparation from $Pd(AcO)_2$	**1.** Simple preparation with stable isolable intermediate (also prepared in situ)	**1.** Simple preparation with stable isolable intermediate
2. Compatible with bulky ligands	**2.** Not compatible with bulkier ligands	**2.** Compatible with bulky ligand and good solubility in solvent	**2.** Compatible with bulky ligand
3. Unstable intermediates	**3.** Poor solubility	**3.** Long life in solution (=1 month)	**3.** Better long life in solution (>1 month)
3. Short life in solution	**3.** Short life in solution		**3.** Avoids carbazole leaving group

2.2.2.1 Synthesis of G1–G4 Buchwald Palladacyclic Precatalysts

The preparation of successive generations of palladacycles is concisely displayed in Scheme 3.8. In 2007 *(63)*, Buchwald and coworkers isolated the stable neutral aryl amine complex that would lead to development of first generation of Buchwald precatalysts **L-Pd-G1** in 2008, with L = SPhos, XPhos or RuPhos *(60)*. The synthesis of these precatalysts could be achieved in two steps, the first involving the generation of thermally sensitive [$PdMe_2$(tmeda)] followed by the oxidative addition of 2-chloroethylamine. Later, Vicente and coworkers developed an alternative C-H activation pathway to **L-Pd-G1** *(64)*.

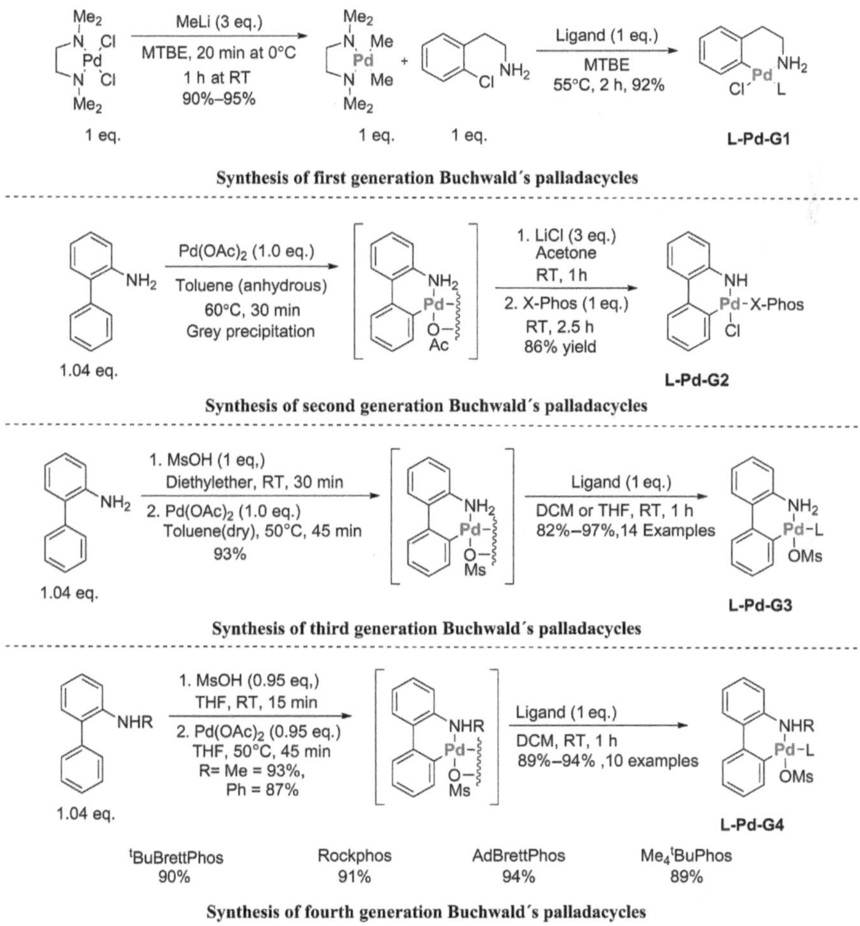

Scheme 3.8 Synthesis of successive Buchwald precatalysts G1–G4.

Synthesis of second generation Buchwald precatalysts **L-Pd-G2** can be attained by the reaction of 2-amino biphenyl with palladium acetate in toluene, followed by ion-exchange with LiCl and addition of the corresponding phosphine ligand *(61)*.

Third generation precatalysts **L-Pd-G3** can be developed by just replacing the chloride ion of **L-Pd-G2** with methanesulfonate ligand. As shown in Scheme 3.8, a μ-OMs dimer is generated from 2-aminobiphenyl via sequential mesylate salt formation and cyclopalladation. The dimeric precursor smoothly yielded the final mononuclear phosphine derivatives in THF as solvent *(41e)*. After reexamining the reaction conditions for this second step, Buchwald's group reported the use of chlorinated solvents as definitive procedure for the incorporation of a wider array of bulky ligands *(45a)*. Similarly, the fourth generation of Buchwald precatalysts was developed by just doing the aryl or alkyl substitution at NH_2 group in 2-aminobiphenyl prior to the next steps displayed in Scheme 3.8 *(62)*.

2.2.2.2 Mode of Activation of Buchwald Palladacyclic Precatalysts

Currently, mechanistic studies give sound information on the intermediate formation and mode of activation of Buchwald precatalysts to yield active LPd(0) species *(45a, 46, 60)*. It involves the abstraction of an N-H proton in preformed complexes **10a** to give the intermediate complex **10b** shown in Fig. 3.9. Deprotonation rate depends upon an acid/base relation between N-H proton and the base employed. Acidity of N-H proton in G1 precatalysts is lower than that of G2-G4 analogous, so a stronger base is required and they are usually unable to activate at ambient temperature. The depicted second step involves the reductive elimination of intermediate

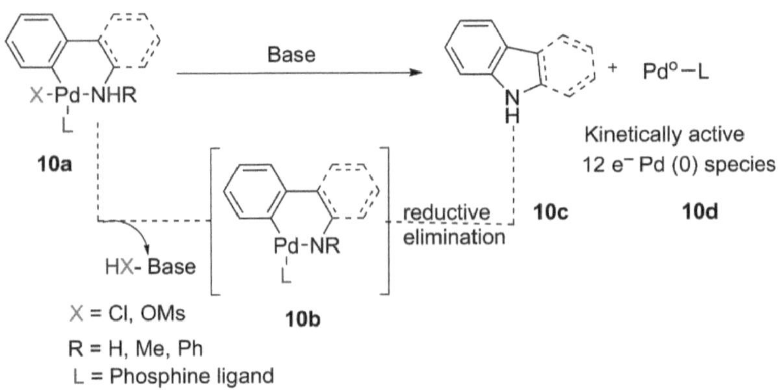

Fig. 3.9 General activation pathway of Buchwald precatalysts G1, G2, G3 and G4.

10b results in formation of active LPd(0) 12 e^- species along with relatively inert indoline (G1) of carbazole (G2,G3). N-substituted carbazole is formed instead when working with G4 precatalysts.

Given the current tremendous applicability of Buchwald-Hartwig amination reaction and the convenient commercial access of G1–G4 palladacycles in several kits and formats, an exhaustive listing of all the reports that have appeared to date would be a difficult task. An impressive work in this sense can be found in recent reviews *(3a, 46)*. We will present here a collection of crucial and updated references.

2.2.2.3 Application of First- and Second-Generation Precatalysts in C—N Bond Forming Reactions

As was mentioned before, in 2008 Buchwald and coworkers described a new class of one-component Pd precatalysts bearing biarylphosphine ligands, later known as first generation precatalysts **L-Pd-G1** *(60)*. They investigated the catalytic performance of **L-Pd-G1** (L = XPhos) in the cross-coupling reaction of 4-chloroanisole and aniline, compared with other traditional Pd sources such as $Pd_2(dba)_3$, $Pd(OAc)_2/PhB(OH)_2$, $[(allyl)PdCl]_2$ as shown in Scheme 3.9. Low loading of **L-Pd-G1** gave complete conversion within 35 minutes, while other sources produced a poor yield. The C—N cross-coupling of electron deficient anilines with unactivated aryl chlorides could also be achieved in good yield with the new prectalalysts, even at low temperatures.

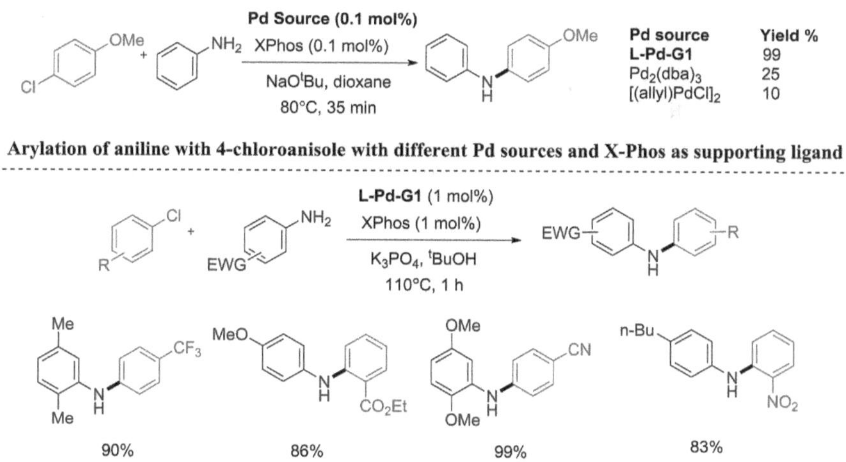

Scheme 3.9 XPhos-Pd- G1 in the arylation of electron poor anilines with unactivated aryl chlorides.

Also, in 2008, Buchwald's group developed a new ligand, BrettPhos, which coordinated to palladium as **L-Pd-G1** precatalyst showed exceptional performance in amination of aryl mesylates *(32)*. Monoarylation of methylamine and other primary aliphatic amines with aryl chlorides, and a high chemoselectivity for the arylation of a primary amine over a secondary amine, were also reported at very low catalyst loadings with high yields.

A general method for the selective monoarylation of acetate esters and aryl methyl ketones with aryl chlorides was reported by Biscoe and Buchwald, using **L-Pd-G1** under mild conditions as precatalysts **(L =**. ^{t}BuXPhos and XPhos) *(65)*.

The excellent catalytic performance of **L-Pd-G1** precatalysts in amination reactions of aryl iodides was soon disclosed, with L = BrettPhos being optimum for primary amines and RuPhos for secondary amines *(66)*. Solvent/NaI effects were also studied and the reactions of heteroarylamines and heteroaryliodides were again conducted in high yields.

Since a BrettPhos-based precatalyst *(32)* was ineffective in reactions involving secondary amines and highly active for the monoarylation of primary amines, and the opposite occurred with the RuPhos based palladacycle *(60, 66)*, in 2010 Fors and Buchwald devised multiligand ligand based palladium catalytic systems to overcome these drawbacks *(67)*. As shown in Scheme 3.10, the mixed catalytic systems BrettPhos-**Pd-G1**/RuPphos

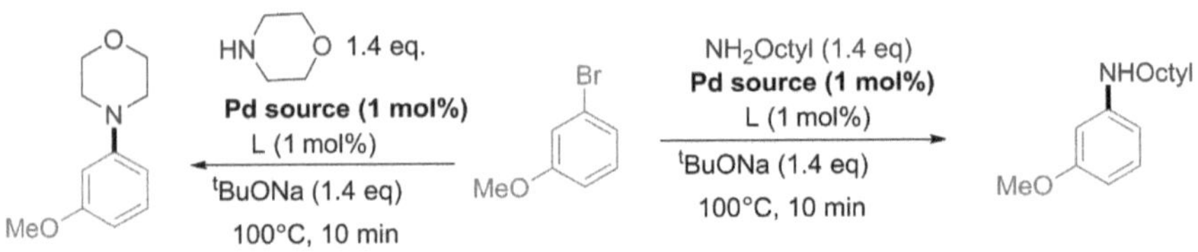

Pd Source	Ligand	GC Yield %
4	-	99
4	1	99
3	-	17
3	2	99

Pd Source	Ligand	Mono: Di	GC Yield %
3	-	99 : 1	99
3	2	97 : 3	94
4	1	96 : 4	92
4	-	97 : 3	30
4	-	89 : 11	77[b]

b = 4 h

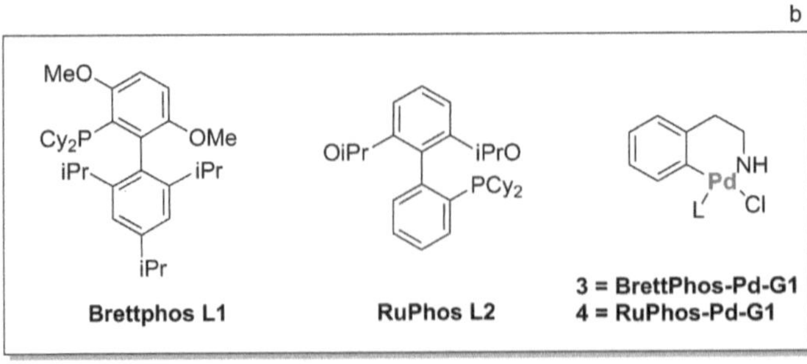

Scheme 3.10 A multiligand based Pd catalyst for C-N cross-coupling reactions.

and RuPhos-**Pd-G1**/BrettPhos exhibited the best properties of each individual catalyst, allowing for the efficient arylation of both primary and secondary amines. The efficiency of the process was thus not dependent on which ligand began bound to Pd.

A mixed catalytic system XPhos-**Pd-G1**/RuPphos was later used in the synthesis of class I phosphoinositide 3-kinases (PI3Ks) inhibitors. A challenging C—N cross-coupling of an elaborated secondary aniline with a sterically hindered heteroaryl chloride was achieved under microwave conditions *(68)*.

Also, in 2010 Buchwald's group proved that its **L-Pd-G1** precatalysts (L = RuPhos, BrettPhos, XPhos or SPhos) were efficient in aminations of unprotected heteroaryl halides *(69)* and C—N bond-forming reactions carried out in a continuous-flow manner (L = BrettPhos and XPhos) *(70)*. A highly efficient and chemoselective method for the construction of arylhydrazines using a broad scope of aryl/heteroaryl chlorides in continuous flow was later reported, and it was applied to the multistep syntheses of unprotected indoles and N1-aryl pyrazoles *(71)*.

A comprehensive study on the combined applicability of RuPhos and BrettPhos **L-Pd-G1** precatalysts in the arylation of a broad range of primary and secondary amines with functionalized aryl and heteroaryl halides was soon reported *(72)*, and a concise "user's guide" about dialkylbiaryl phosphines as valuable ligands for Pd-catalyzed amination, including the role of **L-Pd-G1** precatalysts, was published at the same time *(40)*. Their unique features, including low catalyst loadings, short reaction times, and the unnecessary use of a glovebox, placed these systems among the best at that moment, with promising applications in the synthesis of complex molecules such as pharmaceuticals, natural products, and functional materials. Ruiz-Castillo and Buchwald, in their recent review *(3a)*, brought together updated relevant reports.

An early example of this applicability was reported by Ley and coworkers, that in their continuous synthesis of Imatinib (Gleevec) (Fig. 3.10), a myeloid leukemia drug, included in the last N-arylation of aryl bromide step the use of BrettPhos **L-Pd-G1**. *(73)*. Also, in medicinal chemistry this palladacycle has been used in C—N coupling reactions of haloimidazole derivatives *(74)*, in the palladium-catalyzed amination of *N-free* 2-chloro-7-azaindole (Scheme 3.11) *(75)* and designing a protocol for the facile synthesis of 4-aryl and alkyl substituted, N^6-alkylated pyridazine-3,6-diamines *(76)*. (See Fig. 3.11.)

On the part of RuPhos **L-Pd-G1,** it was used by Brummond and coworkers to introduce linear dialkylamines and cyclic amines in naphtalene

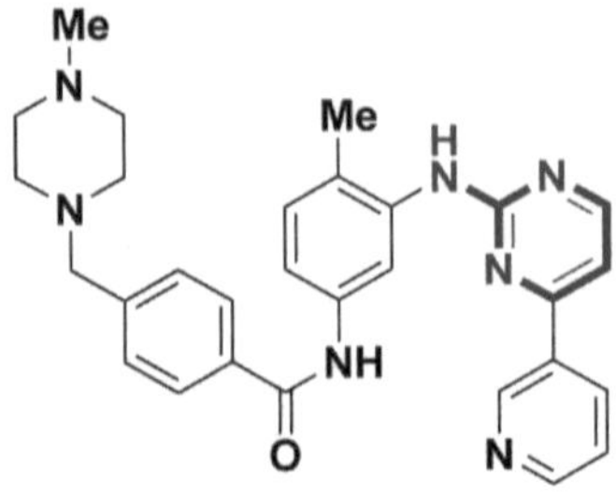

X = Br = 69%
Cl = 84%

Imatinib

Fig. 3.10 Cancer medication developed by Novartis.

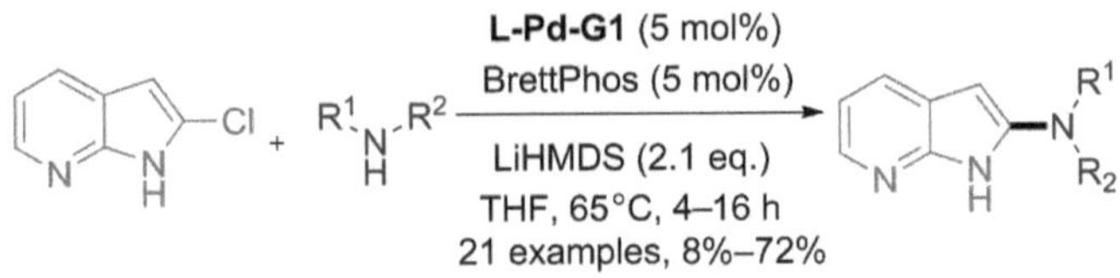

Amine (R_1 ; R_2)= H, Butyl, Bn, Piperazine,Et,Piperidine

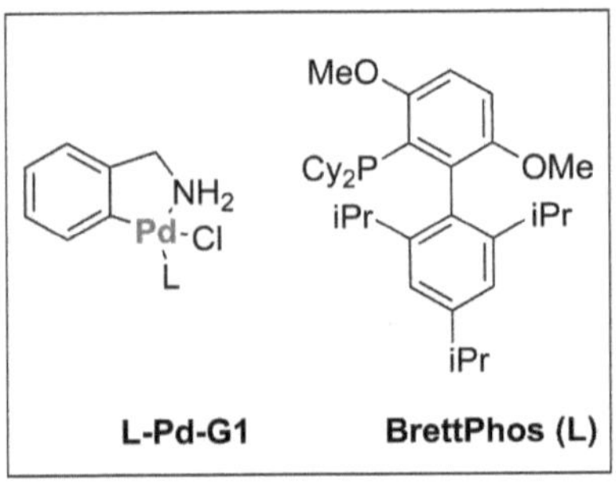

Scheme 3.11 Amination of 2-chloroazaindole with various primary and secondary amines using BrettPhos -Pd-G1.

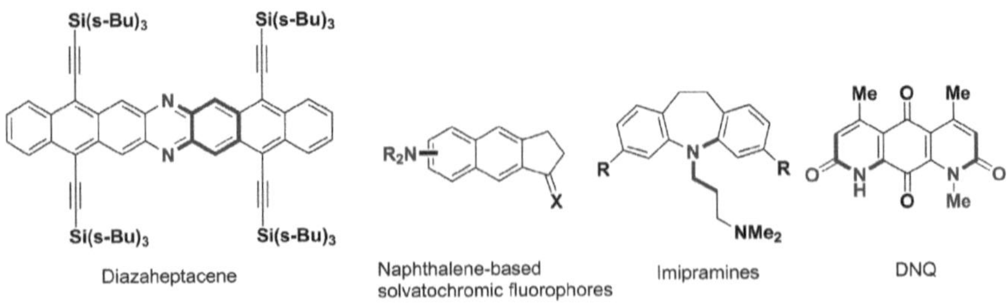

Fig. 3.11 Challenging new materials and biologically active compounds obtained with RuPhos Pd-G1 precatalyst.

rings by Pd-catalyzed C—N coupling, targeting fluorescent markers for biological applications *(77)*. This precatalyst was also employed in the synthesis of new potential OLED materials via the intermolecular cyclization of primary anilines, when Bunz and coworkers prepared the first stable diazaheptacene *(78)*.

During their nonracemic synthesis of diabetes pharmaceutical candidate GK-GKRP disruptor, RuPhos **L-Pd-G1** also had a crucial role *(79)*. Also, the first-generation XPhos palladium precatalyst found application in the synthesis of biologically active compounds via the coupling of primary alkylamines; specifically, Jensen and coworkers prepared a series of 3,7-disubstituted analogues of the tricyclic antidepressant Imipramine *(80)*. The synthesis of (−)-Epi- Indolactam V by an intramolecular Buchwald–Hartwig C—N coupling cyclization reaction was also achieved with XPhos L-Pd-G1 *(81)*.

Regarding second-generation precatalysts that incorporate a much more acidic NH_2 at the 2-aminobiphenyl palladacycle backbone, they proved to be readily-activated generating active LPd(0) at room temperature with weak bases. These precatalysts have found use in several cross-coupling reactions such as the first general method for the Suzuki-Miyaura coupling of unprotected, 5-membered heterocycles, Suzuki couplings with aryl- and alkyl-BF3K salts, borylations of aryl halides, alkynylations in continuous flow, and C-H arylation *(82)*. Palladium-catalyzed chemoselective α-arylation of methyl sulfones with aryl chlorides was also recently reported using Buchwald's 2nd generation palladium dimer and several ligands *(83)*.

Besides being poorly soluble in organic solvents, **L-Pd-G2** are not stable in solution for a long time, and important bulky ligands in C—N bond formation such as BrettPhos and related tBuXPhos, tButBrettPhos, and RockPhos cannot be incorporated. In fact, scarce reports describe the application of **L-Pd-G2** precatalysts in Buchwald-Hartwig reactions. For example, Hergenrother and coworkers developed an efficient multistep route to prepare the potential anticancer agent DNQ. Two intramolecular amidation reactions were key steps of this synthesis and were achieved with XPhos **L-Pd-G2** *(84)*. The efficient synthesis of benzene-fused 6/7-membered amides via Xphos Pd G2 catalyzed intramolecular C—N bond formation was also recently reported *(85)*, and Molander and coworkers employed the same palladacycle in the course of their studies for the synthesis of highly functionalized trifluoromethyl alkenes, using α-trifluoromethyl-β-silyl alcohols *(86)*.

2.2.2.4 Application of Third- and Fourth-Generation Precatalysts in C—N Bond Forming Reactions

In their 2013 cutting-edge article *(41e)*, Buchwald and coworkers introduced a new generation of palladacycle precatalysts for C—C and C—N cross-coupling reactions. Based on the 2-aminobiphenyl scaffold as the previous second generation, but replacing the chloride with a mesylate labile group, this easy to prepare **L-Pd-G3** family exhibited long solution life, could be activated under mild conditions using a weak base at room temperature, and were able to accommodate even bulky L phosphines. The performance of the new generation G3 was compared with that of classical precursors and previous palladacycle-based precatalysts in several cross-coupling reactions, and the preliminary results allowed the authors to anticipate its widespread use. That was the case, and a perspective article by Alami, Messaoudi, and coworkers *(46)* compiled the achievements made in several C-C and C-heteroatom bonds forming reactions using these G3- precatalysts during the initial years 2013 and 2014. For detailed information about examples in this period, which will not be addressed in depth here, please refer to this perspective article and to the 2016 specific review on C—N cross-coupling reactions by Ruiz-Castillo and Buchwald *(3b)*.

As mentioned above, in 2013, Bruno and Buchwald revisited the synthesis of their **L-Pd-G3** precatalysts and achieved the incorporation of bulkier phosphines *(45a)*. Best results in the coupling of functionalized aryl chlorides with various amides were obtained with L = tBuBrettPhos, which also displayed high chemoselectivity toward amides versus alcohol and N—H azoles (Scheme 3.12). Jui and Buchwald also used the tBuBrettPhos palladacycle in the regio- and chemocontrolled synthesis of N-arylbenzimidazoles *(87)*.

The selective monoarylation of ammonia with a wide range of aryl and challenging heteroaryl halides such as diazines was also reported, with the more conformational-rigid bulky ligands Me_3(OMe)XPhos or AdBrettPhos showing the best results *(88)* (Scheme 3.12).

The amination of aryl halides involving ammonium salts and base as an alternative to anhydrous gaseous ammonia or ammonia solutions was investigated by Green and Hartwig in 2014 *(89)*. Among several palladium precatalysts that had already been used in the arylation of ammonia, the dimeric chloride or mesylate bridged G3 palladacycles were also chosen for the screening of three ammonium salts in the amination of p-chlorotoluene.

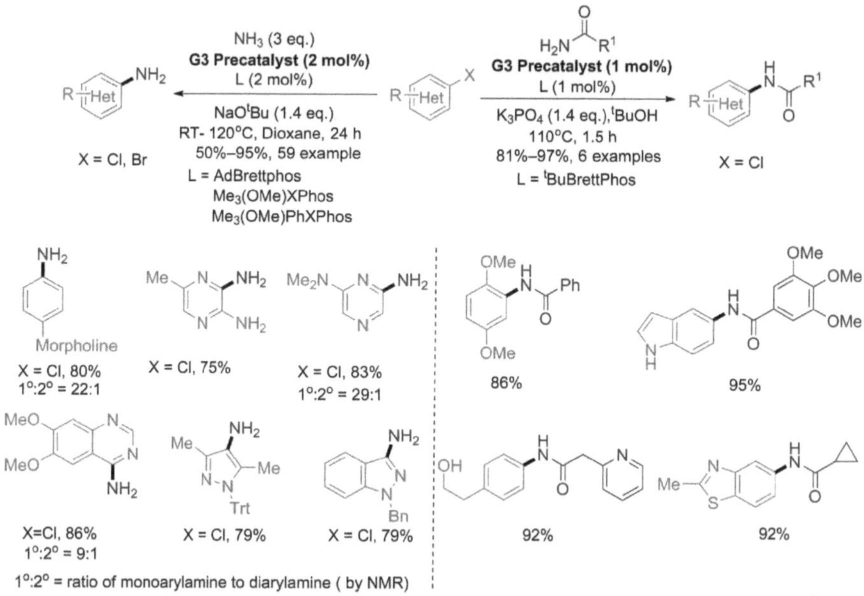

Scheme 3.12 Selective monoarylation of ammonia with aryl chlorides (left). Chemoselectivity against amides *vs* alcohol/azoles (right).

The regioselective synthesis of benzimidazolones via cascade C—N coupling of monosubstituted ureas was also achieved using the G3-BrettPhos palladacycle *(90)*, as shown in Scheme 3.13

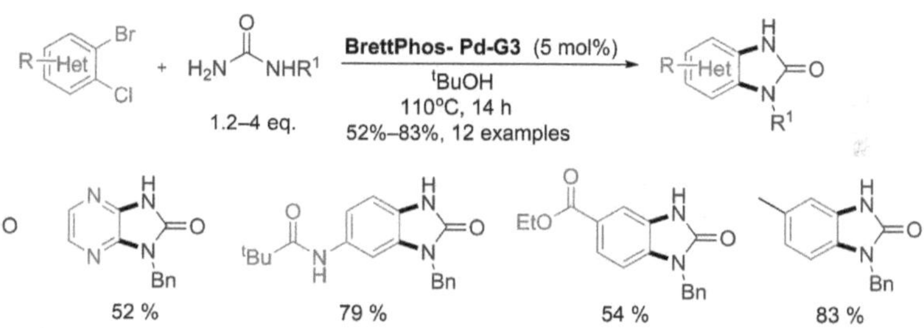

Scheme 3.13 Regioselective synthesis of benzimidazolones via cascade C-N coupling.

In the course of their preliminary pharmacological evaluation of enantiomeric morphinans, Neumeyer and coworkers employed the BrettPhos precatalyst for the synthesis of enantiomerically pure (−)-MCL-609 and (+)- MCL-740 (Scheme 3.14) *(91)*.

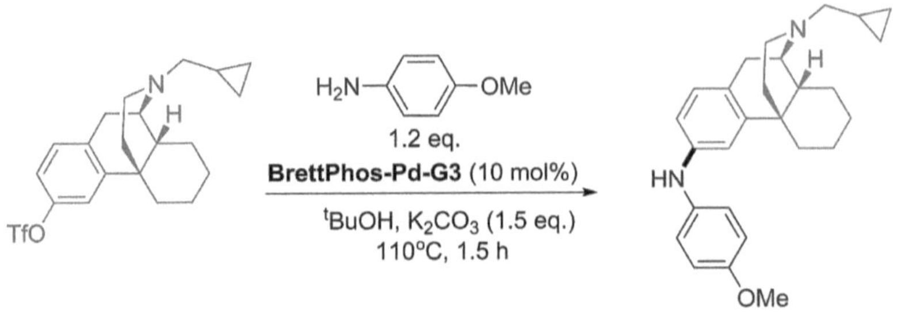

Scheme 3.14 Application of G3 precatalyst in total synthesis of MCL enantiomer.

Buchwald's group also explored the **XPhos** G3 precatalyst in continuous flow processes *(92)*. They specifically targeted the multistep synthesis of 1-substituted benzotriazoles from chloronitrobenzenes and amines under sequential multiphase condition.

The third generation of precatalysts has also found application in the synthesis of new materials via multiple intermolecular couplings of secondary anilines, as the RuPhos palladacycle was used in the synthesis of π-expanded diketopyrrolopyrroles that were promising fluorophores for bio-imaging *(93)*.

In 2015, Buchwald and coworkers reported the Pd-catalyzed arylation of very hindered α,α,α-trisubstituted primary amines with several (hetero)aryl chlorides and bromides. This work included the first example of hybrid (alkyl)aryl ligands (CyPhCPhos and t-BuPhCPhos), which once coordinated to Pd in their corresponding G3 palladacycles improved the results in these specific couplings when compared to its analogues with identical alkyl- or arylphosphorus substituents *(94)*.

Smith and Buchwald demonstrated different approaches for the regioselective synthesis of 2-aminopyrimidines from substituted polychloropyrimidines and chloro-thio-methoxypyrimidines *(95)*. Aryl- and heteroarylamines as substrates required the use of 1% L-Pd-G3/L with L = tBuBrettPhos. Regioselective amination of more nucleophilic dialkylamines did not require Pd catalysis.

A new method to access cyclic peptidomimetics *via* a Pd-catalyzed macroamination was reported, with natural amino acid amines acting as proficient nucleophiles *(96)*. This time, the third generation CPhos precatalyst, which previously succeeded in other hindered amine amination reactions, produced the best results.

Almost simultaneously, a general method for the *N*-arylation of amino acid esters with aryl triflate electrophiles was reported by King and

Buchwald, who conducted reaction optimization by design of experiment (DOE) analysis using JMP software *(97)*. The use of *t*-BuBrettPhos Pd G3 or G4 precatalysts was allowed to work under mild reaction conditions, which resulted in minimal racemization of the amino acid ester. This method was the first synthetic application of the *t*-BuBrettPhos Pd-G4 precatalyst, although related G4 precatalysts with BrettPhos and RuPhos were previously tested in the arylation of primary amines, secondary amines, and primary amides when this fourth generation was first reported *(62)*.

Alami, Messaoudi, and coworkers reported in 2016 a protocol for the bioconjugation of cysteine-containing peptides, using Xantphos Pd-G3 precatalyst *(98a)*. The detection of trace amounts of a by-product from the unexpected reaction of Pd-G3-precatalyst with the thiol function in the cysteine allowed them to envisage a route to 2′-S-linked 2-aminobiphenyls under mild reaction conditions. Subsequently they recently disclosed a new methodology to synthesize 1-aminobiphenyl thioglycosides through the activation of the XantPhos Pd-G3 with various glycosylthiols in very mild conditions *(98b)*. Direct selective functionalization at the residual aniline group was then achieved in the presence of Cs_2CO_3 at 100°C, via Buchwald-Hartwig reaction with the same precatalyst.

As mentioned in the introduction, a new ligand (EPhos) for palladium catalyzed C—N cross-coupling reactions was designed by using mechanistic considerations. The corresponding G3 and G4 palladacycles were successfully used as precatalysts in the synthesis of 2-(hetero)arylaminooxazoles and 4-(hetero)arylaminothiazoles with NaOPh as a mild homogeneous base *(17b)*. Palladacyles used in this study are displayed in Fig. 3.12.

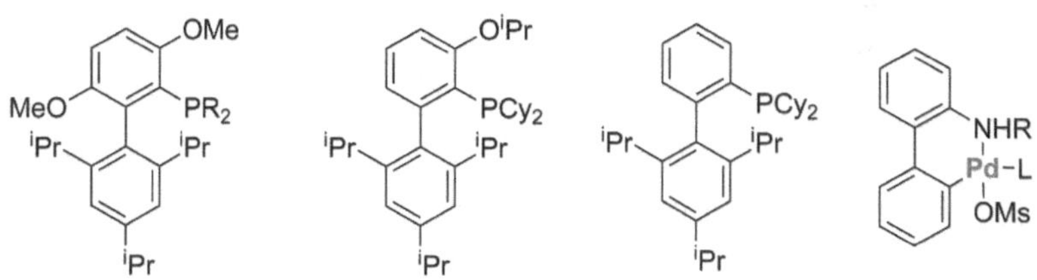

L1 : R = Cy (BrettPhos)
L2 : R = ^{t}Bu (^{t}BuBrettPhos)

L3 : R = EPhos

L4 : R = XPhos

P1 : L = **L1**, R = Me
P2 : L = **L2**, R = H
P3 : L = **L3**, R = Me
P4 : L = **L4**, R = Me

Fig. 3.12 Examples of G3 and G4 palladacycles in the synthesis of 2-(hetero) arylaminooxazoles and 4-(hetero)arylaminothiazoles.

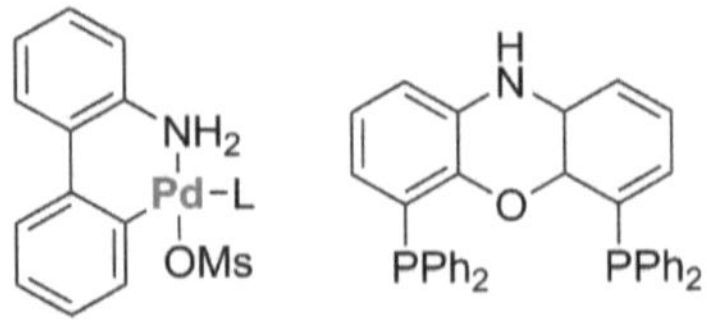

L-Pd-G3 NIXANTPHOS

Fig. 3.13 Nixantphos and its Pd-G3 palladacycle.

The newer G3 NIXANTPHOS-based precatalyst (Fig. 3.13) for coupling of sterically hindered secondary anilines outperformed similar Xantphos system and various mono- and bidentate phosphine ligands using unactivated aryl chloride substrates. Mao, Walsh, and coworkers reported good to excellent yields with as low as 0.05 mol% palladium loading, and also explored morpholine and several anilines as nucleophiles *(38)*.

To conclude this section, it is worth mentioning the comparative study with a large variety of commercially available phosphine ligands and types of Pd catalysts that Spindler and coworkers published in 2015 *(99)*. In addition to Suzuki-Miyaura cross-coupling, they were evaluated in Buchwald-Hartwig amination and Buchwald amidation reactions. Outstanding performance of the **L-Pd-G3** family with respect to low catalyst loadings and high chemoselectivities was reported. This industry perspective again gives an idea of the current relevance and the widespread use of Buchwald's G1–G4 palladacycles, an overview of which we have presented in this chapter.

We should mention here that the application of palladium-based oxidative addition complexes as precatalysts for C—N, C-O, and C-F cross-coupling reactions with a variety of (hetero)arenes has brought a new Buchawald's generation to the stage (Fig. 3.14). It has been labelled G6, and these nonpalladacyclic complexes are a new alternative, particularly useful in the case of the bulkiest biarylphosphine ligands *(100)*.

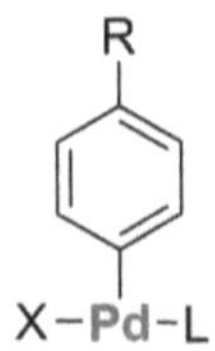

L-Pd-G6

Fig. 3.14 Oxidative addition complexes as precatalysts.

2.3 Group 3: Palladacycles With NHC Ligands

The use of N-heterocyclic carbenes (NHCs) in palladium-catalyzed cross-coupling reactions as alternative ligands to ubiquitous tertiary phosphines has experienced a tremendous development since the isolation of a first stable NHC by Arduengo in 1991 *(101)*. Well-defined NHC containing palladium(II) complexes are no longer in "their infancy", as quoted by Marion and Nolan in his 2008 excellent review *(102)*, but sharing a prominent place besides the best catalysts in cross-coupling reactions *(44, 103)*. Typically, NHC-Pd(II) complexes are classified into five main classes: (1) palladium dimers with bridging halogens; (2) palladacycles; (3) palladium acetates and acetylacetonates; (4) π-allyl complexes; and (5) the PEPPSI family of precatalysts (PEPPSI = Pyridine-Enhanced Precatalyst Preparation Stabilization and Initiation). The synthesis, advantages, and applications in cross-coupling chemistry of those well-defined complexes are outside the scope of this chapter and have been thoroughly addressed in the review articles above mentioned. These different accompanying ligands have been described as "throwaway ligands", merely a protecting shell that must be easily discarded going from the Pd(II) precatalyst to the Pd(0) true catalyst under the reaction conditions *(102, 103c)*.

Regarding the impact of NHC-containing palladacycles in cross-coupling reactions, and specifically applied in Buchwald-Hartwig amination, they have played a relevant role, *(102, 103b)* which will be covered next, although lately other families such as π-allyl complexes and the PEPPSI precatalysts have attracted more attention *(104)*.

In 2003, and stimulated by the promising results obtained at that time with palladacycle/PR_3 systems in cross-coupling reactions, Nolan and coworkers combined the donating properties of NHC's IMes and IPr with the stability of a palladacycle backbone *(49)*. The synthesis of the new complexes was achieved by easy cleavage of dimeric halide-bridged precursors, as shown in Scheme 3.15. The authors investigated the catalytic activity ofNHC based palladacyclic complexes in the cross-coupling of aryl chlorides or triflates with a broad scope of substrates such as heterocyclic alkylamines, dialkylamines, aryl-alkylamines, and primary amines. The use of palladacycles as labile leaving groups to generate active NHC-Pd(0) species was among the first examples of this strategy *(105)* and a mechanism for its activation was proposed *(106)*.

The analogous palladacyclic complexes incorporating the saturated version of the NHC-ligands IMes and IPr were prepared in 2008 by a similar synthetic route *(107)*. Their compared activity in the Buchwald-Hartwig

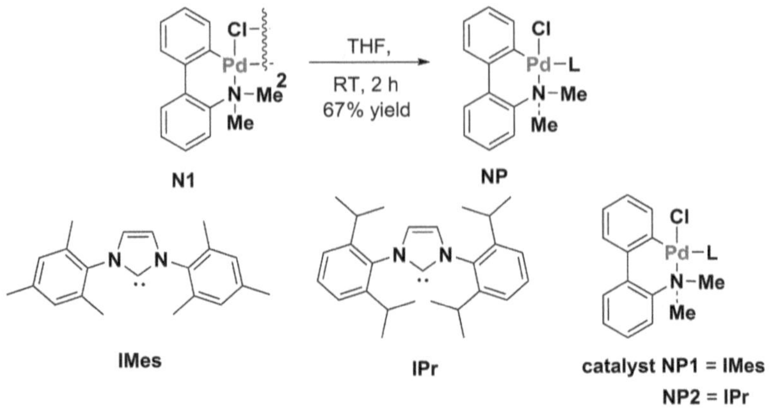

Scheme 3.15 Synthesis of NHC based palladacycles.

amination of several unactivated and diverse aryl chlorides was explored without noticeable differences among saturated/unsaturated derivatives (Scheme 3.16).

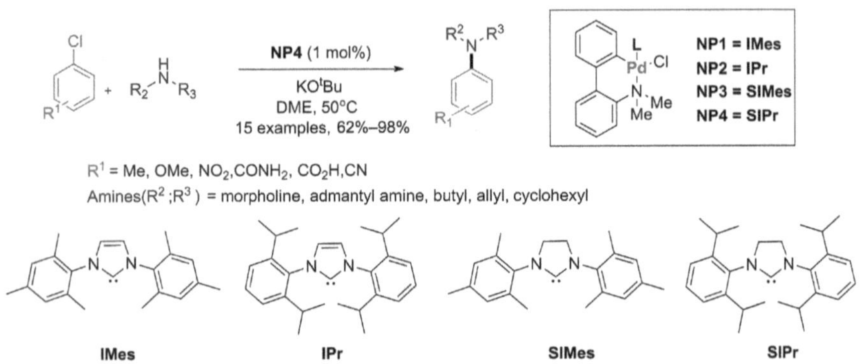

Scheme 3.16 NHC-palladacycles incorporating saturated/insaturated ligands.

There was a previous report by Wu and coworkers of a palladacyclic carbene adduct catalyzing Buchwald-Hartwig amination *(108)*. The authors explored the activity of their IPr-cyclopalladated ferrocenylimines (Scheme 3.17) in the amination of aryl chlorides with a broad range of amines. In 2013, they expanded the scope of their catalytic system to the amination of chloropyridines with primary and secondary amines, including sterically hindered primary amines and alkyl amines with catalyst loadings of 1 mol% *(109)*.

As mentioned in the Group 2 section, Xu and coworkers reported the synthesis of phosphine adducts of cyclopalladated ferrocenylpyridine obtained by bridge-splitting chloride dimers, and explored their performance as

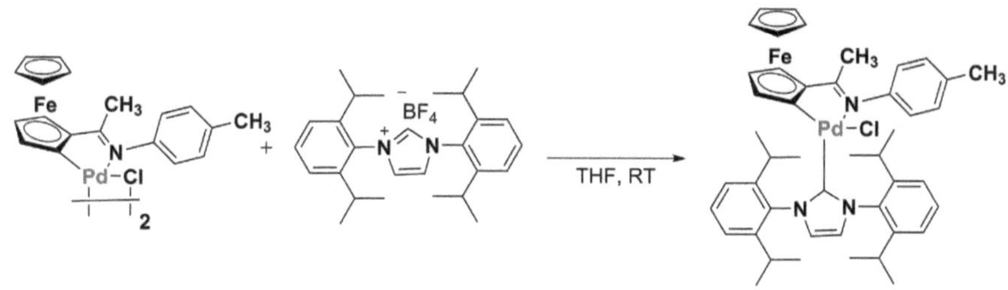

Scheme 3.17 Synthesis of iminoferrocenyl palladacycles.

reusable catalyst in Buchwald-Hartwig amination reactions *(56)*. In 2010, they conducted similar research on analogous NHC's complexes (Scheme 3.18) and examined their application as catalysts for the Suzuki and Buchwald-Hartwig amination and the reusability of the system [Pd(Cl)(NHC)(palladacycle)]/PET-400 *(110)*.

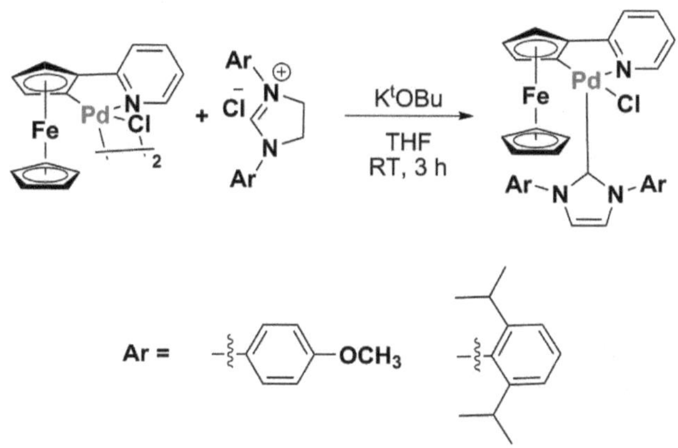

Scheme 3.18 Synthesis of ferrocenylpyridine cyclopalladated complexes.

Robust Pd-NHC acenaphthoimidazolylidene palladacycles have been developed recently by Tu *et al.* *(111)* (Fig. 3.15). High catalytic activity and chemoselectivity in the amination of diverse N-heteroaryl chlorides at low catalyst loading was reported, and different primary and secondary

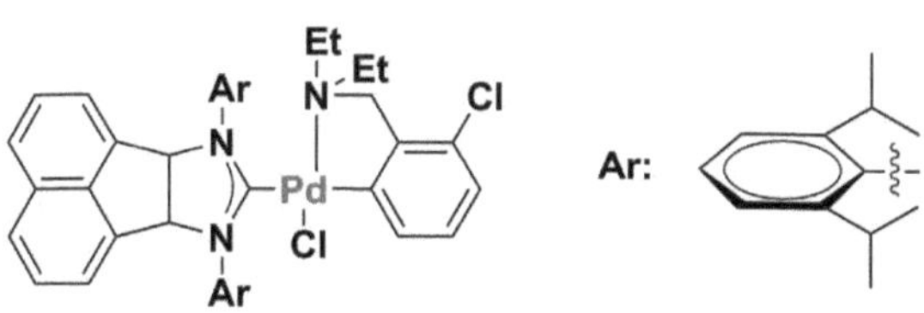

Fig. 3.15 Pd-NHC acenaphthoimidazolylidene palladacycles.

amines were compatible with this catalytic system. The protocol efficiency was highlighted as it could be extended to the direct amination of N-heteroaryl chlorides by an amide and a hydrazine derivative, and used in the synthesis of rosiglitazone, a clinical drug for diabetes mellitus.

Yorimitsu and coworkers have recently developed an efficient protocol for the amination of diaryl sulfoxides with anilines or alkyl amines with the Signacycle A1 palladacyclic complex displayed below (Scheme 3.19). The mild conditions of the new protocol made it compatible with a broad range of functional groups and halogen moieties. Regioselective amination of unsymmetrical diaryl sulfoxide was also achieved *(112)*.

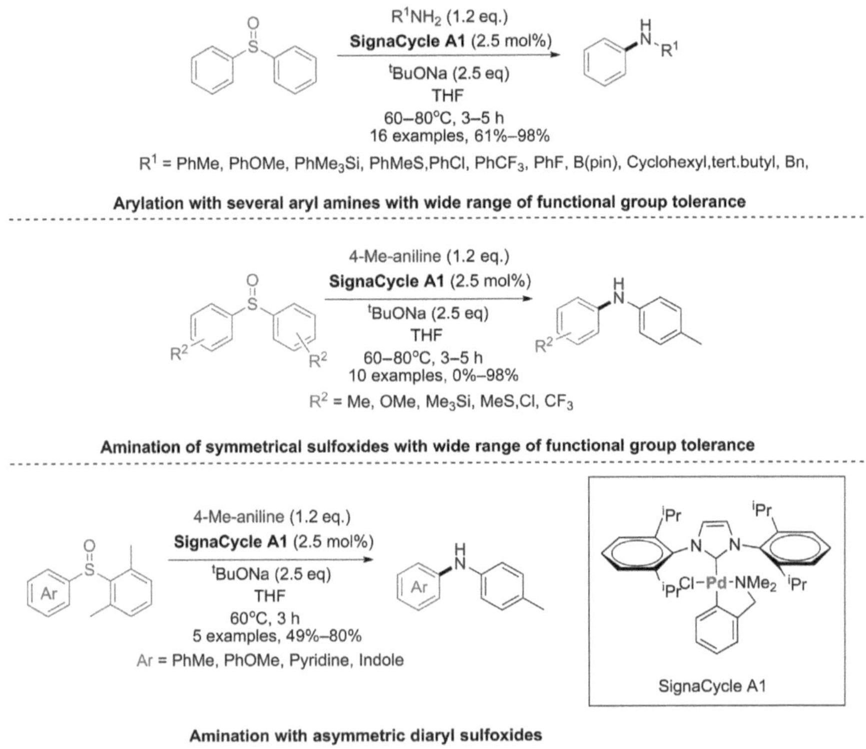

Scheme 3.19 Palladium-catalyzed amination of aryl sulfoxides.

3. PALLADACYCLES IN NEGISHI COUPLING

3.1 Introduction

The Negishi coupling, in its wider version of a palladium- or nickel-catalyzed coupling of organometals containing Zn, Al or Zr with various

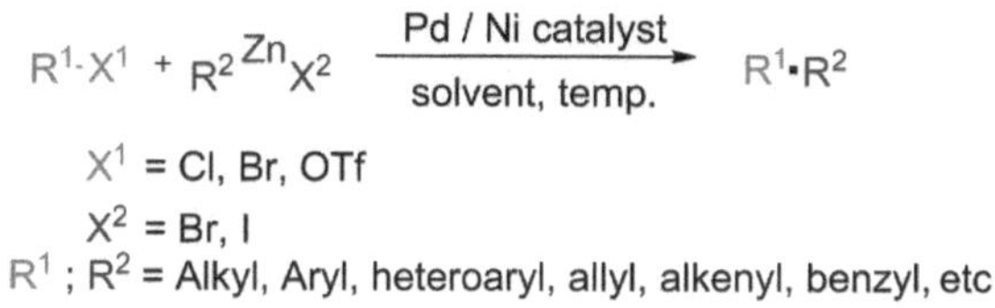

Fig. 3.16 General Negishi reaction.

halide-containing moieties (aryl, vinyl, benzyl, or allyl), is nowadays deservedly seen as a key reaction in the modern formation of carbon-carbon bonds (Fig. 3.16). Entering the new millennium, the Negishi coupling was already well recognized as a "name reaction" in reference works such as the second edition of *"Organic Syntheses based on Name Reactions"* or *"The Merck Index"*, and a decade later professor Negishi received his shared Nobel Prize in Chemistry *(2b, 3d, 113)*.

This current status is well beyond the initial scope reported back in 1977 by Negishi and coworkers, when the preparation of biaryl and diarylmethanes from aryl- and benzylic zinc reagents was achieved using [Ni $(PPh_3)_4$] and [$PdCl_2(PPh_3)_2$] as catalysts (Scheme 3.20) *(114)*.

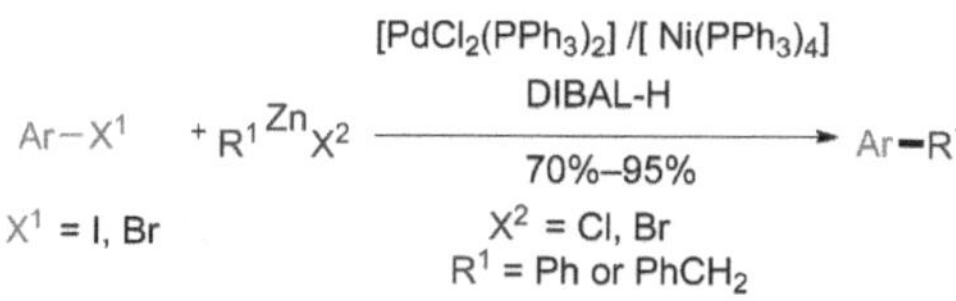

Scheme 3.20 Ni/Pd catalyzed Neghishi reaction.

However, were those early critical findings that the Negishi group made between 1976 and 1978 the ones that contributed the most to establish the basis for the present knowledge of Pd-catalyzed cross-coupling. The original articles from this period and an overview about the early development of the field can be found in Chapter 15 of "Metal Catalyzed Cross-Coupling Reactions. 2nd edition" *(2b)*. This perspective approach was also documented in the 2002 special issue published in *Journal of Organometallic Chemistry (115)* and more recently revisited on the occasion of the Nobel Prize *(113)* and Chapter 3 in the 2014 book "Metal-Catalyzed Cross-Coupling Reactions and More" *(3d)*. A comprehensive discussion organized by specific Negishi cross-coupling types beyond regular aryl-aryl coupling (aryl-alkenyl, alkenyl-aryl, alkenyl-alkenyl, alkylation allylation, and others) is thoroughly presented in the 2002 Handbook of Organopalladium Chemistry for Organic Synthesis

(116) Since then, updated recent developments have been collected in subsequent general references *(2b, 3d, 117)*, or those more specific devoted to alkenylation *(118)* pharmaceutical applications *(119)*, or total synthesis *(120)*, clearly showing the present relevance of Negishi coupling as a topical subject. Most of the above mentioned references cover general aspects of the Negishi coupling that will not be addressed here such as the so-called changeable parameters: *(2b)* metal countercation or specifically organozinc reagents *(117, 119)*, leaving groups, ligands, Pd source, or cocatalysts among others. The high compatibility of organozinc reagents with sensitive functional groups *(121)*, their improved preparation which has widened the scope and applicability of the Negishi reaction *(122)*, and the mechanism involving the Negishi coupling has also been reviewed there. A simplified version displaying a second transmetallation step that would explain dehalogenation and undesired homocoupled product is shown in Fig. 3.17. Detailed studies of the mechanism have recently been presented by Espinet *(123)* and Koszinowski *(124)* groups.

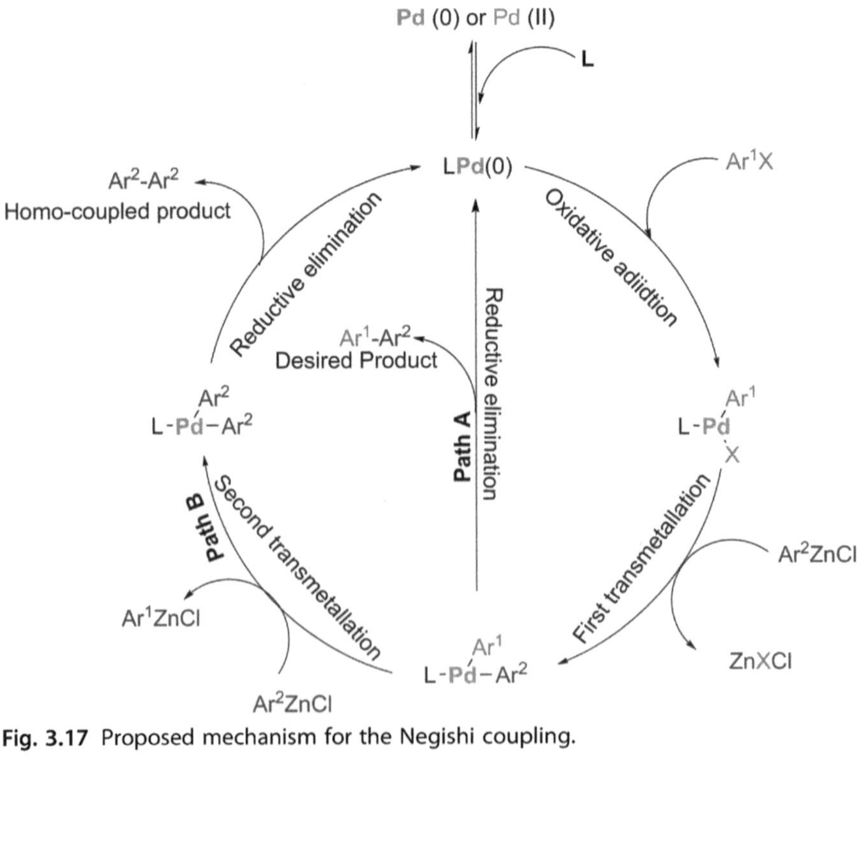

Fig. 3.17 Proposed mechanism for the Negishi coupling.

Regarding the palladium source, just a few Pd(0) and Pd(II) complexes have been widely employed besides the original [$PdCl_2(PPh_3)_2$] and [Pd $(PPh_3)_4$] for both conventional or high activation requirements. Those including chelating diphosphines such as [$PdCl_2$(dppf)] or [$PdCl_2$(DPEphos)] or those involving Pd($^tBu_3P)_2$ o$Pd_2(dba)_3$ are found in a vast majority of reported examples. $Pd(AcO)_2$ or the well-defined PEPPSI-IPr precatalyst have also been profusely used *(125)*. The lack of references to palladacycles as the catalysts of choice for the Negishi coupling in the consulted key review documents *(1b, 2b, 3d, 113, 118, 120)* is thus notable. The Buchwald G1–G4 family of palladacycles applied to the Negishi coupling came into the picture in 2013 *(126)*, and they have been considered the main contribution of palladacycles to this field. *(44c, 117, 119)*. In parallel, specialized work regarding palladacycles *(43, 44d, 50, 127, 128)* hardly mentions its use as catalysts for the Negishi coupling, with only one clear reference found *(50)* that reports an example before the advent of Buchwald's palladacycles. This example is the early application of Herrmann-Beller catalyst in the Negishi coupling, collected in a review article under the generic title "Application of palladacycles in Heck type reactions" *(47b)*. This will be considered the first example in the following section, where we present a chronological account of the reports on palladacycles as catalysts for the Negishi coupling.

3.2 Uses of Palladacycles in the Negishi Coupling

In the above mentioned early review by Herrmann and coworkers, the authors put together the results obtained during the period 1997–1999 using the acetate-bridged phosphapalladacycle, *trans*-di(μ-acetato)-bis[*o*-(di-*o*-tolylphosphino)benzyl] dipalladium(II), as cross-coupling catalyst. Mainly uses in Heck arylation of olefins with aryl halides were reported, but also in Suzuki, Stille, Sonogashira, Grignard, and the Negishi coupling reactions. Both the chemistry of phospha-palladacycles and the mechanism operating during cross-coupling reactions were addressed in detail, and the advantages of the palladacyclic system over conventional in situ alternative options were highlighted. Regarding the Negishi coupling, four examples using 2 mol% catalyst, $ZnMe_2$, or PhZnBr and aryl bromides or chlorides were reported, with GC yields in the range 76%–99% *(47b)*.

More than a decade later, Frech and coworkers *(129)* reported the pincer complex with aliphatic core as active catalyst for the Negishi coupling of a wide variety of aryl bromides with several diarylzinc reagents in N-methylpyrrolidone (NMP) or dioxane as solvent and short reaction times, as exemplified in Scheme 3.21.

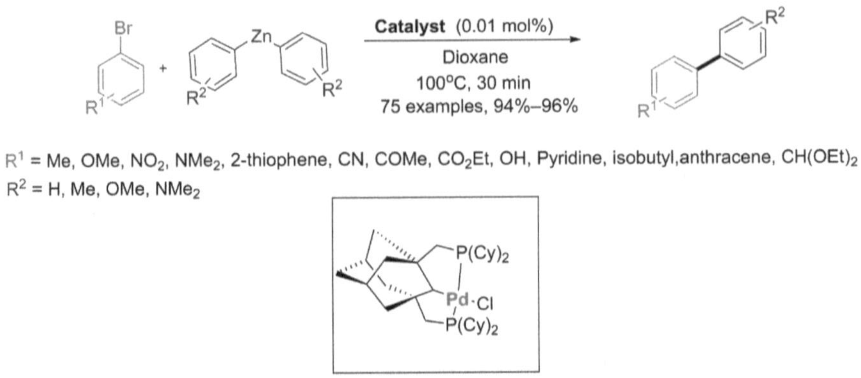

Scheme 3.21 Application of palladium pincer complexes in the Neghishi reaction.

Then in 2013 Buchwald and coworkers applied their third generation palladacycle catalyst to the challenging Negishi cross-coupling of heteroaryl zinc reagents with heteroaryl halides/pseudohalides at ambient temperature or with low catalyst loadings. *(126)* Special attention was paid to pyrazole and imidazole-containing substrates and polyfluoroaryl zinc reagents (see Scheme 3.22 below).

This family of palladacycles ligated by dialkylbiarylphosphine ligands (with XPhos particularly active) proved to generate more efficiently LPd (0) species, when compared with $Pd(OAc)_2$ and $[Pd_2dba_3]$. In these Negishi studies, the size of the substituents on the nonphosphorus-containing ring of the ligand clearly determined the catalyst activity, as the authors had previously found for other cross-coupling reactions.

Almost simultaneously in 2013 two new articles expanded the scope of the XPhos palladacycle in Negishi coupling reactions. Knochel and Buchwald *(130)* developed two methods for the preparation of solid 2-pyridylzinc nucleophiles as alternatives to 2-pyrydil boron reagents in cross-coupling reactions. Specifically, a dioxanate complex was demonstrated to be an effective nucleophile under the Negishi cross-coupling conditions developed with XPhos palladacycle (Scheme 3.23). The dioxanate complex decomposed after long-term storage and required titration before use, although decomposition products did not diminish the reagent's efficacy.

Soon a completely linear-selective Negishi coupling of 3,3-disubstituted allyl zinc reagents with a representative library of aryl, heteroaryl, and vinyl halides was developed for the first time *(131)* Together with the broad scope and the mild conditions of the reaction to rapid access prenylated arenes, its application in the concise and convergent synthesis of natural product

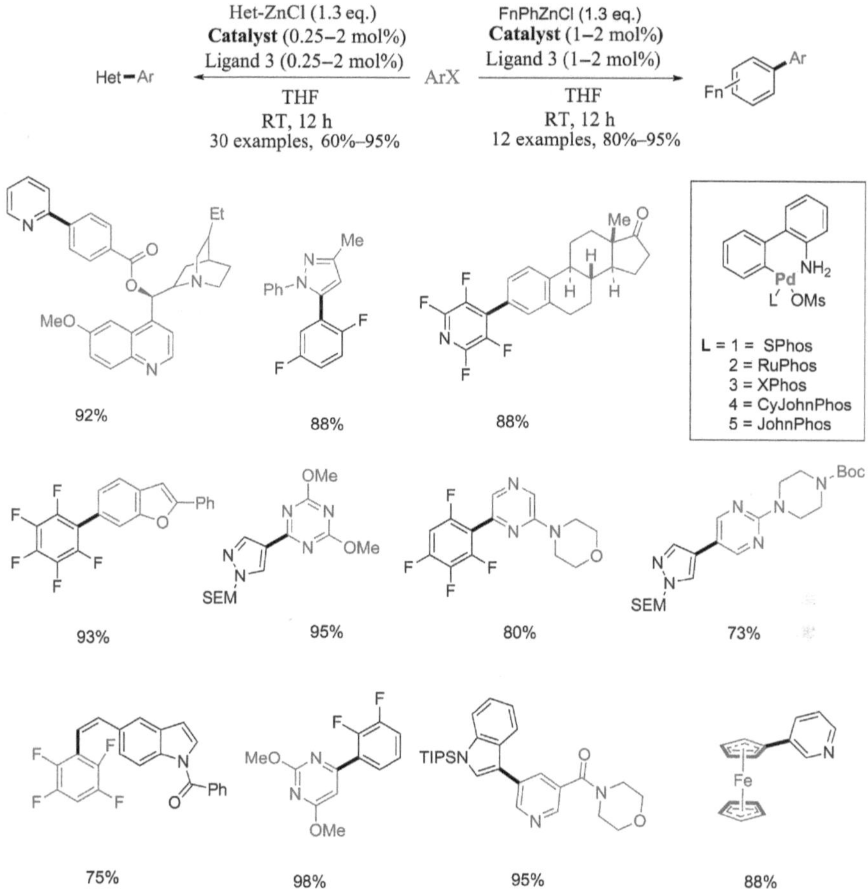

Scheme 3.22 The Negishi cross-coupling of heteroaryl zinc reagents and heteroaryl halides/pseudohalides with L-Pd-G3 precatalysts.

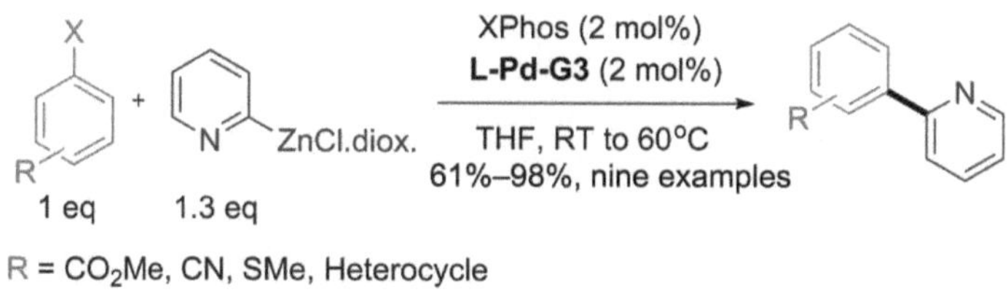

Scheme 3.23 Solid 2-pyridylzinc reagents in the Negishi reaction.

siamenol, and interesting computational studies to provide additional knowledge on the regiochemical controlling parameters were other features highlighted in this article. Again, the catalyst choice was critical, being the precatalyst generated from CPhos the optimal one for this coupling reaction (Scheme 3.24).

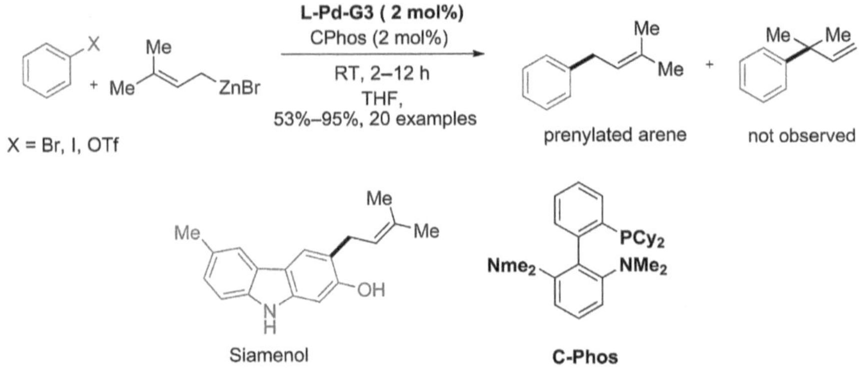

Scheme 3.24 Synthesis of prenylated arenes using a CPhos-Pd-G3 precatalyst.

Buchwald and coworkers expanded their studies on the CPhos palladacycle as catalyst for the highly selective cross-coupling of secondary alkylzinc reagents with heteroaryl halides *(132)*. The mild conditions of the reaction and some examples of the broad spectrum of heteroaryl substrates tested are displayed in Scheme 3.25. Several results using $Pd(AcO)_2$ as palladium source and a series of new CPhos-related biarylphosphine ligands bearing a 2,6-bis(dimethylamino)phenyl group were also reported.

Also, in 2014, Vilarrasa's group checked with moderate success the XPhos G3 palladacycle and its analogue from first generation, bearing a 2-(2-aminoethyl)phenyl) orthometallated backbone, among many representative catalysts such as $Pd(PPh_3)_4$, Pd_2dba_3/Xantphos, Pd_2dba_3/DPE-phos, Pd_2dba_3/XPhos and Pd_2dba_3/RuPhos in a synthetic approach to palmerolides via Negishi cross-coupling *(133)*.

Inspired by Lipshutz's results with in situ formation of alkyl zinc reagents, subsequently cross-coupled with aryl bromides under micellar conditions with a palladium catalyst *(134)*, Buchwald and coworkers developed an improved system for this micelle-enhanced Negishi aqueous cross-coupling of alkyl halides with aryl electrophiles *(135)*. Among other fourth generation precatalysts, a new one incorporating a new biaryl(dialkyl)phosphine ligand (VPhos) showed optimal results in combination with a simple and effective surfactant system (octanoic acid/sodium octanoate). This new catalyst/surfactant system produced good yields in the $C(sp^3)$-$C(sp^2)$ cross-coupling of a broad spectrum of (hetero)aryl halides with little excess of nonaromatic heterocyclicbromides, as presented in Scheme 3.26.

Stradiotto and coworkers *(136)* used XPhos/Pd-G3 plus additional XPhos as Negishi catalyst in a combined protocol for the challenging construction of tetra-substituted trifluoromethylpyrazoles. The new methodology

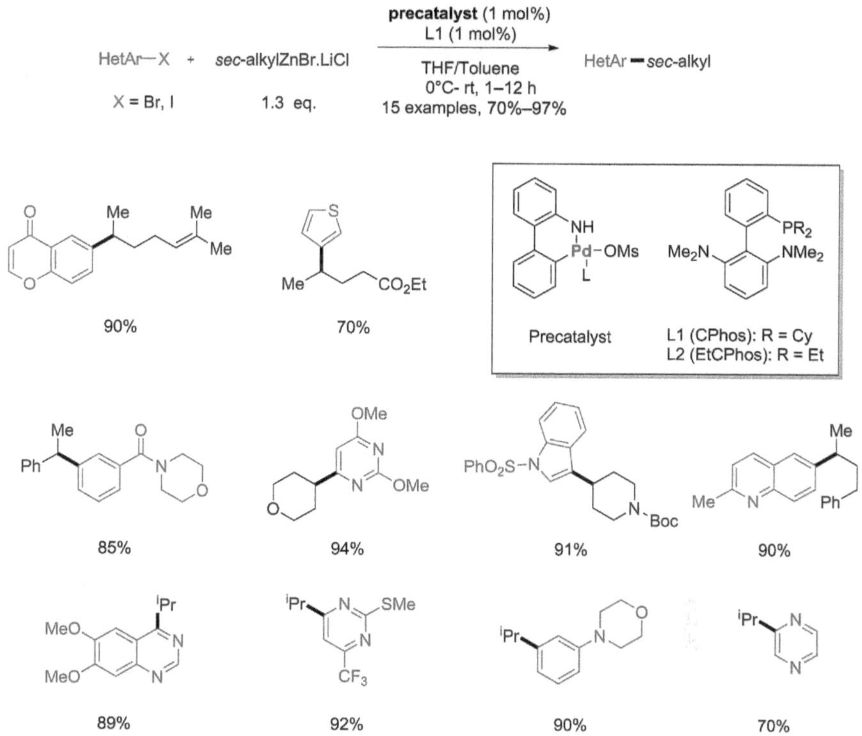

Scheme 3.25 Highly selective palladium-catalyzed cross-coupling of secondary alkylzinc reagents with heteroaryl halides.

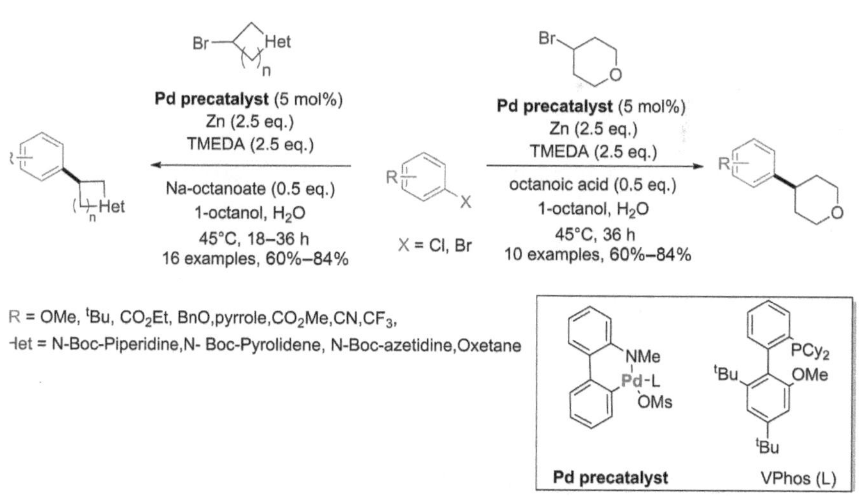

Scheme 3.26 Aqueous Lipshutz-Negishi cross-coupling of alkyl halides with aryl electrophiles using a new precatalyst based on VPhos.

involved as a previous step the halogenation of 3-substituted 1-phenyl-5-trifluoromethylpyrazoles at the 4- position, as shown in Scheme 3.27.

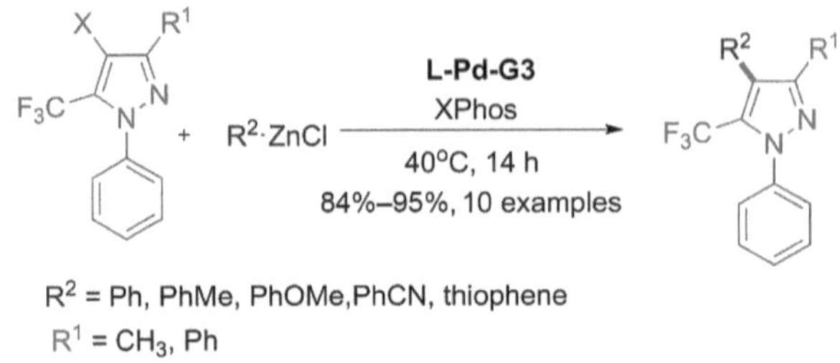

Scheme 3.27 XPhos-Pd-G3 in the synthesis of 1-phenyl-5-trifluoromethylpyrazoles.

A continuous-flow procedure involving a metalation, zincation, and Negishi cross-coupling sequence was developed to yield functionalized 2-fluorobiaryl products *(137)*. Again, the third generation palladacycle pre-catalyst bearing XPhos ligand was chosen for this system, which produced the desired arylated products in short times and high yields thanks to an excellent control of the temperature during the metalation step. Scheme 3.28 shows the three-step sequence to synthesize 2-fluorobiaryls.

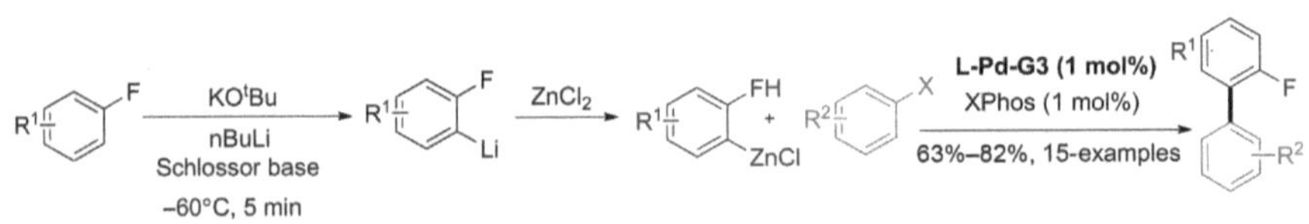

Scheme 3.28 Synthesis of 2-flurobiaryls with an L-Pd-G3 / XPhos catalyst system.

The palladacycle precatalyst XPhos-Pd-G2 in 2 mol% loading was found to be optimal for the highly diastereoselective α-arylation of substituted cyclobutyl nitriles with several aryl, heteroaryl, and alkenyl halides (Scheme 3.29). This is claimed to be the first general example of α-arylation of a nitrile, producing preferentially the *trans*- diastereomer. Further derivatization of the aryl cyclobutyl nitriles was also explored *(138)*.

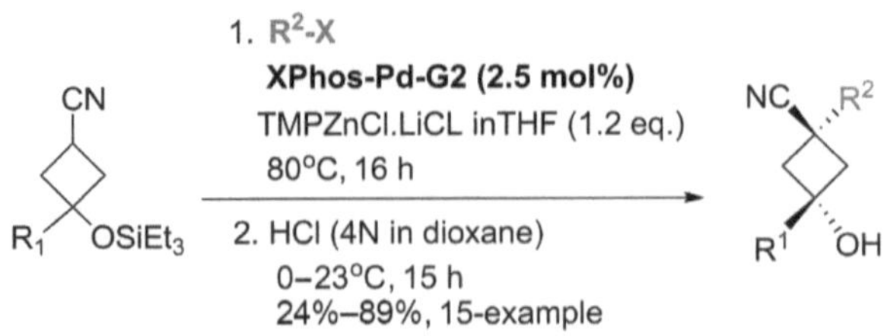

Scheme 3.29 XPhos-Pd-G3 catalyzed α-arylation of substituted cyclobutyl nitriles.

An interesting comparison between CPhos-containing analogous palladacycles from G3, G4, and G5 families as catalysts in Negishi coupling was conducted recently *(139)*. The target reaction was the Negishi coupling of α-CF_3 oxiranyl zincates with aryl bromides or chlorides to access 2-aryl-2-trifluoromethyloxiranes in optically pure form. In addition, a continuous-flow generation of organozincate was linked to the Negishi coupling in a batch reactor with excellent results and broad substrate scope.

Also, in 2017 N-hetero-ortho-phenylenes (N-oPs) containing pyridine moieties and terminal hydroxy groups were obtained by Negishi cross-coupling reactions between bis(chloropyridyl)biphenyl and a methoxymethoxy-protected phenylzinc reagent using XPhos-Pd-G4 as precatalyst (Scheme 3.30) *(140)*.

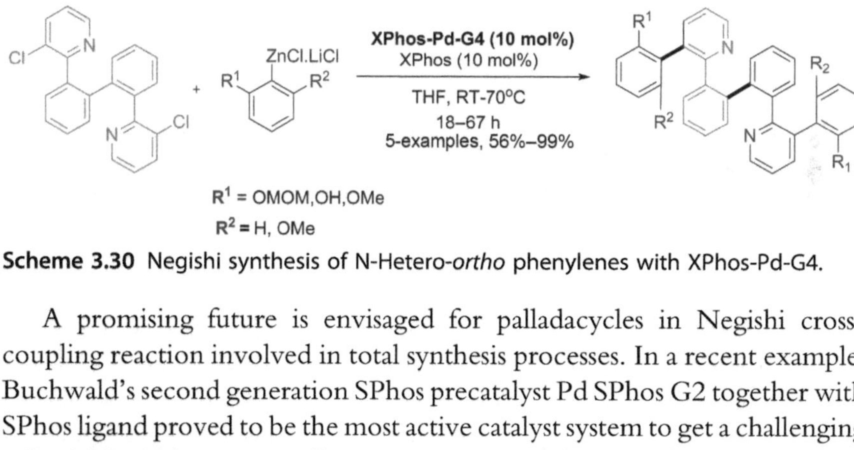

Scheme 3.30 Negishi synthesis of N-Hetero-*ortho* phenylenes with XPhos-Pd-G4.

A promising future is envisaged for palladacycles in Negishi cross-coupling reaction involved in total synthesis processes. In a recent example, Buchwald's second generation SPhos precatalyst Pd SPhos G2 together with SPhos ligand proved to be the most active catalyst system to get a challenging sp2-p3 Negishi cross-coupling constructing of the crucial C15-C16 bond that connects the arene with the decalin subunit in the total synthesis of (+)-stachyflin (Fig. 3.18) *(141)*.

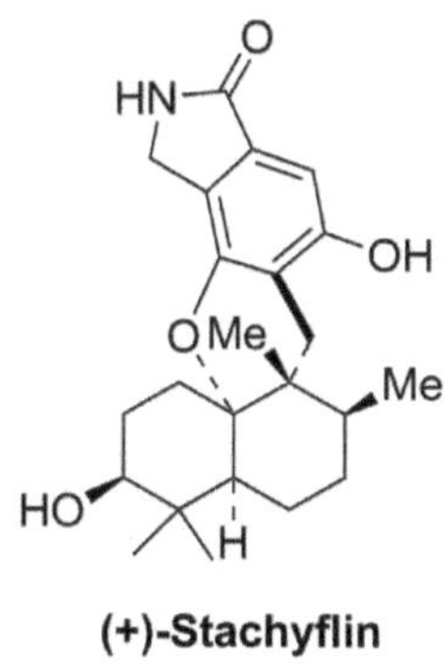

Fig. 3.18 Antiviral meroterpenoid.

4. CONCLUSION

Nowadays the Pd-catalyzed C-N bond forming reactions have become an essential tool for the synthesis of a huge variety of compounds which incorporate N-containing functional groups. Among others, the synthesis of agrochemicals and new organic materials or the pharmaceutical industry are some fields that have benefitted most from the growth and development of this synthetic technology. Preformed palladium precatalysts, specifically palladacycles, have revealed as valuables one-component palladium sources that readily generate the active LnPd(0) species, sometimes outperforming traditional systems with separate Pd and ligand sources. Indeed, the rational ligand/precatalyst design that has accompanied the development of Buchwald-Hartwig amination has produced several generations of reliable systems that work under mild, efficient, and user friendly conditions. Arguably this has been a crucial factor for the universal adoption of Buchwald-Hartwig methodology, including those reactions that have used palladacycles as precatalysts.

On the basis of current commercial availability of Buchwald's palladacycles (G1–G4) and its many practical advantages, in the near future one can anticipate its widespread use in challenging synthetic transformations applied to different disciplines, thus broadening the scope of palladium catalyzed C-N and other cross-coupling reactions.

That would be the case with the Pd-catalyzed Negishi reaction, which has itself evolved as one of the most general methods for C-C bond formation and has already made tremendous contributions in many areas of synthetic organic chemistry. The current and growing availability of Zn organometallics will also foster the expanding of Negishi coupling in the coming years. Although palladacycles have only made a modest contribution as precatalysts in the Negishi reaction so far, greater significance in the field can be expected after the recent advent of G1–G4 palladacycles.

ACKNOWLEDGMENTS

We thank Dr. A. R. Kapdi for helpful suggestions during the preparation of this chapter. J.L.S. wants to acknowledge project CTQ2015-67927-R del 708 Programa Estatal de Investigacion, Desarrollo e Innovacion Orientada a los Restos de la Sociedad, and Agencia de Ciencia y Technologia de la Region de Murcia for project no. 19278/PI/14.

REFERENCES

1. (a) Wu, X.-F.; Anbarasan, P.; Neumann, H.; Beller, M. *Angew. Chem. Int. Ed.* **2010**, *49*, 9047; (b) Seechurn, J. C. C. C.; Kitching, M. O.; Colacot, T. J.; Snieckus, V. *Angew. Chem. Int. Ed.* **2012**, *51*, 5062 and references therein; (c) Devendar, P.; Qu, R.-Y.; Kang, W.-M.; He, B.; Yang, G.-F. *J. Agric. Food Chem.* **2018**, *66*, 2630.
2. (a) Diederich, F., Stang, P. J., Eds.; *Metal-Catalyzed Cross-Coupling Reactions*, Wiley-VCH: Weinheim: Germany, 1998; (b) Jiang, L.; Buchwald, S. L. In *Metal Catalyzed Cross-Coupling Reactions*, 2nd ed.; de Meijere, A., Diederich, F., Eds; John Wiley & Sons: New York, 2004.
3. (a) Ruiz-Castillo, P.; Buchwald, S. L. Chem. Rev. **2016**, *116*, 12564; (b) Mishra, V.; Chundawat, T. S. Curr. Org. Synth. **2018**, *15*, 208; (c) Heravi, M. M.; Kheilkordi, Z.; Zadsirjan, V.; Heydari, M.; Malmir, M. J. Organomet. Chem. **2018**, *17*, 861; (d) de Meijere, A.; Bräse, S.; Oestreich, M. *Metal Catalyzed Cross-Coupling Reactions and More;* Wiley-VCH: Weinheim, 2014;.
4. (a) Beletskaya, I. P.; Cheprakov, A. V. *Organometallics* **2012**, *31*, 7753; (b) Richardson, J.; Ruble, J. C.; Love, E. A.; Berritt, S. *J. Org. Chem.* **2017**, 82, 3741; (c) Inoue, F.; Kashihara, M.; Yadav, M. R.; Nakao, Y. *Angew. Chem. Int. Ed.* **2017**, *56*, 13307.
5. (a) Paul, F.; Patt, J.; Hartwig, J. F. J. Am. Chem. Soc. **1994**, *116*, 5969; (b) Guram, A. S.; Buchwald, S. L. *J. Am. Chem. Soc.* **1994**, *116*, 7901; (c) Guram, A. S.; Rennels, R. A.; Buchwald, S. L. *Angew. Chem. Int. Ed.* **1995**, *34*, 1348; (d) Louie, J.; Hartwig, J. F. *Tetrahedron Lett.* **1995**, *36*, 3609.
6. (a) Smith, M. B.; March, J. *Advanced Organic Chemistry: Reactions, Mechanism and Structure;* Wiley-Intersscience: New York, 2000; (b) van Leeuwen, P.; Kamer, P.; Reek, J.; Dierkes, P. Chem. Rev. **2000**, *100*, 2741.
7. Bunnett, J. F.; Zahler, R. E. *Chem. Rev.* **1951**, *49*, 237.
8. (a) Heaney, H. Chem. Rev. **1962**, *62*, 81; (b) Razzuk, A.; Biehl, E. R. J. Org. Chem. **1987**, *52*, 2619.
9. Lindley, J. *Tetrahedron* **1984**, *40*, 1433.
10. Kosugi, M.; Kameyama, M.; Migita, T. *Chem. Lett.* **1983**, 927.
11. Louie, J.; Hartwig, J. F. Angew. Chem. Int. Ed. **1996**, *35*, 2359.
12. (a) Hartwig, J. F. In *Handbook of Organopalladium Chemistry for Organic Synthesis;* Negishi, E., Ed; Wiley: New York, 2002; (b) Muci, A. R.; Buchwald, S. L. *Top. Curr. Chem* **2002**, *219*, 131.
13. Hesp, K. D.; Genovino, J. RSC Drug Discovery Series No. 52. In *The Royal Society of Chemistry*, Blakemore, D.; Doyle, P.; Fobian, Y., Eds.; Synthetic Methods in Drug Discovery; 2016; Vol. 1.
14. (a) Hartwig, J. *Organotransition Metal Chemistry: From Bonding to Catalysis*, University Science Books: Sausalito, CA, 2010; (b) Sunesson, Y.; Limé, E.; Nilsson Lill, S. O.; Meadows, R. E.; Norrby, P.-O. *J. Org. Chem.* **2014**, *79*, 11961 and references therein.
15. (a) Hoi, K. H.; Calimsiz, S.; Froese, R. D. J.; Hopkinson, A. C.; Organ, M. G. *Chem. Eur. J.* **2012**, *18*, 145; (b) McMullin, C. L.; Bastian, R.; Besora, M.; Orpen, G. A.; Harvey, J. N.; Fey, N. *J. Mol. Catal. A: Chem.* **2010**, *324*, 48.
16. (a) Hamann, B. C.; Hartwig, J. F. *J. Am. Chem. Soc.* **1998**, *120*, 3694; (b) Shekhar, S.; Hartwig, J. F. *Organometallics* **2007**, *26*, 340; (c) Arrechea, P. L.; Buchwald, S. L. *J. Am. Chem. Soc.* **2016**, *138*, 12486.
17. (a) Dennis, J. M.; White, N. A.; Liu, R. Y.; Buchwald, S. L. J. Am. Chem. Soc. **2018**, *140*, 4721; (b) Olsen, P. K.; Arrechea, P. L.; Buchwald, S. L. *Angew. Chem. Int. Ed.* **2017**, *56*, 10569; (c) Hu, H.; Qu, F.; Gerlach, D. L.; Shaughnessy, K. H. ACS Catal. **2017**, 7, 2516.

18. (a) Wolfe, J. P.; Wagaw, S.; Buchwald, S. L. *J. Am. Chem. Soc.* **1996**, *118*, 7215; (b) Harris, M. C.; Geis, O.; Buchwald, S. L. *J. Org. Chem.* **1999**, *64*, 6019.
19. Driver, M. S.; Hartwig, J. F. *J. Am. Chem. Soc.* **1996**, *118*, 7217.
20. (a) Kranenburg, M.; van der Burgt, Y. E. M.; Kamer, P. C. J.; van Leeuwen, P. W. N. M. *Organometallics* **1995**, *14*, 3081; (b) Kranenburg, M.; Delis, J. G. P.; Kamer, P. C. J.; van Leeuwen, P. W. N. M.; Vrieze, K.; Veldman, N.; Spek, A. L.; Goubitz, K.; Fraanje, J. *J. Chem. Soc., Dalton Trans.* **1997**, 1839.
21. Sadighi, J. P.; Harris, M. C.; Buchwald, S. L. *Tetrahedron Lett.* **1998**, 39, 5327.
22. (a) Yin, J.; Buchwald, S. L. *Org. Lett.* **2000**, *2*, 1101; (b) Yin, J.; Buchwald, S. L. *J. Am. Chem. Soc.* **2002**, *124*, 6043.
23. (a) Yamamoto, T.; Nishiyama, M.; Koie, Y. *Tetrahedron Lett.* **1998**, *39*, 2367; (b) Nishiyama, M.; Yamamoto, T.; Koie, Y. Tetrahedron Lett. **1998**, *39*, 617; (c) Hartwig, J. F.; Kawatsura, M.; Hauck, S. I.; Shaughnessy, K. H.; Alcazar-Roman, L. M. *J. Org. Chem.* **1999**, *64*, 5575; (d) Watanabe, M.; Nishiyama, M.; Yamamoto, T.; Koie, Y. *Tetrahedron Lett.* **2000**, *41*, 481; (e) Lee, S.; Jørgensen, M.; Hartwig, J. F. Org. Lett. **2001**, *3*, 2729; (f) Stambuli, J. P.; Kuwano, R.; Hartwig, J. F. *Angew. Chem. Int. Ed.* **2002**, *41*, 4746.
24. (a) Hama, T.; Liu, X. X.; Culkin, D. A.; Hartwig, J. F. J. Am. Chem. Soc. **2003**, *125*, 11176; (b) Kataoka, N.; Shelby, Q.; Stambuli, J. P.; Hartwig, J. F. *J. Org. Chem.* **2002**, *67*, 5553.
25. (a) Huang, J.; Grasa, G.; Nolan, S. P. Org. Lett. **1999**, (1), 1307; (b) Stauffer, S. R.; Lee, S. W.; Stambuli, J. P.; Hauck, S. I.; Hartwig, J. F. Org. Lett. **2000**, (2), 1423; (c) Marion, N.; Navarro, O.; Kelly, R. A.; Nolan, S. P. *Synthesis* **2003**, 2590; (c) Navarro, O.; Marion, N.; Mei, J.; Nolan, S. P. *Chem. Eur. J.* **2006**, *12*, 5142.
26. (a) Zapf, A.; Ehrentraut, A.; Beller, M. *Angew.* Chem. Int. Ed. **2000**, *39*, 4153; (b) Ehrentraut, A.; Zapf, A.; Beller, M. J. M. *Cat. A* **2002**, *515*, 182; (b) Tewari, A.; Hein, M.; Zapf, A.; Beller, M. *Tetrahedron* **2005**, *61*, 9705.
27. (a) Urgaonkar, S.; Nagarajan, M.; Verkade, J. G. *Org. Lett.* **2003**, *5*, 815; (b) Urgaonkar, S.; Nagarajan, M.; Verkade, J. G. *J. Org. Chem.* **2003**, *68*, 452; (c) Urgaonkar, S.; Verkade, J. G. J. Org. Chem. **2004**, *69*, 9135; (d) Urgaonkar, S.; Xu, J. H.; Verkade, J. G. J. Org. Chem. **2003**, *68*, 8416.
28. (a) Shen, Q.; Ogata, T.; Hartwig, J. F. *J. Am. Chem. Soc.* **2008**, *130*, 6586; (b) Shen, Q.; Hartwig, J. F. *Org. Lett.* **2008**, *10*, 4109.
29. (a) Old, D. W.; Wolfe, J. P.; Buchwald, S. L. J. Am. Chem. Soc. **1998**, *120*, 9722; (b) Wolfe, J. P.; Tomori, H.; Sadighi, J. P.; Yin, J. J.; Buchwald, S. L. *J. Org. Chem.* **2000**, *65*, 1158.
30. (a) Wolfe, J. P.; Buchwald, S. L. *Angew. Chem. Int. Ed.* **1999**, *38*, 2413; (b) Wolfe, J. P.; Tomori, H.; Sadighi, J. P.; Yin, J.; Buchwald, S. L. J. Org. Chem. **2000**, *65*, 1158.
31. (a) Billingsley, K.; Buchwald, S. L. J. Am. Chem. Soc. **2007**, *129*, 3358; (b) Huang, X.; Anderson, K. W.; Zim, D.; Jiang, L.; Klapars, A.; Buchwald, S. L. J. Am. Chem. Soc. **2003**, *125*, 6653.
32. Fors, B. P.; Watson, D. A.; Biscoe, M. R.; Buchwald, S. L. *J. Am. Chem. Soc.* **2008**, *130*, 13552.
33. (a) Milne, J. E.; Buchwald, S. L. *J. Am. Chem. Soc.* **2004**, *126*, 13028; (b) Charles, D.; Schultz, P.; Buchwald, S. L. *Org. Lett.* **2005**, 7, 3965.
34. Strieter, E. R.; Buchwald, S. L. *Angew. Chem. Int. Ed.* **2006**, *45*, 925.
35. (a) Lundgren, R. J.; Peters, B. D.; Alsabeh, P. G.; Stradiotto, M. *Angew. Chem. Int. Ed.* **2010**, *49*, 4071; (b) Alsabeh, P. G.; Stradiotto, M. *Angew. Chem. Int. Ed.* **2013**, *52*, 7242.
36. Doherty, S.; Knight, J. G.; McGrady, J. P.; Ferguson, A. M.; Ward, N. A. B.; Harrington, R. W.; Clegg, W. *Adv. Synth. Catal.* **2010**, *352*, 201.

37. (a) Withbroe, G. J.; Singer, R. A.; Sieser, J. E. *Org. Process Res. Dev.* **2008**, *12*, 480; (b) Crawford, S. M.; Lavery, C. B.; Stradiotto, M. *Chem. Eur. J.* **2013**, *19*, 16760.
38. Mao, J.; Zhang, J.; Zhang, S.; Walsh, P. *J.* Dalton Trans. **2018**, *47*, 8690.
39. Ma, Y.-N.; Li, S.-X.; Yang, S.-D. Acc. Chem. Res. **2017**, *50*, 1480.
40. Surry, D. S.; Buchwald, S. L. *Chem. Sci.* **2011**, *2*, 27.
41. (a) Seechurn, J. C. C. C.; Li, H.; Colacot, T. J. *In New Trends in Cross-Coupling: Theory and Applications;* RSC: Cambridge, 2015; (b) Amatore, C.; Broeker, G.; Jutand, A.; Khalil, F. *J. Am. Chem. Soc.* **1997**, *119*, 5176; (c) Fairlamb, I. J. S.; Kapdi, A. R.; Lee, A. F. *Org. Lett.* **2004**, *6*, 4435; (d) Wei, C. S.; Davies, G. H. M.; Soltani, O.; Albrecht, J.; Gao, Q.; Pathirana, C.; Hsiao, Y.; Tummala, S.; Eastgate, M. D. *Angew. Chem. Int. Ed.* **2013**, *52*, 5822; (e) Bruno, N. C.; Tudge, M. T.; Buchwald, S. L. *Chem. Sci.* **2013**, *4*, 916.
42. (a) Stolyarov, I. P.; Demina, L. I.; Cherkashina, N. V. *Russ. J. Inorg. Chem.* **2011**, *56*, 1532; (b) Zalesskiy, S. S.; Ananikov, V. P. *Organometallics* **2012**, *31*, 2302; (c) Carole, W.; Bradley, J.; Sarwar, M.; Colacot, T. *Org. Lett.* **2017**, *21*, 5472.
43. Dupont, J.; Consorti, C. S.; Spencer, J. *Chem. Rev.* **2005**, *105*, 2527.
44. (a) Li, H.; Seechurn, J. C. C. C.; Colacot, T. *J. ACS Catal.* **2012**, *2*, 1147; (b) Gildner, P. G.; Colacot, T. J. *Organometallics* **2015**, *34*, 5497; (c) Hazari, N.; Melvin, P. R.; Beromi, M. M. Nat. Rev. Chem. **2017**, *1*, 25; (d) Dupont, J., Pfeffer, M., Eds *Palladacycles: Synthesis, Characterization and Applications*, Wiley-VCH: Weinheim, 2008.
45. (a) Bruno, N. C.; Buchwald, S. L. *Org. Lett.* **2013**, *15*, 2876; (b) Bedford, R. B.; Cazin, C. S. J.; Coles, S. J.; Horton, G. T. *Organometallics* **2003**, 22, 987; (c) Bedford, -R. B.; Blake, M. E. Adv. Synth. Catal. **2003**, *345*, 1107; (d) Zim, D.; Buchwald, S. L. Org. Lett. **2003**, *5*, 2413.
46. Bruneau, A.; Roche, M.; Alami, M.; Messaoudi, S. ACS Catal. **2015**, *5*, 1386 and references therein.
47. (a) Beller, M.; Fischer, H.; Herrmann, W. A.; Ofele, K.; Brossmer, C. Angew. Chem. Int. Ed. Engl. **1995**, *34*, 1848; (b) Herrmann, W. A.; Bohm, V. P. W.; Reisinger, C.-P. J. Organomet. Chem. **1999**, *576*, 23.
48. (a) Schnyder, A.; Indolese, A. F.; Studer, M.; Blaser, H.-U. Angew. Chem. Int. Ed. **2002**, *41*, 3668; (b) Blaser, H.-U.; Indolese, A. F.; Naud, F.; Nettekoven, U.; Schnyder, A. *Adv. Synth. Catal.* **2004**, *346*, 1583.
49. Viciu, M. S.; Kelly, R. A., III; Stevens, E. D.; Naud, F.; Studer, M.; Nolan, S. P. *Org. Lett.* **2003**, *5*, 1479.
50. Bedford, R. Chem. Commun. **2003**, 1787.
51. (a) Herrmann, W. A.; BroBmer, C.; Ofele, K.; Reisinger, C.-P.; Priermeier, T.; Belier, M.; Fischer, H. *Angew. Chem. Int. Ed. Engl.* **1995**, *34*, 1844; (b) Herrmann, W. A.; Broßmer, C.; Reisinger, C.-P.; Riermeier, T. H.; Öfele, K.; Beller, M. Chem. Eur. J **1997**, *3*, 1357.
52. Beller, M.; Reiemeter, T. H.; Reisinger, C.-P.; Herrmann, W. A. *Tetrahedron Lett.* **1997**, *38*, 2073.
53. a) Christmann, U.; Vilar, R.; White, A. J. P.; Williams, D. J. *Chem. Commun.* **2004**, 1294; (b) Christmann, U.; Pantazis, D. A.; Benet-Buchholz, J.; McGrady, J. E.; Maseras, F.; Vilar, R. *J. Am. Chem. Soc.* **2006**, *128*, 6376.
54. Kirlikovali, K. O.; Cho, E.; Downard, T. J.; Grigoryan, L.; Han, Z.; Hong, S.; Jung, D.; Quintana, J. C.; Reynoso, V.; Ro, S.; Shen, Y.; Swartz, K.; Sahakyan, E. T.; Wixtrom, A. I.; Yoshida, B.; Rheingold, A. L.; Spokoyny, A. M. *Dalton Trans.* **2018**, *47*, 3684.
55. Nettekoven, U.; Naud, F.; Schnyder, A.; Blaser, H.-U. *Synlett* **2004**, (14), 2549.
56. Xu, C.; Wang, Z.-Q.; Fu, W.-J.; Lou, X.-H.; Li, Y.-F.; Cen, F.-F.; Ma, H.-J.; Ji, B.-M. *Organometallics* **2009**, *28*, 1909.

57. Wu, Y. J.; Yang, F.; Zhang, J. L.; Cui, X. L.; Gong, J. F.; Song, M. P.; Li, T. S. *Chin. Sci. Bull.* **2010**, *55*, 2784.
58. Avila-Sorrosa, A.; Estudiante-Negrete, F.; Hernández-Ortega, S.; Toscano, R. A.; Morales-Morales, D. *Inorg. Chim. Acta* **2010**, *363*, 1262.
59. Xu, C.; Wang, Z.-Q.; Zhang, Y.-P.; Dong, X.-M.; Hao, X.-Q.; Fu, W.-J.; Ji, B.-M.; Song, M.-P. *Eur. J. Inorg. Chem.* **2011**, 4878.
60. Biscoe, M. R.; Fors, B. P.; Buchwald, S. L. J. Am. Chem. Soc. **2008**, *130*, 6686.
61. Kinzel, T.; Zhang, Y.; Buchwald, S. L. J. Am. Chem. Soc. **2010**, *132*, 14073.
62. Bruno, N. C.; Niljianskul, N.; Buchwald, S. L. *J. Org. Chem.* **2014**, *79*, 4161.
63. Biscoe, M. R.; Barder, T. E.; Buchwald, S. L. *Angew. Chem. Int. Ed.* **2007**, *46*, 7232.
64. Vicente, J.; Saura-Llamas, I.; Olivia-Madrid, M.; Garcia-Lopez, J. *Organometallics* **2011**, *30*, 4624.
65. Biscoe, M. R.; Buchwald, S. L. *Org. Lett.* **2009**, *11*, 1173.
66. Fors, B. P.; Davis, N. R.; Buchwald, S. L. J. Am. Chem. Soc. **2009**, *131*, 5766.
67. Fors, B. P.; Buchwald, S. L. J. Am. Chem. Soc. **2010**, *132*, 15914.
68. De Turiso, F. G.-L.; Shin, Y.; Brown, M.; Cardozo, M.; Chen, Y.; Fong, D.; Hao, X.; He, X.; Henne, K.; Hu, Y.-L.; Johnson, M. G.; Kohn, T.; Lohman, J.; McBride, H. J.; McGee, L. R.; Medina, J. C.; Metz, D.; Miner, K.; Mohn, D.; Pattaropong, V.; Seganish, J.; Simard, J. L.; Wannberg, S.; Whittington, D. A.; Yu, G.; Cushing, T. D. *J. Med. Chem.* **2012**, *55*, 7667.
69. (a) Henderson, J. L.; McDermott, S. M.; Buchwald, S. L. Org. Lett. **2010**, *12*, 4438; (b) Henderson, J. L.; Buchwald, S. L. Org. Lett. **2010**, *12*, 4442.
70. Naber, J. R.; Buchwald, S. L. Angew. Chem. Int. Ed. **2010**, *49*, 9469.
71. DeAngelis, A.; Wang, D.-H.; Buchwald, S. L. Angew. Chem. Int. Ed. **2013**, *52*, 615.
72. Maiti, D.; Fors, B. P.; Henderson, J. L.; Buchwald, S. L. *Chem. Sci.* **2011**, *2*, 57.
73. Hopkin, M. D.; Baxendale, I. R.; Ley, S. V. *Chem. Commun.* **2010**, *46*, 2450.
74. Stec, M. M.; Andrews, K. L.; Bo, Y.; Caenepeel, S.; Liao, H.; McCarter, J.; Mullady, E. L.; San Miguel, T.; Subramanian, R.; Tamayo, N.; Whittington, D. A.; Wang, L.; Wu, T.; Zalameda, L. P.; Zhang, N.; Hughes, P. E.; Norman, M. H. *Bioorg. Med. Chem. Lett.* **2015**, *25*, 4136.
75. Plas, A.; Martin, C.; Joubert, N.; Viaud-Massuard, M.-C. Org. Lett. **2015**, *17*, 4710.
76. Wlochal, J.; Bayley, A. *Tetrahedron Lett.* **2015**, *56*, 6791.
77. Benedetti, E.; Kocsis, L. S.; Brummond, K. M. *J. Am. Chem. Soc.* **2012**, *134*, 12418.
78. Engelhart, J. U.; Tverskoy, O.; Bunz, U. H. F. *J. Am. Chem. Soc.* **2014**, *136*, 15166.
79. Bourbeau, M. P.; Ashton, K. S.; Yan, J.; St. Jean, D. J., Jr. *J. Org. Chem.* **2014**, *79*, 3684.
80. Christensen, H.; Schjøth-Eskesen, C.; Jensen, M.; Sinning, S.; Jensen, H. H. *Chem. Eur. J.* **2011**, *17*, 10618.
81. Mari, M.; Bartoccini, F.; Piersanti, G. J. Org. Chem. **2013**, *78*, 7727.
82. Bruno, N. C.; Buchwald, S. L. *The Strem Chemiker;* 2014; p. 27and references therein.
83. Gao, G.; Zheng, B.; Fu, Y.; Li, M.; Wang, B.; Chen, X.-Z.; Zhang, Y.-Y.; Liu, J.-J.; Hou, S.-C.; Walsh, P. W. *Asian J. Org. Chem.* **2017**, *6*, 654.
84. Bair, J. S.; Palchaudhuri, R.; Hergenrother, P. J. J. Am. Chem. Soc. **2010**, *132*, 5469.
85. Xu, Z.; Li, K.; Zhai, R.; Liang, T.; Guia, X.; Zhang, R. RSC Adv. **2017**, 7, 51972.
86. Phelan, J. P.; Wiles, R. J.; Lang, S. B.; Kelly, C. H. B.; Molander, G. A. *Chem. Sci.* **2018**, *9*, 3215.
87. Jui, N. T.; Buchwald, S. L. *Angew. Chem. Int. Ed.* **2013**, *52*, 11624.
88. Cheung, C. W.; Surry, D. S.; Buchwald, S. L. Org. Lett. **2013**, *15*, 3734.
89. Green, R. A.; Hartwig, J. F. *Org. Lett.* **2014**, *16*, 4388.
90. Ernst, J. B.; Tay, N. E.; Jui, N. T.; Buchwald, S. L. *Org. Lett.* **2014**, *16*, 3844.
91. Sromek, A. W.; Provencher, B. A.; Russell, S.; Chartoff, E.; Knapp, S. I.; Bidlack, J. M.; Neumeyer, J. L. *ACS Chem. Neurosci.* **2014**, *5*, 93.
92. Chen, M.; Buchwald, S. L. *Angew.* Chem. Int. Ed. **2013**, *52*, 4247.

93. Grzybowski, M.; Hugues, V.; Blanchard-Desce, M.; Gryko, D. T. *Chem. Eur. J.* **2014**, *20*, 12493.
94. Ruiz-Castillo, P.; Blackmond, D. G.; Buchwald, S. L. *J. Am. Chem. Soc.* **2015**, *137*, 3085.
95. Smith, S. M.; Buchwald, S. L. *Org. Lett.* **2016**, *18*, 2180.
96. Hopkins, B. A.; Smith, G. F.; Sciammetta, N. *Org. Lett.* **2016**, *18*, 4072.
97. King, S. M.; Buchwald, S. L. *Org. Lett.* **2016**, *18*, 4128.
98. (a) Al-Shuaeeb, R. A. A.; Kolodych, S.; Koniev, O.; Delacroix, S.; Erb, S.; Nicolaÿ, S.; Cintrat, J.-C.; Brion, J.-D.; Cianférani, S.; Alami, M.; Wagner, A.; Messaoudi, S. *Chem. Eur. J.* **2016**, *22*, 11365; (b) Al-Shuaeeb, R. A. A.; Dejean, C.; Alami, M.; Messaoudi, S. *Chem. Asian J.* **2017**, *12*, 3114.
99. Spindler, F.; Furer, P.; Rotzler, J. *Chim. Oggi Chem. Today* **2015**, *33*, 26.
100. Ingoglia, B. T.; Buchwald, S. L. *Org Lett.* **2017**, *19*, 2853.
101. Arduengo, A. J., III; Harlow, R. L.; Kline, M. *J. Am. Chem. Soc.* **1991**, *113*, 361.
102. Marion, N.; Nolan, S. P. *Acc. Chem. Res.* **2008**, *41*, 1440.
103. (a) Wolfgang, A.; Herrmann, W. A.; Öfele, K.; Preysing, D. V.; Schneider, S. K. *J. Organomet. Chem.* **2003**, *687*, 229; (b) Fortman, G. C.; Nolan, S. P. *Chem. Soc. Rev.* **2011**, *40*, 5151; (c) Chartoire, A.; Nolan, S. P. RSC Catalysis Series No. 21. In *New Trends in Cross-Coupling: Theory and Applications,* Colacot, T. J., Ed.; The Royal Society of Chemistry, 2015 (chapter 4).
104. (a) Tian, X.; Lin, J.; Zou, S.; Lv, J. W.; Huang, Q. F.; Zhu, J.; Huang, S. P.; Wang, Q. W. *J. Organomet. Chem.* **2018**, *861*, 125; (b) Zhang, Y.; Lavigne, G.; Lugan, N.; César, V. *Chem. Eur. J.* **2017**, *23*, 13792; (c) Shi, S. C.; Szostak, M. Chem. Commun. **2017**, *53*, 10584; d) Lan, X.-B.; Li, Y.; Li, Y.-F.; Shen, D.-S.; Ke, Z.; Liu, F.-S. *J. Org. Chem.* **2017**, *82*, 2914; d) Khadra, A.; Mayer, S.; Organ, M. G. *Chem. Eur. J.* **2017**, *23*, 3206.
105. O'Brien, C. J.; Kantchev, E. A. B.; Valente, C.; Hadei, N.; Chass, G. A.; Lough, A.; Hopkinson, A. C.; Organ, M. G. *Chem. Eur. J.* **2006**, *12*, 4743.
106. Navarro, O.; Marion, N.; Oonishi, Y.; Kelly, R. A., III; Nolan, S. P. J. Org. Chem. **2006**, *71*, 685.
107. Broggi, J.; Clavier, H.; Nolan, S. P. *Organometallics* **2008**, *27*, 5527.
108. Li, J.; Cui, M.; Yu, A.; Wu, Y. *J. Organomet. Chem.* **2007**, *692*, 3732.
109. Mu, B.; Lia, J.; Wu, Y. *Appl. Organometal. Chem.* **2013**, *27*, 537.
110. Xu, C.; Li, H.-M.; Liu, H.; Zhang, Z.-Q.; Wanga, Z.-Q.; Fu, W.-J.; Zhang, Y.-Q. Inorg. Chim. Acta **2012**, *386*, 22.
111. Deng, Q.; Zhang, Y.; Zhu, H.; Tu, T. *Chem. Asian J.* **2017**, *12*, 2364.
112. Yoshida, Y.; Otsuka, S.; Nogi, K.; Yorimitsu, H. *Org. Lett.* **2018**, *20*, 113.
113. Negishi, E. *Angew. Chem. Int. Ed.* **2011**, *50*, 6738.
114. (a) Negishi, E.; King, A. O.; Okukado, N. J. Org. Chem. **1977**, *42*, 1821; (b) King, A. O.; Okukado, N.; Negishi, E. J. Chem. Soc. Chem. Commun. **1977**, 683.
115. Tamao, K.; Hiyama, T.; Negishi, E., Eds.; *Special Issue on 30 Years of the Cross-coupling Reactions,* 2002 *J. Organomet. Chem.*, *653*, 1, specifically 34-40 by Negishi, E.
116. Negishi, E.; deMeijere, A. *Handbook of Organopalladium Chemistry for Organic Synthesis.* Wiley-Interscience: New York, 2002.
117. Haas, D.; Hammann, J. M.; Greiner, R.; Knochel, P. *ACS Catal.* **2016**, *6*, 1540.
118. Negishi, E.; Hu, Q.; Huang, Z.; Qian, M.; Wang, G.; Brown, H. C. A. *Aldrichim.Acta* **2005**, *38*, 71.
119. Limberakis, C. RSC Drug Discovery Series No. 52. In *Synthetic Methods in Drug Discovery,* Blakemore, D., Doyle, P., Fobian, Y., Eds The Royal Society of Chemistry, 2016; *Vol. 1* (chapter 2).
120. Heravi, M. M.; Hashemi, E.; Nazari, N. *Mol. Divers* **2014**, *18*, 441.
121. Manolikakes, G.; Dong, Z.; Mayr, H.; Li, J.; Knochel, P. *Chem. Eur. J.* **2009**, *15*, 1324.

122. (a) Haag, B.; Mosrin, M.; Ila, H.; Malakhov, V.; Knochel, P. *Angew. Chem.* **2011**, *123*, 9968; (b) Bernhardt, S.; Manolikakes, G.; Kunz, T.; Knochel, P. *Angew. Chem.* **2011**, *123*, 9372; (c) Stathakis, C. I.; Bernhardt, S.; Quint, V.; Knochel, P. *Angew. Chem.* **2012**, *124*, 9563.
123. (a) del Pozo, J.; Salas, G.; Álvarez, R.; Casares, J. A.; Espinet, P. *Organometallics* **2016**, 35, 3604; (b) Gioria, E.; Martinez-Ilarduya, J. M.; Espinet, P. *Organometallics* **2014**, *33*, 4394.
124. Boeck, K.; Feil, J. E.; Karaghiosoff, K.; Koszinowski, K. *Chem. Eur. J.* **2015**, *21*, 5548.
125. (a) Valente, C.; Belowich, M. E.; Hadei, N.; Organ, M. G. Eur. J. Org. Chem. **2010**, *23*, 4343; (b) Valente, C.; Çalimsiz, S.; Hoi, K. H.; Mallik, D.; Sayah, M.; Organ, M. G. *Angew. Chem. Int. Ed.* **2012**, *51*, 3314.
126. Yang, Y.; Oldenhuis, N. J.; Buchwald, S. L. *Angew. Chem. Int. Ed.* **2013**, *52*, 615.
127. Beletskaya, I. P.; Cheprakov, A. V. *J. Organomet. Chem.* **2004**, *689*, 4055.
128. Ratti, R. *Can. Chem. Trans.* **2014**, (2), 467.
129. Gerber, R.; Blacque, O.; Frech, C. M. *Dalton Trans.* **2011**, *40*, 8996.
130. Colombe, J. R.; Bernhardt, S.; Stathakis, C.; Buchwald, S. L.; Knochel, P. *Org. Lett.* **2013**, *15*, 5754.
131. Yang, Y.; Mustard, T. J. L.; Cheong, P. H.-Y. *Angew. Chem. Int. Ed.* **2013**, *52*, 14098.
132. Yang, Y.; Niedermann, K.; Han, C.; Buchwald, S. L. Org. Lett. **2014**, *16*, 4638.
133. Carrillo, J.; Gómez, A.; Costa, A. M.; Fernández, P.; Isart, C.; Sidera, M.; Vilarrasa, J. Tetrahedron Lett. **2014**, *55*, 4623.
134. (a) Krasovskiy, A.; Duplais, C.; Lipshutz, B. H. *J. Am. Chem. Soc.* **2009**, *131*, 15592; (b) Duplais, C.; Krasovskiy, A.; Lipshutz, B. H. *Organometallics* **2011**, *30*, 6090.
135. Bhonde, V. R.; OÏNeill, B. T.; Buchwald, S. L. *Angew. Chem. Int. Ed.* **2016**, *55*, 1849.
136. Wiethan, C.; Rosa, W. C.; Bonacorso, H. G.; Stradiotto, M. *Org. Biomol. Chem.* **2016**, *14*, 2352.
137. Roesner, S.; Buchwald, S. L. *Angew. Chem. Int. Ed.* **2016**, *55*, 10463.
138. Dalziel, M. E.; Chen, P.; Carrera, D. E.; Zhang, H.; Gosselin, F. Org. Lett. **2017**, *19*, 3446.
139. Zhang, H.; Buchwald, S. L. *J. Am. Chem. Soc.* **2017**, *139*, 11590.
140. Tokoro, Y.; Ohtsuka, N.; Kusakabe, A.; Fukuzawa, S. Eur. J. Org. Chem. **2017**, 2353.
141. Haut, F.-L.; Speck, K.; Wildermuth, R.; Möller, K.; Mayer, P.; Magauer, T. *Tetrahedron* **2018**, *74*, 3348.

CHAPTER 4

Introduction of Water-Solubility in Palladacycles and Their Catalytic Applications

Kevin H. Shaughnessy
Department of Chemistry & Biochemistry, The University of Alabama, Tuscaloosa, AL, United States

Contents

1. INTRODUCTION

1.1 Aqueous-Phase Catalysis

Catalyst systems typically either work under homogeneous (Fig. 4.1A) or heterogeneous (Fig. 4.1B) conditions. In homogeneous catalysis, the catalyst is a soluble species, typically a molecular complex, whereas in heterogeneous catalysis the catalyst is insoluble in the reaction medium. The development of modern metal-catalyzed transformations in organic synthesis has largely relied on the development of homogeneous metal-ligand complexes that can promote reactions with high activity and selectivity. Important classes of reactions, such as cross-coupling of organic halides, olefin metathesis, asymmetric hydrogenation, and hydroformylation primarily rely on molecular metal ligand complexes that are soluble in the reaction medium. The ability to finely tune the ligand environment of a metal complex has allowed

Palladacycles
https://doi.org/10.1016/B978-0-12-815505-9.00005-6
© 2019 Elsevier Inc.
All rights reserved.

the development of catalysts capable of controlling regio-, diastereo-, and enantioselectivity in reactions with high degrees of precision. Similar precise control is generally not possible in heterogeneous catalyst systems, where the nature of the active catalyst species is often ill-defined.

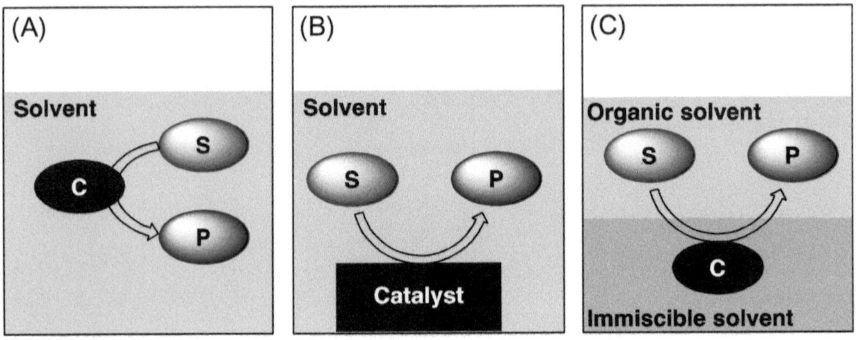

Fig. 4.1 Modes of catalysis. (A) Homogeneous catalysis; (B) Heterogeneous catalysis; (C) Biphasic catalysis.

Heterogeneous catalysts offer a significant advantage over homogeneous catalysts, particularly when applied on an industrial scale. Heterogeneous catalysts can be readily separated from liquid or gaseous organic products by simple filtration. Heterogeneous catalysts are also ideally suited for flow chemistry where reagents are passed through a catalyst bed. Many large volume chemical processes rely on heterogeneous catalysts.

There has been a long-standing interest in finding ways to marry the best properties of homogeneous and heterogeneous catalysts by developing metal-ligand complexes that are constrained in a separable phase. Efforts to heterogenize metal complexes by attaching them to solid supports have met with mixed success *(1)*. An alternative approach that has received significant attention is to utilize two immiscible liquid phases (Fig. 4.1C): an organic phase in which the substrates and products are dissolved and an immiscible phase in which the catalyst is retained. In this way, the catalyst can be a soluble species, yet it can be readily separated from the organic product stream. Examples of organic immiscible phases that have been explored include water, fluorous solvents, and ionic liquids *(2)*. Of these alternative solvents, water has received the most attention due to its low toxicity, nonflammability, and relatively low cost *(3)*.

Aqueous biphasic catalysis was first introduced on an industrial scale by Rhône-Poulenc with the development of an aqueous-phase process for the hydroformylation of propene catalyzed by a Rh/TPPTS catalyst system (TPPTS = trisodium tris(3-sulfonatophenyl)phosphine) *(4)*. In this process, the butanal product can be decanted from the catalyst-containing aqueous phase, allowing the expensive rhodium catalyst to be recycled. The concept of aqueous-biphasic catalysis using water-soluble metal complexes has been applied to nearly all classes of metal-catalyzed reactions *(3c,5)*. Palladium-catalyzed cross-coupling is particularly well-suited for aqueous-biphasic catalysis *(6)*. Organopalladium compounds are generally stable in water under neutral or basic conditions. Large numbers of water-soluble ligand systems have been developed to support aqueous-phase palladium-catalyzed cross-coupling reactions.

1.2 Approaches to Introducing Water-Solubility in Palladacycles

As described throughout this book, palladacycles are valuable precatalysts for a wide range of catalytic processes. Palladacycles tend to be robust, air-stable complexes that can provide high turnover numbers in catalytic processes. With the development of aqueous-phase catalysis, there has been interest in applying palladacycles as catalysts in water. One approach is to use hydrophobic palladacycles for on-water reactions *(7)*. Alternatively, water-soluble palladacycles can be used to perform aqueous-biphasic catalysis (in-water). Two different strategies have been used in the synthesis of water-soluble palladacycles. The most common approach is to modify the organic ligand of the palladacycle with hydrophilic substituents, such as ionizable species or poly(ethylene glycol) (PEG). A second strategy is for the palladium center to be cationic with a noncoordinating counteranion. In this approach, the organic ligand can be hydrophobic with the water-solubility provided by the separated ion pair.

Ionizable substituents, such as carboxylic acids, sulfonic acids, and phenols are most commonly used as water-solubilizing elements for palladacycles. Nájera reported the first examples of a water-soluble palladacycle applied in cross-coupling chemistry *(8)*. The Nájera palladacycle (**1**, Fig. 4.2) is formed from the 4-hydroxyacetophenone oxime. The acidic phenol and oxime OH groups provide water-solubility to the complex, particularly under the basic conditions typically employed in cross-coupling reactions. The solubility of this complex in water has not been reported, however. *N,N*-Dimethyl-4-hydroxybenzylamine (**2**) provides a palladacycle with ionizable phenolic substituent. Palladacycle **3** with a hydroxymethyl substituent is reported to be water soluble, although no data on its solubility is provided *(9)*. The hydroxymethyl group, which is not ionizable under typical aqueous phase conditions, would not be expected to provide significant water-solubility.

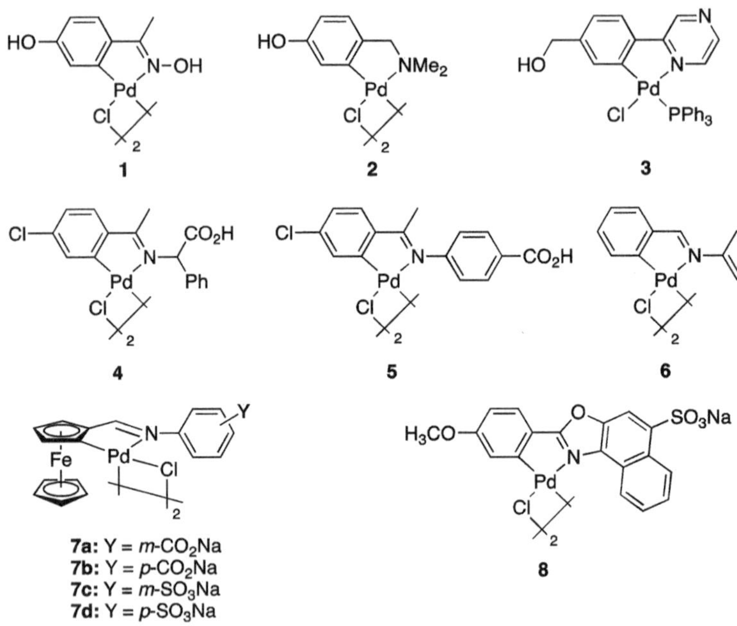

Fig. 4.2 Palladacycles with ionizable hydrophilic substituents.

Sulfonate and carboxylate anions are commonly used to generate water-soluble ligands for aqueous-phase catalysis *(5a)*. Carboxylic acid-functionalized benzaldehyde imine ligands are readily prepared by condensation of aminoacids *(10)* (**4**, Fig. 4.2) or *p*-aminobenzoic acid *(11)* (**5**) with benzaldehyde. Complexation with palladium affords carboxylic acid-substituted palladacycles. A similar approach using 4-aminobenzenesulfonic acid provides a sulfonated benzaldehyde imine palladacycle (**6**) *(12)*. Sulfonated and carboxylated ferrocenylimine ligands have also been used to prepare water-soluble palladacycles (**7**) *(13)*. A readily prepared sulfonated 2-arylnaphthoxazole ligand affords water-soluble palladacycle **8** *(14)*.

Poly (ethylene glycol) is commonly used as a neutral hydrophobic substituent to solubilize compounds in water. The neutral compounds have the advantage of having solubility in moderately polar organic solvents, which makes them easier to prepare and purify than ionic compounds. The PEG-modified complexes still partition into the aqueous phase in organic-water systems. A PEG-substituted 4-aryltriazole palladacycle (**9**) was prepared by Cu-catalyzed cyclization of PEG-azide and an arylalkyne followed by metalation of the 4-aryltriazole with Pd(II) (Fig. 4.3) *(15)*. An SCS-pincer complex with PEG substituents was prepared from PEG-thiolate to give water-soluble pincer complex **10** *(16)*.

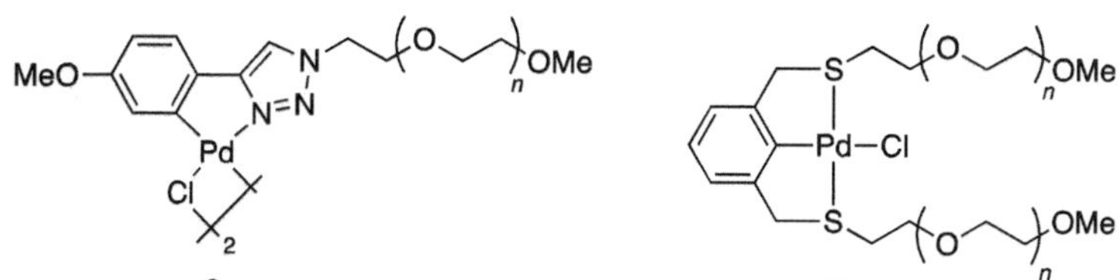

9 **10**

Fig. 4.3 Palladacycles with PEG hydrophilic substituents.

Introducing hydrophilic groups on the ligand requires additional synthetic steps. An alternative approach to preparing water-soluble palladacycle complexes is for the metal center to be cationic with noncoordinating anions. The resulting complexes are water-soluble and have been successfully used as catalyst precursors. Aquo complexes formed by replacing coordinating anions, such as halides, with weakly coordinating anions provide water-soluble palladacycle (**12**, Fig. 4.4) *(17)* and pincer *(18)* (**13**) complexes. Halide abstraction in the presence of other neutral ligands can also provide cationic palladacyclic precatalysts (**14**) *(19)*. Although these precatalyst species are water-soluble, the water-solubility may be lost as these species are converted to the catalytic intermediates.

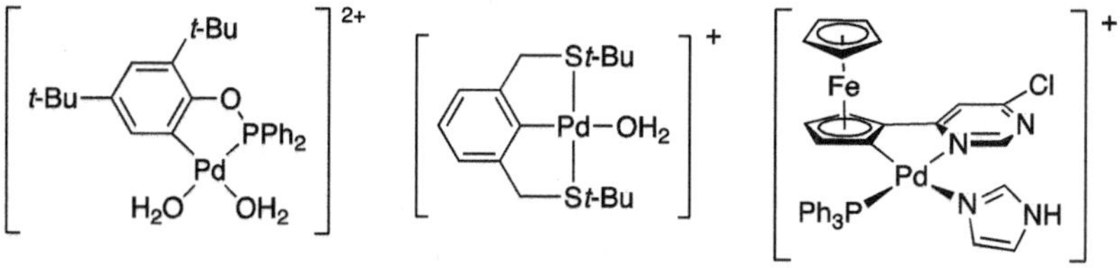

12 **13** **14**

Fig. 4.4 Cationic palladacycle complexes.

2. APPLICATION OF WATER-SOLUBLE PALLADACYCLES IN CATALYTIC PROCESSES

2.1 Oxime-Derived Palladacycles

The first, and still most widely used, hydrophilic palladacyclic precatalyst was developed by the Nájera group *(11,20)*. The Nájera palladacycle is based on an oxime ligand derived from 4′-hydroxyacetophenone (**1**) or 4,4′-dihydroxybenzophenone (**15**), the acetophenone-derived palladacycle **1** being most commonly used (Scheme 4.1). The oxime palladacycles are air- and thermally stable. They are derived from inexpensive materials and can be prepared simply and in high yield from the oxime and $LiPdCl_4$. The exact nature of the active species generated from the oxime palladacycles is not known, but evidence suggests that palladium nanoparticles are the true active species *(21)*. The oxime palladacycles are effective precatalysts for Suzuki, Hiyama, and Heck couplings of aryl halides in aqueous media.

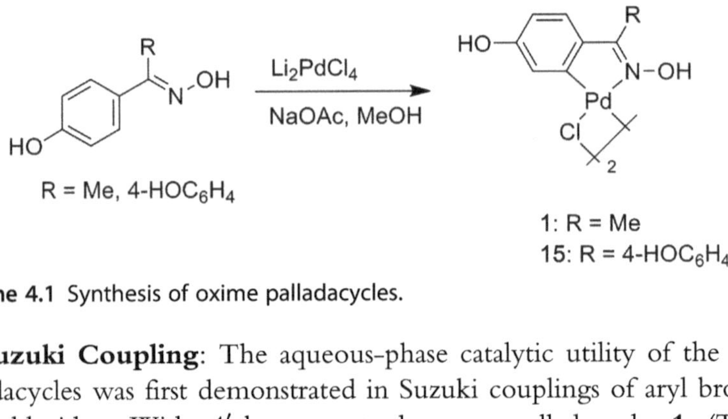

1: R = Me
15: R = 4-HOC_6H_4

Scheme 4.1 Synthesis of oxime palladacycles.

Suzuki Coupling: The aqueous-phase catalytic utility of the oxime palladacycles was first demonstrated in Suzuki couplings of aryl bromides and chlorides. With 4′-bromoacetophenone, palladacycle **1** (TOF = 37,200 h^{-1}) provides a catalyst that is only modestly more active than that derived from $Pd(OAc)_2$ (TOF = 31,300 h^{-1}) *(8b)*. With the less reactive 4′-chloroacetophenone, precatalyst **1** (TOF = 3850 h^{-1}) is much more active than $Pd(OAc)_2$ (TOF = 250 h^{-1}) *(8a)*. Palladacycle **1** provides good yields with a range of structurally diverse aryl bromides using 10^{-3}–10^{-2} mol% palladium at 100°C in water or at room temperature in aqueous methanol (Scheme 4.2). A similar substrate scope was demonstrated for aryl chlorides, but a higher loading of precatalyst **1** (0.01–1 mol%) and the addition of TBAB as an additive was required. TBAB likely serves to stabilize the palladium nanoparticles for the longer reaction times required for coupling of aryl chlorides *(22)*. Palladacycle **1** is also effective in the Suzuki coupling ofbenzylic chlorides with phenyl boronic acid and the coupling of aryl bromides with alkyl boronic acids *(8b)*.

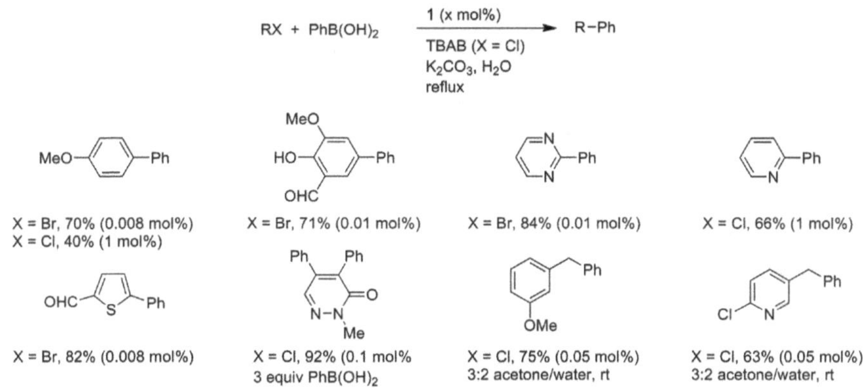

Scheme 4.2 Suzuki coupling of aryl and benzyl halides using precatalyst **1**.

Nájera showed that palladacycle **1** is also an effective precatalyst for the Suzuki coupling of aryl, allyl, and benzyl halides with aryl- and alkenyltrifluoroborates *(23)*. Organotrifluoroborates are attractive alternatives to boronic acids because they are typically more crystalline and easier to purify. Palladacycle **1** in combination with TBAB provides high yields for the coupling of aryl chlorides with aryl trifluoroborates in water at 100°C using either conventional or microwave heating *(23a)*. Benzylic and allylic chlorides are more reactive than aryl chlorides and can be coupled at temperatures ranging from 25°C to 50°C in aqueous acetone. High selectivity is achieved when both sp^2 C—Cl bonds and allylic/benzylic chlorides are present in the substrate (Scheme 4.3). Similar results are achieved with vinyl trifluoroborates *(23b)*. The palladacycle **1**/TBAB system can be used for four reaction cycles before significant degradation in activity occurs. Recycling was demonstrated with 4′-bromoacetophenone, cinnamyl chloride, and benzyl chloride coupling with phenyl trifluoroborate. Palladium leaching into the crude organic product ranged from 50 to 100 ppm/cycle. Palladacycle **1** showed modest activity for the Suzuki coupling of aryl imidazolesulfonates in aqueous methanol (Scheme 4.4). Higher yields are achieved with oxime **16** derived from 4,4′-dichlorobenzophenone oxime *(24)*.

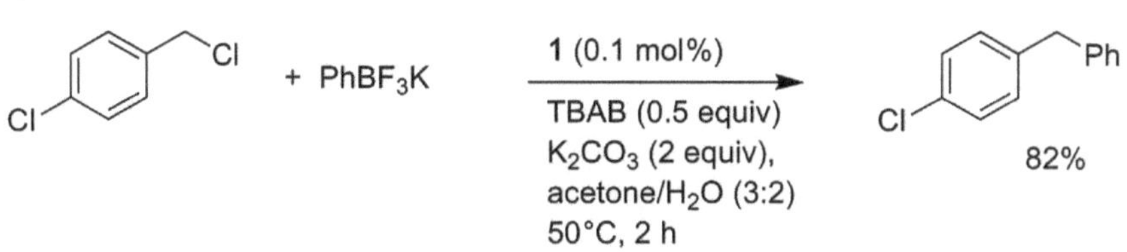

Scheme 4.3 Chemoselective Suzuki coupling of 4-chlorobenzyl chloride.

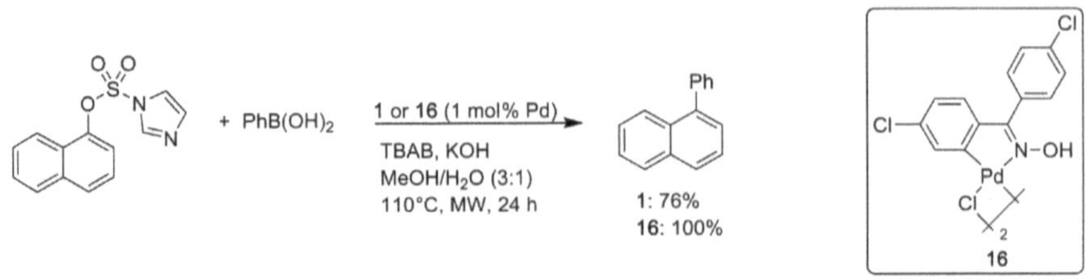

Scheme 4.4 Suzuki coupling of imidazolesulfonates using oxime palladacycles.

Hiyama Coupling: Oxime palladacycle **1** is an effective catalyst for the Hiyama coupling of vinylsiloxanes and aryl halides *(25)*. Nájera reported a fluoride-free methodology using sodium hydroxide as the reaction promoter. Both $Pd(OAc)_2$ and **1** provide similar yields, with **1** affording slightly higher yields in most cases. This result suggests that the active species is likely palladium nanoparticles derived from **1** or $Pd(OAc)_2$. The palladacycle precatalyst provided modestly better recycling than $Pd(OAc)_2$ *(25b)*. In the coupling of 4-bromoacetophenone with trimethyl vinylsiloxane, complex **1** gave yields above 95% for six reaction cycles (Scheme 4.5). In comparison, the $Pd(OAc)_2$-derived catalyst lost activity in the fifth cycle (50% yield). The $Pd(OAc)_2$-derived catalyst also began producing acetophenone as a byproduct in the second reaction cycle, compared to the fourth cycle for complex **1**.

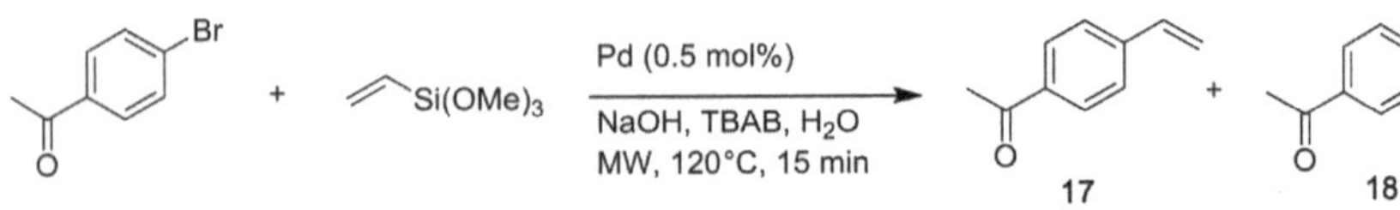

	Pd(OAc)$_2$			1		
Cycle	17 (%)	18 (%)	Pd leaching (ppm)	17 (%)	18 (%)	Pd leaching (ppm)
1	99	0	3.6	99	0	3.6
2	99	3	3.6	99	0	3.6
3	90	7	ND	99	0	ND
4	99	16	76.4	99	5	54.5
5	50	21	192	95	4	90.5
6				99	8	ND
7				42	11	219

Scheme 4.5 Recycling in Hiyama coupling of aryl bromides using $Pd(OAc)_2$ or complex **1**.

Heck Coupling: The Nájera palladacycle (**1**) catalyzes the Heck coupling of aryl iodides and bromides with electron-deficient alkenes *(26)* and styrene derivatives *(27)*. Aryl iodides afford high yields using 0.01 mol% **1** in water at 120°C. In the case of aryl bromides, a mixed solvent of water and DMA is required along with addition of TBAB. Complex **1** is a more effective precatalyst than $Pd(OAc)_2$ under these conditions. Disubstituted alkenes are also effectively arylated with complex **1**. *E*-Disubstituted alkenes provide diastereomeric mixtures of alkene products favoring the *E*-isomer (Scheme 4.6). The product ratio varied from 63:37 to 100/0 *E*:*Z* depending on the nature of the aryl halide and alkene substrates. Dimethyl itaconate is regio- and stereospecifically arylated on the unsubstituted alkene carbon to afford the *E*-product. Using a two-fold excess of aryl halide to alkene allowed the selective β,β-diarylation of electron deficient alkenes. Palladium(II) acetate was a significantly less active precatalyst for the diarylation reaction.

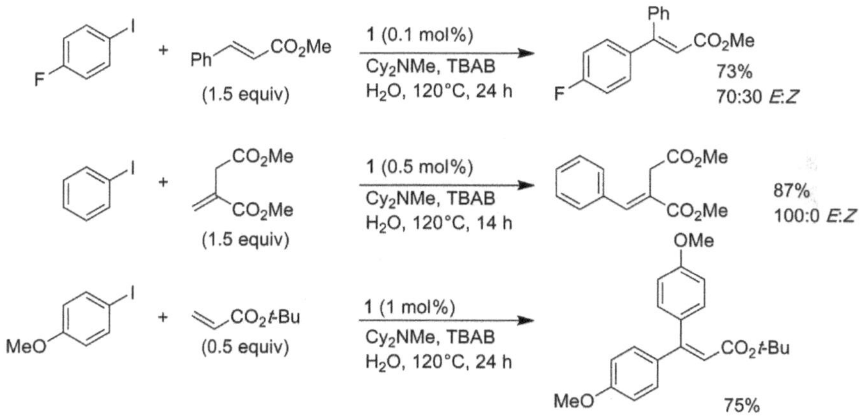

Scheme 4.6 Mono- and diarylation of alkenes using palladacycle **1**.

The Heck coupling of aryl halides with allylic alcohols affords aryl ketones as the major product along with smaller amounts of allylic alcohol products *(28)*. The coupling of iodobenzene with 3-buten-2-ol using 0.1 mol% **1** in water afforded a 98% yield of a mixture of 4-phenyl-2-butanone, 3-phenyl-2-butanone, 4-phenyl-3-buten-2-ol, and 3-phenyl-3-buten-2-ol in a 44:1.5:3.5:1 ratio (Scheme 4.7). A variety of 2-arylethyl ketones and aldehydes were prepared by this methodology in 55%–90% yield. In the case of 2-methyl-3-buten-2-ol, a tertiary alcohol, the allylic product is obtained along with a diene byproduct formed by dehydration of the alcohol. Coupling of 2-iodoaniline with allylic alcohol in water affords quinoline along with dimeric compound **19** (Scheme 4.8). Compound **19** is believed to form by nucleophilic addition of the 1,2-dihydroquinoline intermediate to quinoline.

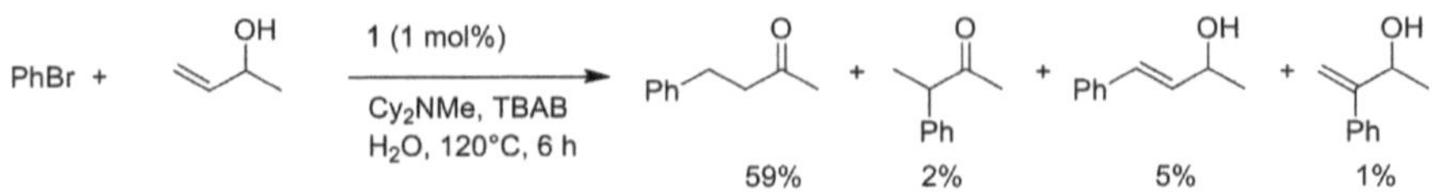

Scheme 4.7 Heck arylation of 3-buten-2-ol using palladacycle **1**.

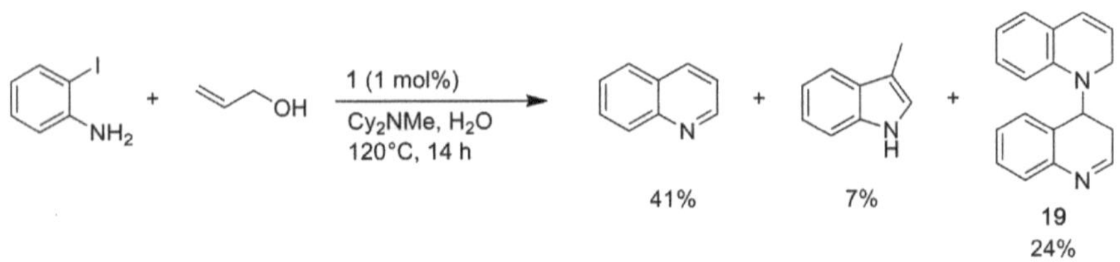

Scheme 4.8 Heck condensation of 2-iodoaniline and allyl alcohol to quinoline.

1.1.1 Supported oxime palladacycles

Fluorous derivatives of the Nájera palladacycle have been prepared and shown to be highly robust catalyst systems in aqueous solvents. The ligand is prepared by addition of fluoroalkyl groups to the hydroxyl substituents of 4,4'-dihydroxybenzophenone, followed by conversion to the oxime and metalation with palladium to afford palladacycle **20** (Scheme 4.9) *(29)*. High yields are obtained in the Suzuki coupling of aryl bromides and aryl-, vinyl-, and alkylboronic acids using 0.05 mol% **20** with TBAB in water at 140°C with microwave heating. Complex **20** could be used for 5 reaction cycles with the yield for each step above 90%, although the reaction time was increased from 2 min for cycle 1 to 7 min for the fifth cycle. In each cycle, palladacycle **20** was recovered by column chromatography using fluorous-phase silica to give approximately 90% recovery. Palladium leaching into the product was extremely low (0.03 ppm) compared to nonfluorinated palladacycle **1** (50–100 ppm) *(23b)*. The significantly lower leaching seen with **20** compared to **1** may be due to the high efficiency of the separation of the organic product from the fluorinated complex using fluorous-phase silica chromatography. Complex **20** showed similarly high activity and recyclability in the Songashira and Stille couplings in water *(29b)*. In contrast, complex **20** was ineffective for the Heck coupling in water or water/DMF, although it did provide good yields in dry DMF.

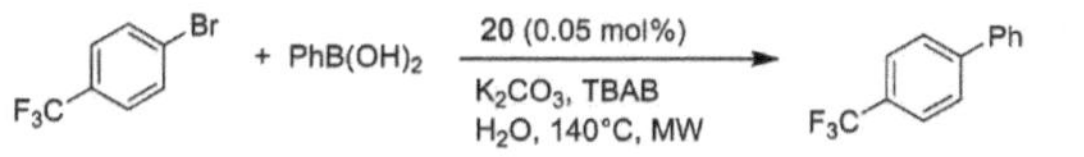

Cycle	Time	Yield (%)	Pd leaching (ppm)
1	2	98	0.031
2	3	96	0.033
3	4	95	0.027
4	6	95	0.023
5	7	92	0.027

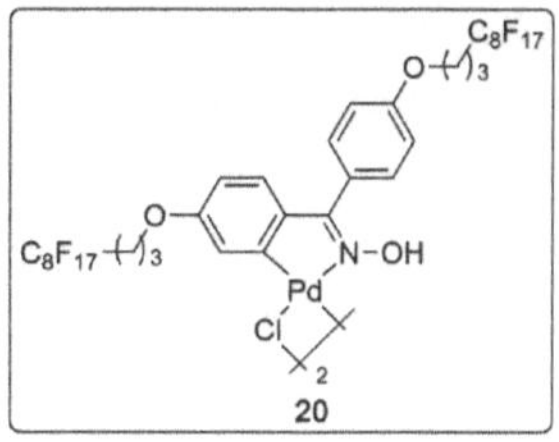

Scheme 4.9 Recycling of catalyst **20** in a Suzuki coupling reaction.

Fluorinated palladacycle **20** is an effective catalyst for carbonylations of aryl iodides and bromides. Carbonylative Suzuki couplings can be achieved using 1 mol% **20** in water at 140°C (microwave) using $Mo(CO)_6$ as the carbon monoxide source (Scheme 4.10) *(30)*. Diaryl ketones are obtained in 60%–85% yields under these conditions. Complex **20** is also effective for alkoxycarbonylation in alcohol solvents and aminocarbonylation in water. Catalyst recycling was not as efficient for these reactions as for the Suzuki coupling. In each case, a single charge of catalyst was used for five reaction cycles. Yields decreased by approximately 10% from the first to last cycle. The decrease in yield/cycle was accompanied by a similar decrease in the amount of palladium catalyst recovered after each cycle. Leaching into the product was very low (<0.1 ppm), however. Annulation of internal alkynes with *ortho*-substituted aryl halides provided access to a variety of carbo- and heterocyclic products in water using palladacycle **20** *(31)*.

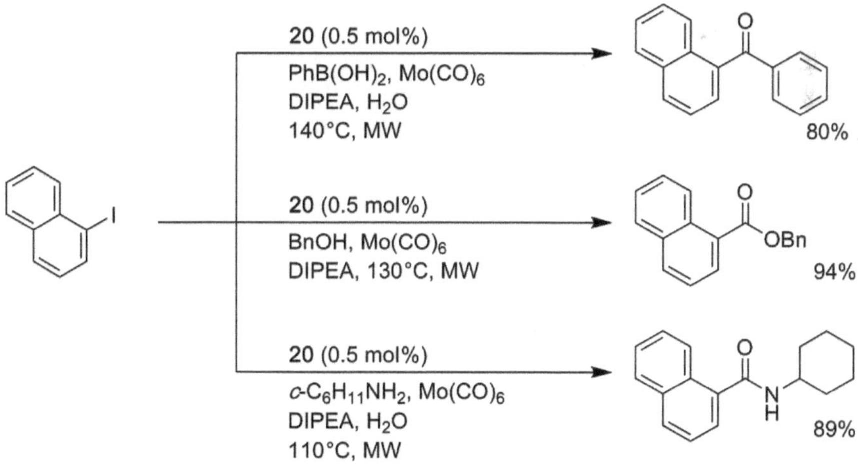

Scheme 4.10 Carbonylative coupling reactions using palladacycle **20**.

1.1.2 Solid-supported oximes

Solid-supported analogs of homogeneous catalysts offer the potential to achieve the best of both homogeneous and heterogeneous catalysis. By supporting organometallic complexes via flexible tethers on solid supports, they have the potential to behave as in solution, yet can be readily separated from the reaction solvent by filtration. In practice, these systems have generally not been industrially viable due to lower activity and catalyst leaching *(1)*.

An ω-alkenyl analog of the Nájera oxime palladacycle (**21**) has been supported on a thiol-modified silica support (Scheme 4.11) *(32)*. The resulting material shows good activity for the Suzuki coupling of aryl bromides and even nonactivated aryl chlorides in water at 100°C using 5 mol% palladium. Modest conversion was even achieved with a highly activated aryl fluoride. The solid catalyst could be used for eight reaction cycles with no loss in yield. Oxime palladacycles supported on MCM-41, polystyrene, and poly(ethylene glycol bismethacrylate) were much less active.

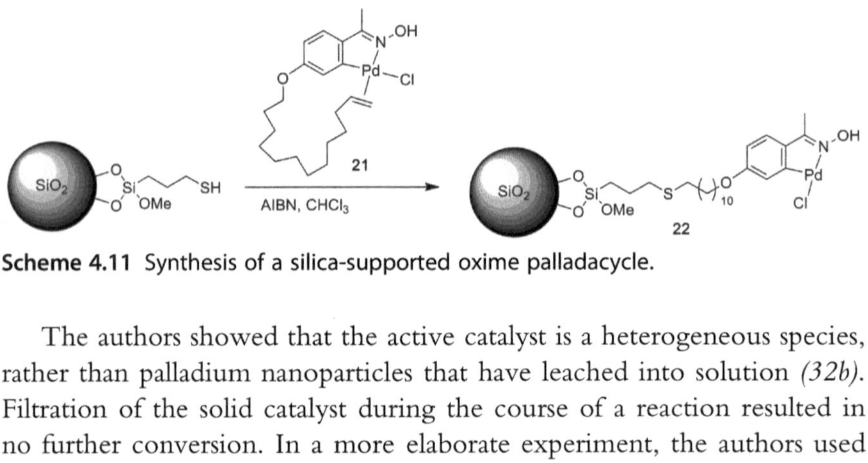

Scheme 4.11 Synthesis of a silica-supported oxime palladacycle.

The authors showed that the active catalyst is a heterogeneous species, rather than palladium nanoparticles that have leached into solution *(32b)*. Filtration of the solid catalyst during the course of a reaction resulted in no further conversion. In a more elaborate experiment, the authors used a three-phase test in which both solid supported and soluble aryl halides were used (Scheme 4.12). If the catalyst is truly heterogeneous, then the

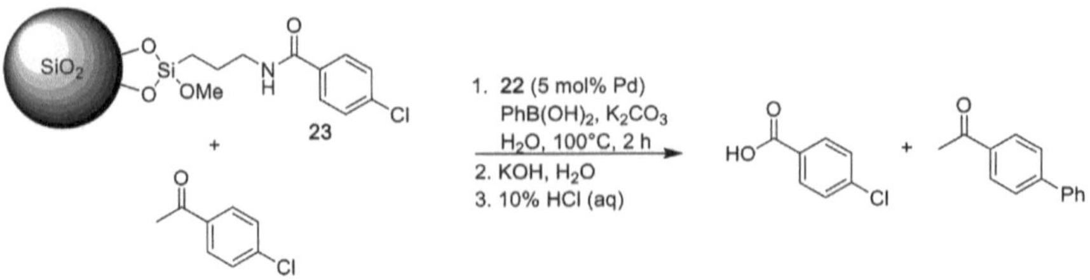

Scheme 4.12 Three-phase test of the heterogeneity of catalyst **22**.

supported halide should not react. When supported aryl chloride **23** and 4-chloroacetophenone were reacted with phenylboronic acid using supported catalyst **22**, no conversion was seen for the supported aryl chloride. The results suggest that the active species remains on the solid support throughout the reaction.

Lee reported an alternative polystyrene-supported oxime palladacycle (**24**, Fig. 4.5) that was effective for the Suzuki coupling of aryl bromides in water/DMF (2:1) at 50°C under microwave irradiation using 1 mol% palladium *(33)*. Activated aryl chlorides could be coupled at 100°C. The catalyst could be used for five reaction cycles with no change in product yield in the coupling of 4-bromoanisole and phenylboronic acid.

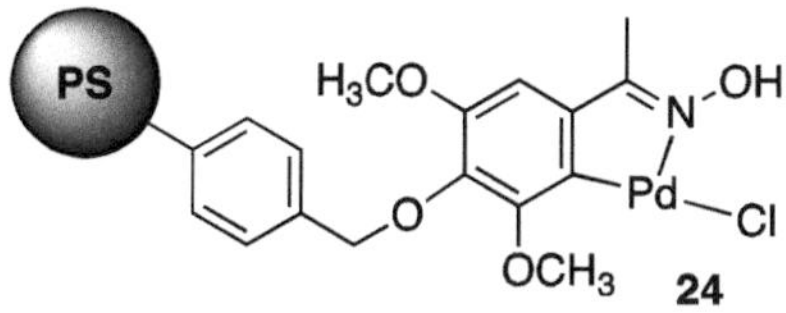

24

Fig. 4.5 Polystyrene supported oxime palladacycle.

2.2 Imine-Derived Palladacycles

Palladacycles derived from aryl imines are readily prepared in a similar manner to oxime palladacycles. The imine ligand provides a higher degree of modification, as the N-aryl substituents can also be readily varied. Imine-based palladacycles derived from aminoacid and aminosulfonic acids can be readily prepared to afford water soluble palladacycles. Grannell prepared the first example of a carboxylate-substituted imine palladacycle (**4**, Fig. 4.2) *(10)*. Imines were prepared by the condensation of benzaldehyde derivatives with amino acid esters. The resulting imines were palladated to afford palladacycles with ester functionality. In principle, these could be converted to carboxylates by hydrolysis, but this was not demonstrated. Grannell then reported a palladacycle prepared from the Schiff base formed from benzaldehyde and *p*-aminobenzoic acid (**5**, Fig. 4.2) *(34)*. Complexation with palladium resulted in the *N,C*-palladacycle, rather than coordination to the carboxylate moiety. The resulting complex is insoluble in neutral water, but soluble at pH 10. No catalytic application of these palladacycles has been reported.

The Shaughnessy group prepared a similar water-soluble imine from sodium *p*-aminobenzenesulfonate (**6**, Fig. 4.6) *(12)*. Several water-soluble benzylamine-derived palladacycles were also prepared containing sulfonate or hydroxyl substituents. Hydroxy-substituted palladacycle **2** was insoluble in neutral water, but dissolved in basic water and polar aprotic solvents (acetonitrile and DMSO). Sulfonated palladacycles **6**, **25**, and **26** were soluble in neutral water. Complexes **2** and **6** are insoluble in methanol and other organic solvents. Complexes **2** and **6** provided effective catalysts for the Suzuki coupling of aryl bromides in combination with *t*-Bu-Amphos. Palladacycles **25** and **26** provided less active catalysts. High yields could be achieved with loadings of **2**/*t*-Bu-Amphos as low as 0.02 mol%. The catalyst derived from **6**/*t*-Bu-Amphos was highly recyclable (Scheme 4.13). The same aqueous solution of **6**/*t*-Bu-Amphos could be used for 11 reaction cycles before the yield began to deteriorate. The catalyst derived from **2** and **6** showed no inhibition in the presence of elemental mercury, suggesting that the active species was a molecular palladium complex. Analysis of the reaction of palladacycle **6** with *t*-Bu-Amphos in the presence of phenylboronic acid in basic water resulted in clean formation of Pd(*t*-Bu-Amphos)$_2$.

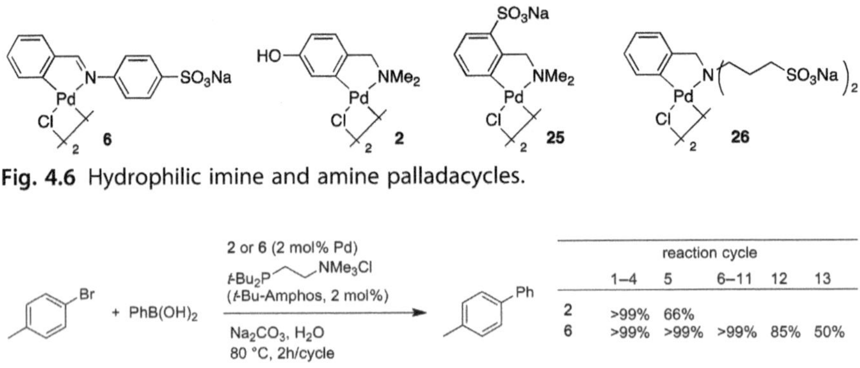

Fig. 4.6 Hydrophilic imine and amine palladacycles.

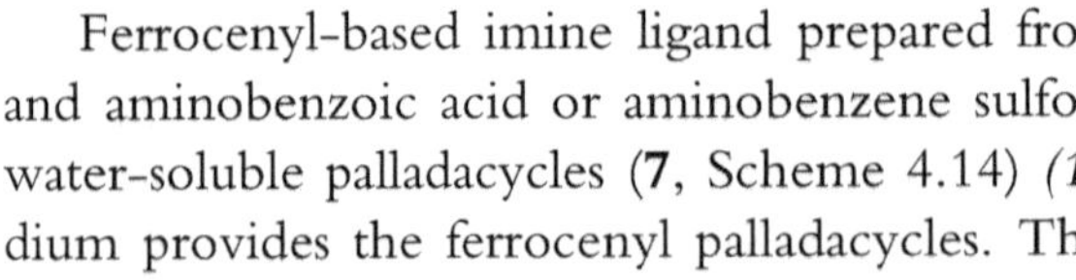

	reaction cycle				
	1–4	5	6–11	12	13
2	>99%	66%			
6	>99%	>99%	>99%	85%	50%

Scheme 4.13 Recycling of precatalysts **27** and **30** with *t*-Bu-Amphos.

Ferrocenyl-based imine ligand prepared from ferrocene carbaldehyde and aminobenzoic acid or aminobenzene sulfonic acid provide access to water-soluble palladacycles (**7**, Scheme 4.14) *(13)*. Metalation with palladium provides the ferrocenyl palladacycles. The *meta*- and *para*- isomers of both the sulfonated and carboxylated complexes were effective catalysts (0.5 mol%) for Suzuki coupling of aryl bromides in water with PEG-2000 at 60°C. Using ethanol as a cosolvent allows the catalyst loading to be lowered to 0.1 mol%. Complex **7d** could be used in six reaction cycles before a significant decrease in yield occurred. The reaction times were increased from 3 to 10 hours after the third cycle, which shows that there was some degradation in catalyst activity with each cycle.

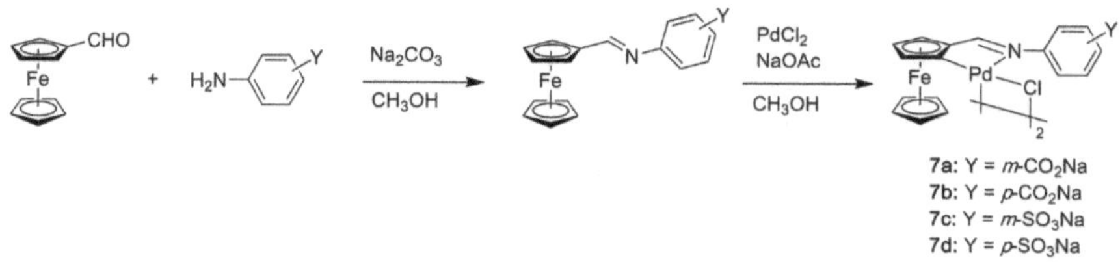

Scheme 4.14 Synthesis of ferrocene-based hydrophilic palladacycles.

2.3 2-Arylheterocycle-Derived Palladacycles

2-Arylheterocycles, such as 2-phenylpyridine, are common ligands for palladacyclic complexes. This class of palladacycles are easily prepared and highly robust. The 2-arylheterocycle provides a modular base for introducing water-solubilizing functionality. Wu reported a palladacycle derived from a sulfonated 2-arylnaphthoxazole ligand (**8**, Fig. 4.2) *(14)*. The 2-arylnaphthoxazole ligand is easily prepared by condensation of benzaldehyde derivatives with 1-amino-2-naphthol (Scheme 4.15) *(35)*. Complex **8** (0.1 mol%) is an effective catalyst for the Suzuki coupling of aryl bromides in water at room temperature. Moderate to good yields were achieved with challenging heteroaryl bromide substrates. Complex **8** shows low activity for coupling of unactivated aryl chlorides. The authors did not report efforts to recycle the aqueous catalyst solution or the level of palladium leaching into the organic products.

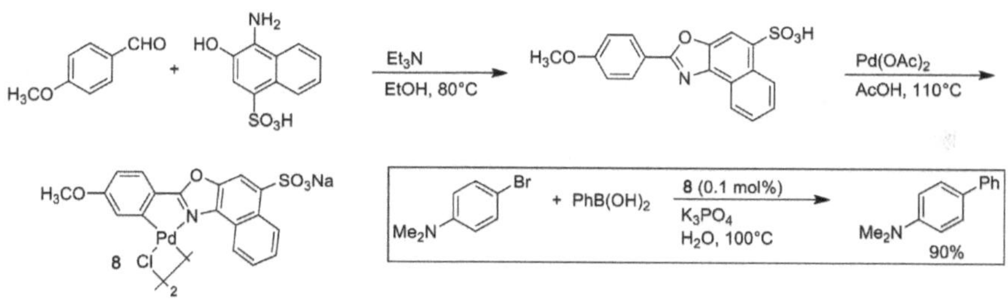

Scheme 4.15 Synthesis of sulfonated 2-arylnaphthoxazole palladacycle **8**.

Copper-catalyzed cycloaddition of PEG azide and an aryl alkyne provides an efficient entry to PEG-modified 4-aryltriazole ligands (Scheme 4.16) *(15)*. Metalation with palladium affords complex **9**, which showed high productivity for Suzuki couplings of aryl iodides, bromides, and chlorides in water at 100°C. High conversion is achieved with palladium loadings as low as 1×10^{-4} mol% (9.8×10^5 turnovers). Addition of elemental mercury completely deactivated the catalyst, suggesting that complex **9** decomposes to palladium nanoparticles, which act as the active catalyst. A wide range of aryl bromides were coupled with TONs ranging from 10^3 to 10^5. Notably, 2- and 3-bromopyridine was coupled using 1×10^{-3} mol% palladium. Complex **9** (1×10^{-4} mol% Pd) showed modest recyclability. Yields above 90% were achieved for three cycles, after which they dropped dramatically. Reaction times were doubled with each cycle indicating a loss of activity with each cycle. Although only a few cycles were achieved, the catalyst loading was much lower than other systems that report high yields over more cycles. Complex **9** also showed good activity for the Sonogashira coupling of aryl bromides and phenylacetylene in refluxing water.

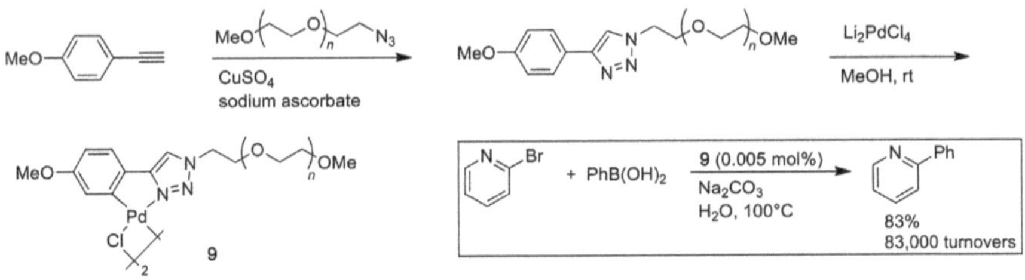

Scheme 4.16 Synthesis and application of PEG-modified palladacycle **9**.

Hydroxymethyl-substituted 2-arylpyrazine-derived palladacycles complexed to PPh_3 (**3**, Fig. 4.2) or S-Phos (**27**, Scheme 4.17) are reported to be water-soluble *(9)*. No data regarding the water-solubility is provided, however. The hydroxymethyl substituent would be expected to provide limited hydrophilic character, particularly with highly hydrophobic phosphine ligands coordinated. S-Phos-coordinated complex **27** (0.2 mol%) is effective for the Suzuki coupling of unactivated aryl chlorides in water. In contrast, $Pd(OAc)_2$/S-Phos gave little conversion under the same conditions. Complex **27** is the only palladacycle system that has shown broad applicability to aryl chlorides in water. Complex **27** is also an effective catalyst for the Hartwig-Buchwald coupling of aryl chlorides and aniline derivatives in water at 100°C. Buchwald has reported on-water amination of aryl chlorides using a hydrophobic palladacycle coordinated to X-phos *(36)*.

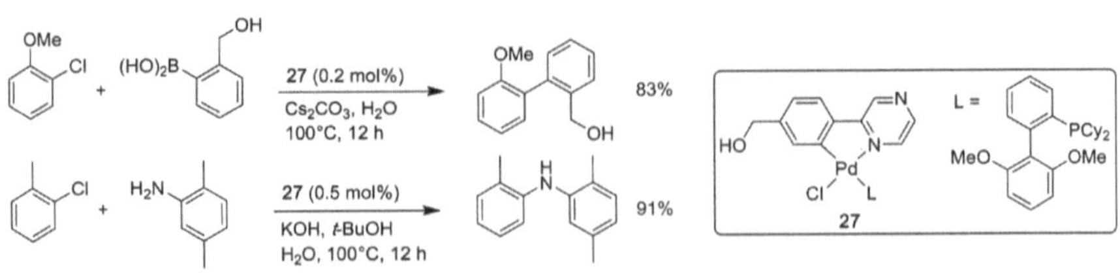

Scheme 4.17 Suzuki and Hartwig-Buchwald coupling in water catalyzed by palladacycle-phosphine complex **27**.

4-Ferrocenylpyridimine-based palladacycle **14** (Fig. 4.4) does not have any hydrophilic substituents on the organic ligand *(19)*. Rather the water-solubility is provided by the charge separated cationic palladium center. Complex **14** (0.5 mol%) is an effective catalyst for the Suzuki coupling of aryl bromides in water at 100°C. Unactivated aryl chlorides give low conversion unless tricyclohexylphosphine (PCy_3) is added to the reaction system. The combination of **14** and PCy_3 affords a 78% yield for the coupling of 4-chlorotoluene and 4-hydroxymethylphenylboronic acid.

2.4 Phosphorus-Derived Palladacycles

Phosphorus-based palladacycles are highly effective precatalysts for a range of cross-coupling reactions in organic solvents *(37)*. Eppinger reported the use of a phosphinite-derived palladacycle for the Suzuki coupling of aryl bromides in water (Scheme 4.18) *(17)*. Complex **28** (0.02 mol%) provides good yields for the Suzuki coupling of aryl bromides in water at 30°C. The organic products were recovered by filtration and were reported to be spectroscopically and analytically pure without additional purification. The water solubility of **28**, or the catalytic species derived from it, is unknown. The degree of palladium contamination in the organic products was not reported.

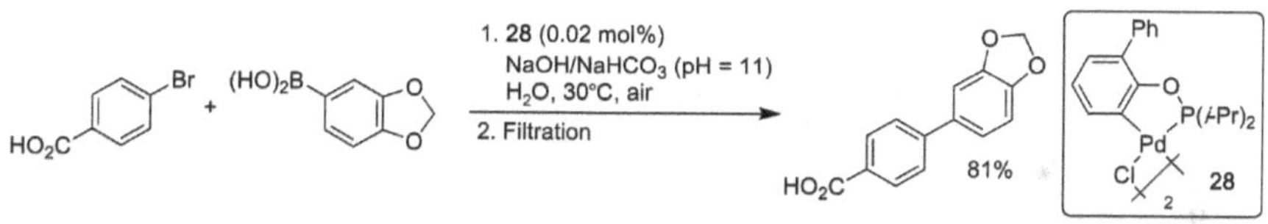

Scheme 4.18 On-water Suzuki coupling catalyzed by palladacycle **28**.

2.5 Water-Soluble Palladium Pincer Complexes

Pincer complexes have received some attention as catalyst precursors for cross-coupling reactions. The tridentate coordination of pincer ligands almost certainly requires conversion of the pincer complex into a lower coordination active species. Most likely, pincer complexes serve as precursors to palladium nanoparticles. Berbreiter reported the use of a PEG- modified SCS pincer complex (**10**) for Heck couplings *(16)*. Complex **10** gave modest yields in the Heck coupling of aryl iodides with activated alkenes at 150°C in DMA under microwave heating. The reactions could be performed in water using hydrophilic substrates. Catalyst recycling was demonstrated in a thermomorphic solvent system (1:9:20 H_2O/DMA/heptane) that is two phases at room temperature but becomes homogeneous at higher temperature. Yields of 49%, 98%, 88%, and 78% were achieved over four cycles using the same aqueous/DMA solution of the catalyst. The low yield in cycle one is due to product loss in the aqueous/DMA layer.

Cationic SCS palladacycle aquo complex **13** has been applied to aqueous phase cross-coupling reactions. Complex **13** was evaluated in the Suzuki coupling of 3-halobenzoic acid with phenylboronic acid or tetraphenylborate *(18)*. Tetraphenylborate gave slightly higher rates of conversion than phenylboronic acid. The catalyst productivity was found to reach a peak around pH 10 and then decrease at higher pH. The pK_a of the aquo ligand in **13** is 10.7, which seems to correlate with the observed maximum. Hydroxide plays both promoting and inhibiting roles in the Suzuki coupling, so similar dependency on [$^{-}$OH] is seen in systems without an acidic proton *(38)*. Thus, the observed pH dependence may not be solely due to the protonation state of the precatalyst. Complex **13** is inactive in Stille and Heck couplings *(39)*.

PCP-pincer complex **29** provides modest activity in the Stille coupling, but is inactive in the Heck coupling (Scheme 4.19) *(39)*. In contrast, bis(aquo) palladacycle **30** was significantly more productive. Precatalyst **30** provides a TON of 134,000 for the Suzuki coupling of 3-bromobenzoic acid, 50,000 for the Stille coupling of 3-iodobenzoic acid, and 22,000 for the Heck coupling of 3-iodobezoic acid. All reactions were carried out at 100°C in water. Elemental mercury had no effect on the Suzuki coupling using complex **30**, suggesting that the active species is not metallic palladium.

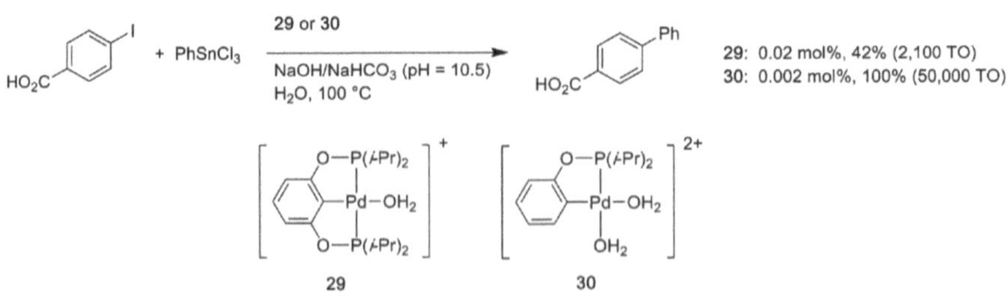

Scheme 4.19 Comparison of pincer and palladacycle complexes for Stille coupling in water.

Hydrophilic *N*-heterocyclic carbene-derived pincer complexes provide effective catalysts for aqueous-phase Suzuki couplings. Pincer **31a** (Scheme 4.20) with a CNC coordination environment is prepared by condensation of an *N*-alkylimidazole with 2,6-dibromoisonicotinic acid followed by metalation with palladium *(40)*. Hydrolysis under acidic conditions provides the more hydrophilic triacid form (**31b**). Both **31a** and **31b** gave high conversions for Suzuki coupling of 4-bromoacetophenone and phenyl boronic acid down to 10^{-3} mol% catalyst. Complex **31b** gave a 93% yield at 10^{-4} mol%, whereas complex **31a** gave no conversion at this loading. Both complexes performed comparably with a range of aryl bromides, although complex **31b** was more effective with electron-rich aryl bromides. Complexes **31a** and **31b** were only modestly recyclable. The catalytic activity was completely inhibited by addition of elemental mercury, whereas addition of triphenylphosphine had no effect. These results suggest a heterogeneous active species.

31a: R = Et
31b: R = H

32

Scheme 4.20 Hydrophilic NHC-derived pincer complexes.

Tu reported a related CNC-pincer complex, but with benzimidazolylidine substituents and only one carboxylate group to provide water solubility (**32**, Scheme 4.20) *(41)*. Complex **32** affords high yields in aqueous-phase Suzuki couplings of aryl bromides using 0.005 mol% Pd. In contrast to complex **31**, little inhibition of **32** is observed with elemental mercury or poly(vinylpyrrolidine), which suggests a molecular species is the active catalyst. Complex **32** provides modest activity for the Heck coupling of aryl iodides *(42)*.

3. CONCLUSION

Palladacycles are attractive precatalysts as they are typically air- and moisture-stable and can often provide high turnover numbers. These properties make hydrophilic palladacycles useful precatalysts for aqueous-biphasic catalysis. Numerous examples have been reported that are highly recyclable with limited leaching of palladium to the organic product. The nature of the catalyst species derived from palladacycles is often poorly understood. In many cases, evidence suggests that the palladacycles decompose to palladium nanoparticles which serve as the true catalyst. In other cases, evidence supports a true homogeneous catalyst. Further work is needed to understand the mechanisms by which the palladacyclic catalysts operate.

REFERENCES

1. Hübner, S.; de Vries, J. G.; Farina, V. Why Does Industry Not Use Immobilized Transition Metal Complexes as Catalysts? *Adv. Synth. Catal.* **2016**, *358*, 3–25.
2. (a)Sheldon, R. A. Green Solvents for Sustainable Organic Synthesis: State of the Art. *Green Chem.* **2005**, 7, 267–278; (b)Liu, S.; Xiao, J. Toward Green Catalytic Synthesis—Transition Metal-Catalyzed Reactions in Non-conventional Media. *J. Mol. Catal. A Chem.* **2007**, *270*, 1–43; (c)Pollet, P.; Davey, E. A.; Ureña-Benavides, E. E.; Eckert, C. A.; Liotta, C. L. Solvents for Sustainable Chemical Processes. *Green Chem.* **2014**, *16*, 1034–1058.
3. (a)Li, C.-J.; Chen, L. Organic Chemistry in Water. *Chem. Soc. Rev.* **2006**, *35*, 68–82; (b)Butler, R. N.; Coyne, A. G. Water: Nature's Reaction Enforcers Comparative Effects for Organic Synthesis "In-Water" and "On-Water". *Chem. Rev.* **2010**, *110*, 6302–6337; (c)Dunn, P. Water as a Green Solvent for Pharmaceutical Applications. In: *Handbook of Green Chemistry—Green Solvents;* Li, C.-J., Ed.; Vol. 5; John Wiley & Sons: Weinheim, 2010; pp 363–383.
4. (a)Kuntz, E. G. FR 2,314,910, 1975.(b)Kuntz, E. G. FR 2,349,562, 1976.; (c) Kuntz, E. G. Homogeneous catalysis… in water. *Chemtech* **1987**, *17*, 570–575
5. (a)Shaughnessy, K. H. Hydrophilic Ligands and Their Application in Aqueous-Phase Metal-Catalyzed Reactions. *Chem. Rev.* **2009**, *109*, 643–710; (b)Simon, M.-O.; Li, C.-J. Green chemistry oriented organic synthesis in water. *Chem. Soc. Rev.* **2012**, *41*, 1415–1427.
6. (a)Polshettiwar, V.; Decottignies, A.; Len, C.; Fihri, A. Suzuki-Miyaura Cross-Coupling Reactions in Aqueous Media: Green and Sustainable Syntheses of Biaryls. *ChemSusChem* **2010**, *3*, 502–522; (b)Li, B.; Dixneuf, P. H. sp^2 C–H Bond Activation in Water and Catalytic Cross-Coupling Reactions. *Chem. Soc. Rev.* **2013**, *42*, 5744–5767; (c)Levin, E.; Ivry, E.; Diesendruck, C. E.; Lemcoff, N. G. Water in N-Heterocyclic Carbene-Assisted Catalysis. *Chem. Rev.* **2015**, *115*, 4607–4692; (d)Shaughnessy, K. H. Greener Approaches to Cross-Coupling. In: *New Trends in Cross-Coupling*: Theory and Application; Colacot, T. J., Ed.; Royal Society of Chemistry: Cambridge, 2015; pp 645–696.
7. (a)Narayan, S.; Muldoon, J.; Finn, M. G.; Fokin, V. V.; Kolb, H. C.; Sharpless, K. B. "On Water": Unique Reactivity of Organic Compounds in Aqeuous Suspension. *Angew. Chem. Int. Ed.* **2005**, *44*, 3275–3279; (b)Chanda, A.; Fokin, V. V. Organic Synthesis "On Water"*Chem. Rev.* **2009**, *109*, 725–748; (c)Mase, N.; Barbas, C. F., III In Water, on Water, and by Water: Mimicking Nature's Aldolases With Organocatalysis and Water. *Org. Biomol. Chem.* **2010**, *8*, 4043–4050.
8. (a)Botella, L.; Najera, C. A Convenient Oxime-Carbapalladacycle-Catalyzed Suzuki Cross-Coupling of Aryl Chlorides in Water. *Angew. Chem. Int. Ed.* **2002**, *41*,

179–181; (b)Botella, L.; Nájera, C. Cross-Coupling Reactions With Boronic Acids in Water Catalysed by Oxime-Derived Palladacycles. *J. Organomet. Chem.* **2002**, *663*, 46–57.

9. Han, X.; Li, H.-M.; Xu, C.; Xiao, Z.-Q.; Wang, Z.-Q.; Fu, W.-J.; Hao, X.-Q.; Song, M.-P. Water-Soluble Palladacycles Containing Hydroxymethyl Groups: Synthesis, Crystal Structures and Use as Catalysts for Amination and Suzuki Coupling of Reactions. *Transit. Met. Chem.* **2016**, *41*, 403–411.
10. Albert, J.; Magali Cadena, J.; González, A.; Granell, J.; Solans, X.; Font-Bardia, M. Cyclopalladation of Schiff Bases From Phenylalanine and 2-Phenylglicine. *J. Organomet. Chem.* **2002**, *663*, 277–283.
11. Alonso, D. A.; Botella, L.; Nájera, C.; Pacheco, M. C. Synthetic Applications of Oxime-Derived Palladacycles as Versatile Catalysts in Cross-Coupling Reactions. *Synthesis* **2004**, ;1713–1718.
12. Huang, R.; Shaughnessy, K. H. Water-Soluble Palladacycles as Precursors to Highly Recyclable Catalysts for the Suzuki Coupling of Aryl Bromides in Aqueous Solvents. *Organometallics* **2006**, *25*, 4105–4112.
13. Li, Y.; Fu, Z.; Rao, L.; Wang, L.; Li, R.; Li, T.; Wu, Y. Water-Soluble and Recyclable Cyclopalladated Ferrocenylimine for Suzuki Coupling Reaction. *J. Chin. Chem. Soc. (Weinheim, Ger.)* **2014**, *61*, 397–403.
14. Li, H.; Wu, Y. Water-Soluble 2-Arylnaphthoxazole-Derived Palladium(II) Complexes as Phosphine-Free Catalysts for the Suzuki Reaction in Aqueous Solvent. *Appl. Organomet. Chem.* **2008**, *22*, 233–236.
15. Zhang, G.; Zhang, W.; Luan, Y.; Han, X.; Ding, C. PEG Click-Triazole Palladacycle: An Efficient Precatalyst for Palladium-Catalyzed Suzuki-Miyaura and Copper-free Sonogashira Reactions in Neat Water. *Chin. J. Chem.* **2015**, *33*, 705–710.
16. Bergbreiter, D. E.; Furyk, S. Microwave-Promoted Heck Reactions Using an Oligo (ethyleneglycol)-Bound SCS Palladacycle Under Thermomorphic Conditions. *Green Chem.* **2004**, *6*, 280–285.
17. Marziale, A. N.; Faul, S. H.; Reiner, T.; Schneider, S.; Eppinger, J. Facile Palladium-Catalyzed Suzuki-Miyaura Coupling in Air and Water at Ambient Temperature. *Green Chem.* **2010**, *12*, 35–38.
18. Nakai, H.; Ogo, S.; Watanabe, Y. pH-Dependent Cross-Couplng Reactions of Water-Soluble Organic Halides with Organoboron Compounds Catalyzed by the Organometallic Aqua Complex $[(SCS)Pd^{II}(H_2O)]^+$ ($SCS = C_6H_3$-2,6-$(CH_2SBu^t)_2$). *Organometallics* **2002**, *21*, 1674–1678.
19. Li, H.-M.; Xu, C.; Hao, X.-Q.; Li, Z.; Wang, Z.-Q.; Fu, W.-J.; Song, M.-P. Cationic Cyclopalladated Ferrocenylpyrimidine Complex as an Efficient Catalyst for the Synthesis of Substituted Biarylmethanol by the Suzuki Reaction in Water. *Inorg. Chim. Acta* **2013**, *404*, 236–240.
20. Nájera, C. Oxime-Derived Palladacycles: Applications in Catalysis. *ChemCatChem* **2016**, *8*, 1865–1881.
21. Alonso, D. A.; Nájera, C. Oxime-Derived Palladacycles as Source of Palladium Nanoparticles. *Chem. Soc. Rev.* **2010**, *39*, 2891–2902.
22. Reetz, M. T.; Westerman, E. Phosphane-Free Palladium-Catalyzed Coupling Reactions: The Decisive Role of Pd Nanoparticles. *Angew. Chem. Int. Ed.* **2000**, *39*, 165–168.
23. (a)Alacid, E.; Nájera, C. First Cross-Coupling Reaction of Potassium Aryltrifluoroborates with Organic Chlorides in Aqueous Media Catalyzed by an Oxime-Derived Palladacycle. *Org. Lett.* **2008**, *10*, 5011–5014; (b)Alacid, E.; Nájera, C. Palladium-Catalyzed Cross-Coupling Reactions of Potassium Alkenyltrifluoroborates with Organic Halides in Aqueous Media. *J. Org. Chem.* **2009**, *74*, 2321–2327.

24. Cívicos, J. F.; Alonso, D. A.; Nájera, C. Oxime-Palladacycle-Catalyzed Suzuki-Miyaura Arylation and Alkenylation of Aryl Imidazolesulfonates under Aqueous and Phosphane-Free Conditions. *Eur. J. Org. Chem.* **2012**, ;3670–3676.
25. (a)Alacid, E.; Nájera, C. The First Fluoride-Free Hiyama Reaction of Vinylsiloxanes Promoted by Sodium Hydroxide in Water. *Adv. Synth. Catal.* **2006**, *348*, 2085–2091; (b)Alacid, E.; Nájera, C. Aqueous Sodium Hydroxide Promoted Cross-Coupling Reactions of Alkenyltrialkoxysilanes Under Ligand-Free Conditions. *J. Org. Chem.* **2008**, *73*, 2315–2322.
26. (a)Botella, L.; Nájera, C. Controlled Mono and Double Heck Reactions in Water Catalyzed by an Oxime-Derived Palladacycle. *Tetrahedron Lett.* **2004**, *45*, 1833–1836; (b)Botella, L.; Nájera, C. Mono- and β,β-double-Heck Reactions of α,β-unsaturated Carbonyl Compounds in Aqeuous Media. *J. Org. Chem.* **2005**, *70*, 4360–4369.
27. Botella, L.; Nájera, C. Synthesis of Methylated Resveratrol and Analogues by Heck Reactions in Organic and Aqueous Solvents. *Tetrahedron* **2004**, *60*, 5563–5570.
28. Alacid, E.; Nájera, C. Arylation of Allyl Alcohols in Organic and Aqueous Media Catalyzed by Oxime-Derived Palladacycles: Synthesis of β-arylated Carbonyl Compounds. *Adv. Synth. Catal.* **2007**, *349*, 2572–2584.
29. (a)Susanto, W.; Chu, C.-Y.; Ang, W. J.; Chou, T.-C.; Lo, L.-C.; Lam, Y. Development of a Fluorous, Oxime-Based Palladacycle for Microwave-Promoted Carbon-Carbon Coupling Reactions in Aqueous Media. *Green Chem.* **2012**, *14*, 77–80; (b)Susanto, W.; Chu, C.-Y.; Ang, W. J.; Chou, T.-C.; Lo, L.-C.; Lam, Y. Fluorous Oxime Palladacycle: A Precatalyst for Carbon-Carbon Coupling Reactions in Aqueous and Organic Medium. *J. Org. Chem.* **2012**, *77*, 2729–2742.
30. Ang, W. J.; Lo, L.-C.; Lam, Y. Expedient Carbonylation of Aryl Halides in Aqueous or Neat Condition. *Tetrahedron* **2014**, *70*, 8545–8558.
31. Ang, W. J.; Tai, C.-H.; Lo, L.-C.; Lam, Y. Palladium-Catalyzed Annulation of Internal Alkynes in Aqueous Medium. *RSC Adv.* **2014**, *4*, 4921–4929.
32. (a)Baleizão, C.; Corma, A.; García, H.; Leyva, A. An Oxime-Carbapalladacycle Complex Covalently Anchored to Silica as an Active and Reusable Heterogeneous Catalyst for Suzuki Cross-Coupling in Water. *Chem. Commun.* **2003**, 606–607; (b)Baleizão, C.; Corma, A.; García, H.; Leyva, A. Oxime Carbapalladacycle Covalently Anchored to High Surface Area Inorganic Supports or Polymers as Heterogeneous Green Catalysts for the Suzuki Reaction in Water. *J. Org. Chem.* **2004**, *69*, 439–446.
33. Park, S.-J.; Cho, H.-J.; Lee, S.-M.; Lee, Y.-S. Microwave-Assisted C-C Coupling Reaction Using Polymer-Supported Electron-Rich Oxime Palladacycles in Aqueous Condition. *Tetrahedron Lett.* **2017**, *58*, 2670–2674.
34. Albert, J.; D'Andrea, L.; Bautista, J.; González, A.; Granell, J.; Font-Bardia, M.; Calvet, T. Unexpected Structure of Palladacycles Containing Carboxylated Schiff Bases. Synthesis of Some Water-Soluble Metallacycles. *Organometallics* **2008**, *27*, 5108–5117.
35. Li, H.; Wei, K.; Wu, Y.-J. A Convenient Synthesis of 2-Arylnaphtho[1,2-d]oxazole Derivatives Promoted by Triethylamine. *Chin. J. Chem.* **2007**, *25*, 1704–1709.
36. Huang, X.; Anderson, K. W.; Zim, D.; Jiang, L.; Klapars, A.; Buchwald, S. L. Expanding Pd-Catalyzed C-N Bond-Forming Processes: The First Amidation of Arylsulfonates, Aqueous Amination, and Complementarity with Cu-Catalyzed Reactions. *J. Am. Chem. Soc.* **2003**, *125*, 6653–6655.
37. Bedford, R. B. Palladacyclic Catalysts in C-C and C-Heteroatom Bond-Forming Reactions. *Chem. Commun.* **2003**, ;1787–1796.
38. (a)Amatore, C.; Jutand, A.; Le Duc, G. Kinetic Data for the Transmetalation/Reductive Elimination in Palladium-Catalyzed Suzuki–Miyaura Reactions: Unexpected Triple Role of Hydroxide Ions Used as Base. *Chem. Eur. J.* **2011**, *17*, 2492–2508; (b)Amatore, C.; Le Duc, G.; Jutand, A. Mechanism of Palladium-Catalyzed Suzuki–Miyaura Reactions: Multiple and Antagonistic Roles of Anionic "Bases" and Their Countercations. *Chem. Eur. J.* **2013**, *19*, 10082–10093.

39. Ogo, S.; Takebe, Y.; Uehara, K.; Yamazaki, T.; Nakai, H.; Watanabe, Y.; Fukuzumi, S. pH-Dependent C-C Coupling Reactions Catalyzed by Water-Soluble Palladacyclic Aqua Catalysts in Water. *Organometallics* **2006**, *25*, 331–338.
40. Inés, B.; San Martin, R.; Moure, M. J.; Domínguez, E. Insights into the Role of New Palladium Pincer Complexes as Robust and Recyclable Precatalysts for Suzuki-Miyaura Couplings in Neat Water. *Adv. Synth. Catal.* **2009**, *351*, 2124–2132.
41. Tu, T.; Feng, X.; Wang, Z.; Liu, X. A Robust Hydrophilic Pyridine-Bridged bis-Benzimidazolylidene Palladium Pincer Complex: Synthesis and Its Catalytic Application Towards Suzuki-Miyaura Couplings in Aqueous Solvents. *Dalton Trans.* **2010**, *39*, 10598–10600.
42. Wang, Z.; Feng, X.; Fang, W.; Tu, T. Efficient Aqueous-Phase Heck Reaction Catalyzed by a Robust Hydrophilic Pyridine-Bridged Bisbenzimidazolylidene-Palladium Pincer Complex. *Synlett* **2011**, 951–954.

CHAPTER 5

Palladacyclic Complexes as Efficient Catalysts for C—H Bond Functionalization Reactions

Raysa Khan, John Spencer
Department of Chemistry, School of Life Sciences, University of Sussex, Brighton, East Sussex, United Kingdom

Contents

1. INTRODUCTION: C—H ACTIVATION BACKGROUND

C—H activation with subsequent functionalization in situ offers an approach to introduce substitutes into molecules where a previous multi-step can be transformed into a synthesis with fewer steps *(1)*. Successful C—H functionalization methods need to surpass a number of critical challenges in addition to cleaving very strong bonds. Additionally, the successful catalysts in these processes need to fulfill *inter alia* the following criteria:

- display selectivity for a particular kind of C—H bond in a system,
- remain stable and activated in the presence of necessary oxidants, other coordinating functional groups in the product or solvent, and
- be resistant to further functionalization i.e., overoxidation of the products.

Palladacycles
https://doi.org/10.1016/B978-0-12-815505-9.00006-8
© 2019 Elsevier Inc.
All rights reserved.

Palladacycles offer a unique opportunity for site-selective, atom-, and step-economical synthesis *(2)*. Pd-catalysed C—H activation methodologies are remarkably versatile in terms of the substrate and product scope and can be

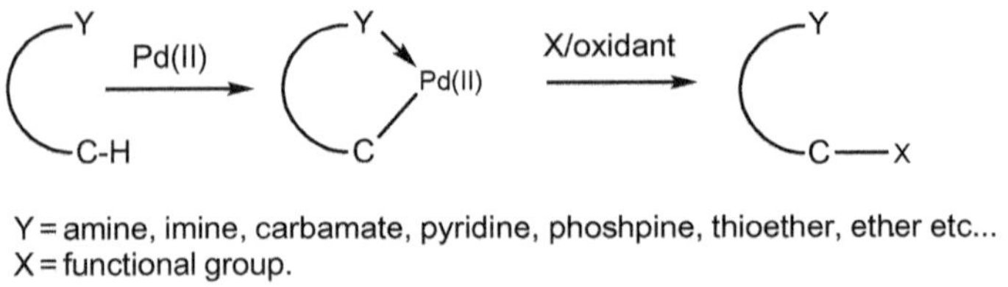

Scheme 5.1 Role of palladacycles in ortho C—H functionalization.

used for the synthesis of both simple and complex molecules *(3)*. Pd catalysts are compatible with a variety of conditions and can be employed with a wide range of functional groups *(4)* (Scheme 5.1).

A wide range of experimental and computational studies are available on the reactivity patterns of Pd complexes in C—H functionalization; concerted metalation deprotonation is a common reaction pathway, which has been treated in detail elsewhere *(5–7)*. The catalytic C—H functionalization reactions involve a number oxidation states Pd(0), Pd(I), Pd (II) and Pd(IV), and the catalytic cycles are commonly identified as $Pd^{0/II}$ and $Pd^{II/IV}$ pathways; others have also shown the involvement of Pd(III) *(8, 9)*. Numerous organometallic palladacycles have been isolated and characterized for most of these oxidation states *(10, 11)*.

In this review, we will highlight selected examples of the various roles of palladacycles in synthesis, a field that has grown immensely since our comprehensive review of the area in 2005. Inevitably, numerous deserving examples are not cited due to space restrictions. We apologize for any such omissions since examples have been selected for illustrative purposes rather than attempting to give an exhaustive account.

2. MECHANISTIC STUDIES

A number of Pd(II)/(IV) pathways utilize iodonium salts as oxidants *(12–14)*. These can be either symmetrical or be unsymmetrical and contain a mesityl group as a leaving group (which leaves as the aryl iodide). This is obviously limited in terms of atom economy but is a convenient route to ortho-arylated substrates (Scheme 5.2).

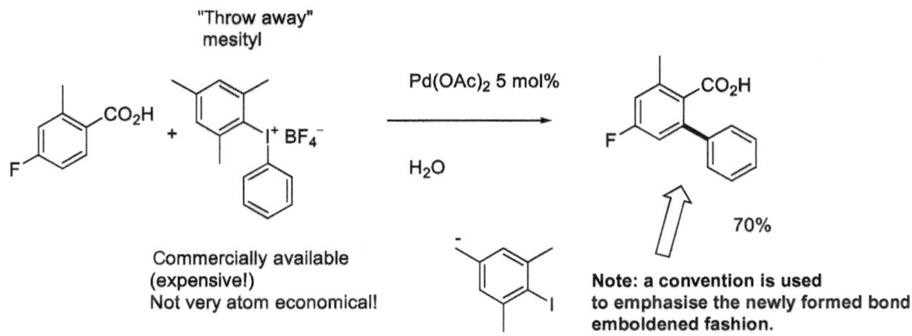

Scheme 5.2 Ortho-arylation of a benzoic acid derivative *(15)*.

A typical catalytic cycle for the above transformation involves Pd(II)/(IV) catalysis (Scheme 5.3).

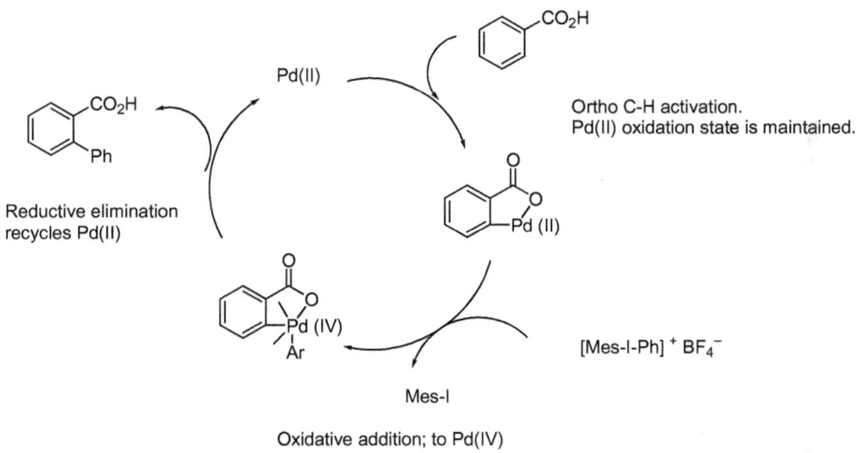

Scheme 5.3 Proposed catalytic cycle.

Similar methodology enables one to deuterate aromatic molecules in the ortho position, presumably via Pd(II) catalysis *(16)*. Note, double C—H activation can occur; often an ortho blocking substituent (e.g., in this case CF_3) is used. In other instances, selectivity is achieved *via* palladacycle ring size, i.e., 5- over 6-membered (Scheme 5.4) *(17)*.

3. SYNTHETIC EXAMPLES

Under certain conditions, double C—H activation is achievable, with benzoquinone *(18)* or sodium persulfate *(19)* acting as oxidant and the arylation occurring via a palladacycle (Scheme 5.5).

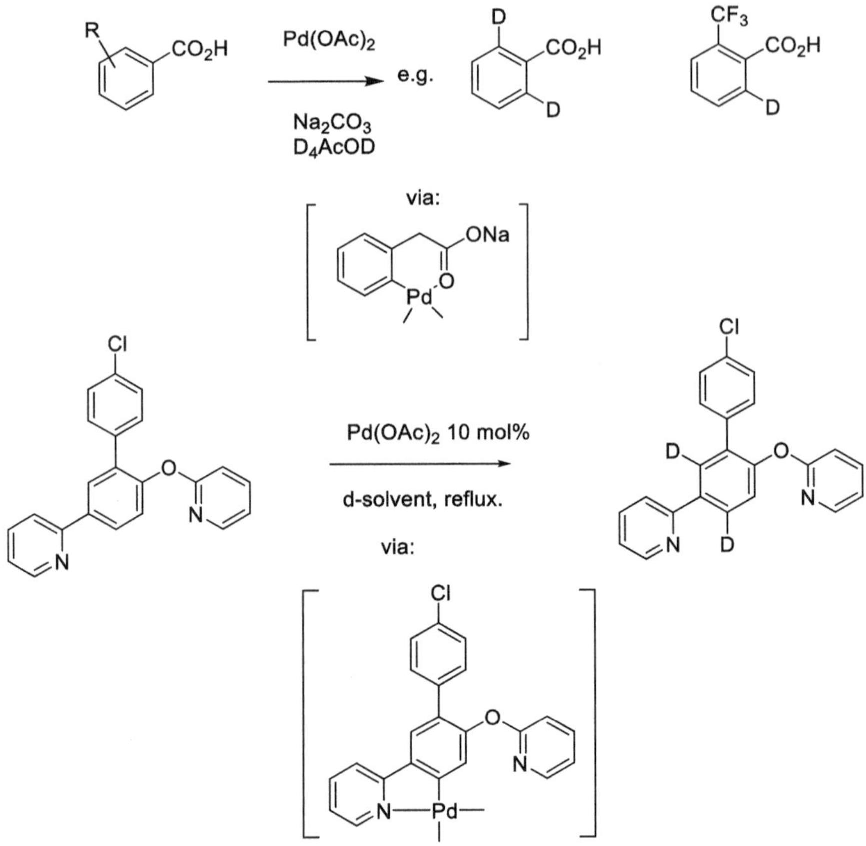

Scheme 5.4 Deuteration via palladacycles.

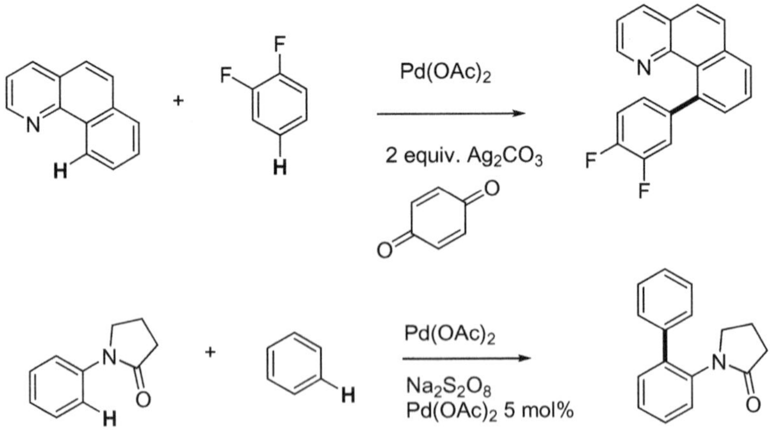

Scheme 5.5 Double C—H activations via a palladacycle.

In oxidative C—H functionalizations, remarkably oxygen can be used as a terminal oxidant under Pd catalysis (Scheme 5.6).

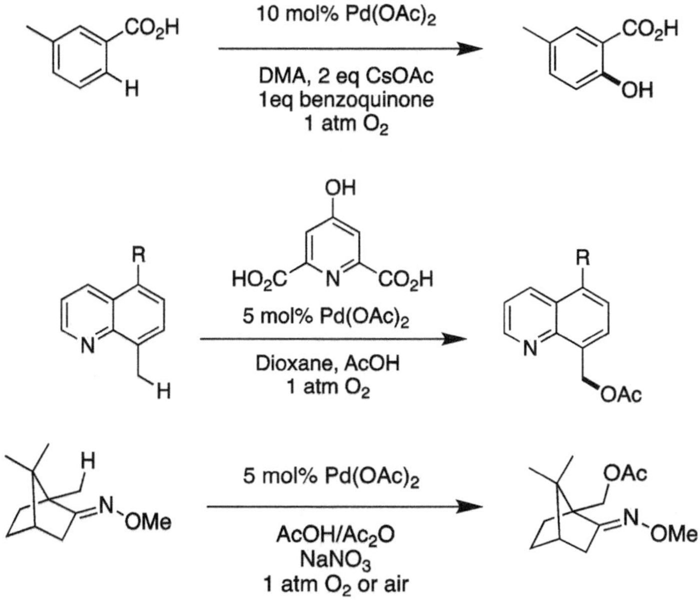

Scheme 5.6 C—H oxygenation via palladacycles.

A similar acetoxylation process has recently been reported to occur in the solid state on graphene oxide *(20)*. Many studies have focused on combining C—H activation chemistry with other transformations in order to achieve optimal results under milder reaction conditions and high atom-economy. Sanford *et al.* presented C—H functionalizations of 2-phenylpyridine and other analogues, which occur at room temperature with visible light photocatalysis to afford biaryl products *(21)*. Our work has shown that, under harsher conditions, using fluoroaryl diazonium compounds, competing S_NAr reactions, termed the nuisance effect, can lead to mixtures of the expected fluoroaryl and ether products *(22–24)*. The use of alternative light sources e.g., blue LEDs, visible light, cheaper organic dyes, represents a current area of significant interest *(25–27)* (Scheme 5.7).

A protocol developed by Yu and coworkers employed transient directing group to lead to substituted aliphatic ketones *(28)*. The formation of

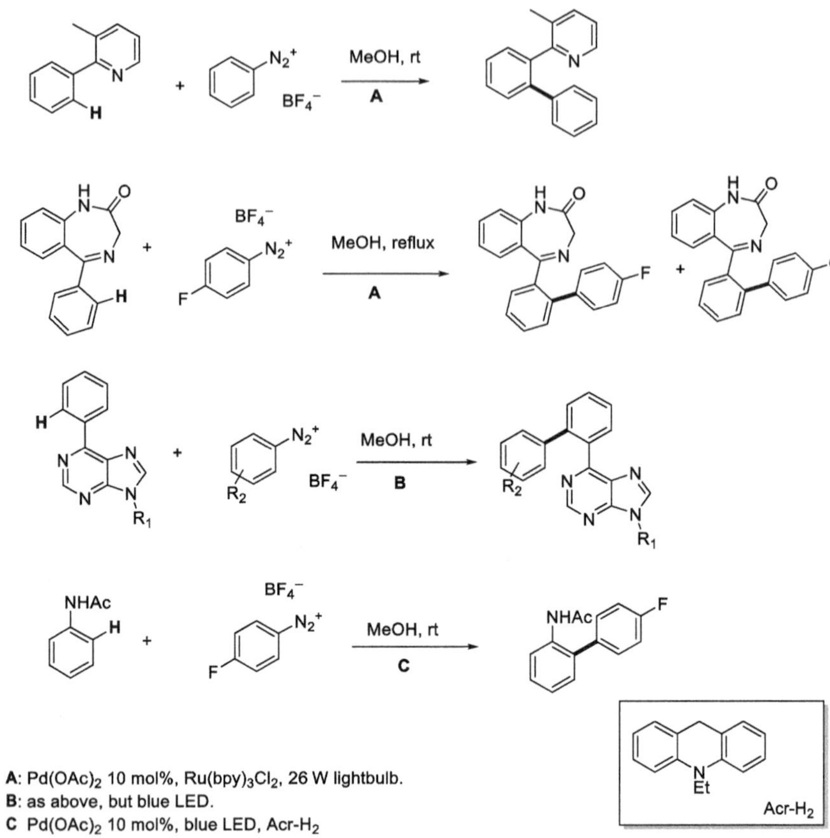

Scheme 5.7 Ortho C—H activation using aryldiazonium salts.

the six-membered Pd complex intermediate with α-benzyl β-alanine directing group was found to be crucial for the C—H bond activation. Good enantioselectivities have been reported for benzylic arylations *(29)* (Scheme 5.8).

Over-arylation may be avoidable by the use of blocking groups or steric hindrance. In some cases, however, it may be desirable and polyaromatics can be made *(3, 30)* (Scheme 5.9).

Pyrimidines and similar heterocycles act as suitable directing groups for Pd-catalyzed C—H activation *(31)*. Unsymmetrical halogenated intermediates, formed via palladacycles and a CMD process, can be synthesized leading to the divergent construction of novel biphenyl *s*-aryltetrazines *(32)* (Scheme 5.10).

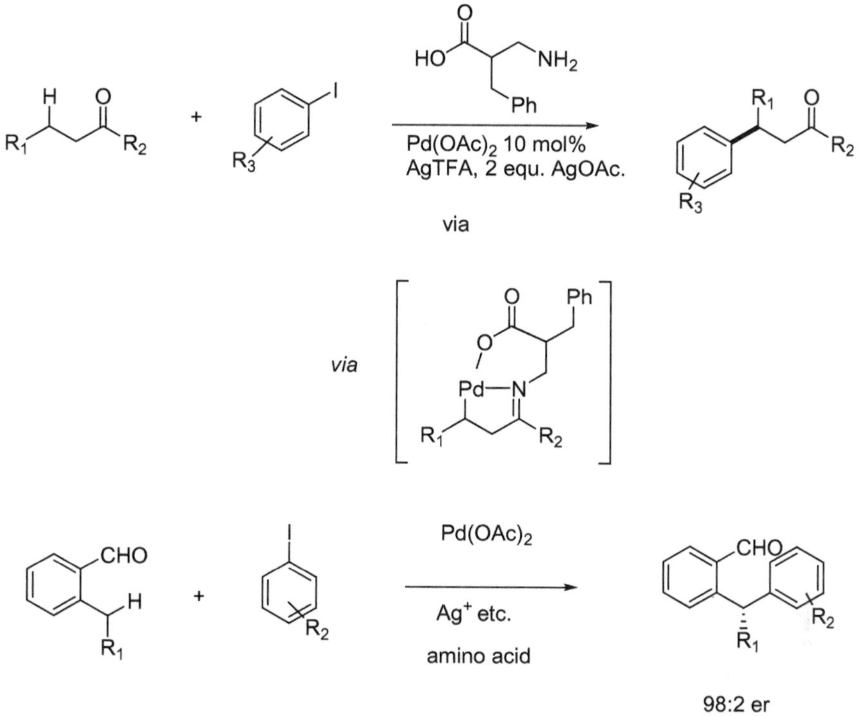

Scheme 5.8 Transient directing group-activated C—H activations.

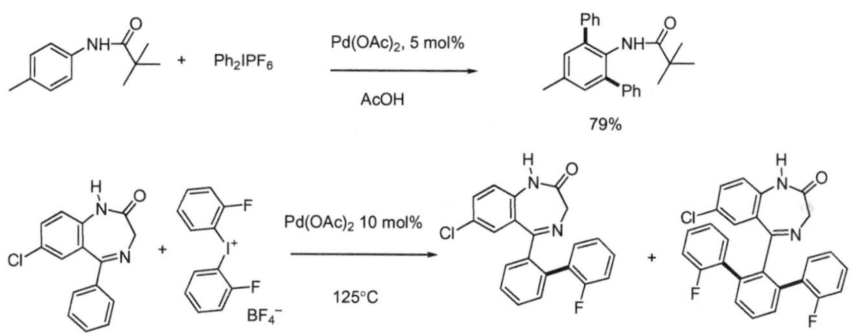

Scheme 5.9 Diarylation processes.

4. LATE STAGE C—H FUNCTIONALIZATION OF DRUG-LIKE MOLECULES AND NATURAL PRODUCTS

The above discussion highlights a few selected metal catalyzed C—H functionalization protocols *(33–35)*. Many other C—H activation systems have been developed in this field including using other metals *(36–40)*. The use of C—H

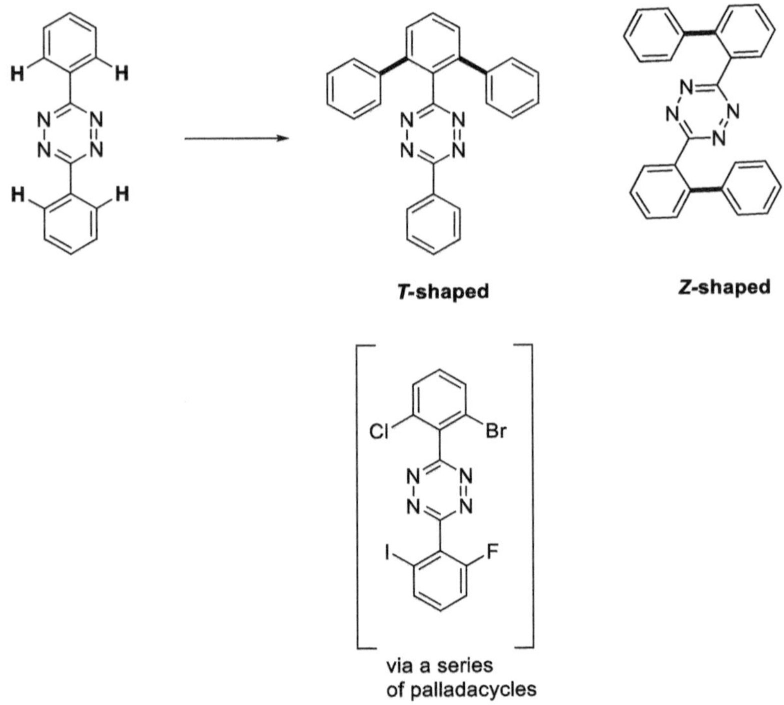

Scheme 5.10 Shape-selective synthesis.

functionalization protocols has enabled the synthesis of many complex molecules with a greener approach. Late-stage C—H functionalization protocols, in which libraries of complex bioactive molecules can be synthesized efficiently, has greatly contributed in speeding up the process of lead and hit-to-lead optimization in medicinal chemistry, as well as natural product synthesis, and diversification *(41)*. Direct C—H functionalizations are continuing to advance and therefore should continue to greatly contribute to energy efficient, highly selective, atom-economic transformations for the synthesis of organic molecules and bioactive compounds in the future *(42–46)* (Scheme 5.11).

5. CARBON-HETEROATOM BOND FORMATION

As stated above, we are unable to highlight comprehensively the advances in palladacycle chemistry; hence, this section gives an unbiased focus on another area of palladacycle chemistry. Palladacycles can be involved in C—Ge or C—Si *(47)*, C—S *(48)*, or C—O, bond formations for example *(49)*. C—N bond formations have been recently covered comprehensively *(50–53)* (Scheme 5.12).

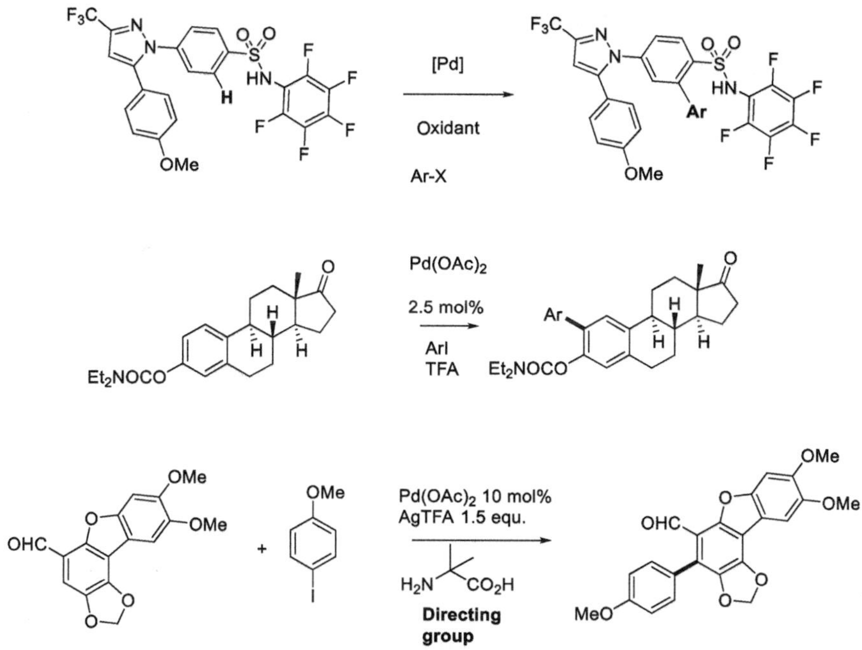

Scheme 5.11 Functionalization of complex molecules via palladacycles.

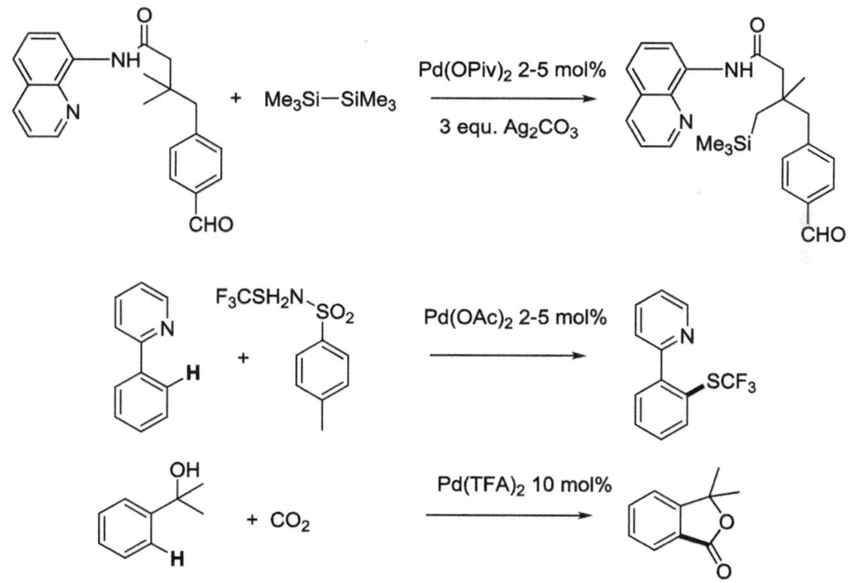

Scheme 5.12 C-heteroatom bond formations.

6. META C—H FUNCTIONALIZATIONS

Many of the earlier examples have focussed on the ortho-modification of molecules utilizing the chelating ability of directing groups to functionalize in the 2-position. Novel examples of meta-functionalization are now becoming more commonplace due to careful tethered ligand design, which can direct toward the meta position *(54–58)*. This is a fantastic extension to add to the arsenal of palladacycle-mediated transformations (Scheme 5.13).

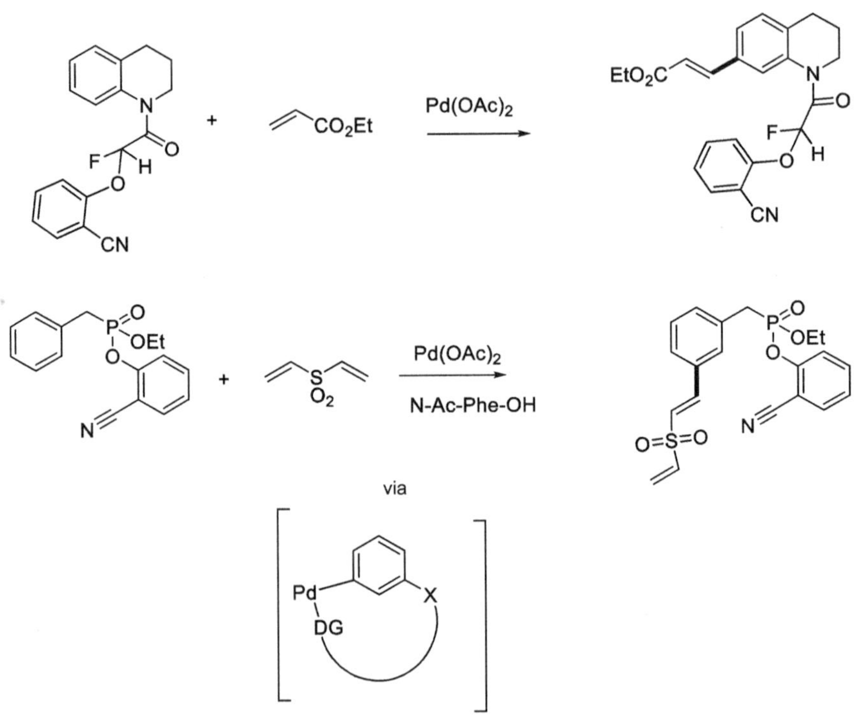

Scheme 5.13 Meta-activations.

The use of CO_2 as a transient director is an ingenious means of attaching an aromatic group formally in the meta position *(59)* (Scheme 5.14).

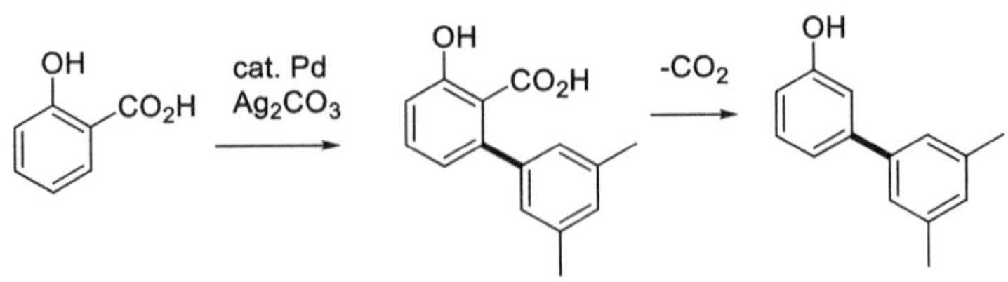

Scheme 5.14 CO_2 as a transient director group.

Alternatively, a Catellani-type intermediate *(60)* can be used to direct a meta CH activation *(55)* (Scheme 5.15).

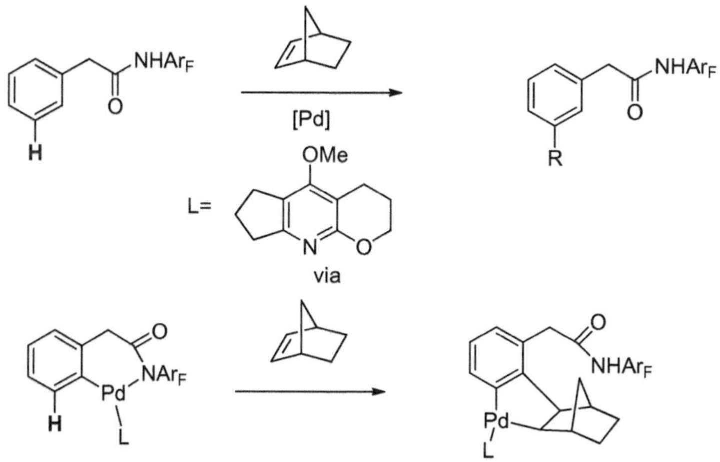

Scheme 5.15 Meta C—H activations using Catelani-type chemistry.

7. TRANSANNULAR C—H ACTIVATIONS

Recent advances in the activation of drug-like scaffolds include the use of transannular C—H activation chemistry (Scheme 5.16).

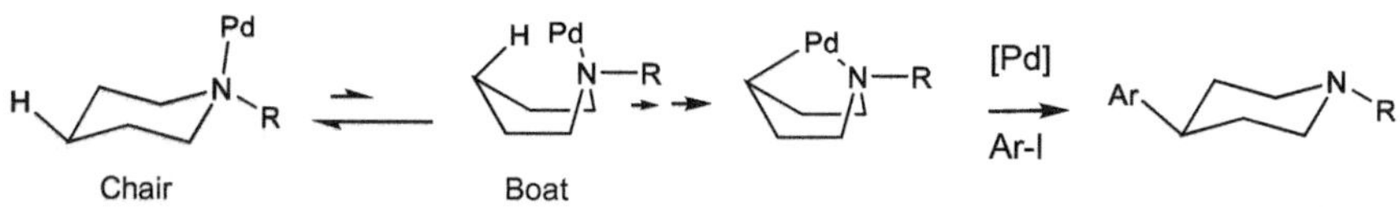

Scheme 5.16 Chair-boat transformations in C—H activations.

Sanford's group showed the use of such a chair-boat transformation in the synthesis of high value drug-like molecules *(61)* (Scheme 5.17).

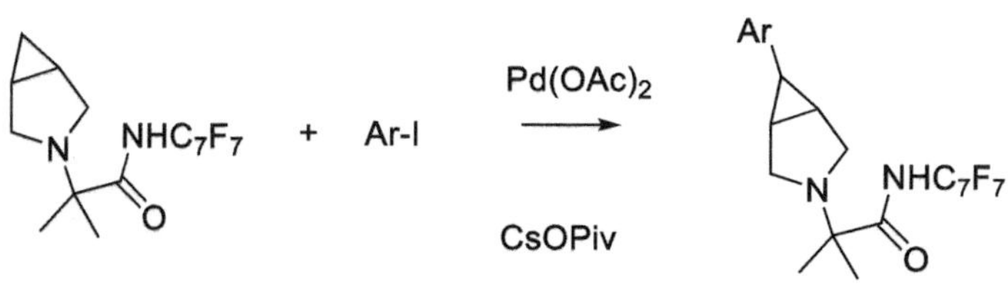

Scheme 5.17 Transannular C—H activation.

Others are showing the utility of a tether in providing high C-sp^3-character molecules *(62)* (Scheme 5.18).

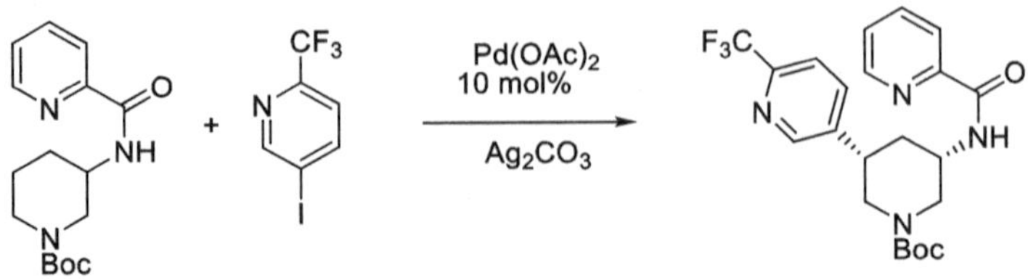

Scheme 5.18 Functionalized piperidines.

Finally, chemistry involving a 4-membered palladacyclic intermediate led to interesting aziridine scaffolds *(63)* (Scheme 5.19).

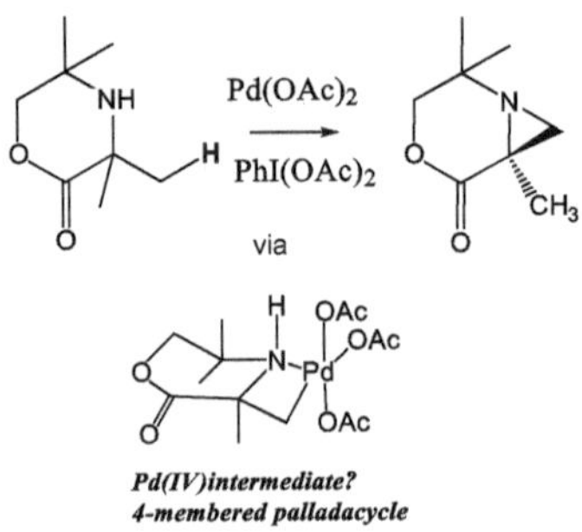

Scheme 5.19 Aziridine synthesis.

8. CONCLUSIONS AND FUTURE PERSPECTIVES

The wealth of examples illustrated in this chapter serves to show just how far the field of C—H functionalization has advanced in just over a decade. Previous reviews tend to focus on palladacycles in Pd(II)/(0) cycles, often as precatalysts in Suzuki-Miyaura or Heck-type chemistry *(10, 11)*. The increased understanding of the mechanistic and theoretical aspects of palladacycle chemistry, new types of reactivity *(64)*, a broader scope of substrates, improved ligand design, and the application toward the synthesis of "escape from flatland" drug-like entities *(65, 66)*, often using late stage functionalization chemistry, makes these reactions more step and atom economical. Despite these advances, much of this chemistry still employs high molecular weight aryl iodides or iodonium salts, for example, high catalyst loadings of 10 mol% or greater, meaning that these reactions are probably more "substoichiometric" than catalytic. A push toward higher turnover numbers, more double C—H activation type chemistry, with improved regioselective outcomes would be a vital hurdle to overcome, which is certainly achievable.

REFERENCES

1. Li, J. J., Ed.; *C-H Bond Activation in Organic Synthesis*, CRC Press, 2017.
2. Dupont, M.; Jairton, P., Eds.; *Palladacycles: Synthesis, Characterization and Applications*, Wiley-VCH Verlag GmbH & Co. KGaA, 2008.
3. Functionalization, C. C.À.H. B.; Neufeldt, S. R.; Sanford, M. S. *Accounts Chem. Res.* **2012**, *45*, 936–946.
4. Dupont, J.; Consorti, C. S.; Spencer, J. *Chem. Rev.* **2005**, *105*, 2527–2571.
5. Davies, D. L.; Macgregor, S. A.; McMullin, C. L. *Chem. Rev.* **2017**, *117*, 8649–8709.
6. Lapointe, D.; Markiewicz, T.; Whipp, C. J.; Toderian, A.; Fagnou, K. *J. Org. Chem.* **2011**, *76*, 749–759.
7. Yang, Y. F.; Hong, X.; Yu, J. Q.; Houk, K. N. *Acc. Chem. Res.* **2017**, *50*, 2853–2860.
8. Mian, M. R.; Iguchi, H.; Takaishi, S.; Murasugi, H.; Miyamoto, T.; Okamoto, H.; Tanaka, H.; Kuroda, S. I.; Breedlove, B. K.; Yamashita, M. *J. Am. Chem. Soc.* **2017**, *139*, 6562–6565.
9. Powers, D. C.; Ritter, T. *Nat. Chem.* **2009**, *1*, 302–309.
10. Dupont, J.; Pfeffer, M.; Spencer, J. *Eur. J. Inorg. Chem.* **2001**, *2001*, 1917–1927.
11. Bedford, R. B. *Chem. Commun.* **2003**, 1787–1796.
12. Merritt, E. A.; Olofsson, B. *Angew. Chem. Int. Ed.* **2009**, *48*, 9052–9070.
13. Yusubov, M. S.; Maskaev, A. V.; Zhdankin, V. V. *Arkivoc* **2011**, *2011*, 370–409.
14. Lindstedt, E.; Ghosh, R.; Olofsson, B. *Org. Lett.* **2013**, *15*, 6070–6073.
15. Wu, Z.; Chen, S.; Hu, C.; Li, Z.; Xiang, H.; Zhou, X. *ChemCatChem* **2013**, *5*, 2839–2842.
16. Ma, S.; Villa, G.; Thuy-Boun, P. S.; Homs, A.; Yu, J. Q. *Angew. Chem. Int. Ed.* **2014**, *53*, 734–737.
17. Yin, D. W.; Liu, G. *J. Org. Chem.* **2018**, *83*, 3987–4001.
18. Hull, K. L.; Sanford, M. S. *J. Am. Chem. Soc.* **2007**, *129*, 11904–11905.
19. Yeung, C. S.; Zhao, X.; Nadine, B.; Dong, V. M. *Chem. Sci.* **2010**, *1*, 331–336.
20. Zhang, Y.; Zhao, Y.; Luo, Y.; Xiao, L.; Huang, Y.; Li, X.; Peng, Q.; Liu, Y.; Yang, B.; Zhu, C.; Zhou, X.; Zhang, J. *Org. Lett.* **2017**, *19*, 6470–6473.
21. Kalyani, D.; McMurtrey, K. B.; Neufeldt, S. R.; Sanford, M. S. *J. Am. Chem. Soc.* **2011**, *133*, 18566–18569.
22. Khan, R.; Boonseng, S.; Kemmitt, P. D.; Felix, R.; Coles, S. J.; Tizzard, G. J.; Williams, G.; Simmonds, O.; Harvey, J. L.; Atack, J.; Cox, H.; Spencer, J. *Adv. Synth. Catal.* **2017**, *359*, 3261–3269.
23. Liang, L.; Xie, M. S.; Wang, H. X.; Niu, H. Y.; Qu, G. R.; Guo, H. M. *J. Org. Chem.* **2017**, *82*, 5966–5973.
24. Jiang, J.; Zhang, W. M.; Dai, J. J.; Xu, J.; Xu, H. J. *J. Org. Chem.* **2017**, *82*, 3622–3630.
25. Douglas, J. J.; Sevrin, M. J.; Stephenson, C. R. J. *Org. Process Res. Dev.* **2016**, *20*, 1134–1147.
26. Majek, M.; Jacobi Von Wangelin, A. *Acc. Chem. Res.* **2016**, *49*, 2316–2327.
27. Pramanik, M. M. D.; Rastogi, N. *Chem. Commun.* **2016**, *52*, 8557–8560.
28. Zhu, R. Y.; Li, Z. Q.; Park, H. S.; Senanayake, C. H.; Yu, J. Q. *J. Am. Chem. Soc.* **2018**, *140*, 3564–3568.
29. Zhang, F.-L.; Hong, K.; Li, T.-J.; Park, T.-J.; Yu, J.-Q. *Science (80-)* **2016**, *350*, 252–256.
30. Khan, R.; Felix, R.; Kemmitt, P. D.; Coles, S. J.; Day, I. J.; Tizzard, G. J.; Spencer, J. *Adv. Synth. Catal.* **2016**, *358*, 98–109.
31. Gupta, S.; Melanson, J. A.; Vaillancourt, L.; Nugent, W. A.; Tanoury, G. J.; Schatte, G.; Snieckus, V. *Org. Lett.* **2018**, *20*, 3745–3748.
32. Mboyi, C. D.; Testa, C.; Reeb, S.; Genc, S.; Cattey, H.; Fleurat-Lessard, P.; Roger, J.; Hierso, J. C. *ACS Catal.* **2017**, 7, 8493–8501.
33. Noisier, A. F. M.; Brimble, M. A. *Chem. Rev.* **2014**, *114*, 8775–8806.
34. Yamaguchi, J.; Yamaguchi, A. D.; Itami, K. *Angew. Chem. Int. Ed.* **2012**, *51*, 8960–9009.

35. McMurray, L.; O'Hara, F.; Gaunt, M. J. *Chem. Soc. Rev.* **2011**, *40*, 1885–1898.
36. He, J.; Hamann, L. G.; Davies, H. M. L.; Beckwith, R. E. J. *Nat. Commun.* **2015**, *6*, 1–9.
37. Hartwig, J. F. *J. Am. Chem. Soc.* **2016**, *138*, 2–24.
38. Wencel-Delord, J.; Glorius, F. *Nat. Chem.* **2013**, *5*, 369–375.
39. Zhu, Y.; Bauer, M.; Ackermann, L. *Chem. Eur. J.* **2015**, *21*, 9980–9983.
40. Chen, W.; Ma, L.; Paul, A.; Seidel, D. *Nat. Chem.* **2018**, *10*, 165–169.
41. Cernak, T.; Dykstra, K. D.; Tyagarajan, S.; Vachal, P.; Krska, S. W. *Chem. Soc. Rev.* **2016**, *45*, 546–576.
42. Zhang, M. Y.; Barrow, R. A. *J. Org. Chem.* **2018**, *83*, 6776–6782.
43. Bedford, R. B.; Brenner, P. B.; Durrant, S. J.; Gallagher, T.; Méndez-Gálvez, C.; Montgomery, M. *J. Org. Chem.* **2016**, *81*, 3473–3478.
44. Blakemore, D. C.; Castro, L.; Churcher, I.; Rees, D. C.; Thomas, A. W.; Wilson, D. M.; Wood, A. *Nat. Chem.* **2018**, *10*, 383–394.
45. Haines, B. E.; Berry, J. F.; Yu, J. Q.; Musaev, D. G. *ACS Catal.* **2016**, *6*, 829–839.
46. Dai, H. X.; Stepan, A. F.; Plummer, M. S.; Zhang, Y. H.; Yu, J. Q. *J. Am. Chem. Soc.* **2011**, *133*, 7222–7228.
47. Deb, A.; Singh, S.; Seth, K.; Pimparkar, S.; Bhaskararao, B.; Guin, S.; Sunoj, R. B.; Maiti, D. *ACS Catal.* **2017**, *7*, 8171–8175.
48. Kesavan, A.; Chaitanya, M.; Anbarasan, P. *Eur. J. Org. Chem.* **2018**, *2018*, 3276–3279.
49. Song, L.; Zhu, L.; Zhang, Z.; Ye, J.-H.; Yan, S.-S.; Han, J.-L.; Yin, Z.-B.; Lan, Y.; Yu, D.-G. *Org. Lett.* **2018**, *20*, 3776–3779.
50. Bandara, H. M. D.; Jin, D.; Mantell, M. A.; Field, K. D.; Wang, A.; Narayanan, R. P.; Deskins, N. A.; Emmert, M. H. *Catal. Sci. Technol.* **2016**, *6*, 5304–5310.
51. Berzina, B.; Sokolovs, I.; Suna, E. *ACS Catal.* **2015**, *5*, 7008–7014.
52. Timsina, Y. N.; Gupton, B. F.; Ellis, K. C. *ACS Catal.* **2018**, *8*, 5732–5776.
53. Sharma, A.; Hartwig, J. F. *Nature* **2015**, *517*, 600–604.
54. Leow, D.; Li, G.; Mei, T. S.; Yu, J. Q. *Nature* **2012**, *486*, 518–522.
55. Wang, X. C.; Gong, W.; Fang, L. Z.; Zhu, R. Y.; Li, S.; Engle, K. M.; Yu, J. Q. *Nature* **2015**, *519*, 334–338.
56. Bera, M.; Sahoo, S. K.; Maiti, D. *ACS Catal.* **2016**, *6*, 3575–3579.
57. Jayarajan, R.; Das, J.; Bag, S.; Chowdhury, R.; Maiti, D. *Angew. Chem. Int. Ed.* **2018**, *57*, 7659–7663.
58. Dey, A.; Agasti, S.; Maiti, D. *Org. Biomol. Chem.* **2016**, *14*, 5440–5453.
59. Luo, J.; Preciado, S.; Larrosa, I. *J. Am. Chem. Soc.* **2014**, *136*, 4109–4112.
60. Della Ca, N.; Fontana, M.; Motti, E.; Catellani, M. *Acc. Chem. Res.* **2016**, *49*, 1389–1400.
61. Topczewski, J. J.; Cabrera, P. J.; Saper, N. I.; Sanford, M. S. *Nature* **2016**, *531*, 220–224.
62. Van Steijvoort, B. F.; Kaval, N.; Kulago, A. A.; Maes, B. U. W. *ACS Catal.* **2016**, *6*, 4486–4490.
63. McNally, A.; Haffemayer, B.; Collins, B. S. L.; Gaunt, M. J. *Nature* **2014**, *510*, 129–133.
64. Yu, J.; Lv, W.; Cheng, G. *Org. Lett.* **2018**, *20*, 4732–4735.
65. Lovering, F.; Bikker, J.; Humblet, C. *J. Med. Chem.* **2009**, *52*, 6752–6756.
66. Hernando, E.; Villalva, J.; Martínez, Á.M.; Alonso, I.; Rodríguez, N.; Gómez Arrayás, R.; Carretero, J. C. *ACS Catal.* **2016**, *6*, 6868–6882.

FURTHER READING

67. Chaumontet, M.; Piccardi, R.; Audic, N.; Hitce, J.; Peglion, J. L.; Clot, E.; Baudoin, O. *J. Am. Chem. Soc.* **2008**, *130*, 15157–15166.
68. Deprez, N. R.; Sanford, M. S. *J. Am. Chem. Soc.* **2009**, *131*, 11234–11241.
69. Mukherjee, A.; Sen, T. K.; Ghorai, P. K.; Mandal, S. K. *Organometallics* **2013**, *32*, 7213–7224.

CHAPTER 6

Multicomponent Reaction Sequences Using Palladacyclic Complexes

Helena C. Malinakova
Department of Chemistry, The University of Kansas, Lawrence, KS, United States

Contents

1. INTRODUCTION

Multicomponent reactions, defined as synthetic protocols that join together three or more substrates in a highly regio- and stereoselective manner to deliver structurally complex organic molecules, have seen a dramatic rise in applications in all fields of organic synthesis. The methodology afforded impressively efficient solutions to challenging problems in the total synthesis of natural products *(1)*. Synthetic methodologies based on multicomponent reactions proved to be ideally suited for the diversity-oriented synthesis of combinatorial libraries aimed toward drug-discovery goals *(2)*. Various

Palladacycles
https://doi.org/10.1016/B978-0-12-815505-9.00007-X
© 2019 Elsevier Inc.
All rights reserved.

modes of catalysis, including organocatalysis *(3)*, transition metal catalysis *(4)*, as well as dual-catalysis of various types *(5)*, were employed for the design of multicomponent reactions. Palladium catalysis, encompassing a number of historically significant Nobel prize-winning transformations, has played a prominent role in the development of novel multicomponent reactions *(6)*. An easy access to three different oxidation states of palladium, e.g., Pd(0), Pd(II), and Pd(IV), and the extensive knowledge of how different types of auxiliary ligands and additives could select between alternative catalytic cycles available to the given set of substrates *(6)*, are key features responsible for the extensive use of palladium catalysts in multicomponent reactions *(7)*.

An important subset of the palladium-catalyzed multicomponent reactions features a cyclopalladated complex as the key component or a reactive intermediate. For the purpose of this review, *cyclopalladated complexes*, often also called *palladacycles*, have been defined as organopalladium species featuring at least one carbon-palladium bond (where the group R is formally considered a mononegative, two electron donor), and either a second carbon-palladium bond or a coordination bond between the palladium center and a heteroatom X (X = O, N, S, P where X is formally considered a neutral two electron donor) (Fig. 6.1) *(8)*.

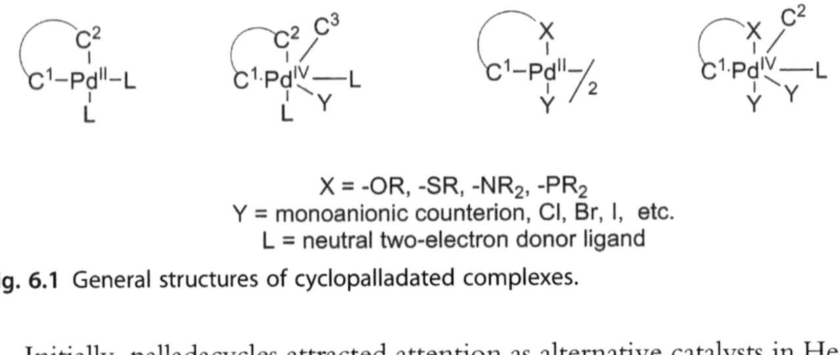

Fig. 6.1 General structures of cyclopalladated complexes.

Initially, palladacycles attracted attention as alternative catalysts in Heck reactions, and their mode of action was often traced to acting as a source of highly active colloidal Pd^0 nanoparticles *(9)*. Since then, however, palladacycles have been shown to operate as key intermediates in complex catalytic cycles while playing diverse roles and containing palladium in all accessible oxidation states (Pd^{II}, Pd^{IV}). In this review, we highlight the most notable multicomponent reactions reported over the past decade (after 2008) that feature palladacycles in all different mechanistic roles.

The material will be divided into three broad areas, including:

(1) *Multicomponent reactions using palladium catalyst with a cyclopalladated auxiliary ligand.* Herein, the auxiliary ligand may or may not remain attached to the palladium center throughout the entire catalytic cycle. Thus, reactions where the cyclopalladated complex plays a role of a stable Lewis acidic center, or reactions where the cyclopalladated complex acts as a precatalyst, are presented.

(2) *Multicomponent reactions involving in situ palladation of the substrate.* Herein, protocols featuring a regiocontrolled C—H activation via palladation of one of the substrates followed by functionalization, are presented.

(3) *Multicomponent reactions with a stoichiometric cyclopalladated complex.* Herein, reactions featuring multiple small molecule insertions into stoichiometric palladacycles, and leading to organic products, are presented.

Selected mechanistic studies that are relevant to elucidating the role played by cyclopalladated complexes in carbon-carbon bond formation, are discussed where appropriate.

2. MULTICOMPONENT REACTIONS USING PALLADIUM CATALYST WITH A CYCLOPALLADATED AUXILIARY LIGANDS

2.1 Recent Applications of Herrmann's Palladacycle

Herrmann-Beller palladacycle (HBP, Scheme 6.1) was among the first palladacycles recognized for catalytic activity superior to traditional combinations of phosphine ligands with Pd(II) or Pd(0) complexes in the Heck reactions *(10)*. Recently, an excellent performance of Herrmann-Beller palladacycle as the precatalyst in a dual-catalytic multicomponent reaction was reported by Lautens *(11)*. The reaction between *ortho*-iodobenzyl azides, terminal alkynes and internal alkynes catalyzed by CuI and the Herrmann-Beller palladacycle afforded a variety of tricyclic triazoles (Scheme 6.1).

Initial attempts to employ $Pd(OAc)_2$ with added PPh_3 afforded only complex mixtures of products. Deactivation of the Cu(I) center by the free phosphine ligand was deemed to be responsible for this observation. Notably, when the traditional catalytic system was replaced with the Herrmann-Beller palladacycle, along with pivalic acid as an additive facilitating the C—H activation, the triazole products were obtained in good

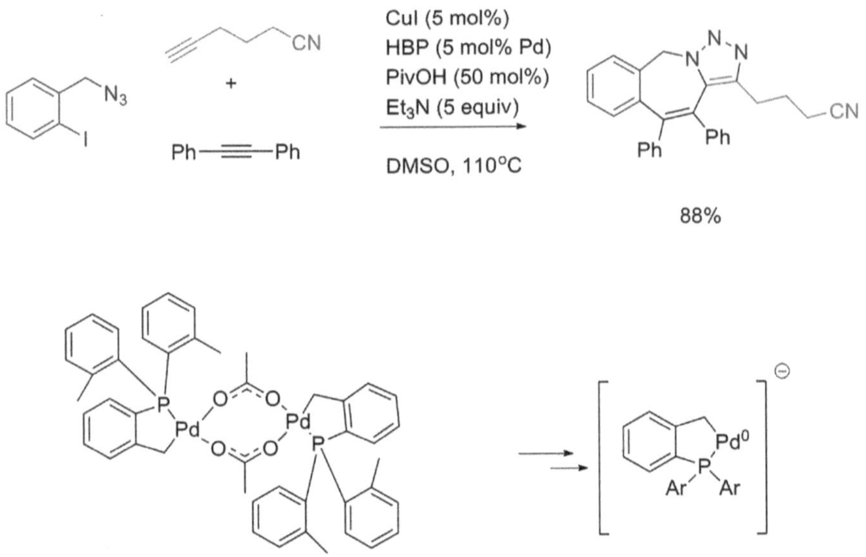

HBP (Herrmann-Beller palladacycle)

Scheme 6.1 Synthesis of tricyclic triazoles using the Herrmann-Beller palladacycle as catalyst.

(50%–80%) yields (Scheme 6.1). The presence of the cyclopalladated phosphine ligand (e.g., the C-Pd(II)-P cycle) was found to be critical to the success of the catalysis. When cyclopalladated complexes introduced by Buchwald (Generation I and III) featuring cyclopalladated amines (e.g., the C-Pd(II)-N cycle) with X-Phos monodentate ligands were used as catalysts, the triazoles were not produced, as the requisite carbopalladation of the internal alkyne was inhibited. The mechanism of the three-component protocol was rationalized by two independent catalytic cycles. Initially Cu(I)-catalyzed [3+2] cycloaddition between the azide and the terminal alkynes takes place. Subsequently, the cycloaddition product enters the Pd(0)/Pd(II)-catalytic cycle, and undergoes oxidative addition at the Csp^2—I bond, carbopalladation of the internal alkynes, and an intramolecular electrophilic palladation of the C—H bond in the triazole ring, giving rise to an eight-membered C-Pd(II)-C palladacycle that generates the tricyclic triazole product after reductive elimination. Experiments with deuterated substrates supported the electrophilic palladation pathway. To appreciate the unique features of the Herrmann-Beller palladacycle, it is relevant to refer to prior studies which suggested that an in situ reduction of the

Herrmann's catalyst gave rise to an active monoanionic $[C\text{-}Pd(0)]^-$ — complex retaining the aryl-Pd bond (Scheme 6.1) *(12)*. The superior activity of the Herrmann's catalyst in comparison to traditional $Pd(II)/PR_3$ systems toward oxidative addition into aryl bromide bonds was demonstrated by a three-component aminocarbonylation of aryl and heteroaryl halides using $[Mo(CO)_6]$ as a convenient source of the CO, and aryl and heteroaryl amines to deliver a diverse ensemble of valuable amide building blocks (Scheme 6.2) *(13)*.

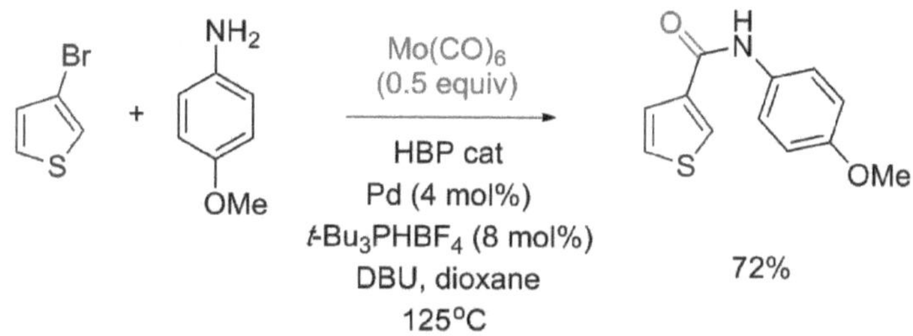

Scheme 6.2 Aminocarbonylation using $Mo(CO)_6$ and Herrmann-Beller palladacycle as catalyst.

The aminocarbonylation of heteroaryl bromides (2-bromothiophene, 3-bromothiophene, 3-bromobenzothiophene, 2-bromopyridine) with $Mo(CO)_6$ and aromatic and heteroaromatic amines required the use of the Herrmann's catalyst, along with $t\text{-}Bu_3PHBF_4$ (DBU, dioxane) (Scheme 6.2). In contrast, the aminocarbonylation of the more reactive aryl iodide could be successfully realized using $Pd(OAc)_2$ (dioxane, DBU) in the absence of an added phosphine ligand.

2.2 Palladium-Catalyzed Cross-Coupling Polymerization

In principle, polymerization represents an extreme extension of a multi-component reaction as defined herein. Palladium-catalyzed cross-coupling polymerization of bifunctional (AB) monomers possessing a complementary electrophilic group A (C—Br bond) and a nucleophilic group B (C—Bpin bond) has recently received much attention as a promising methodology for the preparation of valuable new materials *(14)*. For a successful cross-coupling polymerization, it is critical to identify an initiator that easily generates active Pd(0) species capable of delivering uniform "capping groups" at the origin of each polymer chain. Furthermore, reaction conditions that favor a well-controlled chain-growth polymerization involving an intramolecular transfer of the Pd(0) to the next electrophilic site at the end of the growing polymer chain, must be established in order to produce polymers with a narrow molecular weight distribution. Cyclopalladated complexes have been used with success as the initiators of such polymerization processes. In a study reported by Hu, a palladium catalyst featuring a cyclopalladated acetanilide provided the "capping aromatic group" at the origin of the polymer chain, and served as the source of the active Pd(0) species *(15)*. The aim of the study was to evaluate the effect of replacing the established *t*-Bu_3P phosphine ligand in the structure of the cyclopalladated acetanilide catalyst with the tris-adamantylphosphine (Ad_3P) ligand on the outcomes of the polymerization of 7-bromo-9,9-dihexylfluoren-2-yl boronic acid pinacol ester and its derivatives (Scheme 6.3). The ligand change resulted in a somewhat slower polymerization, but the process was well controlled, delivering products with narrow distributions of molecular weights, as well as a well-controlled capping group (established by MALDI analysis). A linear relationship between the conversion and the length of the polymer chain was demonstrated confirming the operation of the desired chain-growth polymerization.

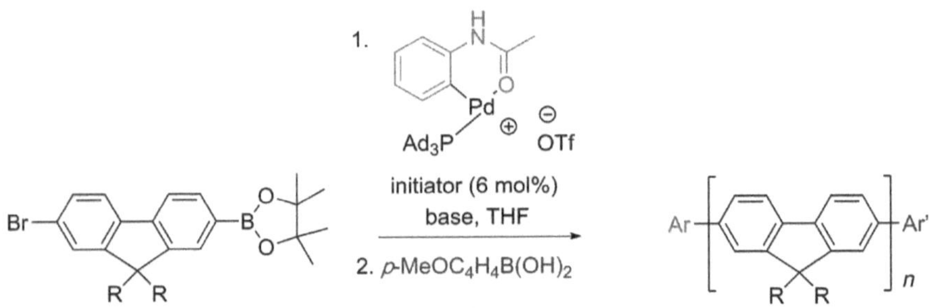

Scheme 6.3 Cyclopalladated acetanilide as the catalyst for cross-coupling polymerization.

Choi has reported a significant advance in the cross-coupling polymerization for the synthesis of polythiophenes, namely poly(3-hexyl)thiophene or poly(3-ethyl)thiophene, as well as a block copolymer of these two 5-bromo-4-*n*-alkyllthien-2-yl-boronate monomers (Scheme 6.4) *(16)*.

Two problems hampering the prior polymerization protocols were addressed in this work. The application of MIDA-boronates (*N*-methylimidodiacetic acid boronates) rather than boronate esters, avoided the competing protodeboronation, and led to a generally high degree of polymerization. A slow initiation observed with the previously used phenylpalladium bromide complexes featuring *t*-Bu_3P or various

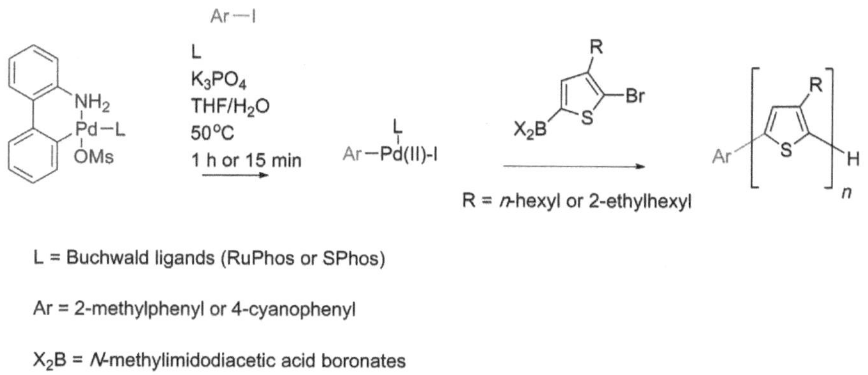

Scheme 6.4 Cyclopalladated complexes with Buchwald ligands as precatalysts for the preparation of polythiophenes from MIDA-boronates.

N-heterocyclic carbenes as ligands L (LPhPd(II)Br), was addressed by employing a more active precatalyst. Thus, mixing catalytic amounts of the Buchwald (Generation III) cyclopalladated complex bearing the S-Phos or Ru-Phos ligands with aryl iodides and base (K_3PO_4) in THF-H_2O for 15–60 min prior to the addition of the monomers resulted in much faster initiation. Under these conditions, the cyclopalladated precatalysts undergo rapid reduction to active Pd(0) species which gives rise to the aryl(L)Pd(II)I catalyst responsible for the introduction of the aryl (Ar) "capping groups" at the origin of the growing polymer chains (Scheme 6.4). The main advantage of the cyclopalladated complex was the rapid generation of the active LPd(0) species. Furthermore, control experiments indicated that the palladacycle-based precatalyst favored more strongly the intramolecular transfer of the Pd(0) catalyst via the intramolecular oxidative addition to the second Csp^2-Br site in the same molecule of the substrate, thus supporting the chain-growth polymerization.

2.3 Cyclopalladated Catalysts Remaining in Pd(II) Oxidation State

Complexes featuring a cyclopalladated auxiliary ligand have been used in protocols in which the bonds between the auxiliary ligand and the Pd(II) center remain intact throughout the entire reaction, and the oxidation state of palladium does not change. In these transformations, the palladium center operates either as a Lewis acidic site, activating the coordinated organic substrates, or mediates the formation of new C—C or C—X bonds via transmetallation and/or migratory insertion events *(6)*. Although interconversion between Pd(II) and Pd(IV) oxidation states would allow for the cyclopalladated ligand to remain attached to the Pd(II)/Pd(IV) centers, synthetic protocols relying on such catalytic cycles have so far exclusively featured one of the substrates as the cyclopalladated ligand *(17)*.

A palladium-catalyzed three-component coupling reaction between anhydrides (acetic anhydride or benzoic anhydride); amino acids, and enones afforded functionalized 1,5-diones via the Michael addition of azalactones, formed in situ by the condensation of the amino acid with the anhydride (Scheme 6.5) *(18)*. The regioselectivity of electrophilic attack at either the C-2 or the C-4 (α-carbon to the carbonyl group) could be controlled by properly matching the choice of either the monopalladated or the bispalladated ferrocene-derived chiral nonracemic catalysts to the selected anhydride (Scheme 6.5).

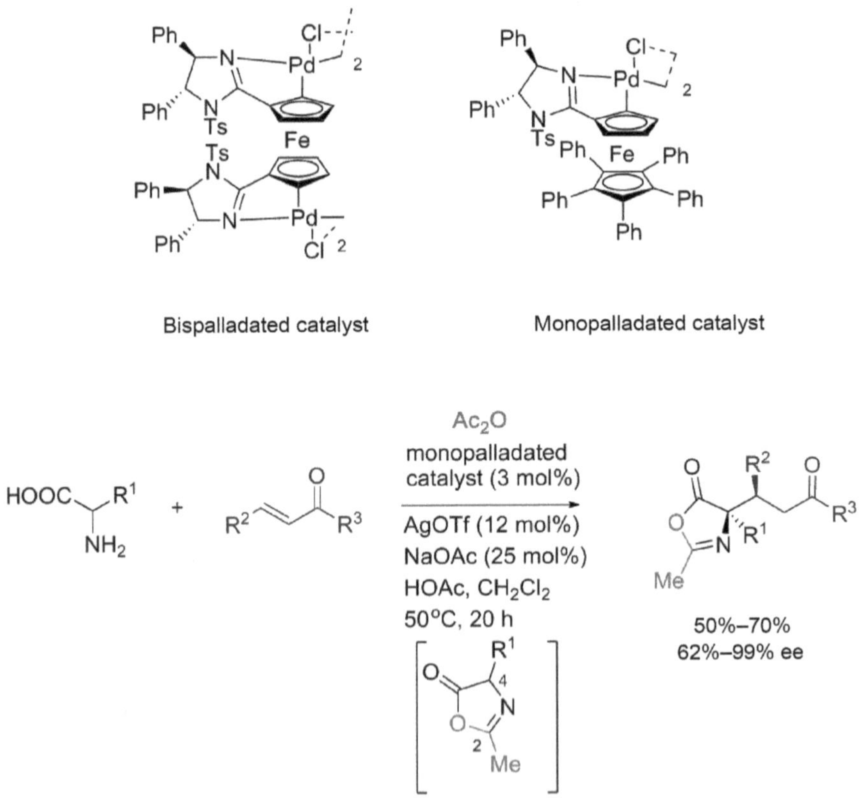

Scheme 6.5 Cyclopalladated catalysts for asymmetric synthesis of azalactone 1,5-diones from acetic anhydride, amino acids and enones.

Under the reaction conditions (AcOH, NaOAc), azalactones formed from racemic amino acids undergo reversible enolization and/or deprotonation yielding achiral nucleophiles capable of reacting with the Michael acceptor via either the C-2 or the C-4 positions (Scheme 6.5). When the bispalladated ferrocene catalyst was used in reactions with acetic anhydride, mixtures of C2 and C4 Michal adducts were obtained. To alleviate this problem, excess of benzoic anhydride had to be employed to achieve selective functionalization at the sterically more accessible C-4 position of the azalactone. This protocol proved to be quite cumbersome, particularly due to the challenging removal of excess benzoic anhydride. After replacing the *bispalladated complex* with an analogous *monopalladated catalyst*, selective C-4 functionalization of the azalactone generated from acetic anhydride was easily realized. The protocol was much more practical considering the relative ease of removal of excess acetic anhydride. Furthermore, excellent enantioselectivity was noted in the final Michael addition products. However, the same monopalladated catalyst did not perform well in reactions with benzoic anhydride.

These observations were rationalized by noting the difference in reactivity between the weakly nucleophilic and more sterically hindered azalactone generated from benzoic anhydride, and the more nucleophilic and less sterically hindered azalactone derived from acetic anhydride. The Ph-substituted azalactone requires bimetallic activation via coordination of both the lactone and the enone each to a separate Pd(II) center in the bimetallic catalyst. The transition state for the bimetallic pathways that minimizes steric hindrance caused by the Ph group strongly favors the attack on the C-4 position in the azalactone. Analogous transition state for the less sterically hindered azalactone derived from acetic anhydride does not lead to a regioselective transformation, and the lack of selectivity in reactions of the Me-substituted azalactone catalyzed by the bispalladacycle was explained by a competition between the bimetallic and monometallic pathways. Consequently, the optimized monopalladated catalyst featuring a sterically encumbered bottom ring of the ferrocene, makes only the monometallic activation pathway available, and for that reason reactions with Me-substituted azalactones proceed with excellent selectivities for functionalization at the C-4 position of the azalactone, as well as high enantiomeric purity of the 1,5-diones (Scheme 6.5).

Seeking a more efficient catalyst for a Pd(II)-catalyzed three-component coupling of boronic acids, allenes, and imines for the preparation of substituted homoallylic amines (Scheme 6.6), the performance of

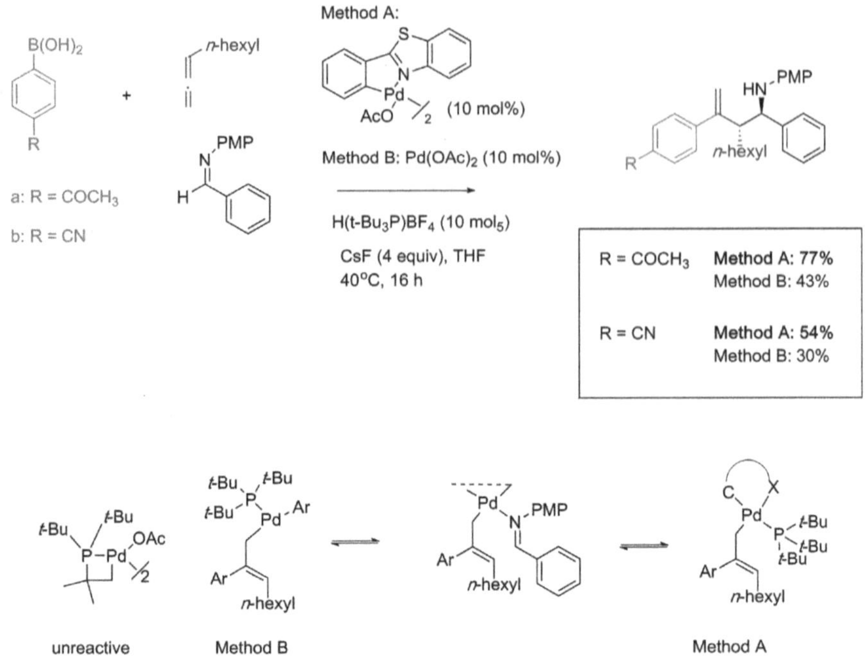

Scheme 6.6 Superior performance of a cyclopalladated complex vs Pd(OAc) $_2$in three-component coupling of boronic acids and allenes with imines.

cyclopalladated complexes with diverse auxiliary ligands in the presence or absence of free t-Bu$_3$P ligand (Fig. 6.2) was compared to the originally developed Pd(OAc)$_2$/t-Bu$_3$P catalyst (Scheme 6.6) *(19)*.

The key step in the catalytic cycle for this three-component coupling reaction is the nucleophilic attack of the Pd(II)-bonded allyl fragment on the imine coordinated to the Pd(II) center. Selective formation of antihomoallylic amines supports the proposed closed transition state for the nucleophilic allyl transfer (Scheme 6.6) *(20)*. Transmetallation of the aryl group from boron to palladium, followed by migratory insertion of the allene, gives rise to the allylpalladium complex. Computational study by Szabo confirmed that the η-1 bonding of the allyl fragment to the Pd(II) center was necessary for the nucleophilicity of the substituted terminus of the allyl fragment *(21)*. In agreement with this mechanistic proposal, sterically bulky monodentate phosphines (t-Bu$_3$P) used along with Pd(OAc)$_2$ were found to be competent catalysts for this transformation, presumably favoring the requisite η-1 bonding of the allyl fragment by steric demands of the ligand (Method B, Scheme 6.6).

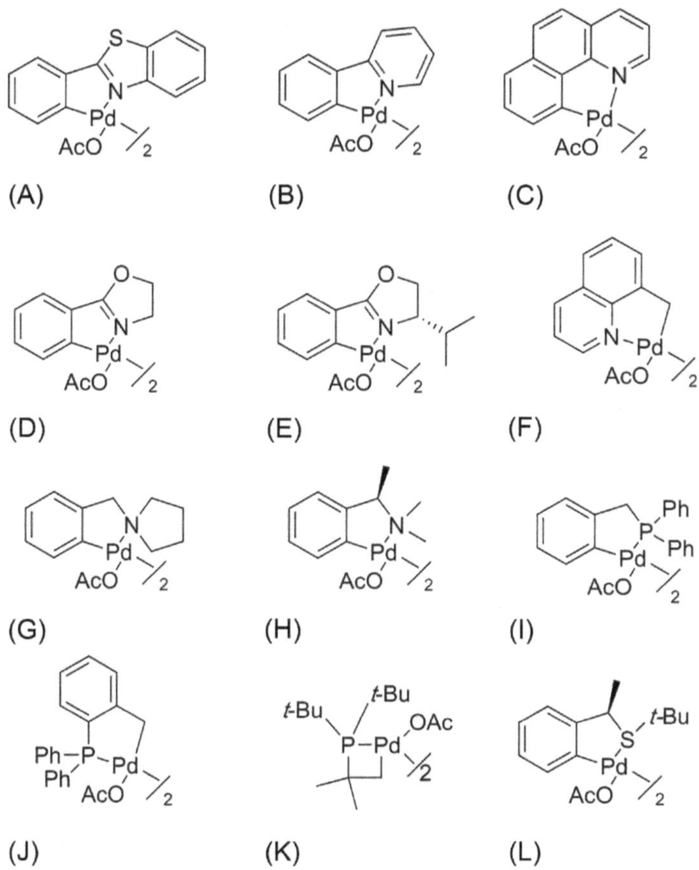

Fig. 6.2 Cyclopalladated complexes used as catalysts in the three-component coupling of boronic acids, allenes and imines.

Comparison of the catalytic activity of diverse cyclopalladated complexes (Fig. 6.2) indicated that the presence of a free sterically bulky phosphine ligand (added as $H(t\text{-}Bu_3P)BF_4$ in 1:1 mol ratio) along with the cyclopalladated complex (10 mol% Pd, added as a dimer) was necessary for optimum performance. Palladacycles featuring sterically encumbered N-heteroatom (X = N), either sp^2-hybridized (=N—) or sp^3-hybridized (—NR_2), proved to be the most active catalysts. The complex shown in Scheme 6.6 (Method A) was selected for exploration of the reaction scope with respect to substituted boronic acids. In most cases, the catalytic system used in Method A showed improved performance in comparison to the traditional catalyst used in Method B (Scheme 6.6). Examples of highest gains

(34% and 24% for R = COMe and R = CN) in product yields for two electron-deficient boronic acids are shown in Scheme 6.6. The improved performance is likely related to the presence of a relatively electron rich Pd(II) center and the steric bulk around the N-atom favoring the requisite η-1-coordination of the allyl group, both features contributing to increased nucleophilicity of the allyl fragment. In contrast, palladacycles featuring P and S heteroatoms (X = P, S) did not shown improved activity, a finding consistent with the expected lower electron density on the Pd(II) center. It is notable that a four-membered palladacycle that is likely to be formed by cyclopalladation of the *t*-Bu_3P ligand under the reaction conditions of Method A (complex K in Fig. 6.2 and Scheme 6.6), was found to be inactive as a catalyst, shedding some light on the reasons for subpar performance of the $Pd(OAc)_2$/*t*-Bu_3P catalyst in this transformation. Only minimal asymmetric induction (2% and 3% ee) was observed when palladacycles with chiral nonracemic cyclopalladated auxiliary ligands were employed (Fig. 6.2). ^{31}P NMR monitoring was performed on reactions stoichiometric in palladium performed under the conditions of Method A and Method B. The two systems gave rise to distinct signals in the ^{31}P NMR, ruling out the possibility that the cyclopalladated complexes simply acted as precatalysts. Under the conditions of a stoichiometric Method A, signals at 50.9 ppm (a new Pd(II) complex with ligand L) and 63.4 ppm (free *t*-Bu_3P ligand) were detected, whereas the stoichiometric conditions in Method B yielded signals at −9.2 ppm (corresponding to the signal reported for palladacycle K, Fig. 6.2) and 84.8 ppm (presumably the $Pd(0)L_2$, L = *t*-Bu_3P) *(19)*.

An interesting mechanistic study described a palladium(II) catalyzed addition of arylboron derivatives to aldehydes via a pathway that did not involve a boron to palladium transmetallation. The process employed a palladacycle with a cyclopalladated auxiliary ligand $P(OAr)_3$ as the catalyst (Scheme 6.7) *(22)*.

The authors reported that under scrupulously anhydrous conditions, bromoaryl aldehydes reacted with dried boronic acids (boroxines) and dried K_3PO_4 in toluene to afford the corresponding bromoaryl alcohols in excellent yields. No byproducts arising from a C^1—Br/C^2—B cross-coupling were present, and no formation of Pd(0) indicative of the catalyst decomposition was detected. Consequently, catalyst loading could be lowered to as low as 0.0005 mol%. Furthermore, electron deficient boronic acids, known to be sluggish in transmetallation, proved to be effective nucleophiles. Extensive control experiments confirmed that the anhydrous conditions were indeed responsible for inhibiting the transmetallation and

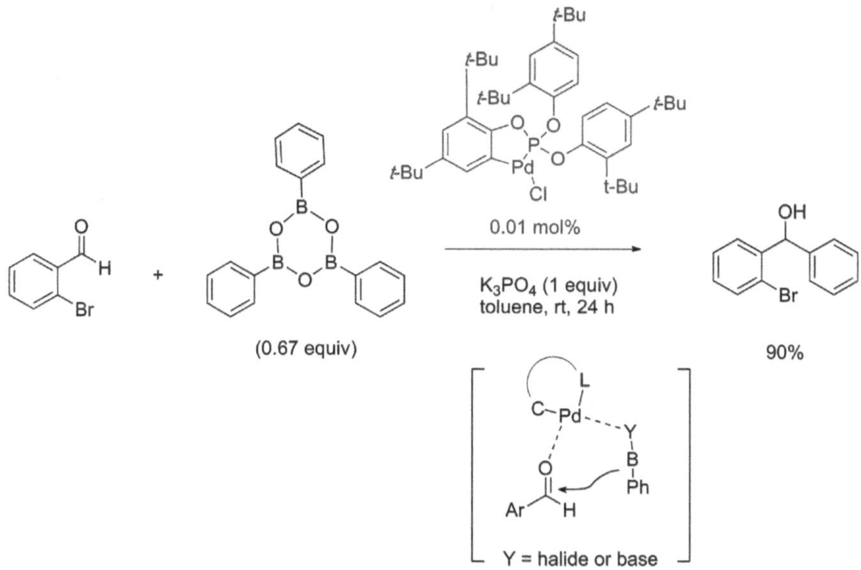

Scheme 6.7 Mechanism of arylation of aldehydes with boroxines catalyzed by a cyclopalladated complex.

therefore the cross-coupling pathway. However, the base (K_3PO_4) was required for good yields of the alcohols, and thus the transition state was proposed to feature coordination of the boroxine to the base as well as the base to the palladium center (Scheme 6.7). The findings of this study have the potential to further expand the development of novel carbon-carbon bond-forming avenues for multicomponent coupling reactions catalyzed by cyclopalladated complexes with auxiliary palladated ligands.

3. MULTICOMPONENT REACTIONS INVOLVING IN SITU PALLADATION OF THE SUBSTRATE

Catalytic cycles in which a cyclopalladated substrate operates as a catalytically active intermediate can be found in an ever-growing number of synthetically powerful multicomponent coupling reactions. The catalytic cycles utilize interconversions between all the oxidation states available to palladium, e.g., Pd(0)/Pd(II)/Pd(IV). Selective C—H activation, as well as regio- and stereo-selective olefin difunctionalizations, and construction of polycyclic systems, represent synthetically most valuable outcomes of these methodologies.

3.1 Reactions Featuring Norbornene as a Component of the Cyclopalladated Reactive Intermediate

A three-component coupling of iodoarenes, alkyl halides, and either olefins, alkynes, or boronic acids, catalyzed by palladium, and mediated by stoichiometric norbornene additive is an excellent example of a process relying on cyclopalladated complexes with palladium in oxidation states both Pd(II) and Pd(IV). The protocol was developed by Catellani, and initially used for the preparation of contiguously trisubstituted aromatic rings (Scheme 6.8) *(23)*.

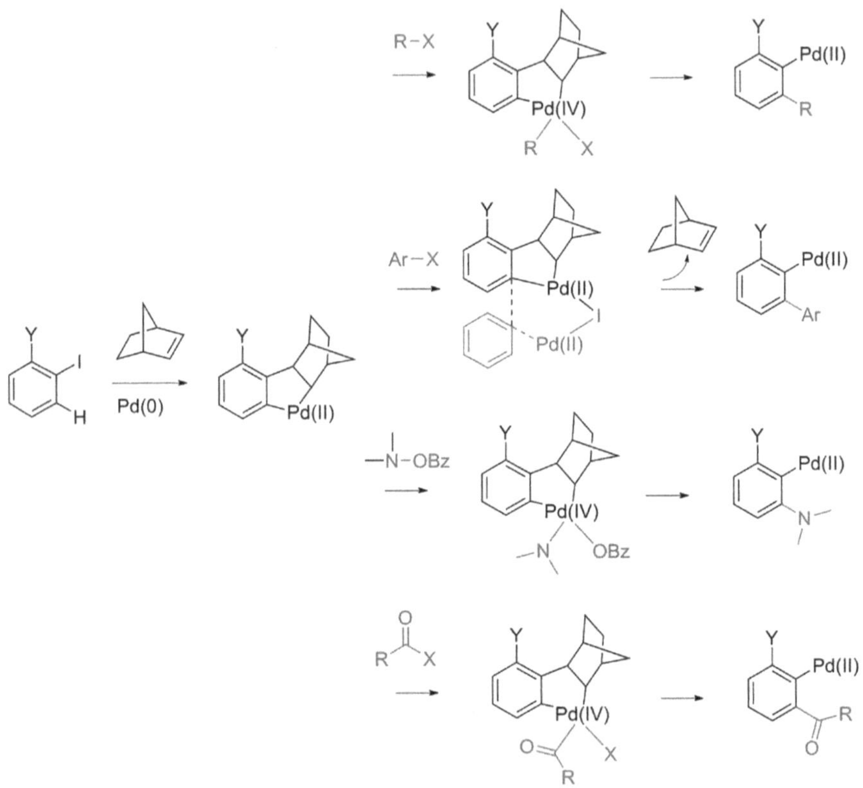

Scheme 6.8 Initial steps in catalytic cycles of novel variants of the Catellani protocol.

The palladacycle in Pd(II) oxidation state is formed via oxidative addition of Pd(0) into the aryl iodide and a migratory insertion of the norbornene, followed by directed palladation of the ortho-position in the aromatic ring. Alkyl iodide is less reactive toward oxidative addition to Pd(0), and will therefore preferentially oxidize the pallada(II)cycle to a pallada(IV)cycle. The pallada(IV)cycle then undergoes reductive elimination to form a new

C—C bond, deinsertion of norbornene, and the catalytic cycle is concluded by the reaction of the third component (olefin, alkyne or boronic acid) with the acyclic Pd(II) intermediate to recover the Pd(0) catalyst and release the bifunctionalized aromatic ring. The stoichiometric quantities of the norbornene additive are needed to favor the migratory insertion step.

Variants of the original protocol employ substituted aryl bromides instead of the alkyl iodide component (Scheme 6.8) *(24)*. In these cases, opening of the originally formed pallada(II)cycle is more likely to involve an aryl-aryl exchange via transmetallation between two arylpalladium(II) centers, rather than oxidation to a Pd(IV) palladacycle (Scheme 6.8). Impressive applications of this methodology to the syntheses of heterocycles were reported (Scheme 6.9) *(24)*.

In recent years, diversity of components employed in this methodology

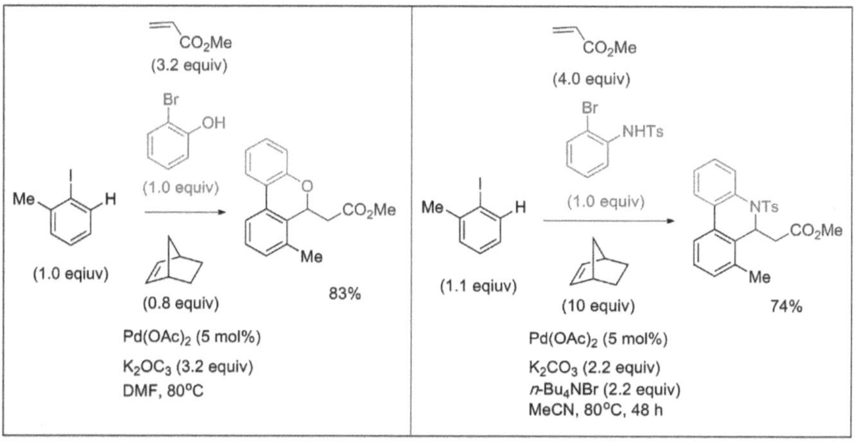

Scheme 6.9 Applications of the Catellani protocol to the synthesis of heterocycles with an aryl–aryl bond.

has been greatly expanded. Thus, transformations featuring amination of the C—H bond in the position *ortho* to the C—I bond in aryl iodides by suitable amine derivatives, along with final functionalization of the original C—I bond via Suzuki reaction, borylation, hydrazone insertion, or a Heck reaction, were reported (Scheme 6.10) *(25)*. The amine derivatives were proposed to open the norbornene-containing palladacycle either via oxidative addition to yield a Pd(IV) complex (Scheme 6.8), or by an electrophilic substitution. Alternatively, acyl derivatives were used to replace the amine components, and open of the pallada(II)cycles via oxidative addition to generate Pd(IV) intermediates (Scheme 6.8) *(26)*.

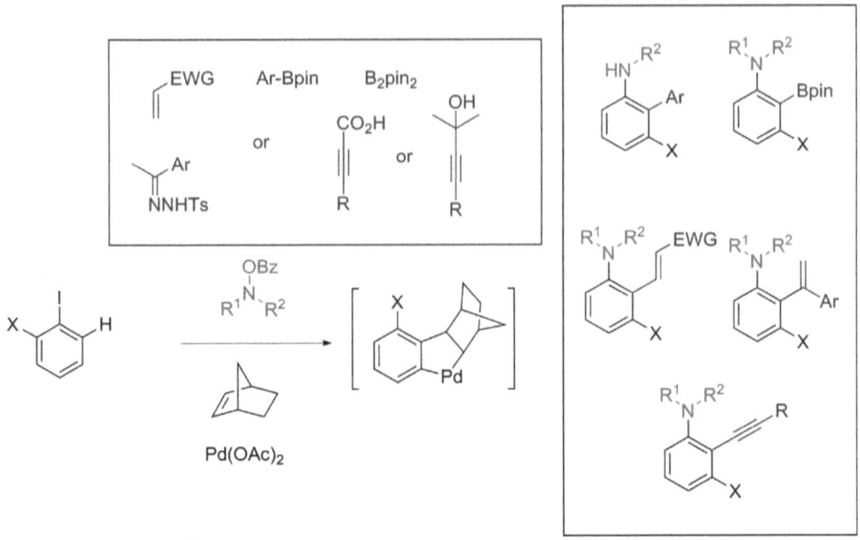

Scheme 6.10 Application of amines as components in the Catellani protocol.

Impressive applications of Catellani's three-component coupling to natural product synthesis are also known (Scheme 6.11) *(27)*.

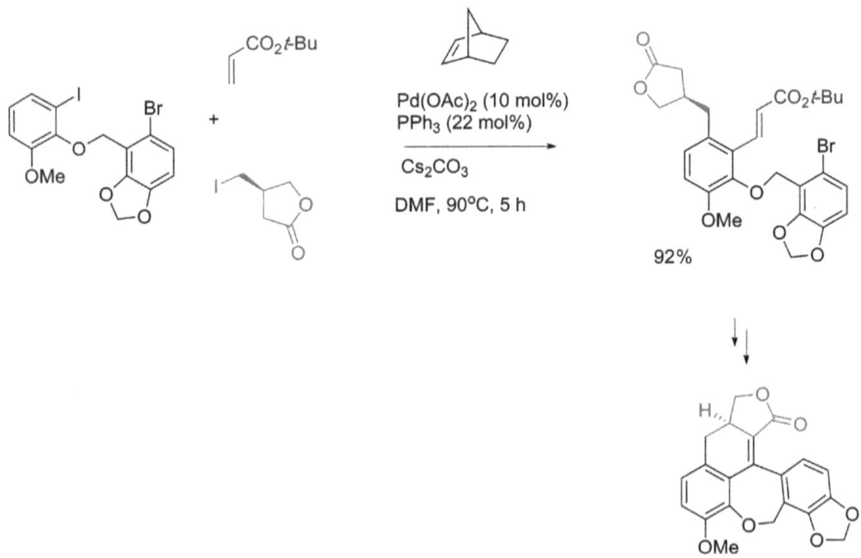

Scheme 6.11 Catellani protocol in a natural product synthesis.

A study relying on DFT calculations was designed to elucidate the mechanism of Pd-catalyzed cyclotrimerization of enantiopure halonorbornenes

(Scheme 6.12). Under palladium catalysis ($Pd(OAc)_2$; PPh_3, n-Bu_4NOAc, Na_2CO_s) enantiopure halonorbornes reacted to afford C3 or C3v symmetric syn-tris(norborneno)benzenes with a high selectivity (Scheme 6.12) *(28)*.

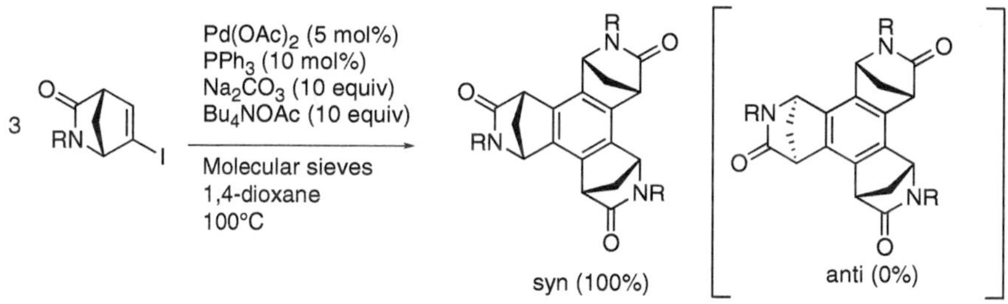

Scheme 6.12 Palladium catalyzed cyclotrimerization of enantiopure halonorbornenes.

DFT calculations revealed that the final ring closure involved a *disrotatory electrocyclization* of a trienylpalladium intermediate followed by HPdX elimination. The activation energy for the ring closure of the trienyl organic compound lacking the C-Pd bond proved to be actually higher. The trienylpalladium complex was generated by sequential *regioselective* migratory insertion of the halonorborne into the Csp^2-Pd bond, and HX elimination.

Notably, a reversible opening and formation of a palladacycle via a palladation of a Csp^2-H bond determined the stereoselectivity of the trimerization (Scheme 6.13). The palladacycle served as a common intermediate on the pathway toward the *syn* or *anti* products, depending on the nature of the "symmetry breaking" event during the palladacycle-opening step (Scheme 6.13) *(28)*.

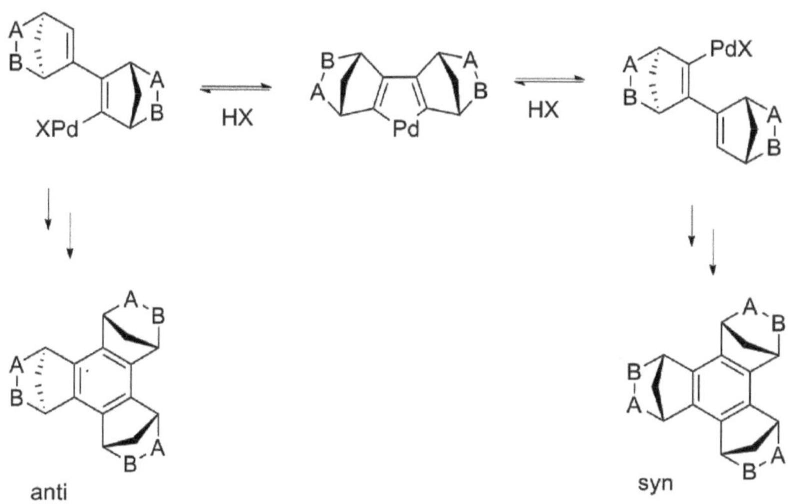

Scheme 6.13 Reversible C–H palladation in the stereoselective cyclotrimerization of halonorbornenes.

3.2 Reactions Featuring Chelation-Directed Difunctionalization of Unsaturated Substrates Via Cyclopalladated Intermediates

Recent studies by Engle described the development of regio- and stereoselective protocols for difunctionalization of butenoic and pentenoic acid derivatives *(29, 30, 31)*. Appropriately designed directing groups in the carboxylic acid derivatives coordinate to the Pd(II) center, as well as to the olefin (Scheme 6.14). The subsequent attack of carbon- or nitrogen-based nucleophiles generates either five-membered or six-membered cyclopalladated complexes (Pd(II)) as the key intermediates. The pallada(II)cycles can then react with organohalides (R-I or R-Br, R = methyl, vinyl, aryl, alkynyl) via oxidative addition to form palladium(IV) metalacycles that undergo reductive elimination to establish the second new bond. Thus, the bisfunctionalization involves Pd(II)/Pd(IV) cycle *(29, 30)*. Alternatively, the pallada(II)cycles can react with a second component, e.g., bispinacolborane (B_2pin_2), via transmetallation and reductive elimination to afford the bisfunctionalized organic products. In this case, a stoichiometric oxidant must be present to close the Pd(II)/Pd(0) catalytic cycle *(31)*.

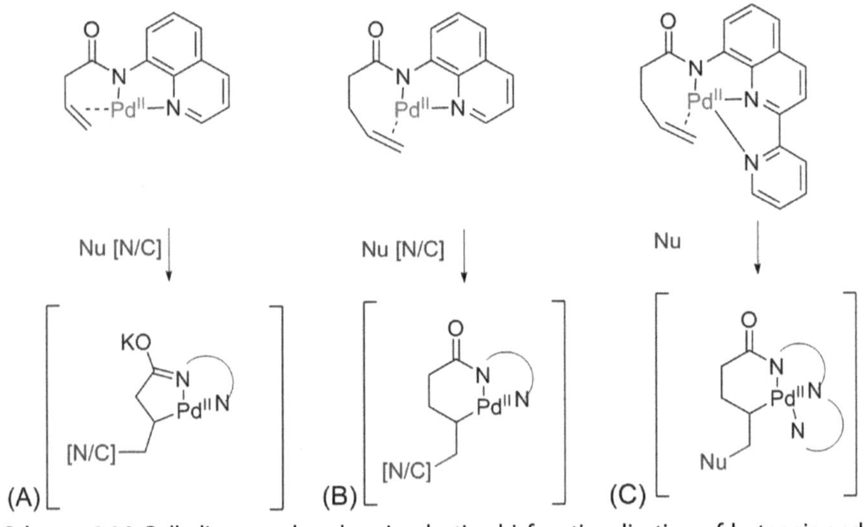

Scheme 6.14 Palladium-catalyzed regioselective bisfunctionalization of butenoic and pentenoic acid derivatives.

The stability of the pallada(II)cycles plays a crucial role in the multicomponent reactions. In contrast to the relatively long-lived five-membered palladacycles (part (a), Scheme 6.14), the six-membered palladacycles with the bidentate directing group (parts (b), Scheme 6.14) are relatively unstable, and rapidly decompose via β-hydride elimination. Thus, only components reacting at high rates with these six-membered intermediate, e.g., B_2pin_2, could be used successfully for functionalization of pentenoic acids *(31)*. To solve this problem, a tridentate pincer-type auxiliary ligand, that was able to sufficiently stabilize the six-membered palladacycle, was introduced (part (c), Scheme 6.14). However, to date this directing group was only applied for a regiocontrolled two-component hydrofunctionalization of the pentenoic acids (Scheme 6.15) *(32)*. The regiodivergent protocol was controlled by the attachment of the amide directing group either to the carboxylic acid, or to the amine group in the amino acid substrate (Scheme 6.15).

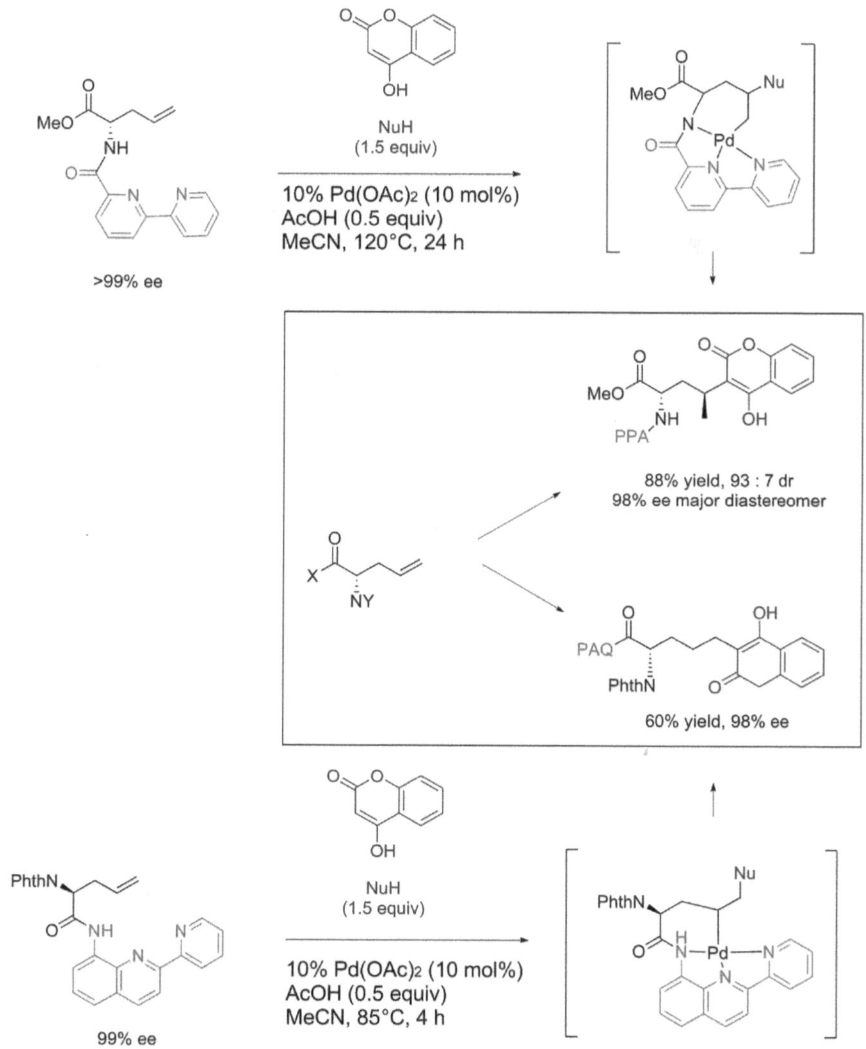

Scheme 6.15 Palladium-catalyzed regiodivergent hydrofunctionalization of pentenoic acid derivatives.

In the first application of this concept, a three-component difunctionalization of butenoic acids with the bidentate directing group was reported, featuring both carbon-based nucleophiles and electrophiles (Scheme 6.16) *(29)*.

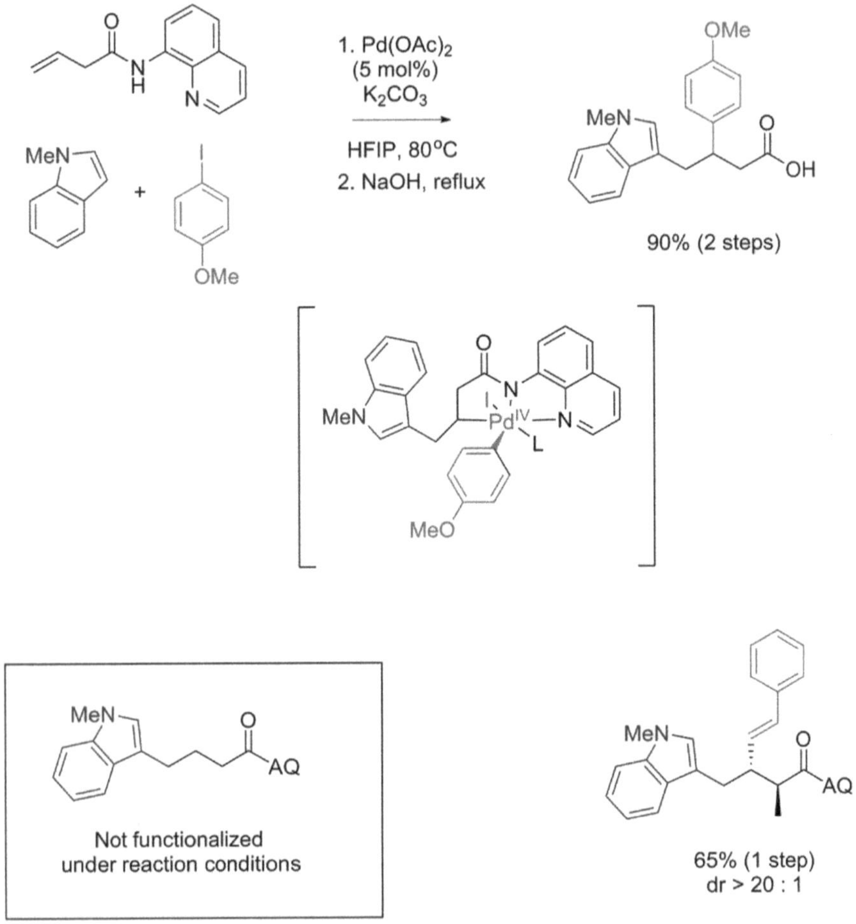

Scheme 6.16 Palladium-catalyzed bisfunctionalization of butenoic acids with two carbon nucleophiles.

Control experiments indicated that the second C—C bond is formed during the opening of the pallada(IV)cycle via the reductive elimination, since the monofunctionalized addition product (Scheme 6.16) did not react further under the reaction conditions. Furthermore, the α methyl group, which would be expected to be favored for C—H activation following the initial hydrofunctionalization with HNu, proved to be unreactive (Scheme 6.16) *(29)*. A subsequent report extended the scope of the three-component coupling to N-based nucleophiles (R_2NH = amides, sulfonamides, and N-heterocycles) and a broader variety of organohalide electrophiles, including vinyl iodides, methyl iodide, and alkynyl bromides. Examples of difunctionalization of terminally substituted olefins (internal olefins) were provided, as well as reactions with substrates mono- or disubstituted at the α carbon (Scheme 6.17) *(30)*. More extensive mechanistic studies, including DFT calculations and "Reaction Progress Kinetic Analysis" (RPKA), were a part of this work. The mechanistic studies supported the proposal that the reductive elimination from a Pd(IV) intermediate was the rate-determining step, a finding consistent with lowest reactivity observed empirically for electron-deficient aryl iodides *(30)*.

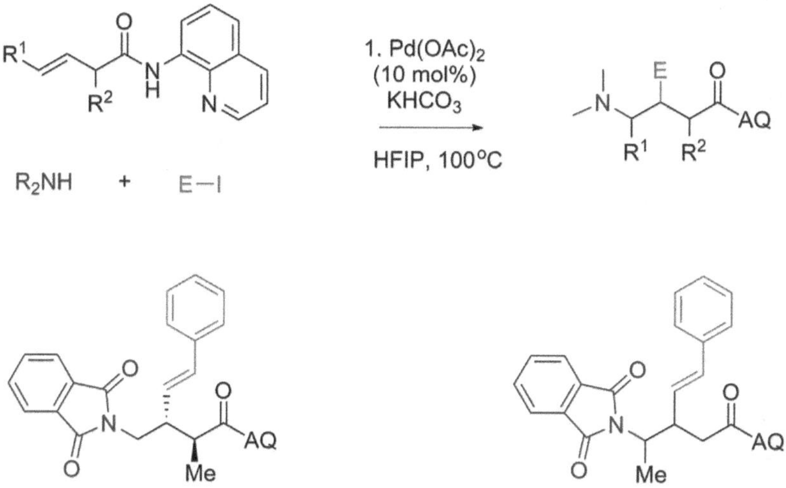

92%
dr = 5 : 1

30%
dr = 1.1 : 1

Scheme 6.17 Expanded scope of the palladium-catalyzed bisfunctionalization of substituted butenoic acids.

The most advanced iteration of this methodology to date described a regio- and stereoselective carbo- or amino-borylation *(31)*. The three-component reactions employed both substituted butenoic and, for the first time, also pentenoic acids that featured the bidentate auxiliary group. The pentenoic acids could react successfully via the unstable six-membered palladacycle due to the high rates of the transmetallation with the boron derivatives. However, the presence of substituents on the double bond in the pentenoic acids was not tolerated (Scheme 6.18). As stated above, the catalytic cycle involves Pd(II)/Pd(0) interconversions, and requires the presence of a stoichiometric oxidant (*vide supra*) *(31)*. The possibility to remove the auxiliary directing group via oxidation of the C—B bond followed by the formation of a lactone was ascertained, highlighting the synthetic potential of the new methodology.

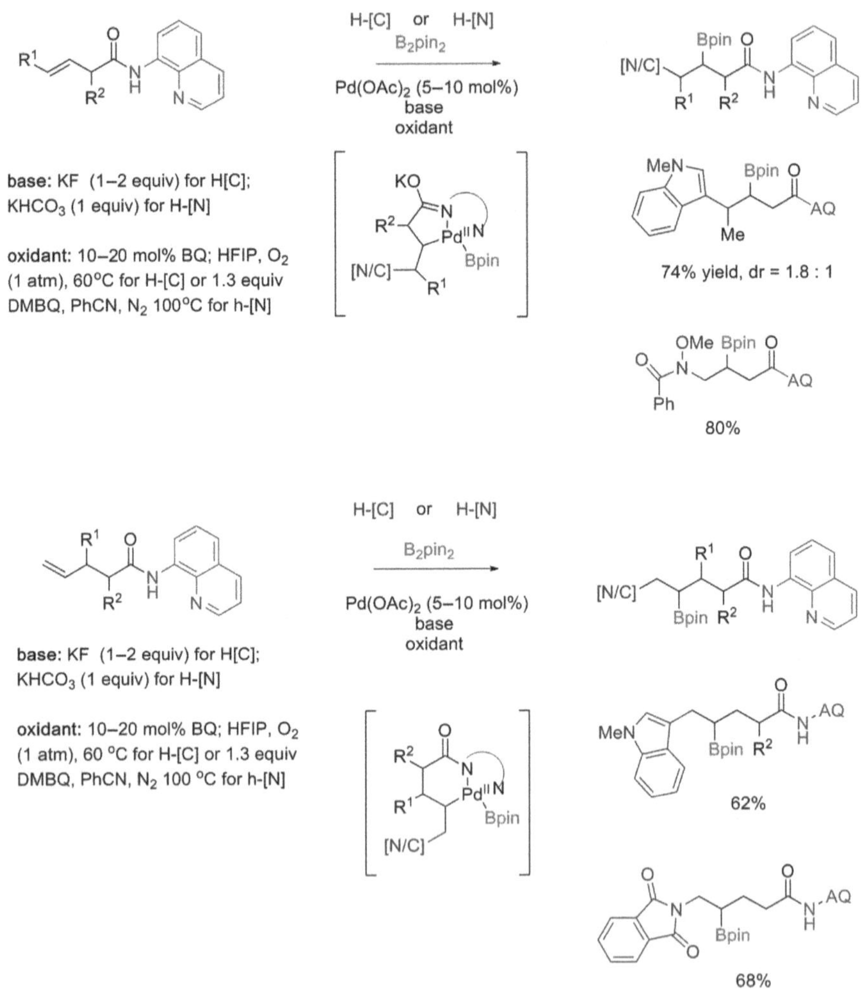

Scheme 6.18 Palladium-catalyzed carbo- or amino-borylation of butenoic and pentenoic acid derivatives.

A similar mechanistic concept was applied toward palladium-catalyzed difunctionalization of allenes with aryl iodides and sulfonyl hydrazones via Pd(0)/Pd(II) catalytic cycle (Scheme 6.19) *(33)*.

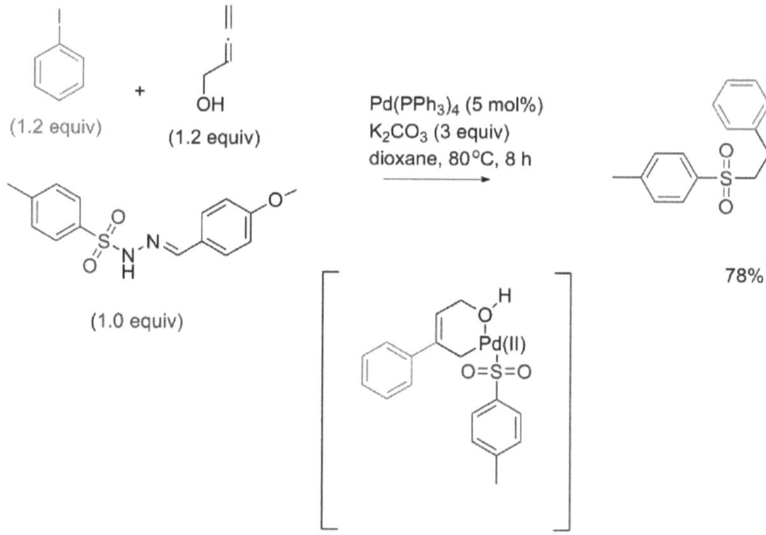

Scheme 6.19 Palladium-catalyzed bisfunctionalization of allenes with aryl iodides and sulfonyl hydrazones.

The presence of an unprotected hydroxyl group in the allene proved to be critical for the success of the reaction. Coordination of the hydroxyl group of the allene to the Pd(II) center in the Ar-Pd(II)-I intermediate increased the reactivity of the allene, and directed the regio- and stereochemical outcome. Formation of a six-membered cyclopalladated intermediate was favored, ultimately leading to the (*Z*)-α-hydroxymethyl allylic sulfones (Scheme 6.19). The protocol relies on the "masked functionality concept". Under the basic reaction conditions, sulfonyl hydrazones decompose to diazonium compounds and subsequently into sulfonyl anions that displace the iodide in the cyclopalladated complex. Reductive elimination generates the C—S bond and closes the Pd(II)/Pd(0) catalytic cycle (Scheme 6.19).

3.3 Reactions Featuring Diverse Components for Construction of Polycyclic Targets Via Palladacyclic Intermediates

A recently reported palladium-catalyzed three-component synthesis of diverse aromatic amides via carbonylation is notable for the use of two stable reagents with "masked functionalities", namely $Mo(CO)_6$ (see also

Scheme 6.2, *vide supra*) as the "CO equivalent", and nitroarenes as the source of the "ArN" component (Scheme 6.20) *(34)*.

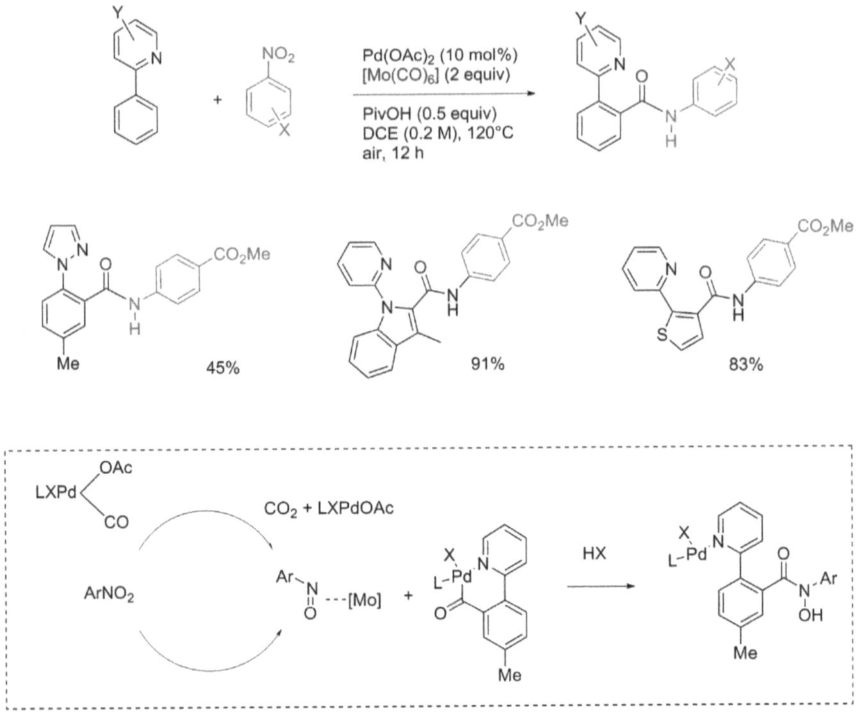

Scheme 6.20 Palladium-catalyzed synthesis of aryl and heteroaryl amides using $Mo(CO)_6$ and nitroarenes as the "CO" and the "ArN" components.

Studies including deuterium-labeling and control experiments supported the mechanistic rationale for this protocol. First, pyridine-directed cyclopalladation (C—H ortho activation) generates a cyclopalladated complex acting as the active catalyst. Then, CO liberated from $Mo(CO)_6$, undergoes insertion into the C—Pd bond to afford an acylpalladium(II) intermediate. In a separate cycle, a palladium(II) complex bearing the CO ligand, e.g., LXPd(CO)OAc is involved in the reduction of the nitro group in $Ar'NO_2$ into the nitroso group in Ar′N=O with concomitant oxidation of CO to CO_2. The resulting aryl nitroso compound coordinates to the Mo center, and acts as an N-nucleophile, attacking the electrophilic carbon in the acylpalladium ArCO-Pd(II) complex to yield Ar′N(OH)-CO-Ar organic product, and Pd(0) as the departing leaving group. The organic product is then reduced to N-H group by either Mo or Pd catalysts to afford the final amides. Control experiments ruled out the involvement of isocyanates in this process.

Palladium-catalyzed difunctionalization of 2-iodobiphenyls with primary alkyl chlorides and terminal monosubstituted olefins was proposed to feature a Pd(0)/Pd(II)/Pd(IV) cycle and a cyclopalladated intermediate (Scheme 6.21). The optimum reaction conditions ($Pd(OAc)_2$ 10 mol%, KOAc (4 equiv), K_2CO_3 in DMF) also required the presence of a reductant (isopropanol IPA, 2 equiv), presumably to accelerate the reduction of Pd(II) to Pd(0) *(35)*. A series of control experiments was performed, and results confirmed that a symmetrical *dibenzopalladacyclopentadiene Pd(II)* palladacycle featuring the C-Pd(II)-C functionality was indeed the crucial intermediate. This palladacyclopentadiene reacted selectively with the alkyl chloride via oxidative addition to form a Pd(IV) palladacycle. Reductive elimination followed by an intermolecular Heck reaction afforded the three-component coupling products (Scheme 6.21). No reaction took place when a product of the Heck reaction between the biaryl iodide and the olefin was treated under the reaction conditions. Thus, the olefin-directed C-H activation followed by alkylation of an acyclic Pd(II) species with the alkyl chloride, could be ruled out.

The scope of the protocol proved to be quite broad with respect to all the components, including primary alkyl chlorides bearing a variety of additional functional groups (amide, epoxide, nitrile, ester, and ketones), as well as monosubstituted terminal olefins including styrenes, heteroaryl-substituted vinyls, and Michael acceptors.

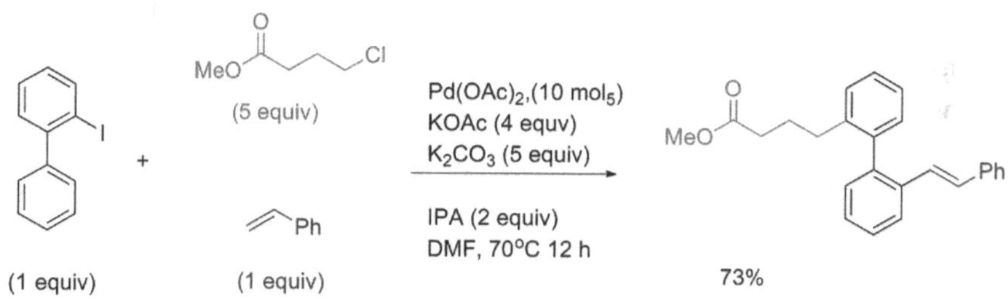

Scheme 6.21 Palladium-catalyzed three-component coupling of iodobiphenyls with aliphatic chlorides and olefins.

Construction of spirocenters in polycyclic structures represents one of the most difficult synthetic challenges. Most frequently employed approaches rely on intramolecular Heck reactions, or cycloadditions to exocyclic olefins. Two recent reports demonstrate how the application of these methodologies in the context of multicomponent reactions achieves a dramatic increase of complexity in just a single operation.

Yang reported a three-component coupling of an aryl iodide with an unsaturated side-chain with two equivalents of a different aromatic iodide component to afford spirocyclic indoles (Scheme 6.22) *(36)*.

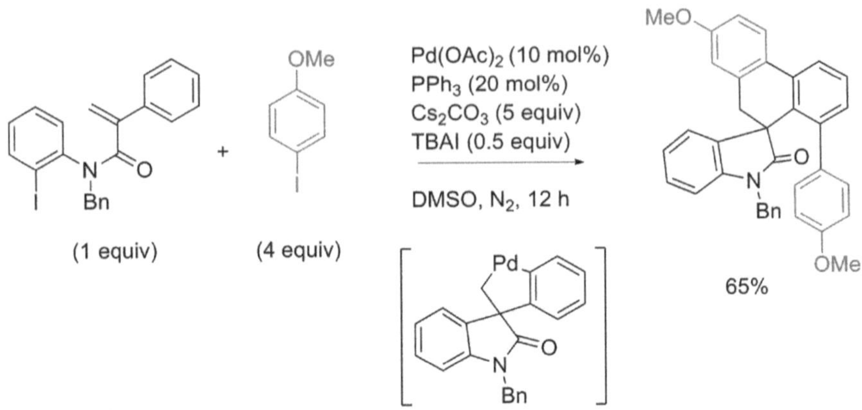

Scheme 6.22 Preparation of spirocyclic indoles.

The five-membered palladacycle intermediate (Scheme 6.22) generated by an intramolecular Heck reaction followed by palladation of the aromatic ring, was proposed as the key intermediate. Opening of this pallada(II)cycle by the first equivalent of the aryl iodide, either via Pd(IV) or Pd(II) intermediates, creates the first new aryl-aryl bond between the two different aryl iodide substrates. The entire sequence involving the generation and subsequent opening of a five-membered pallada(II)cycle is then repeated in the second *ortho*-position in the aromatic ring of the amide, leading to the attachment of the second equivalent of the aryl iodide (Scheme 6.22). The catalytic cycle is then terminated by the palladation of the C—H bond in the aromatic ring yielding an unstable six-membered palladacycle that rapidly collapses via reductive elimination to deliver the final five-membered ring in the spiroindole products (Scheme 6.22). A stable stoichiometric analog of the initially formed five-membered palladacycle featuring two PPh_3 ligands was isolated, and shown to be a competent intermediate under the reaction conditions. Control experiments confirmed that the second equivalent of the aryl iodide is indeed attached via a palladacyclic intermediate, rather than via a pathway involving amide-directed palladation of the aromatic ring. An independently synthesized analog to the spirocyclic amide product bearing only a single aryl-aryl bond failed to undergo the second arylation under the reaction conditions. Overall, three remote C—H activation events took place and a total of four new C—C bonds were generated in a single operation.

Tetrahydroquinolines with a spiroheterocyclic ring were prepared via a formal multicomponent coupling reaction, executed as a one-pot three-step sequential protocol *(37)*. *N*-alkylation of the amine with a benzyl bromide took place in step 1 and was followed in step 2 by an intramolecular Heck reaction yielding an exocyclic olefin. The in situ generated olefin then reacted with aryl nitrile *N*-oxides in a [3 + 2] cycloaddition as step 3 (Scheme 6.23). Although only 22%–28% combined yields for the three-step sequence were realized, the outcomes were deemed satisfactory, considering that a significant increase in the product complexity was achieved. The work is relevant here, since a stable palladacycle with Pd(II) coordinated to the amine group was isolated in 55%–66% yields from an analogous protocol stoichiometric in palladium (Scheme 6.23). Subsequently, the palladacycle was shown to yield the proposed Heck product featuring the exocyclic double bond (Scheme 6.23).

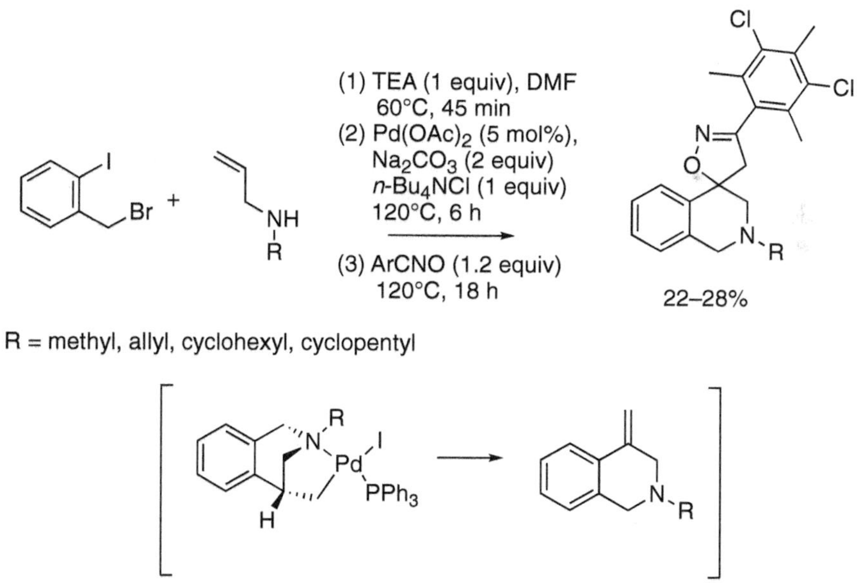

Scheme 6.23 Preparation of spirocyclic tetrahydroquinolines.

4. MULTICOMPONENT REACTIONS WITH A STOICHIOMETRIC CYCLOPALLADATED COMPLEX

The challenges encountered in the construction of medium-size rings continue to inspire applications of cyclopalladated complexes as stoichiometric building blocks that can be used to synthesize cyclic organic products. Examples of two studies of this type are provided in Schemes 6.24 and 6.25.

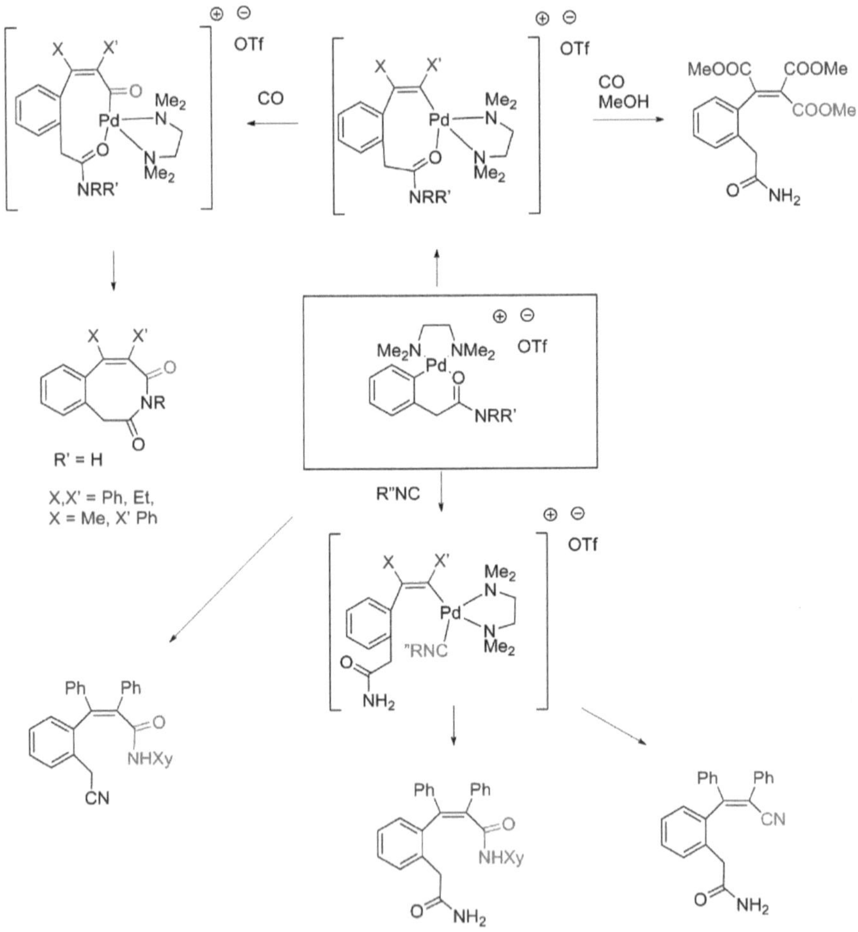

Scheme 6.24 Preparation of benzo[*d*]azocine-2,4(1H,3H)-diones from stable pallada(II) cycles.

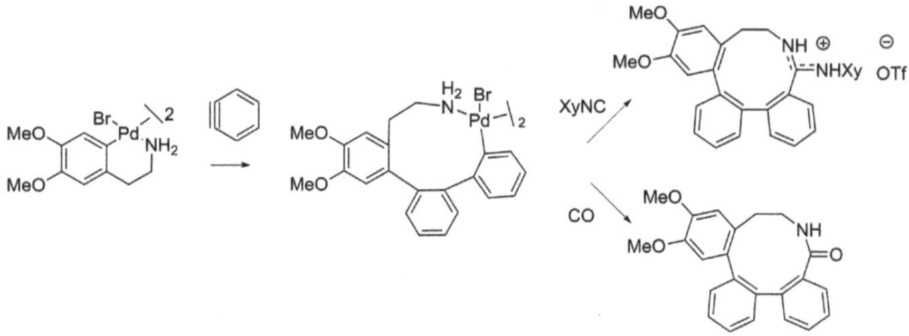

Scheme 6.25 Preparation of a stable 10-membered pallada(II)cycle for reactions with CO and isonitriles.

Using cyclopalladated arylamide complexes stabilized by the TMEDA auxiliary ligand, various disubstituted aryls could be produced via sequential migratory insertions of internal alkynes followed by the addition of isonitriles (R″NC) or CO in methanol. Synthetically the most valuable outcome was the pathway leading to eight-membered rings in benzo[d]azocine-2,4 (1H,3H)-diones via sequential insertion of internal alkynes followed by the insertion of carbon monoxide (Scheme 6.24) *(38)*.

In a similar process, the reaction of stable palladated arylamine with two equivalents of benzyne afforded stable and isolable 10-membered palladacycles. The palladacycles were used to prepare the corresponding 10-membered heterocycles via the migratory insertions of CO or isonitrile (XyNC) followed by reductive elimination (Scheme 6.25) *(39)*.

5. SUMMARY

Studies described herein clearly demonstrate the utility of palladacyclic complexes as stoichiometric substrates, precatalysts, catalysts, and in situ generated catalytic intermediates in synthetically powerful multicomponent reaction sequences, including controlled polymerizations. The unique properties of the palladacyclic complexes include an adjustable balance between stability and reactivity, and the possibility to selectively mediate complex catalytic cycles featuring all the accessible oxidation states of palladium. In the future, the emergence of even more complex multicomponent reaction sequences can be expected, including multicatalytic systems using palladacycles complexes along with a range of different types of catalysts.

REFERENCES

1. Smietana, M.; Benedetti, E.; Bressy, C.; Arseniyadis, S. Multicomponent Reactions in Natural Product Synthesis. In *Efficiency in Natural Product Synthesis,* Huang, P.-Q.; Yao, Z.-J.; Hsung, R. P., Eds.; John Wiley & Sons, Inc: Hoboken, 2018; pp 319–344 (chapter 8).
2. Lutz, W. The application of multi-component reactions in drug discovery. Curr. Med. Chem. **2002**, *9*, 2085–2093.
3. Gasperi, T.; Orsini, M.; Vetica, F.; De Figueiredo, R. M. Organocatalytic Asymmetric Multicomponent Reactions. In *Multicomponent Reactions: Concepts and Applications for Design and Synthesis,* Herrera, R. P.; Marques-Lopez, E., Eds.; John Wiley & Sons, Inc: Hoboken, 2015; pp 16–71 (chapter 2).
4. Ros, A.; Matas, I. Metal-Catalyzed Multicomponent Reactions. In: *Multicomponent Reactions: Concepts and Applications for Design and Synthesis;* Herrera, R. P.; Marques-Lopez, E., Eds.; John Wiley & Sons, Inc: Hoboken, 2015; pp 72–126 (chapter 3).
5. (a)Qian, Y.; Jing, C.; Shi, T.; Ji, J.; Tang, M.; Zhou, J.; Zhaim, C.; Hu, W. Dual Catalysis in Highly Enantioselective Multicomponent Reaction With Water: An Efficient Approach to Chiral b-amino-a-hydroxy Acid Derivatives. *Chem. Cat. Chem.* **2011**, *3*, 653–656; (b) Abbiati, G.; Rossi, E. Silver and Gold-Catalyzed Multicomponent Reactions. Müller, T. J. J. Guest Ed.; Beilstein J. Org. Chem. 2014**,** 10*,* 481-513.
6. Malleron, J.-L.; Fiaud, J.-C.; Legros, J.-Y., Eds.; *Handbook of Palladium-Catalyzed Organic Reactions: Synthetic Aspects and Catalytic Cycles*, Academic Press: Cambridge, 1997.
7. (a)Benmahdjoub, S.; Ibrahim, N.; Benmerad, B.; Alami, M.; Messaoudi, S. One-Pot assembly of unsymmetrical biaryl thioglycosides through chemoselective palladium-catalyzed three-component tandem reaction. *Org. Lett.* **2018**, *20*, 4067–4071; (b) Hao, T.-T.; Liang, H.-R.; Ou-Yang, Y.-H.; Yin, C.-Z.; Zheng, X.-L.; Yuan, M.-L.; Li, R.-X.; Fu, H.-Y.; Chen, H. Palladium-Catalyzed Domino Reaction for Stereoselective Synthesis of Multisubstituted Olefins: Construction of Blue Luminogens. *J. Org. Chem.* **2018**, *83*, 4441–4454; (c)Diaz-Marta, A. S.; Tubio, C. R.; Carbajales, C.; Fernandez, C.; Escalante, L.; Sotelo, E.; Guitian, F.; Barrio, V. L.; Gil, A.; Coelho, A. Three-Dimensional Printing in Catalysis: Combining 3D Heterogeneous Copper and Palladium Catalysts for Multicatalytic Multicomponent Reactions. *ACS Catal.* **2018**, *8*, 392–404; (d)Wang, B.; Sun, S.; Yu, J.-T.; Jiang, Y.; Cheng, J. Palladium-Catalyzed Multicomponent Reactions of *o*-alkynylanilines, Aryl iodides, and CO_2 Toward 3,3-diaryl 2,4-Quinolinediones. *Org. Lett.* **2017**, *19*, 4319–4322; (e) Peng, J.; Gao, Y.; Hu, W.; Gao, Y.; Hu, M.; Wu, W.; Ren, Y.; Jiang, H. Palladium-Catalyzed Multicomponent Reaction (MCR) of Propargylic Carbonates With Isocyanides. *Org. Lett.* **2016**, *18*, 5924–5927.
8. (a)Cope, A. C.; Siekman, R. W. Formation of Covalent Bonds From Platinum or Palladium to Carbon by Direct Substitution. *J. Am. Chem. Soc.* **1965**, *87*, 3272–3273; (b) Trofimenko, S. Cyclopalladation Reaction. *Inorg. Chem.* **1973**, *12*, 1215–1221; (c) Vicente, J.; Saura-Llamas, I.; Cuadrado, J.; Ramirea de Arellano, M. C. Ortho-Metalated Primary Amines. 6.1 The First Synthesis of Six-Membered Palladacycles From Primary Amines Containing Electron-Withdrawing Substituents: End of the Limiting Rules of Cope and Friedrich on Cyclopalladation of Benzyl- and Phenethylamines. *Organometallics* **2003**, *22*, 5513–5517.
9. (a)Dupont, J.; Consorti, C. C.; Spencer, J. The Potential of Palladacycles: More Than Just Precatalysts. *Chem. Rev.* **2005**, *105*, 2527–2572; (b)Rao, G. K.; Kumar, A.; Kumar, S.; Dupare, U. B.; Singh, A. K. Palladacycles of Thioethers Catalyzing Suzuki–Miyaura C–C Coupling: Generation and Catalytic Activity of Nanoparticles. *Organometallics* **2013**, *32*, 2452–2458.

10. Herrmann, W. A.; Brossmer, C.-P.; Reisinger, C.-P.; Riermeier, T. H.; Öfele, K.; Beller, M. Palladacycles: Efficient New Catalysts for the Heck Vinylation of Aryl Halides. *Chem. Eur. J.* **1997**, *3*, 1357–1364.
11. Qureshi, Z.; Kim, J. Y.; Bruun, T.; Lam, H.; Lautens, M. Cu/Pd-Catalyzed Synthesis of Fully Decorated Polycyclic Triazoles: Introducing C–H Functionalization to Multicomponent Multicatalytic Reactions ($(MC)^2R$). *ACS Catal.* **2016**, *6*, 4946–4952.
12. (a)Böhm, V. P. W.; Herrmann, W. A. Mechanism of the Heck Reaction Using a Phosphapalladacycle as the Catalyst: Classical Versus Palladium(IV) Intermediates. *Chem. Eur. J.* **2001**, 7, 4191–4197; (b)Trzeciak, A. N.; Ziółkowski, J. J. Structural and Mechanistic Studies of Pd-Catalyzed CC Bond Formation: The Case of Carbonylation and Heck Reaction. *Coord. Chem. Rev.* **2005**, *249*, 2308–2322.
13. Begouin, A.; Queiroz, M.-J. R. P. Palladium-Catalysed Multicomponent Aminocarbonylation of Aryl or Heteroaryl Halides With [$Mo(CO)_6$] and Aryl- or Heteroarylamines Using Conventional Heating. *Eur. J. Org. Chem.* **2009**, 2820–2827.
14. Koizumi, T.; Kanbara, T. Cross-Coupling Polymerization. In *Organometallic Reactions and Polymerization;* Osakada, K., Ed.; Lecture Notes in Chemistry; Springer: Berlin, Heidelberg, 2014; Vol. 85.
15. Dong, J.; Guo, H.; Hu, Q.-S. Controlled Pd(0)/Ad_3P-Catalyzed Suzuki Cross-Coupling Polymerization of AB-type Monomers With Ad_3P-Coordinated Acetanilide-Based Palladacycle Complex as Initiator. *ACS Macro Lett.* **2017**, *6*, 1301–1304.
16. Seo, K.-B.; Lee, I.-H.; Lee, J.; Choi, I.; Choi, T.-L. A Rational Design of Highly Controlled Suzuki–Miyaura Catalyst- Transfer Polycondensation for Precision Synthesis of Polythiophenes and Their Block Copolymers: Marriage of Palladacycle Precatalysts with MIDA-Boronates. *J. Am. Chem. Soc.* **2018**, *140*, 4335–4343.
17. (a)Hull, K. L.; Sanford, M. S. Catalytic and Highly Regioselective Cross-Coupling of Aromatic C-H Substrates. *J. Am. Chem. Soc.* **2007**, *129*, 11904–11905; (b)Maleckis, A.; Kampf, J. W.; Sanford, M. S. A Detailed Study of Acetate-Assisted C–H Activation at Palladium(IV) Centers. *J. Am. Chem. Soc.* **2013**, *135*, 6618–6625.
18. Weber, M.; Peters, R. Pd(II)-Catalyzed Regio-, Enantio-, and Diastereoselective 1,4-Addition of Azlactones Formed in situ From Racemic Unprotected Amino Acids and Acetic Anhydride. *J. Org. Chem.* **2012**, 77, 10846–10855.
19. Shiota, A.; Malinakova, H. C. Palladacycles: Effective Catalysts for a Multicomponent Reaction With Allylpalladium(II)-Intermediates. *J. Organomet. Chem.* **2012**, *704*, 9–16.
20. Yus, M.; González-Goez, J.-C.; Foubelo, F. Diastereoselective Allylation of Carbonyl Compounds and Imines: Application to the Synthesis of Natural Products. *Chem. Rev.* **2013**, *113*, 5595–5698.
21. Wallner, O. A.; Szabo, K. J. Origin of the Regio- and Stereoselectivity in Palladium-Catalyzed Electrophilic Substitution Via Bis(allyl)palladium Complexes. *Chem. Eur. J.* **2003**, *9*, 4025–4030.
22. Liao, Y.-X.; Xing, C.-H.; Israel, M.; Hu, Q.-S. A Nontransmetalation Reaction Pathway for Anionic Four-Electron Donor-Based Palladacycle-Catalyzed Addition Reactions of Arylborons With Aldehydes. *Tetrahedron Lett.* **2011**, *52*, 3324–3328.
23. (a)Catellani, M.; Motti, E.; Della Ca, N. Catalytic Sequential Reactions Involving Palladacycle-Directed Aryl Coupling Steps. *Acc. Chem. Res.* **2008**, *41*, 1512–1522; (b)-Della Ca, N.; Fontana, M.; Motti, E.; Catellani, M. Pd/Norbornene: A Winning Combination for Selective Aromatic Functionalization via C–H Bond Activation. Acc. *Chem. Res.* **2016**, *49*, 1389–1400.
24. (a)Motti, E.; Faccini, F.; Ferrari, I.; Catellani, M.; Ferraccioli, R. Sequential Unsymmetrical Aryl Coupling of *o*-substituted Aryl Iodides with *o*-Bromophenols and Reaction with Olefins: Palladium-Catalyzed Synthesis of 6*H*-dibenzopyran Derivatives. *Org. Lett.* **2006**, *8*, 3967–3970; (b)Della Ca, N.; Motti, E.; Catellani, M. Palladium-Catalyzed Synthesis of Selectively Substituted Phenanthridine Derivatives. *Adv. Synth. Catal.* **2008**, *350*, 2513–2516.

25. (a)Zhou, P.-X.; Ye, Y.-Y.; Ma, J.-W.; Zheng, L.; Tang, Q.; Qiu, Y.-F.; Song, B.; Qiu, Z.-H.; Xu, P.-F.; Liang, Y.-M. Palladium-Catalyzed/Norbornene-Mediated Ortho-amination/N-tosylhydrazone Insertion Reaction: An Approach to the Synthesis of Ortho-aminated Vinyl-arenes. *J. Org. Chem.* **2014**, *79*, 6627–6633; (b)Chen, Z.-Y.; Ye, C.-Q.; Zhu, H.; Zeng, X.-P.; Yuan, J.-J. Palladium/Norbornene-Mediated Tandem C−H Amination/C−I Alkenylation Reaction of Aryl Iodides With Secondary Cyclic O-benzoyl Hydroxylamines and Activated Terminal Olefins. *Chem. Eur. J.* **2014**, *20*, 4237–4241; (c)Ye, C.; Zhu, H.; Chen, Z. Synthesis of Biaryl Tertiary Amines Through Pd/norbornene Joint Catalysis in a Remote C−H Amination/Suzuki Coupling Reaction. *J. Org. Chem.* **2014**, *79*, 8900–8905; (d)Shi, H.; Babinski, D. J.; Ritter, T. Modular C−H Functionalization Cascade of Aryl Iodides. *J. Am. Chem. Soc.* **2015**, *137*, 3775–3778; (e)Pan, S.; Ma, X.; Zhong, D.; Chen, W.; Liu, M.; Wu, H. Palladium-Catalyzed One-Pot Consecutive Amination and Sonogashira Coupling for Selective Synthesis of 2-Alkynylanilines. *Adv. Synth. Catal.* **2015**, *357*, 3052–3056; (f)Sun, F.; Gu, Z. Decarboxylative Alkynyl Termination of Palladium-Catalyzed Catellani Reaction: A Facile Synthesis of α-alkynyl Anilines via ortho C−H Amination and Alkynylation. *Org. Lett.* **2015**, *17*, 2222–2225.
26. (a)Zhou, P.-X.; Ye, Y.-Y.; Liu, C.; Zhao, L.-B.; Hou, J.-Y.; Chen, D.-Q.; Tang, Q.; Wang, A.-Q.; Zhang, J.-Y.; Huang, Q.-X.; Xu, P.-F.; Liang, Y.-M. Palladium-Catalyzed Acylation/Alkenylation of Aryl Iodide: A Domino Approach Based on the Catellani-Lautens Reaction. *ACS Catal.* **2015**, *5*, 4927–4931; (b)Huang, Y.; Zhu, R.; Zhao, K.; Gu, Z. Palladium-Catalyzed Catellani Ortho-Acylation Reaction: An Efficient and Regiospecific Synthesis of Diaryl Ketones. *Angew. Chem. Int. Ed.* **2015**, *54*, 12669–12672; (c)Dong, Z.; Wang, J.; Ren, Z.; Dong, G. Ortho-C−H acylation of Aryl Iodides by Palladium/Norbornene Catalysis. *Angew. Chem. Int. Ed.* **2015**, *54*, 12664–12668.
27. Weinstabl, H.; Suhartono, M.; Qureshi, Z.; Lautens, M. Total Synthesis of (+)-Linoxepin by Utilizing the Catellani Reaction. *Angew. Chem. Int. Ed.* **2013**, *52*, 5305–5308.
28. Yamanaka, M.; Morishima, M.; Shibata, Y.; Higashibayashi, S.; Sakurai, H. DFT Studies of Mechanism and Origin of Stereoselectivity of Palladium-Catalyzed Cyclotrimerization Reactions Affording syn- Tris(norborneno)benzenes. *Organometallics* **2014**, *33*, 3060–3068.
29. Liu, Z.; Zeng, T.; Yang, K.-S.; Engle, K. M. β,γ-Vicinal Dicarbofunctionalization of Alkenyl Carbonyl Compounds via Directed Nucleopalladation. *J. Am. Chem. Soc.* **2016**, *138*, 15122–15125.
30. Zhen Liu, Z.; Wang, Y.; Wang, Z.; Zeng, T.; Liu, P.; Engle, K. M. Catalytic Intermolecular Carboamination of Unactivated Alkenes via Directed Aminopalladation. *J. Am. Chem. Soc.* **2017**, *139*, 11261–11270.
31. Liu, Z.; Ni, H.-Q.; Zeng, T.; Engle, K. M. Catalytic Carbo- and Aminoboration of Alkenyl Carbonyl Compounds via Five- and Six-Membered Palladacycles. *J. Am. Chem. Soc.* **2018**, *140*, 3223–3227.
32. O'Duill, M.l.; Matsuura, R.; Wang, Y.; Turnbull, J. L.; Gurak, J. A., Jr.; Gao, D.-W.; Lu, G.; Liu, P.; Engle, K. M. Tridentate Directing Groups Stabilize 6-Membered Palladacycles in Catalytic Alkene Hydrofunctionalization. *J. Am. Chem. Soc.* **2017**, *139*, 15576–15579.
33. Hou, Y.; Shen, Q.; Zhu, L.; Han, Y.; Zhao, Y.; Qin, M.; Gong, P. Palladium-Catalyzed Three-Component Tandem Reaction of Sulfonyl Hydrazones, Aryl Iodides and Allenes: Highly Stereoselective Synthesis of (Z)-α-Hydroxymethyl Allylic Sulfones. *RSC Adv.* **2017**, 7, 50372–50377.
34. Zhou, F.; Wang, D.-S.; Guan, X.; Driver, T. G. Nitroarenes as the Nitrogen Source in Intermolecular Palladium-Catalyzed Aryl C–H Bond Aminocarbonylation Reactions. *Angew. Chem. Int. Ed.* **2017**, *56*, 4530–4534.

35. Chen, D.; Shi, G.; Jiang, H.; Zhang, Y.; Zhang, Y. Sequential Difunctionalization of 2-Iodobiphenyls by Exploiting the Reactivities of a Palladacycle and an Acyclic Arylpalladium Species. Org. Lett. **2016**, *18*, 2130–2133.
36. Luo, X.; Xu, Y.; Xiao, G.; Liu, W.; Qian, C.; Deng, G.; Song, J.; Liang, Y.; Yang, C. Palladium-Catalyzed Tandem Reaction of Three Aryl Iodides Involving Triple C–H Activation. *Org. Lett.* **2018**, *20*, 2997–3000.
37. Beccalli, E. M.; Broggini, G.; Martinelli, M.; Masciocchi, N.; Sottocornola, S. New 4-Spiroannulated Tetrahydroisoquinolines by a One-Pot Sequential Procedure. Isolation and Characterization of σ-Alkylpalladium Heck Intermediates. *Org. Lett.* **2006**, *8*, 4521–4524.
38. Frutos-Pedreño, R.; González-Herrero, P.; Vicente, J. Sequential Insertion of Alkynes and CO or Isocyanides into the Pd–C Bond of Cyclopalladated Phenylacetamides. Synthesis of Eight-Membered Palladacycles, Benzo[d]azocine-2,4(1H,3H)-Diones, and Highly Functionalized Acrylonitrile and Acrylamide Derivatives. *Organometallics* **2012**, *31*, 3361–3372.
39. Oliva-Madrid, M.-J.; Saura-Llamas, I.; Bautista, D.; Vicente, J. Benzyne–benzyne–RNC or CO Triple Sequential Insertion into the Pd–C Bond: Synthesis of Ten-Membered N-heterocycles Through Stable Ten- and Eleven-Membered Palladacycles. *Chem. Commun.* **2013**, *49*, 7997–7999.

CHAPTER 7

Enantioselective Synthesis Using Chiral Palladacycles

Vineeta Soni, Benudhar Punji
Organometallic Synthesis and Catalysis Group, Chemical Engineering Division, CSIR–National Chemical Laboratory (CSIR–NCL), Pune, India

Contents

1. INTRODUCTION

Transition-metal-catalyzed enantioselective synthesis represents an important tool in modern organic chemistry because of the high efficiency and selectivity rendered by the metal catalyst *(1–3)*. The stability and predictability of the metal catalyst in such transformation is of critical importance in order to enantioselectively convert a large amount of substrate molecules to products. In this aspect, palladacycle catalysts have many attractive advantages *(4–9)*. Palladacycles are the palladium compound consisting of a chelating ligand with at least one Pd-carbon covalent bond. This stabilizes the palladacycle and creates unique electronic properties and a steric environment around the palladium center. Additionally, the strong Pd-carbon σ-bond allows the catalytic reactions through higher oxidation states. Thus, very often, palladacycles display unique chemical reactivities that are not feasible with other palladium complexes. Though palladacycles were extensively explored in various traditional catalytic reactions *(4,6)*, their application as catalysts in asymmetric transformation is relatively less explored. In this chapter, we summarize the development of chiral palladacycle catalysts for the enantioselective synthesis of diverse organic scaffolds. In particular, inter- and intramolecular enantioselective arylation, and allylic

Palladacycles
https://doi.org/10.1016/B978-0-12-815505-9.00008-1
© 2019 Elsevier Inc.
All rights reserved.

alkylation catalyzed by chiral palladacycle are elaborated. Further, enantioselective [3,3]-sigmatropic rearrangements of allylic imidates using ferrocenyl and cobalt oxazoline palladacycle have been highlighted. In addition, the discovery and development of several palladacycles for the asymmetric hydrophosphination reaction are discussed. The novelty of this phosphination reaction to synthesize important chiral tertiary phosphines and other bidentate and tridentate ligands has been deliberated.

2. PALLADACYCLES IN ENANTIOSELECTIVE ARYLATION

Enantioselective arylation has been considered as an attractive strategy for the synthesis of organic scaffolds with benzylic stereocenter. In particular, chiral R-branched amines that are generated by the enantioselective arylation of imines represent an essential component of many drug candidates. Thus, general methods for the asymmetric synthesis of aryl amines are of considerable importance. The stereoselective addition of organometallic reagents to imines represents one of the most efficient approaches *(10–14)*. There have been many reports on the arylation of imines using diverse metal-catalysts *(15–23)*. However, palladium/palladacycle catalyzed asymmetric arylation is relatively less explored *(24,25)*. Lu and Zeng have independently demonstrated the palladium catalyzed enantioselective arylation of imino acids and esters employing chiral bidentate N-ligand **8.1** (Scheme 7.1) *(26–28)*. Among various ligands screened, the pyridinyl-oxazoline-based ligand **8.1** afforded moderate to good yield and enantioselectivity. Notably, chiral bisphosphine ligands are not suitable under the demonstrated catalytic conditions. However, Burke has shown that a chiral diphosphane ligand system is suitable for the enantioselective arylation of *N*-tosyl imines *(29)*. Similarly, there are also precedents on the arylation of cyclic ketimines to give chiral α-diaryl alkyl amines that are important components of potent drugs (Scheme 7.2) *(30,31)*.

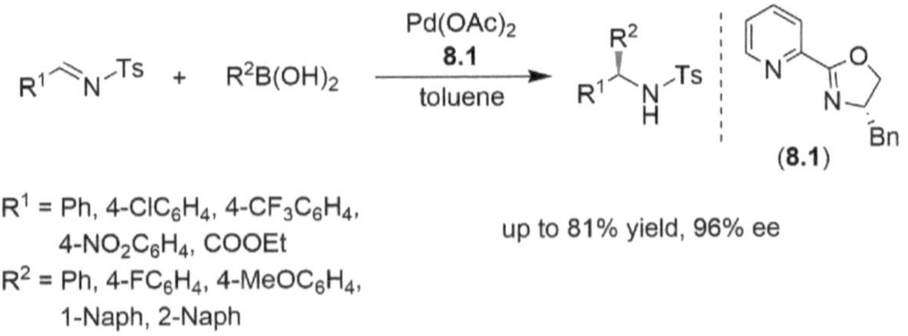

Scheme 7.1 Enantioselective arylation of imines.

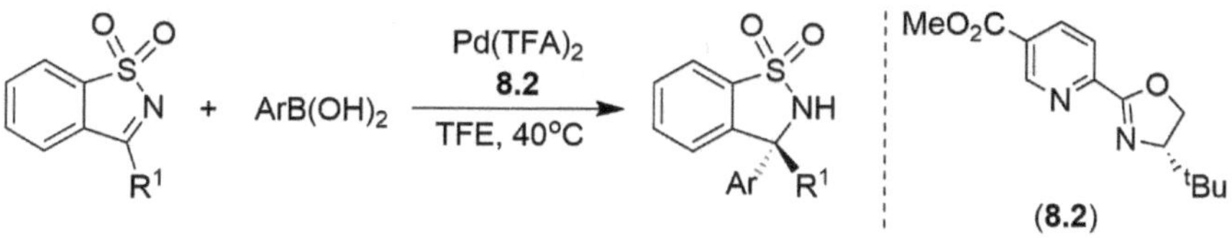

Ar = Ph, 4-MeC_6H_4, 4-$MeOC_6H_4$, 4-ClC_6H_4, 4-FC_6H_4, etc.
R^1 = Me, Et, Bu, ^{i}Pr, etc.

up to 99% yield, 95% ee

Scheme 7.2 Enantioselective arylation of cyclic ketimines.

Shi group has synthesized a series of chiral cationic C2-symmetric *N*-heterocyclic carbene (NHC)Pd^{2+} diaquo complexes derived from 1,1′-binaphthalenyl-2,2′-diamine (BINAM) and H8-BINAM (Fig. 7.1) *(32,33)*. These complexes are used for the enantioselective arylation of *N*-tosylarylimines with arylboronic acids under mild conditions (Scheme 7.3). This method provided easy access to chiral diarylmethylamides in excellent yields and good to high enantioselectivities. The reaction enables the use of diverse *N*-tosylimines and arylboronic acids and allows a broad substrate scope irrelevant to the electron-donating or electron-withdrawing nature of the substituents.

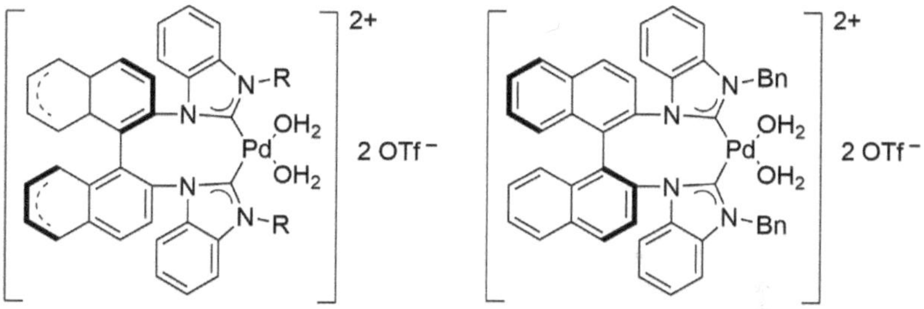

(*R*)-*N*-Me-NHC-Pd^{2+} (**8.3a**): 83%
(*R*)-*N*-Bn-NHC-Pd^{2+} (**8.3b**): 90%
(*R*)-*N*-3,5-Xyl-NHC-Pd^{2+} (**8.3c**): 88%
(*R*)-H_8-*N*-Bn-NHC-Pd^{2+} (**8.3d**): 80%

(*S*)-N-Bn-NHC-Pd^{2+} (**8.3e**): 85%

Fig. 7.1 Chiral cationic NHC-Pd^{2+} diaquo complexes.

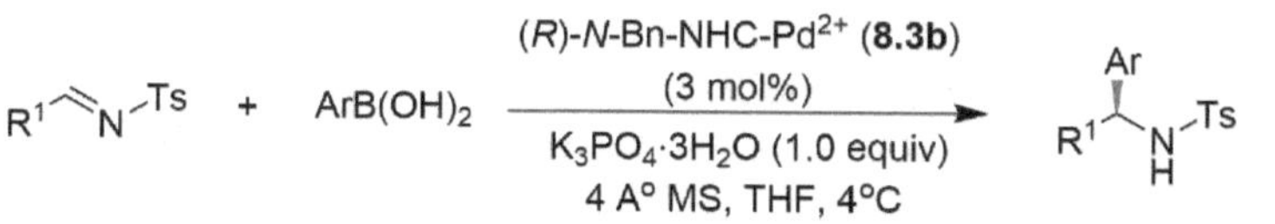

R^1 = Ar, ^{n}Pr, Cy
Ar = Ph, 3-MeC_6H_4, 4-$MeOC_6H_4$, 4-ClC_6H_4, 4-$CF_3C_6H_4$, etc.

up to 99% yield, 94% *ee*
30 examples

Scheme 7.3 Arylation of *N*-tosylarylimines with arylboronic acids using (NHC)Pd.

A readily accessible planar-chiral ferrocene imidazoline palladacycle **8.4** has been demonstrated by Peters and co-workers as a highly efficient catalyst for the formation of *N*-substituted benzylic stereo centers (Scheme 7.4) *(34)*. This chiral palladacycle catalyst accelerates the 1,2-addition of arylboroxines to aliphatic and aromatic imines with high levels of enantioselectivity. Notably, external base was not necessary for the activation of the boroxines while acetate was used as an anionic ligand. The high performance of this method was due to the use of an acetate anionic ligand that made the reaction possible without use of a stoichiometric amount of exogenous base, as base is often necessary for the transmetalation of aryl moiety from boron to palladium. With this methodology, common problems such as aryl-aryl homocoupling as well as imine hydrolysis were fully suppressed. The reaction tolerated crucial functional groups and afforded good to excellent yield of desired products with high enantiocontrol.

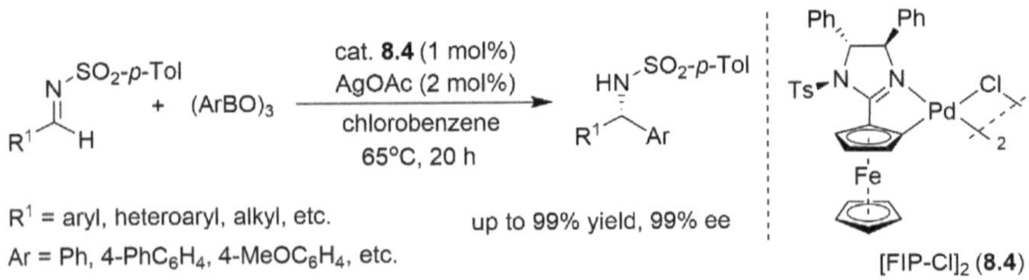

Scheme 7.4 Ferrocenyl-based palladacycle catalyzed arylation of imines.

The enantioselective intramolecular α-arylation of amides to generate oxindoles is an emerging field because the oxindoles are an important substructure in many biologically active compounds. Hartwig demonstrated that chiral carbene-based ligands **8.5** and **8.6** in combination with $Pd(OAc)_2$ can be employed efficiently for the asymmetric synthesis of oxindoles (Scheme 7.5) *(35)*. Later, Glorius synthesized imidazolium palladium complexes and demonstrated the α-arylation of amides with moderate enantioselectivity *(36)*. Though the demonstrated catalysis does not use chiral palladacycles, formation of palladacycle intermediate has been proposed during the arylation reaction.

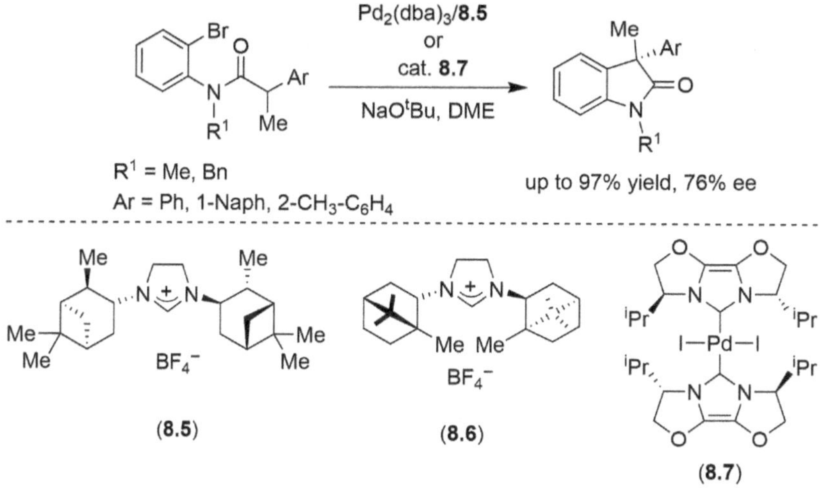

Scheme 7.5 Enantioselective intramolecular arylation of amides.

Similarly, Kundig also synthesized bulky NHC-Chiral ligands from readily available *ortho*-substituted α-alkylbenzylamines. These ligands were used for the palladium-catalyzed asymmetric intramolecular α-arylation of amides with excellent conversion and high enantioselectivities (Scheme 7.6) *(37,38)*. Later, the same group showed the synthesis of substituted spirooxindoles by palladium catalyzed asymmetric intramolecular arylation of amides (Scheme 7.7) *(39)*. Various substituents such as OMe, F, Cl, CF_3 were well tolerated on the amide backbones to yield the desired products in excellent yields with good enantioselectivities. Notably, synthesis of various ring sizes 4–6 spirooxindoles has been shown to employ this methodology.

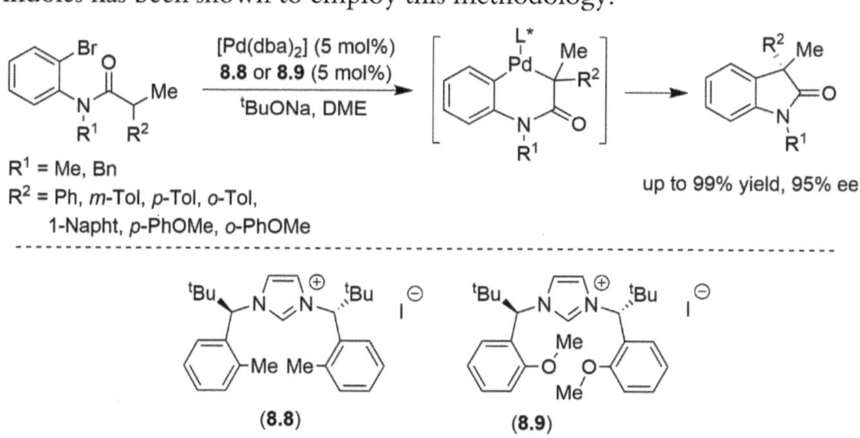

Scheme 7.6 Palladium-catalyzed asymmetric intramolecular α-arylation of amides.

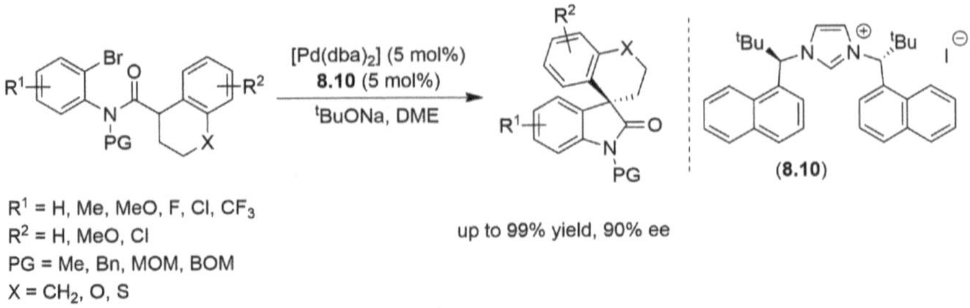

Scheme 7.7 Asymmetric synthesis of spirooxindoles.

3. ASYMMETRIC ALLYLIC ALKYLATION USING PALLADACYCLES

Asymmetric allylation represents one of the important reactions for the selective construction of C—C bond in many crucial synthetic intermediates. Though many metal catalysts are known for this reaction *(40–45)*, palladacycles as catalysts have been given special attention in recent years. The Peters group has developed a series of planar chiral pentaphenylferrocene-based phosphinooxazoline ligands and their palladium complexes for the allylic substitution reaction (Scheme 7.8) *(46)*. The authors have used oxazoline moiety as an efficient ortho directing group, and by the choice of suitable lithiation conditions, both possible diastereomers (for **8.11**) with regard to the planar chirality were accessed. The planar chiral organolithium species was trapped using diarylphosphino chlorides to obtain P,N ligands. The palladium-allyl complexes synthesized from these ligands were demonstrated for the allylic substitution of cyclic allylic acetates, wherein some of them allowed for good enantioselection (up to 91% ee) with the challenging substrates.

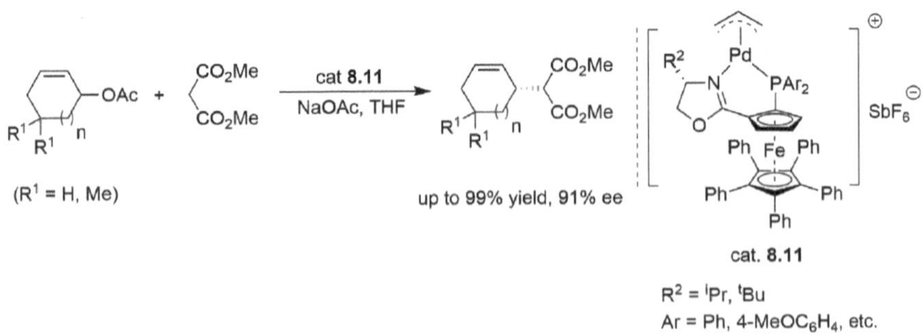

Scheme 7.8 Asymmetric allylic substitution of cyclic allylic acetates.

Stepnicka and co-workers developed a series of chiral phosphinoferrocene carboxamide ligands and tested for the palladium catalyzed enantioselective allylic alkylation of 1,3-diphenylallyl acetate with dimethyl malonate (Scheme 7.9) *(47–49)*. The reaction proceeded with complete conversion and almost 90% ee at room temperature and produced alkylated product as *R*-isomer. The most efficient one is the planar-only chiral ligand (S_p)-2-(diphenylphosphino)-1-(*N*-benzylcarbamoyl)-ferrocene ((*Sp*)-**8.12**) and the cationic (η^3-allyl)palladium(II) complex [Pd-(η^3-1,3-$Ph_2C_3H_3$){(S_p)-(**8.12**)-2*O*,*P*}]ClO_4 ((S_p)-**8.15**).

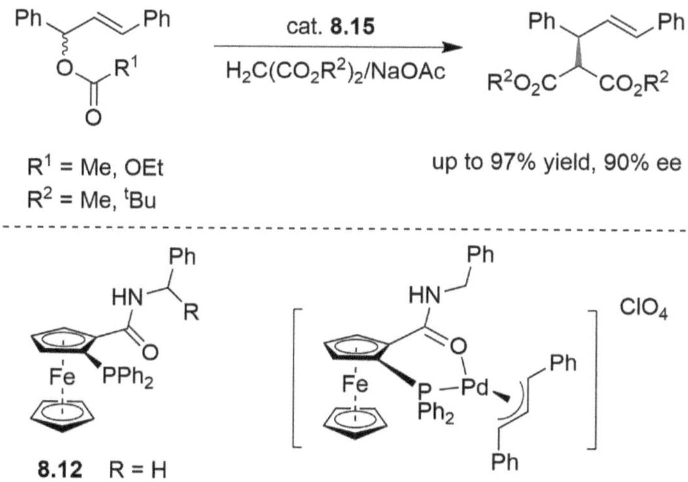

8.12 R = H
8.13 R = (*R*)-Me
8.14 R = (*S*)-Me

8.15

Scheme 7.9 Phosphinoferrocene carboxamides in enantioselective allylic alkylation

In 2009, Bedford and Wingad prepared various BINOL phosphite based chiral palladacycles and investigated their application for the phenylation of cyclohexen-2-one and for the allylation of benzaldehyde (Scheme 7.10) *(50)*. All these complexes resulted in the moderate to good conversion of desired products, however, enantioselectivity was extremely low (<15%). The authors assumed that the breaking up of binuclear palladacycle into monomeric and approach of substrates away from the chiral BINOL groups might be responsible for the poor enantiocontrol. This is consistent with the fact that little improvement of ee was observed with complexes **8.17, 8.18,** and **8.19** compared to **8.16**. The highest ee observed was 14% and 15% for 1,4-conjugate addition and allylation respectively.

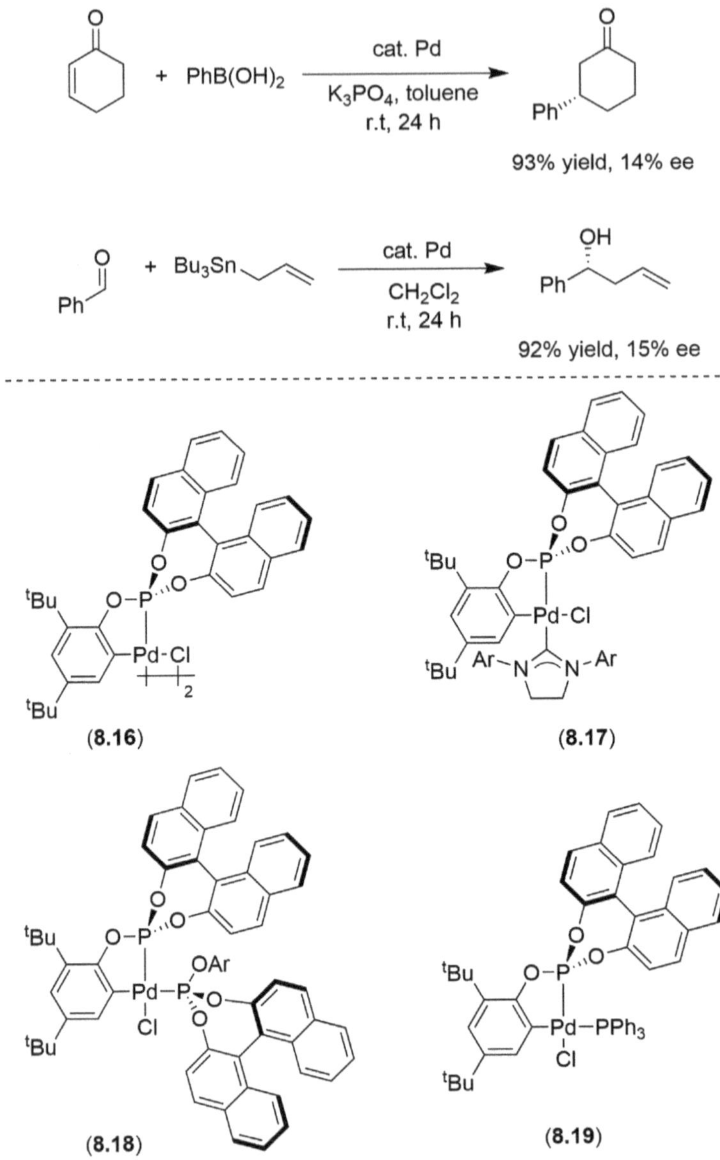

Scheme 7.10 Palladacycle catalyzed 1,4-conjugate addition and allylation reaction.

Further, the same group demonstrated the influence of metahydroxyl functionality on the structure of palladacycle and for the enantiocontrol. For example, the hydroxyl group forms a hydrogen bonding to the adjacent chloride ligand. These complexes showed better enantiocontrol compared to the nonhydroxyl palladacycle for the asymmetric allylation of benzaldehyde with allyl tributyltin (Scheme 7.11) *(51)*. It was found that the hydrogen bonding interactions influence both the isomerism and co-planarity of the metal coordination spheres in the dimer.

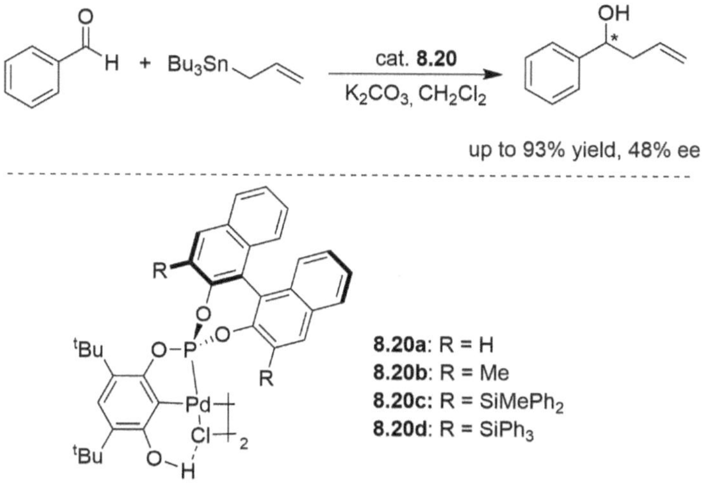

Scheme 7.11 Allylation of benzaldehyde employing metahydroxyl palladacycle.

4. ENANTIOSELECTIVE REARRANGEMENT REACTION

The allylic imidate rearrangement reaction, developed by Overman, is the key step in the overall conversion of allylic alcohol to allylic amines (Scheme 7.12) *(52–54)*. This reaction can be carried out by heating an imidates at high temperature via concerted pericyclic [3,3]-sigmatropic rearrangement to form amide. However, the reaction in the presence of palladium catalyst accelerates the rate in folds and can be performed at mild conditions. Moreover, asymmetric allylic imidate rearrangement could be performed using suitable ligand precursors. Recently, a variety of planar palladacycle complexes have been efficiently employed for this asymmetric reaction *(55,56)*.

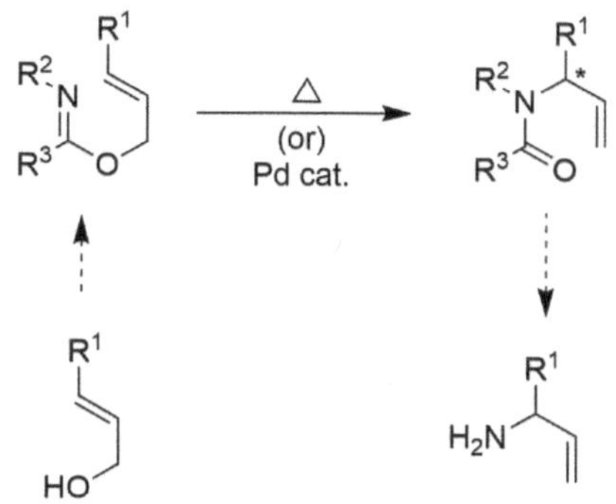

Scheme 7.12 Allylic imidate rearrangement in the conversion of allylic alcohols into allylic amines.

Overman and co-workers have developed a series of cyclopalladated ferrocenyl amines and imines, and evaluated as catalysts for the [3,3]-rearrangement of allylic benzimidates to allylic benzamides (Scheme 7.13) *(57)*. The chloro-bridge species **8.21** was a poor catalyst for the reaction, however, the imine palladacycle **8.22** gave improved ee. A ferrocenyl bis(palladacycle) synthesized by Kang gave high enantioselectivity for the asymmetric aza-Claisen rearrangement of allylic imidates *(58)*.

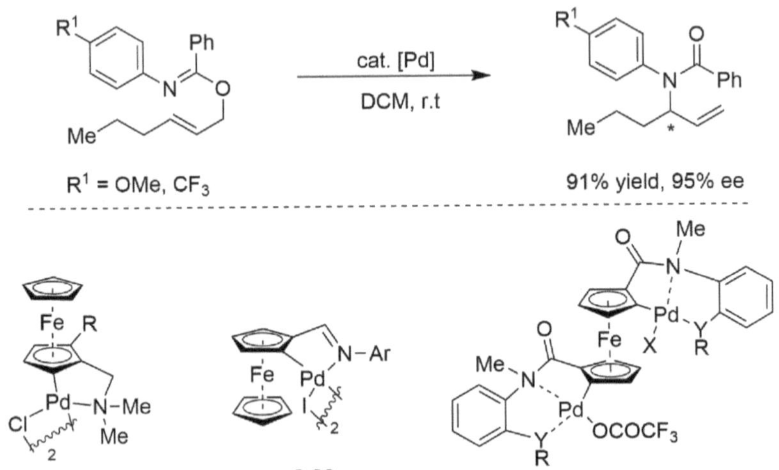

8.21a; R = Me
8.21b; R = $SiMe_3$

8.22

8.23a: YR = SMe
8.23b: YR = OMe
8.23b: YR = O^iPr

Scheme 7.13 Rearrangement reaction by amine and imine-based palladacycles.

Overmann has reported the preparation and evaluation of a small library of ferrocene oxazoline palladacycle (FOP) complexes as asymmetric catalysts for the rearrangement of prochiral allylic imidates *(59,60)*. The catalyst library included various members in which the C3 substituent of the palladated cyclopentadienyl ring is varied. Although substitution at the C3 position is important for catalyst stability, it has little to no effect on the efficiency or selectivity of the catalytic rearrangement. The most readily prepared of this group, the silyl FOP catalysts derived from **8.24** and **8.25** (Fig. 7.2), proved to be the best catalysts for the asymmetric rearrangement of allylic *N*-arylbenzimidates and gave >90% ee in the production of chiral allylic amides from prochiral allylic imidates. Herein, several allylic imidate motifs were prepared and evaluated for asymmetric allylic imidate rearrangements.

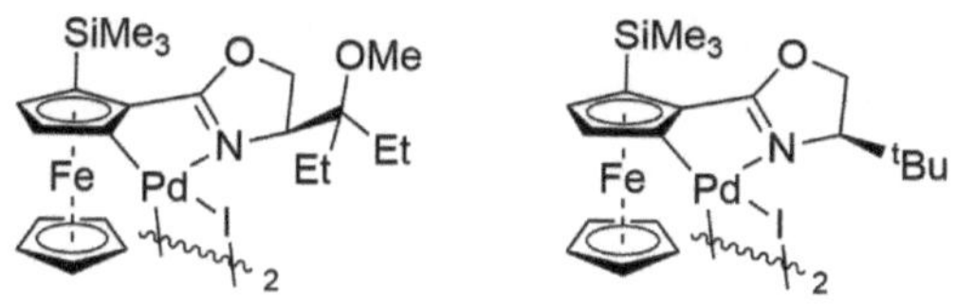

(8.24) (8.25)

Fig. 7.2 Palladacycle catalysts by Overman.

Moyano group has developed ferrocenyl-oxazoline-based chiral cyclopalladated complexes that hold significant promise in the asymmetric aza-Claisen reaction (Scheme 7.14) *(61)*. In particular, the trinuclear complexes **8.26a–c** are able to catalyze the enantioselective rearrangement of allylic imidates to allylic amides and afforded very good enantiocontrol. The use of the (S)-2-phenyl-4-ferrocenyloxazoline-derived trinuclear complex **8.26c** in the [3,3]-sigmatropic rearrangement of (*E*)-3-phenylallyl (N-phenyl)benzimidate gave 49% yield and with an unprecedented high optical purity (90% ee).

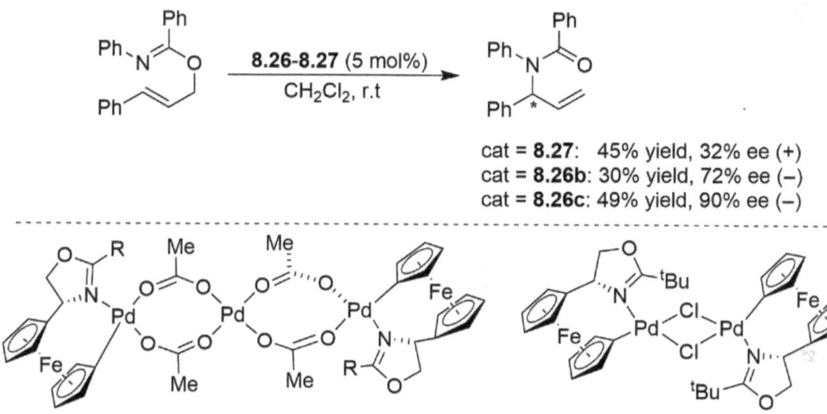

8.26a; R = ^{i}Pr
8.26b; R = ^{t}Bu
8.26c; R = Ph

8.27

Scheme 7.14 Aza-Claisen rearrangement using bi- and tri-nuclear palladacycles.

Although several allylic *N*-arylbenzimidates can undergo rearrangement in excellent enantioselectivities, the resulted products are of limited use as the removal of two nitrogen substituents (benzoyl and 4-methoxyphenyl) is cumbersome. However, allylic *N*-(4-methoxyphenyl)trifluoroacetimidates emerged as promising substrates as the product allylic amides can be converted to the parent allylic amines in good yields. Rearrangement of *E* allylic *N*-(4-methoxyphenyl)trifluoroacetimidates take place in good yields employing catalyst **8.25**, although enantioselectivities were slightly less than those realized with allylic *N*-arylbenzimidates (Scheme 7.15) *(60)*. Similar to oxazoline-based catalysts, the Peters group has synthesized highly efficient ferrocenyl-imidazoline palladacycles (FIPs) catalysts **8.28** for the aza-Claisen rearrangement of *N*-para-methoxyphenyl trifluoroacetimidates, leading to the formation of protected chiral primary allylic amines *(62,63)*. These catalysts showed good activity, enantioselectivity and tolerance for various functional groups in substrates. Rearrangement of *N*-hexyl-trifluoroacetimidates is also efficiently demonstrated by the use of the same catalyst *(64)*.

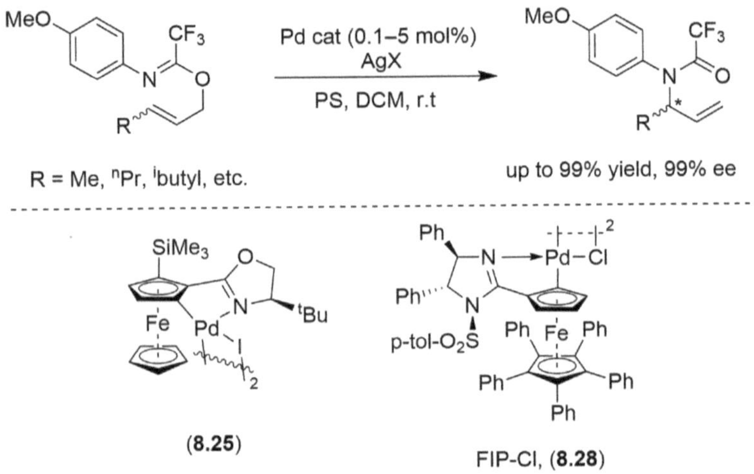

Scheme 7.15 Aza-Claisen rearrangement using FIP-catalyst.

The success of the ferrocenyl-imidazoline palladacycle (FIP) catalysts **8.28** led to the development of bis(palladation) complex **8.29** for the rearrangement of *N*-para-methoxyphenyl trifluoroacetimidates (Scheme 7.16) *(65)*. This catalyst was successful for the rearrangement reaction when R = ^{i}Pr, however, it was inactive when R = Ph. Thus, this methodology was extended to allylic amines containing ester, ketone, ether, silyl ether, and acetal. The use of trifluoroacetimidate substrates has been extended to the asymmetric synthesis of quaternary stereogenic centers with palladacyclic catalyst precursor **8.28** (Scheme 7.17).

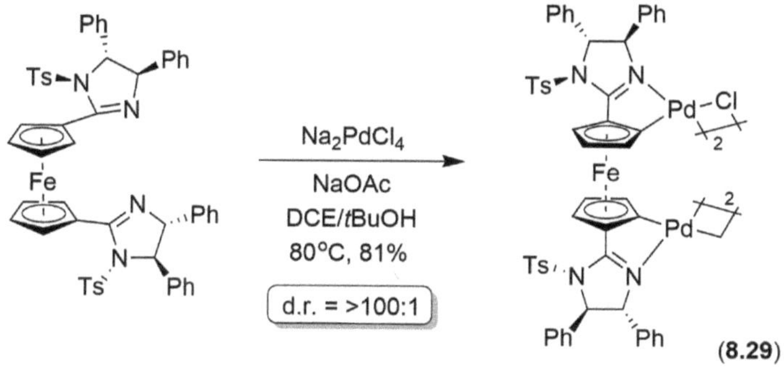

(8.29)

Scheme 7.16 Synthesis of a 1,1′-ferrocenyl-bisimidazoline palladacycle (FBIP).

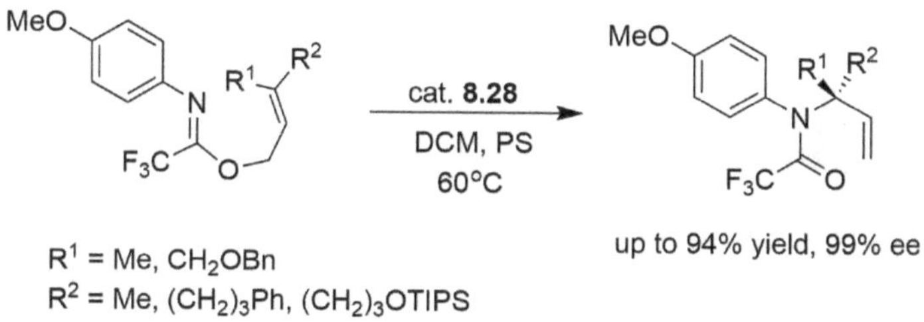

Scheme 7.17 Asymmetric synthesis of quaternary stereogenic centers.

Trichloroacetimidate is an ideal candidate for the asymmetric rearrangement as this compound can be prepared easily by the reaction of an alcohol with trichloroacetonitrile in the presence of base, and the trichloroacetamide obtained on rearrangement may be hydrolyzed under mild reaction conditions to give unprotected amines. Overman and co-workers successfully demonstrated the rearrangement of allylic trichloroacetimidates using catalyst **8.30** (Scheme 7.18) *(66)*. Substrates bearing unbranched R substituents rearranged at room temperature, however, increasing the reaction temperature to 38°C, enabled the use of iso-butyl and cyclohexyl R substituents. Phenyl and tert-butyl R substituents resulted in very poor conversion. A range of functional groups were tolerated in R substituents, such as OAc, CO_2Me, $CH(OCH_2CH_2O)COMe$, OTBDMS, NBn(Boc), and NBn_2 (92%–97% ee). Similarly, Richards has demonstrated the asymmetric rearrangement of allylic imidates employing amine-coordinated palladacycles **8.31** and **8.32** (Fig. 7.3) *(67)*.

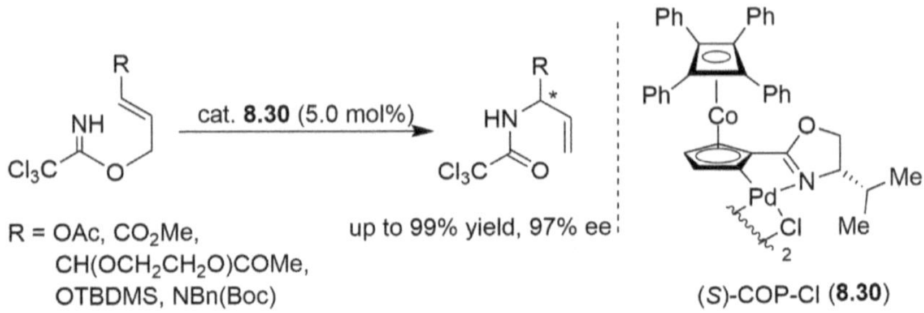

Scheme 7.18 Rearrangement of allylic trichloroacetimidates.

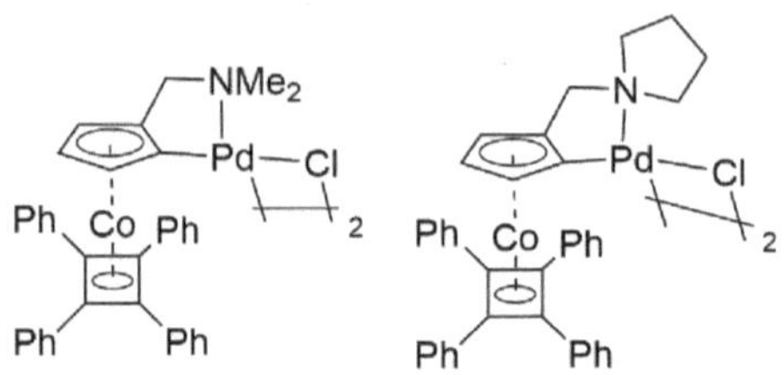

(**8.31**) (**8.32**)

Fig. 7.3 Amine-coordinated palladacycles for rearrangement reaction.

Generally, the aza-Claisen rearrangement generates carboxamide protected amines, and only for trichloro- and trifluoroacetamide products can the protecting groups be readily removed. However, considering the tedious process in preparation and storage of trichloro- and trifluoroacetamide, Peters demonstrated enantioselective rearrangement of allylic carbamates using palladacycle $[PPFOP\text{-}Cl]_2$ (**8.33**) (Scheme 7.19). With this asymmetric methodology, they were capable of transforming achiral allylic alcohols in a single step and with high enantio- and regioselectivity into sulfonyl-protected chiral allylic amines. Herein, in the presence of 3 mol% catalyst, 2 mol% $AgNO_3$, $^iPr_2NEt$ or 1,8-bis(*N,N*-dimethylamino)naphthalene base gave good yield of product with 88% ee *(68,69)*. These reactions have been demonstrated to proceed through a decarboxylative rearrangement of allylic carbamates, which explains the preference for the branched allylic product regioisomers. The allylic carbamates can be generated in situ by addition of the corresponding allylic alcohol to an isocyanate.

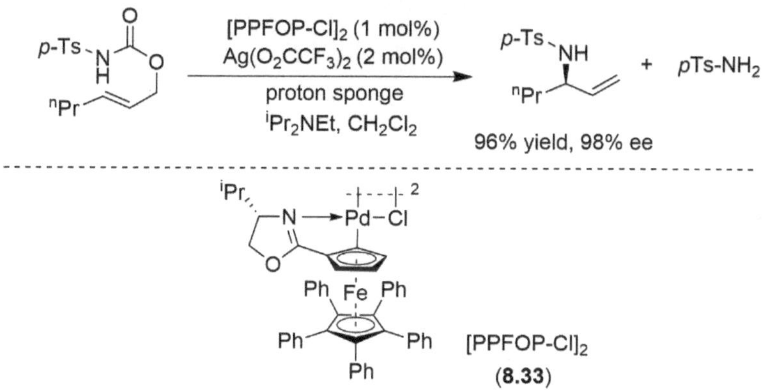

(**8.33**)

Scheme 7.19 Enantioselective allylic carbamate rearrangement.

In addition to ferrocene, iron, and cobalt containing metallocenes chiral palladacycles are also efficient catalysts for the allylic imidate rearrangement *(56)*. The first report on the application of palladacycle **8.30** (COP-Cl) to the allylic imidate rearrangement employed the substrate allylic N-arylbenzimidates (Scheme 7.20). Upon activation of the catalyst with silver trifluoroacetate, the *Z* substrates gave significantly higher enantioselectivities (94%–95% ee), whereas the *E* isomer resulted in low stereocontrol *(70,71)*. Similarly, the same palladium catalyst was employed for the enantioselective [3,3]-sigmatropic rearrangement of (*E*)- and (*Z*)-allyloxy substituted N-heterocycles that generates *N*-allyl N-heterocyclic amides in good yields and high enantioselectivities (up to 96% ee) (Scheme 7.21) *(72)*. Generally, the chiral palladacycle **8.30** (5 mol%) was used as a catalyst with silver(I) trifluoroacetate (10 mol%) at 35–45°C.

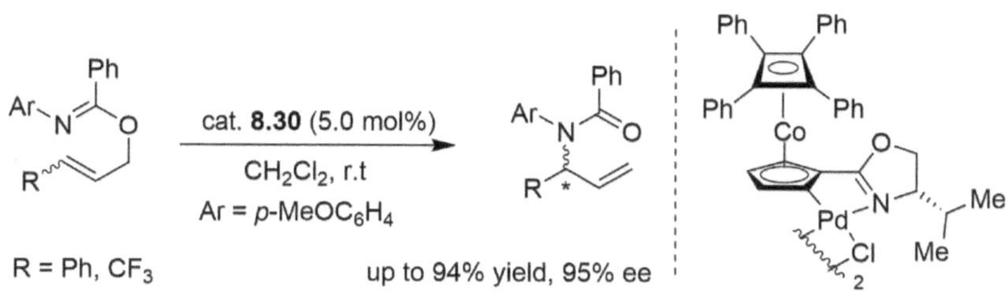

(*S*)-COP-Cl (**8.30**)

Scheme 7.20 Asymmetric rearrangement using catalyst 8.30.

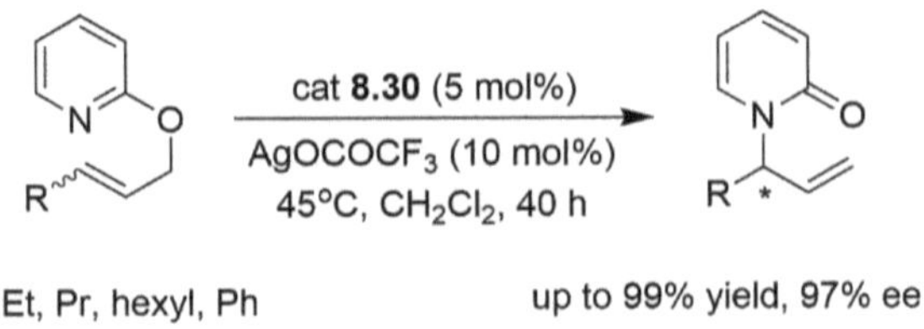

Scheme 7.21 Asymmetric rearrangement of allyloxy pyridine.

Intramolecular aminopalladation of the double bond is undoubtedly a central event in the rearrangement reaction as it would afford numerous nitrogen heterocycles. Overman has shown that the use of ferrocenyloxazoline palladacycle **8.34** can efficiently deliver five-membered nitrogen heterocycles such as 2-oxazolidinones, 2-imidazolidinones, and 2-pyrrolidinones with high stereocontrol (Scheme 7.22) *(73)*.

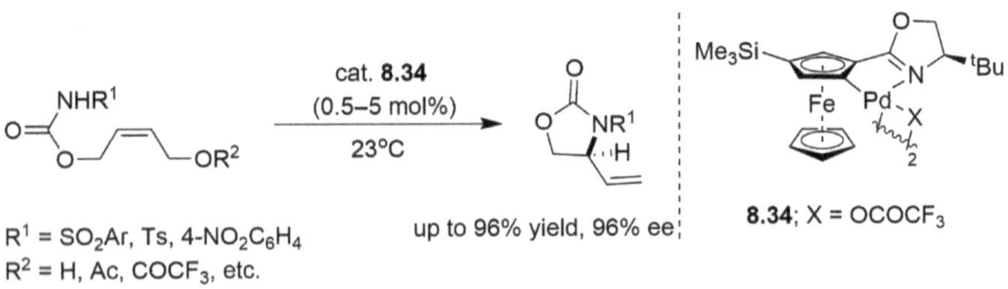

Scheme 7.22 Asymmetric intramolecular aminopalladation.

Nonferrocene planar chiral palladacycles have been developed and efficiently employed for the rearrangement reaction. List and co-workers have found suitable conditions for the rearrangement of N-PMP imidates using various chiral palladacycles (Scheme 7.23) *(74)*. This reaction proceeds through asymmetric counter anion directed palladium catalysis, wherein catalyst **Pd-5** along TRIP-Ag gave good yield of desired product with 96% ee.

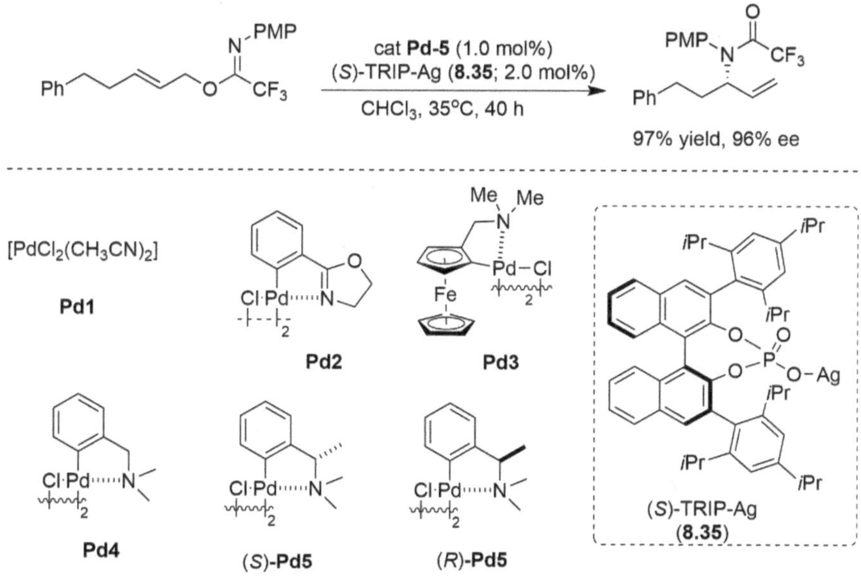

Scheme 7.23 Rearrangement of N-PMP imidates.

Overman has developed a new family of enantiopure C,N-palladacycles (PIN-acac complexes and evaluated as asymmetric catalysts for the reaction of allylic trichloroacetimidates with external nonmetal bound nucleophiles (Scheme 7.24) *(75)*. These air- and moisture-stable complexes were formed in good overall yield in three steps from 2-iodo-1-naphthoic acid and β-amino alcohols. Three PIN complexes were characterized by single-crystal X-ray analysis. Although these PIN palladacycles displayed useful levels of catalytic activity for the allylic substitution of prochiral allylic trichloroacetimidates with carboxylic acid nucleophiles and exclusively provided branched allylic esters in high yield, enantioselectivities were low to moderate. Nonetheless, the accessible and easily varied naphthalene-imidazoline ligands may be useful for the synthesis of related enantiopure axial chiral cyclometalated complexes that might find future applications in asymmetric catalysis. The experimental and computational results showed that with COP catalysts the alkene π-bond of an allylic imidate substrate is preferentially coordinated cis to the carbon ligand of the palladacycle with attack of an external nucleophile occurring from the least sterically hindered face in the quadrant.

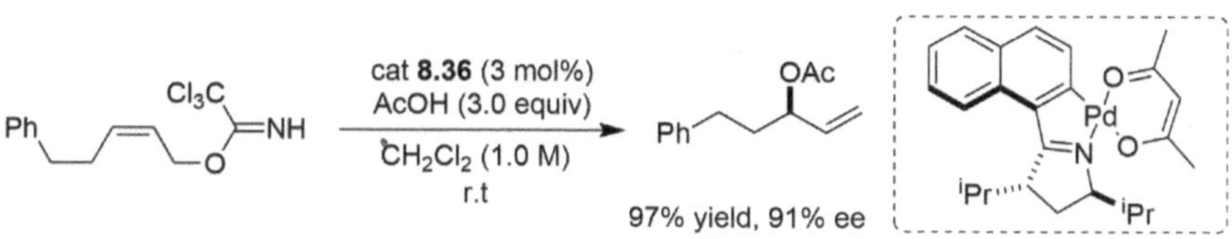

(8.36)

Scheme 7.24 Asymmetric synthesis of branched allylic acetates.

5. PALLADACYCLES IN ENANTIOSELECTIVE HYDROPHOSPHINATION

Chiral phosphines are highly valuable ligands in metal complexes and are widely used in enantioselective organic transformations *(76,77)*. In addition, they are extensively employed in pest control and bioorganic chemistry *(78,79)*. Generally, the chiral phosphines are prepared by resolution or by using a stoichiometric amount of enantiopure substrates *(80)*. Asymmetric hydrophosphination, the stereo-control addition of a P-H bond to unsaturated C—C bond, provides an efficient atom-economy access to chiral phosphines. The catalytic asymmetric synthesis of chiral phosphines has been reported by various metal catalysts based on Ni and Pt. *(81–86)* Recently, chiral palladacycles as catalyst for the asymmetric hydrophosphination reaction have been explored by various research groups. Leung and co-workers have demonstrated the chiral C,N-palladacycle **8.37** for the asymmetric hydrophosphination of aromatic enones, and prepared several chiral tertiary phosphines with high yields and stereoselectivities (Scheme 7.25) *(87,88)*. A range of aromatic enones containing —Cl, —Br, —NO_2, —OH, —OMe functionalities were efficiently employed for the reaction, and resulted with excellent yield and moderate enantioselectivity.

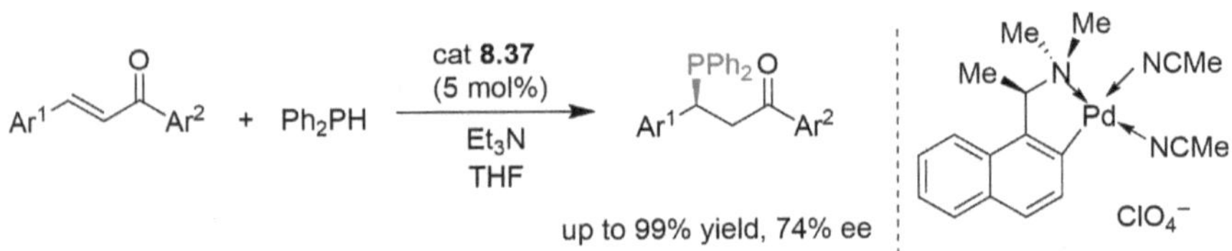

(**8.37**)

Ar^1 = Ph, 2-Naph, 4-ClC_6H_4, 4-BrC_6H_4, 4-$NO_2C_6H_4$, 4-OHC_6H_4
Ar^2 = Ph, 1-Naph, 2-Naph

Scheme 7.25 Asymmetric hydrophosphination of aromatic enones.

When the catalyst **8.37** was used to activate the chiral secondary phosphine R^1R^2PH, unsatisfactory reactivity and enantioselectivity were observed. In order to select a better hydrophosphination catalyst for wider applications, a series of palladacycles **8.38-8.42** as catalyst were synthesized and screened (Scheme 7.26) *(89)*. All the complexes are structurally similar with two coordination sites readily available for catalysis. Complexes

8.38–8.41 are C,N-cyclopalladated complexes, with similar electronic features originating from the σ-donating nitrogen and the π-accepting aromatic carbon. On the other hand, the C,P complex **8.42** is electronically designed to exhibit higher oxophilicity toward substrates because of the presence of both relatively soft carbon and phosphorus donors. The catalysts **8.38–8.41** were employed for hydrophosphination using secondary phosphines (Scheme 7.26), wherein the reaction proceeded at significantly slower rates. However, the C,P complex **8.42** was found to be a better catalyst in this hydrophosphination reaction, with very high reactivity and excellent enantioselectivity (up to 98%). Under optimal conditions, the reaction time was noticeably reduced to 2 h.

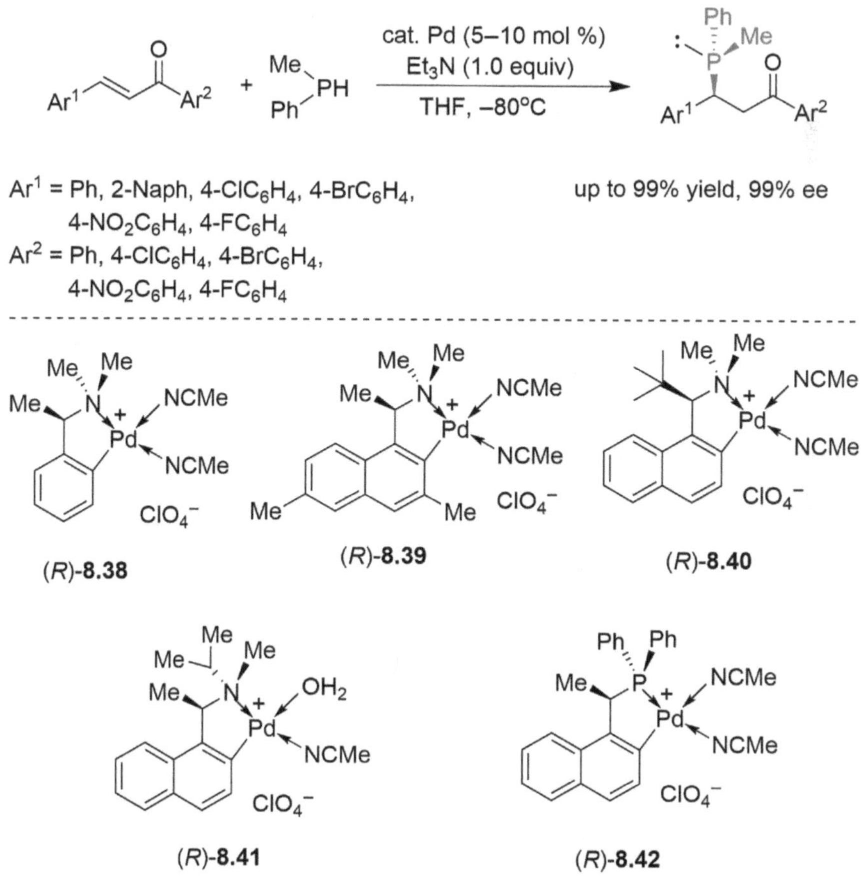

Scheme 7.26 Asymmetric hydrophosphination using secondary phosphines.

The asymmetric hydrophosphination of enones is also demonstrated by Gong and Song employing chiral PCN pincer Pd(II) complexes (Scheme 7.27) *(90)*. Among various complexes synthesized, the phosphinite-imidazoline-based palladacycle with (4S)-phenyl and *N*-Tol-p groups (**8.43**) was found to be the best for reactions of various enones with diphenylphosphine and afforded optically active phosphine derivatives in high yields with excellent enantioselectivities (up to 98% ee). In particular, heteroaryl-containing enones such as 2-alkenoylpyridines that may bind tightly to the catalyst were also tolerated, producing the corresponding pyridine-functionalized chiral phosphine oxides in good yields with good enantioselectivities. In addition, it was shown that the obtained pyridine-functionalized phosphine oxide acted as a $NC_{sp3}O$ pincer preligand in the reaction with $PdCl_2$.

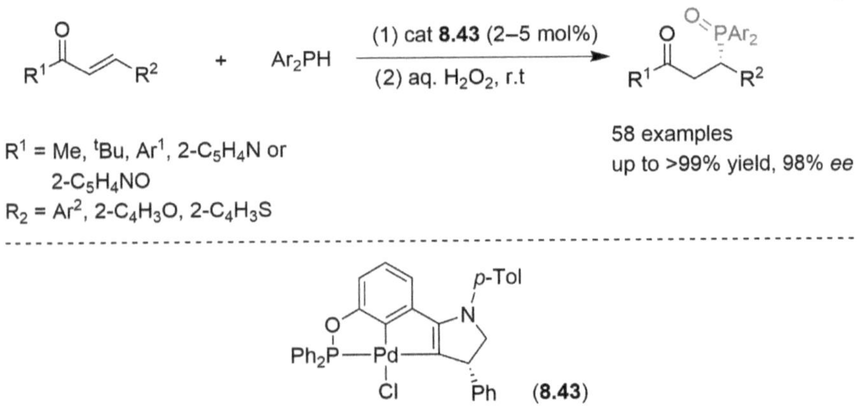

(**8.43**)

Scheme 7.27 Asymmetric hydrophosphination using pincer palladacycle 8.43.

Double hydrophosphination of bis(enones) with $PhPH_2$ has been demonstrated by Leung, wherein the reaction led to the asymmetric intermolecular construction of chiral tertiary *P*-heterocycles (Scheme 7.28) *(91)*. This reaction was catalyzed by palladacycle **8.42** which resulted in the synthesis of a number of chiral *P*-heterocycles with high enantiomeric excess. A mechanism has been proposed for the reaction, wherein stepwise double hydrophosphination of bis(enones) with primary phosphine is predicted. Leung also demonstrated the chemo- and enantioselective addition of diphenylphosphine to α,β–unsaturated imines catalyzed by palladacycle **8.42** (Scheme 7.29) *(92)*. This method exhibited broad functional group tolerance, and a wide range of substrates were efficiently converted into the desired products in high yields with excellent chemo- and enantioselectivities. With this methodology, a series of chiral tertiary enaminophosphines (chiral P,N-ligands) have been synthesized that are potentially useful for further catalysis.

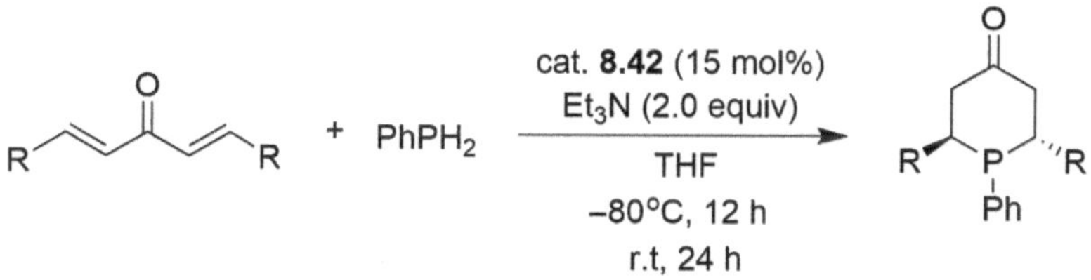

R = Ph, 4-FC_6H_4, 4-ClC_6H_4, 3-ClC_6H_4, 4-BrC_6H_4, 3-BrC_6H_4,

up to 99% yield, 97% ee

Scheme 7.28 Asymmetric hydrophosphination of bis(enones).

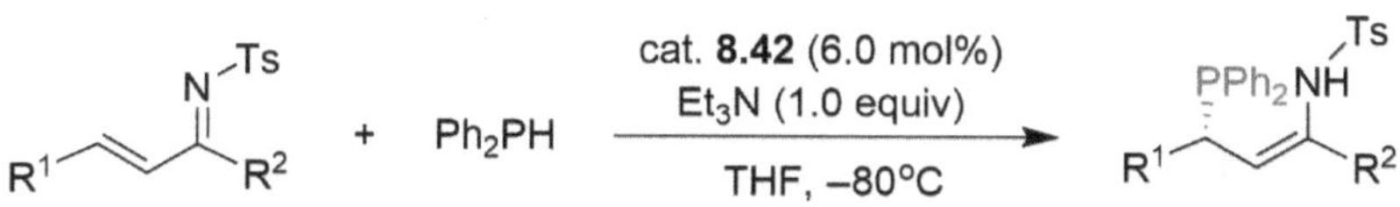

up to 99% yield, 99% ee

R^1 = Ph, 4-FC_6H_4, 4-ClC_6H_4, 3-ClC_6H_4, 2-ClC_6H_4, 4-BrC_6H_4, 3-BrC_6H_4, 4-$CF_3C_6H_4$, 2-Naph

Scheme 7.29 Asymmetric hydrophosphination of *α,β*-unsaturated imines.

The *β,γ*-unsaturated *α*-ketoesters are excellent substrates for a number of reactions due to their superior reactivities *versus* typical *α,β*-unsaturated carbonyls. They are valuable electrophiles in conjugate addition, including oxy- and aza-Michael additions. Notably, the resultant products can be further converted into other synthetically and biologically useful compounds. With this consideration, Leung has shown the enantioselective hydrophosphination of *β,γ*-unsaturated *α*-ketoesters and amides using chiral palladacycle **8.42** (Scheme 7.30) *(93)*. Excellent yields of up to >99% and enantioselectivities of up to 90% have been achieved. This method provides a direct access to important chiral tertiary phosphines that can facilitate the preparation of catalytically versatile P,O and P,N bidentate ligands. Similarly, the same group exclusively reported the hydrophosphination of N-enoyl phthalimides and 4-oxo-enamides which provided an efficient excess to *β*-phosphiniamides and phosphinocarboxamides, respectively *(94,95)*.

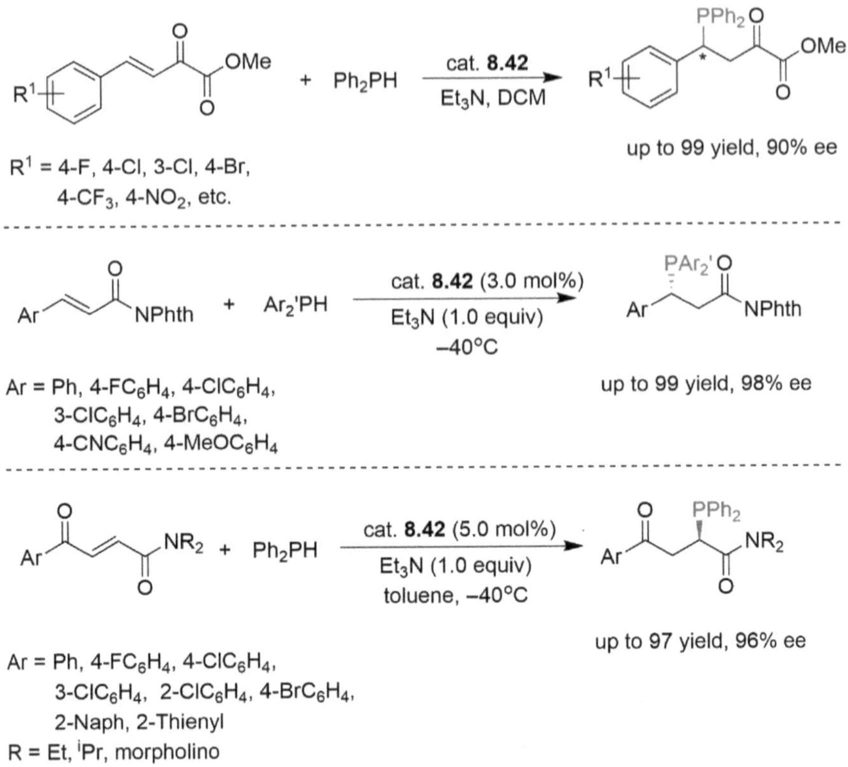

Scheme 7.30 Asymmetric hydrophosphination of β,γ-unsaturated α-ketoesters and amides.

Catalytic asymmetric hydrophosphination of pyridyl-substituted diene has been demonstrated employing a palladacycle pincer catalyst **8.44** in excellent yields (Scheme 7.31) *(96,97)*. Notably, the cyclopalladation of the resulted pyridyl-substituted chiral phosphine sulfide (N—P=S) and oxide (N—P=O) compounds could efficiently afford the asymmetric N—C(sp^3)*—S and N—C(sp^3)*—O pincer complexes. Interestingly, when the new pincer complexes were applied as catalysts in asymmetric hydrophosphination, the newly developed aliphatic pincer catalyst could be recycled over three runs and obtained in large quantities via a one-pot "self-breeding" catalytic protocol.

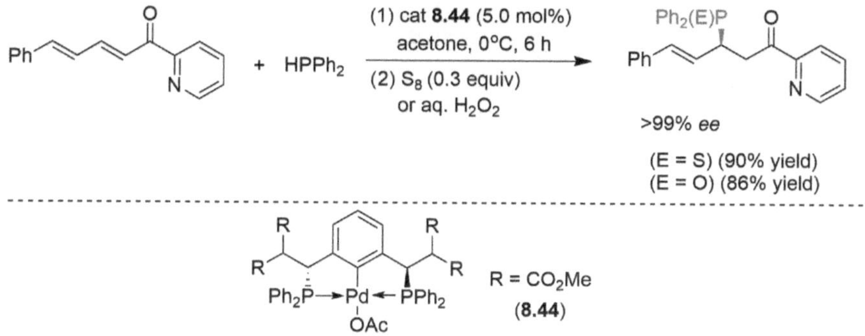

Scheme 7.31 Catalytic hydrophosphination of pyridyl-substituted diene.

Asymmetric hydrophosphination of substituted methylidenemalonate esters with diphenylphosphine that provides a direct access to chiral tertiary phosphines is being demonstrated using palladacycle **8.42** (Scheme 7.32) *(98)*. Leung group has screened three different C,N and C,P palladacycles as catalyst for the hydrophosphination using diphenylphosphine. After screening various reaction conditions, catalyst **8.42** was found to be the best of all. The reaction afforded excellent yield of desired products with high enantioselectivity. Important functionalities such as 2-furan, NH_2, and PhCH=CH are well tolerated under the reaction conditions. Similar to this, enantioselective Michael addition reaction of diphenylphosphine to substituted alkenylisoxazoles has been developed (Scheme 7.32) *(99)*. The reaction proceeded efficiently under mild conditions with high yields and moderate to excellent enantioselectivity. This methodology provided a direct access to a library of chiral tertiary phosphine-functionalized isoxazoles. Notably, the alkenyl having important heterocycles such as 2-furanyl, 2-pyrrolyl, 3-pyridyl, 4-pyridyl are successfully employed for the reaction.

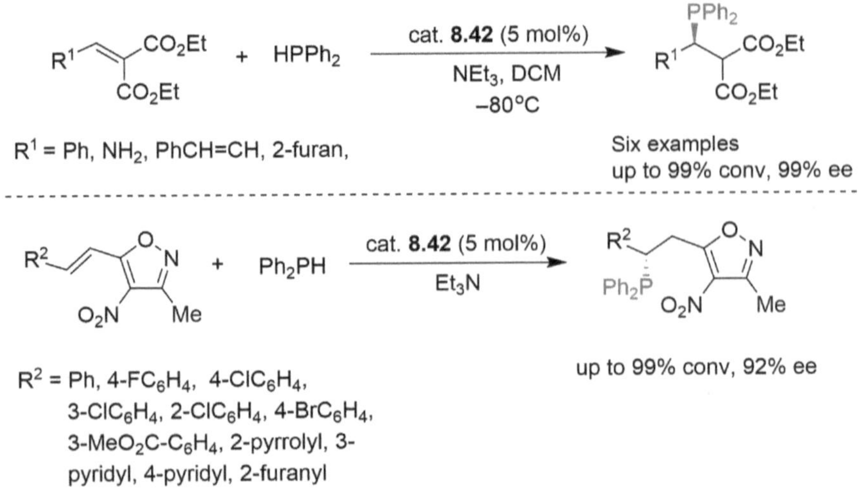

Scheme 7.32 Hydrophosphination of methylidenemalonate esters and alkenylisoxazoles.

6. SUMMARY AND OUTLOOK

Chiral palladacycles have found tremendous catalytic application in the synthesis of enantiomeric pure organic compounds. The present chapter focused on enantioselective arylation and allylic alkylation, [3,3]-sigmatropic rearrangement reaction and asymmetric hydrophosphination of alkenes. In these reactions, well-defined chiral palladacycles have been applied as catalyst. Usually these catalysts performed excellent in term of conversion and enantiocontrol. The activities of palladacycles were effectively tuned to achieve desired enantiomeric compounds in high yields.

Based on the large numbers of application received by palladacycle, it could be assumed that the chiral palladacycle can be extended to other new asymmetric synthesis in the next couple of years. The main targets should be enantioselective direct C-H bond functionalizations leading to the synthesis of various biologically important compounds. Design and synthesis of new chiral palladacycles, with wider application than the existing ones, could be another direction to this field. Another important area is to heterogenize the palladacycle on solid supports to make it a more robust and recyclable catalyst.

REFERENCES

1. Hegedus, L. S. *Transition Metals in the Synthesis of Complex Organic Molecules;* University Science: Mill Valley, CA, 1999.
2. Tsuji, J. Transition Metal Reagents and Catalysts. In *Innovation in Organic Synthesis,* Wiley: Chichester, 2000.
3. Hartwig, J. *Organotransition Metal Chemistry;* University Science: Sausalito, CA, 2010.
4. Dupont, J.; Consorti, C. S.; Spencer, J. The Potential of Palladacycles: More Than Just Precatalysts. *Chem. Rev.* **2005**, *105*, 2527–2572.
5. Albrecht, M.; van Koten, G. Platinum Group Organometallics Based on "Pincer" Complexes: Sensors, Switches, and Catalysts. *Angew. Chem. Int. Ed.* **2001**, *40*, 3750–3781.
6. van der Boom, M. E.; Milstein, D. Cyclometalated Phosphine-Based Pincer Complexes: A Mechanistic Insight in Catalysis, Coordination, and Bond Activation. *Chem. Rev.* **2003**, *103*, 1759–1792.
7. Singleton, J. T. The Uses of Pincer Complexes in Organic Synthesis. *Tetrahedron* **2003**, *59*, 1837–1857.
8. Tsuji, J. *Palladium Reagents and Catalysts. New Perspectives for the 21st Century;* 2004.
9. Szabó, K. J.; Wendt, O. F. B. *Pincer and Pincer-Type Complexes: Applications in Organic Synthesis and Catalysis;* Wiley-VCH: Weinheim, 2014.
10. Bloch, R. Additions of Organometallic Reagents to C-N bonds: Reactivity and Selectivity. *Chem. Rev.* **1998**, *98*, 1407–1438.
11. Hermanns, N.; Dahmen, S.; Bolm, C.; Bräse, S. Asymmetric, Catalytic Phenyl Transfer to Imines: Highly Enantioselective Synthesis of Diarylmethylamines. *Angew. Chem. Int. Ed.* **2002**, *41*, 3692–3694.
12. Akullian, L. C.; Snapper, M. L.; Hoveyda, A. H. Three-Component Enantioselective Synthesis of Propargylamines Through Zr-Catalyzed Additions of Alkyl Zinc Reagents to Alkynylimines. *Angew. Chem. Int. Ed.* **2003**, *42*, 4244–4247.
13. Fagnou, K.; Lautens, M. Rhodium-Catalyzed Carbon-Carbon Bond Forming Reactions of Organometallic Compounds. *Chem. Rev.* **2003**, *103*, 169–196.
14. Li, S.-W.; Batey, R. A. Allylation and Highly Diastereoselective Syn or Anti Crotylation of N-Toluenesulfonylimines Using Potassium Allyl- and Crotyltrifluoroborates. *Chem. Commun.* **2004**, 1382–1383.
15. Hayashi, T.; Yamasaki, K. Rhodium-Catalyzed Asymmetric 1,4-Addition and Its Related Asymmetric Reactions. *Chem. Rev.* **2003**, *103*, 2829–2844.
16. Ueda, M.; Saito, A.; Miyaura, N. Rhodium-Catalyzed Addition of Arylboronic acids to N-Sulfonyl Aldimines. *Synlett* **2000**, *2000*, 1637–1639.
17. Kuriyama, M.; Soeta, T.; Hao, X.; Chen, Q.; Tomioka, K. N-Boc-l-valine-Connected Amidomonophosphane Rhodium(I) Catalyst for Asymmetric Arylation of N-Tosylarylimines With Arylboroxines. *J. Am. Chem. Soc.* **2004**, *126*, 8128–8129.
18. Tokunaga, N.; Otomaru, Y.; Okamoto, K.; Ueyama, K.; Shintani, R.; Hayashi, T. C2-Symmetric Bicyclo[2.2.2]octadienes as Chiral Ligands: Their High Performance in Rhodium-Catalyzed Asymmetric Arylation of N-Tosylarylimines. *J. Am. Chem. Soc.* **2004**, *126*, 13584–13585.
19. Weix, D. J.; Shi, Y.; Ellman, J. A. Diastereoselective and Enantioselective Rh(I)-Catalyzed Additions of Arylboronic Acids to N-tert-Butanesulfinyl and N-Diphenylphosphinoyl Aldimines. *J. Am. Chem. Soc.* **2005**, *127*, 1092–1093.
20. Jagt, R. B. C.; Toullec, P. Y.; Geerdink, D.; de Vries, J. G.; Feringa, B. L.; Minnaard, A. J. A Ligand-Library Approach to the Highly Efficient Rhodium/Phosphoramidite-Catalyzed Asymmetric Arylation of N,N-Dimethylsulfamoyl-Protected Aldimines. *Angew. Chem. Int. Ed.* **2006**, *45*, 2789–2791.

21. Duan, H.-F.; Jia, Y.-X.; Wang, L.-X.; Zhou, Q.-L. Enantioselective Rh-Catalyzed Arylation of N-Tosylarylimines With Arylboronic Acids. *Org. Lett.* **2006**, *8*, 2567–2569.
22. Nakagawa, H.; Rech, J. C.; Sindelar, R. W.; Ellman, J. A. Catalytic Enantioselective Addition of Arylboronic Acids to N-Boc Imines Generated In Situ. *Org. Lett.* **2007**, *9*, 5155–5157.
23. Wang, Z.-Q.; Feng, C.-G.; Xu, M.-H.; Lin, G.-Q. Design of C2-Symmetric Tetrahydropentalenes as New Chiral Diene Ligands for Highly Enantioselective Rh-catalyzed arylation of N-tosylarylimines with arylboronic acids. *J. Am. Chem. Soc.* **2007**, *129*, 5336–5337.
24. He, P.; Lu, Y.; Hu, Q.-S. Phosphinite- and Phosphite-Based Type I Palladacycles as Highly Active Catalysts for Addition Reactions of Arylboronic Acids With Aldehydes, α,β-Unsaturated Ketones, α-Ketoesters, and Aldimines. *Tetrahedron Lett.* **2007**, *48*, 5283–5288.
25. Zhang, Q.; Chen, J.; Liu, M.; Wu, H.; Cheng, J.; Qin, C.; Su, W.; Ding, J. Palladium-Catalyzed Addition of Arylboronic Acids to N-Tosylarylimines. *Synlett* **2008**, *2008*, 935–939.
26. Dai, H.; Yang, M.; Lu, X. Palladium(II)-Catalyzed One-Pot Enantioselective Synthesis of Arylglycine Derivatives From Ethyl Glyoxylate, p-Toluenesulfonyl Isocyanate and Arylboronic Acids. *Adv. Synth. Catal.* **2008**, *350*, 249–253.
27. Dai, H.; Lu, X. Palladium(II)/2,2′-bipyridine-Catalyzed Addition of Arylboronic Acids to N-Tosyl-Arylaldimines. *Tetrahedron Lett.* **2009**, *50*, 3478–3481.
28. Chen, J.; Lu, X.; Lou, W.; Ye, Y.; Jiang, H.; Zeng, W. Palladium(II)-Catalyzed Enantioselective Arylation of α-imino Esters. *J. Org. Chem.* **2012**, 77, 8541–8548.
29. Marques, C. S.; Burke, A. J. Catalytic Enantioselective Addition of Phenylboronic Acid and Phenylboroxine to N-tosylimines: Pd(II) and Rh(I) Catalysis. *Eur. J. Org. Chem.* **2010**, *2010*, 1639–1643.
30. Yang, G.; Zhang, W. A Palladium-Catalyzed Enantioselective Addition of Arylboronic Acids to Cyclic Ketimines. *Angew. Chem. Int. Ed.* **2013**, *52*, 7540–7544.
31. Jiang, C.; Lu, Y.; Hayashi, T. High Performance of a Palladium Phosphinooxazoline Catalyst in the Asymmetric Arylation of Cyclic N-Sulfonyl Ketimines. *Angew. Chem. Int. Ed.* **2014**, *53*, 9936–9939.
32. Ma, G.-N.; Zhang, T.; Shi, M. Catalytic Enantioselective Arylation of N-Tosylarylimines With Arylboronic Acids Using C2-Symmetric Cationic N-Heterocyclic Carbene Pd^{2+} Diaquo Complexes. *Org. Lett.* **2009**, *11*, 875–878.
33. Liu, Z.; Shi, M. Catalytic Asymmetric Addition of Arylboronic Acids to N-Boc Imines Generated in Situ Using C2-Symmetric Cationic N-Heterocyclic Carbenes (NHCs) Pd^{2+} Diaquo Complexes. *Tetrahedron* **2010**, *66*, 2619–2623.
34. Schrapel, C.; Peters, R. Exogenous-Base-Free Palladacycle-Catalyzed Highly Enantioselective Arylation Of Imines With Arylboroxines. *Angew. Chem. Int. Ed.* **2015**, *54*, 10289–10293.
35. Lee, S.; Hartwig, J. F. Improved Catalysts for the Palladium-Catalyzed Synthesis of Oxindoles by Amide α-arylation. Rate Acceleration, Use of Aryl Chloride Substrates, and a New Carbene Ligand for Asymmetric Transformations. *J. Org. Chem.* **2001**, *66*, 3402–3415.
36. Glorius, F.; Altenhoff, G.; Goddard, R.; Lehmann, C. Oxazolines as Chiral Building Blocks for Imidazolium Salts and N-Heterocyclic Carbene Ligands. *Chem. Commun.* **2002**, 2704–2705.
37. Kündig, E. P.; Seidel, T. M.; Jia, Y.-X.; Bernardinelli, G. Bulky Chiral Carbene Ligands and Their Application in the Palladium-Catalyzed Asymmetric Intramolecular α-Arylation of Amides. *Angew. Chem. Int. Ed.* **2007**, *46*, 8484–8487.
38. Jia, Y.-X.; Katayev, D.; Bernardinelli, G.; Seidel, T. M.; Kündig, E. P. New Chiral N-Heterocyclic Carbene Ligands in Palladium-Catalyzed α-Arylations of Amides: Conformational Locking Through Allylic Strain as a Device for Stereocontrol. *Chem. Eur. J.* **2010**, *16*, 6300–6309.

39. Katayev, D.; Jia, Y.-X.; Sharma, A. K.; Banerjee, D.; Besnard, C.; Sunoj, R. B.; Kündig, E. P. Synthesis of 3,3-Disubstituted Oxindoles by Palladium-Catalyzed Asymmetric Intramolecular α-Arylation of Amides: Reaction Development and Mechanistic Studies. *Chem. Eur. J.* **2013**, *19*, 11916–11927.
40. Consiglio, G.; Waymouth, R. M. Enantioselective Homogeneous Catalysis Involving Transition-Metal-Allyl Intermediates. *Chem. Rev.* **1989**, *89*, 257–276.
41. Trost, B. M.; Van Vranken, D. L. Asymmetric Transition Metal-Catalyzed Allylic Alkylations. *Chem. Rev.* **1996**, *96*, 395–422.
42. Trost, B. M.; Crawley, M. L. Asymmetric Transition-Metal-Catalyzed Allylic Alkylations: Applications in Total Synthesis. *Chem. Rev.* **2003**, *103*, 2921–2944.
43. Trost, B. M. Asymmetric Allylic Alkylation, an Enabling Methodology. *J. Org. Chem.* **2004**, *69*, 5813–5837.
44. Helmchen, G. Enantioselective Palladium-Catalyzed Allylic Substitutions With Asymmetric Chiral Ligands. *J. Organomet. Chem.* **1999**, *576*, 203–214.
45. Graening, T.; Schmalz, H.-G. Pd-Catalyzed Enantioselective Allylic Substitution: New Strategic Options for the Total Synthesis of Natural Products. *Angew. Chem. Int. Ed.* **2003**, *42*, 2580–2584.
46. Garcia, M. A.; Frey, W.; Peters, R. Sterically Demanding Planar Chiral P,N Ligands by Diastereoselective Ortho Lithiation of Pentaphenylferrocenyloxazolines and Their Application to Palladium-Catalyzed Substitutions With Cyclic Allylic Acetates. *Organometallics* **2014**, *33*, 1068–1078.
47. Lamač, M.; Tauchman, J.; Štěpnička, P. Preparation of Chiral Phosphinoferrocene Carboxamide Ligands and Their Application to Palladium-Catalyzed Asymmetric Allylic Alkylation. *Organometallics* **2007**, *26*, 5042–5049.
48. Lamač, M.; Císařová, I.; Štěpnička, P. Synthesis, Coordination and Catalytic Utility of Novel Phosphanyl-Ferrocenecarboxylic Ligands Combining Planar and Central Chirality. *Eur. J. Inorg. Chem.* **2007**, *2007*, 2274–2287.
49. Lamač, M.; Císařová, I.; Štěpnička, P. Preparation, Coordination and Catalytic Use of Planar-Chiral Monocarboxylated DPPF Analogues. *New J. Chem.* **2009**, *33*, 1549–1562.
50. Bedford, R. B.; Dumycz, H.; Haddow, M. F.; Pilarski, L. T.; Orpen, A. G.; Pringle, P. G.; Wingad, R. L. Chiral Triaryl Phosphite-Based Palladacycles and Platinacycles: Synthesis and Application to Asymmetric Lewis Acid Catalysis. *Dalton Trans.* **2009**, 7796–7804.
51. Bedford, R. B.; Chang, Y.-N.; Haddow, M. F.; McMullin, C. L. Chiral Palladacycles Based on Resorcinol Monophosphite Ligands: The Role of the Meta-Hydroxyl in Ligand C−H Activation and Catalysis. *Dalton Trans.* **2011**, *40*, 9042–9050.
52. Overman, L. E. Thermal and Mercuric Ion Catalyzed [3,3]-Sigmatropic Rearrangement of Allylic Trichloroacetimidates. 1,3 Transposition of Alcohol and Amine Functions. *J. Am. Chem. Soc.* **1974**, *96*, 597–599.
53. Overman, L. E. A General Method for the Synthesis of Amines by the Rearrangement of Allylic Trichloroacetimidates. 1,3 Transposition of Alcohol and Amine Functions. *J. Am. Chem. Soc.* **1976**, *98*, 2901–2910.
54. Overman, L. E. Allylic and Propargylic Imidic Esters in Organic Synthesis. *Acc. Chem. Res.* **1980**, *13*, 218–224.
55. Nomura, H.; Richards, C. J. Allylic Imidate Rearrangements Catalyzed by Planar Chiral Palladacycles. *Chem. Asian J.* **2010**, *5*, 1726–1740.
56. Cannon, J. S.; Overman, L. E. Palladium(II)-Catalyzed Enantioselective Reactions Using cop Catalysts. *Acc. Chem. Res.* **2016**, *49*, 2220–2231.
57. Cohen, F.; Overman, L. E. Planar-Chiral Cyclopalladated Ferrocenyl Amines and Imines as Enantioselective Catalysts for Allylic Imidate Rearrangements. *Tetrahedron: Asymmetry* **1998**, *9*, 3213–3222.

58. Kang, J.; Yew, K. H.; Kim, T. H.; Choi, D. H. Preparation of bis[palladacycles] and Application to Asymmetric aza-Claisen Rearrangement of Allylic Imidates. *Tetrahedron Lett.* **2002**, *43*, 9509–9512.
59. Donde, Y.; Overman, L. E. High Enantioselection in the Rearrangement of Allylic Imidates With Ferrocenyl Oxazoline Catalysts. *J. Am. Chem. Soc.* **1999**, *121*, 2933–2934.
60. Anderson, C. E.; Donde, Y.; Douglas, C. J.; Overman, L. E. Catalytic Asymmetric Synthesis of Chiral Allylic Amines. Evaluation of Ferrocenyloxazoline Palladacycle Catalysts and Imidate Motifs. *J. Org. Chem.* **2005**, *70*, 648–657.
61. Moyano, A.; Rosol, M.; Moreno, R. M.; López, C.; Maestro, M. A. Oxazoline-Mediated Interannular Cyclopalladation of Ferrocene: Chiral Palladium(II) Catalysts for the Enantioselective aza-Claisen Rearrangement. *Angew. Chem. Int. Ed.* **2005**, *44*, 1865–1869.
62. Weiss, M. E.; Fischer, D. F.; Xin, Z.-q.; Jautze, S.; Schweizer, W. B.; Peters, R. Practical, Highly Active, and Enantioselective Ferrocenyl-Imidazoline Palladacycle Catalysts (FIPs) for the aza-Claisen Rearrangement of N-para-methoxyphenyl Trifluoroacetimidates. *Angew. Chem. Int. Ed.* **2006**, *45*, 5694–5698.
63. Fischer, D. F.; Barakat, A.; Xin, Z.-q.; Weiss, M. E.; Peters, R. The Asymmetric aza-Claisen Rearrangement: Development of Widely Applicable Pentaphenylferrocenyl Palladacycle Catalysts. *Chem.- Eur. J.* **2009**, *15*, 8722–8741.
64. Peters, R.; Xin, Z.-q.; Maier, F. Catalyst Versus Substrate Induced Selectivity: Kinetic Resolution by Palladacycle Catalyzed Allylic Imidate Rearrangements. *Chem. Asian J.* **2010**, *5*, 1770–1774.
65. Jautze, S.; Diethelm, S.; Frey, W.; Peters, R. Diastereoselective Bis-cyclopalladation of Ferrocene-1,1′-diyl bis-imidazolines: Translation of Central via Axial Into Planar Chirality. *Organometallics* **2009**, *28*, 2001–2004.
66. Anderson, C. E.; Overman, L. E. Catalytic Asymmetric Rearrangement of Allylic Trichloroacetimidates. A Practical Method for Preparing Allylic Amines and Congeners of High Enantiomeric Purity. *J. Am. Chem. Soc.* **2003**, *125*, 12412–12413.
67. Cassar, D. J.; Ilyashenko, G.; Ismail, M.; Woods, J.; Hughes, D. L.; Richards, C. J. Enantioselective Synthesis and Application to the Allylic Imidate Rearrangement of Amine-coordinated Palladacycle Catalysts of Cobalt Sandwich Complexes. *Chem. Eur. J.* **2013**, *19*, 17951–17962.
68. Bauer, J. M.; Frey, W.; Peters, R. Asymmetric Cascade Reaction to Allylic Sulfonamides From Allylic Alcohols by Palladium(II)/Base-Catalyzed Rearrangement of Allylic Carbamates. *Angew. Chem. Int. Ed.* **2014**, *53*, 7634–7638.
69. Bauer, J. M.; Frey, W.; Peters, R. Dual Palladium(II)/Tertiary Amine Catalysis for Asymmetric Regioselective Rearrangements of Allylic Carbamates. *Chem. Eur. J.* **2016**, *22*, 5767–5777.
70. Kang, J.; Hyung Kim, T.; Han Yew, K.; Ki Lee, W. The Effect Of Face-Blocking in the Enantioselective aza-Claisen Rearrangement of Allylic Imidates. *Tetrahedron Asymmetry* **2003**, *14*, 415–418.
71. Overman, L. E.; Owen, C. E.; Pavan, M. M.; Richards, C. J. Catalytic Asymmetric Rearrangement of Allylic N-aryl Trifluoroacetimidates. A Useful Method for Transforming Prochiral Allylic Alcohols to Chiral Allylic Amines. *Org. Lett.* **2003**, *5*, 1809–1812.
72. Rodrigues, A.; Lee, E. E.; Batey, R. A. Enantioselective Palladium(II)-Catalyzed Formal [3,3]-Sigmatropic Rearrangement of 2-Allyloxypyridines and Related Heterocycles. *Org. Lett.* **2010**, *12*, 260–263.
73. Overman, L. E.; Remarchuk, T. P. Catalytic Asymmetric Intramolecular Aminopalladation: Enantioselective Synthesis of Vinyl-Substituted 2-Oxazolidinones, 2-Imidazolidinones, and 2-Pyrrolidinones. *J. Am. Chem. Soc.* **2002**, *124*, 12–13.

74. Jiang, G.; Halder, R.; Fang, Y.; List, B. A Highly Enantioselective Overman Rearrangement Through Asymmetric Counteranion-Directed Palladium Catalysis. *Angew. Chem. Int. Ed.* **2011**, *50*, 9752–9755.
75. Cannon, J. S.; Frederich, J. H.; Overman, L. E. Palladacyclic Imidazoline-Naphthalene Complexes: Synthesis and Catalytic Performance in Pd(II)-Catalyzed Enantioselective Reactions of Allylic Trichloroacetimidates. *J. Org. Chem.* **2012**, 77, 1939–1951.
76. Jacobsen, E. N.; Pfaltz, A.; Yamamoto, H. In *Comprehensive Asymmetric Catalysis,* Jacobsen, E. N. Pfaltz, A. Yamamoto, H., Eds; Springer: New York, 1999.
77. Tang, W.; Zhang, X. New Chiral Phosphorus Ligands for Enantioselective Hydrogenation. *Chem. Rev.* **2003**, *103*, 3029–3070.
78. Pietrusiewicz, K. M.; Zablocka, M. Preparation of Scalemic P-chiral Phosphines and Their Derivatives. *Chem. Rev.* **1994**, *94*, 1375–1411.
79. Urig, S.; Fritz-Wolf, K.; Réau, R.; Herold-Mende, C.; Tóth, K.; Davioud-Charvet, E.; Becker, K. Undressing of Phosphine Gold(I) Complexes as Irreversible Inhibitors of Human Disulfide Reductases. *Angew. Chem. Int. Ed.* **2006**, *45*, 1881–1886.
80. Kagan, H. B.; Sasaki, M. *The Chemistry of Organophosphorus Compounds, ed;* F.R. Hartley, John Wiley and Sons: Chichester, 1990.
81. Wicht, D. K.; Kourkine, I. V.; Lew, B. M.; Nthenge, J. M.; Glueck, D. S. Platinum-Catalyzed Acrylonitrile Hydrophosphination Via Olefin Insertion into a Pt-P Bond. *J. Am. Chem. Soc.* **1997**, *119*, 5039–5040.
82. Wicht, D. K.; Kourkine, I. V.; Kovacik, I.; Glueck, D. S.; Concolino, T. E.; Yap, G. P. A.; Incarvito, C. D.; Rheingold, A. L. Platinum-Catalyzed Acrylonitrile Hydrophosphination. P – C Bond Formation via Olefin Insertion Into a Pt – P Bond. *Organometallics* **1999**, *18*, 5381–5394.
83. Kovacik, I.; Wicht, D. K.; Grewal, N. S.; Glueck, D. S.; Incarvito, C. D.; Guzei, I. A.; Rheingold, A. L. Pt(Me-duphos)-Catalyzed Asymmetric Hydrophosphination of Activated Olefins: Enantioselective Synthesis of Chiral Phosphines. *Organometallics* **2000**, *19*, 950–953.
84. Sadow, A. D.; Haller, I.; Fadini, L.; Togni, A. Nickel(II)-Catalyzed Highly Enantioselective Hydrophosphination of Methacrylonitrile. *J. Am. Chem. Soc.* **2004**, *126*, 14704–14705.
85. Sadow, A. D.; Togni, A. Enantioselective Addition of Secondary Phosphines to Methacrylonitrile: Catalysis and Mechanism. *J. Am. Chem. Soc.* **2005**, *127*, 17012–17024.
86. Scriban, C.; Kovacik, I.; Glueck, D. S. A Protic Additive Suppresses Formation ofByproducts in Platinum-Catalyzed Hydrophosphination of Activated Olefins. Evidence for P – C and C – C Bond Formation by Michael Addition. *Organometallics* **2005**, *24*, 4871–4874.
87. Huang, Y.; Pullarkat, S. A.; Li, Y.; Leung, P.-H. Palladium(II)-Catalyzed Asymmetric Hydrophosphination of Enones: Efficient Access to Chiral Tertiary Phosphines. *Chem. Commun.* **2010**, *46*, 6950–6952.
88. Chew, R. J.; Leung, P.-H. Our Odyssey With Functionalized Chiral Phosphines: From Optical Resolution to Asymmetric Synthesis to Catalysis. *Chem. Rec.* **2016**, *16*, 141–158.
89. Huang, Y.; Pullarkat, S. A.; Li, Y.; Leung, P.-H. Palladacycle-Catalyzed Asymmetric Hydrophosphination of Enones for Synthesis of C*- and P*-Chiral Tertiary Phosphines. *Inorg. Chem.* **2012**, *51*, 2533–2540.
90. Hao, X.-Q.; Huang, J.-J.; Wang, T.; Lv, J.; Gong, J.-F.; Song, M.-P. PCN Pincer Palladium(II) Complex Catalyzed Enantioselective Hydrophosphination of Enones: Synthesis of Pyridine-Functionalized Chiral Phosphine Oxides as $NC_{sp3}O$ Pincer Preligands. *J. Org. Chem.* **2014**, *79*, 9512–9530.
91. Huang, Y.; Pullarkat, S. A.; Teong, S.; Chew, R. J.; Li, Y.; Leung, P.-H. Palladacycle-Catalyzed Asymmetric Intermolecular Construction of Chiral Tertiary P-Heterocycles by Stepwise Addition of H – P – H Bonds to bis(enones). *Organometallics* **2012**, *31*, 4871–4875.

92. Huang, Y.; Chew, R. J.; Pullarkat, S. A.; Li, Y.; Leung, P.-H. Asymmetric Synthesis of Enaminophosphines Via Palladacycle-Catalyzed Addition of Ph_2PH to α,β-Unsaturated Imines. *J. Org. Chem.* **2012**, *77*, 6849–6854.
93. Chew, R. J.; Teo, K. Y.; Huang, Y.; Li, B.-B.; Li, Y.; Pullarkat, S. A.; Leung, P.-H. Enantioselective Phospha-Michael Addition of Diarylphosphines to β,γ-Unsaturated α-Ketoesters and Amides. *Chem. Commun.* **2014**, *50*, 8768–8770.
94. Chew, R. J.; Lu, Y.; Jia, Y.-X.; Li, B.-B.; Wong, E. H. Y.; Goh, R.; Li, Y.; Huang, Y.; Pullarkat, S. A.; Leung, P.-H. Palladacycle Catalyzed Asymmetric P−H Addition of Diarylphosphines to N-enoyl Phthalimides. *Chem. Eur. J.* **2014**, *20*, 14514–14517.
95. Chew, R. J.; Li, X.-R.; Li, Y.; Pullarkat, S. A.; Leung, P.-H. Pd-catalyzed Enantiodivergent and Regiospecific Phospha-Michael Addition of Diphenylphosphine to 4-Oxo-enamides: Efficient Access to Chiral Phosphinocarboxamides and Their Analogues. *Chem. Eur. J.* **2015**, *21*, 4800–4804.
96. Yang, X.-Y.; Tay, W. S.; Li, Y.; Pullarkat, S. A.; Leung, P.-H. The Synthesis and Efficient One-pot Catalytic "Self-Breeding" of Asymmetrical NC(sp^3)E-hybridised Pincer Complexes. *Chem. Commun.* **2016**, *52*, 4211–4214.
97. Jia, Y.-X.; Jonathan Chew, R.; Li, B.-B.; Zhu, P.; Li, Y.; Pullarkat, S. A.; Soon Tan, N.; Leung, P.-H. Palladacycle Promoted Base Controlled Regio- and Enantioselective Hydrophosphination of 2-Pyridylacrylate/Amide and the Cytotoxicity of Their Gold Complexes. *Dalton Trans.* **2015**, *44*, 17557–17564.
98. Xu, C.; Jun Hao Kennard, G.; Hennersdorf, F.; Li, Y.; Pullarkat, S. A.; Leung, P.-H. Chiral Phosphapalladacycles as Efficient Catalysts for the Asymmetric Hydrophosphination of Substituted Methylidenemalonate Esters: Direct Access to Functionalized Tertiary Chiral Phosphines. *Organometallics* **2012**, *31*, 3022–3026.
99. Chew, R. J.; Huang, Y.; Li, Y.; Pullarkat, S. A.; Leung, P.-H. Enantioselective Addition of Diphenylphosphine to 3-Methyl-4-nitro-5-alkenylisoxazoles. *Adv. Synth. Catal.* **2013**, *355*, 1403–1408.

CHAPTER 8

Nanoparticular or Colloidal Pathways for Palladacycles-Mediated Catalytic Processes

Jagrut A. Shah, Bhalchandra M. Bhanage
Department of Chemistry, Institute of Chemical Technology, Mumbai, India

Metal nanoparticles (NPs) or colloids have gained considerable attention in the area of molecular homogeneous catalysis or bulk metal/supports-based heterogeneous catalysts *(1–4)*. The interest in these particles is due their small size as well as their high proportion of surface atoms which leads to number of active sites. Their distinctive electronic properties are on the midst or borders of molecular as well as metallic states *(5)*. Furthermore, the surface properties of these materials can be enhanced by supplementing a stabilizer or by employing a suitable support. The nanoparticles in the range of one to a few nanometers in diameter (i.e., those containing a few tens to a few hundred atoms) and usually stabilized by solvent are the most catalytically active particles. Some nanoparticular systems have been applied in catalysis, not only in hydrogenation reactions, but also in the dehydrogenation of amine-boranes, C-C coupling and hydroformylation reactions *(2)*.

Palladacycles are efficient and versatile catalysts for various coupling reactions and well known to give very high turnover numbers. Palladacycles are also considered efficient precatalysts in many organic reactions which act as a Pd(0) NPs reservoir after their initial self-catalytic role. The early reports on the presence of PdNPs/Pd-colloids in palladacycle chemistry brought about an outbreak of activity to bring about advancement of palladacycles and related complexes which could act as reservoirs of highly active PdNPs or Pd-colloids. Notable work amongst them are the reports by Reetz *(6)*, Gladysz *(7, 8)*, DuPont *(9, 10)*, Vries *(11)*, Najera *(12)*, etc. and these studies have confirmed the aforementioned nanoparticular involvement by TEM characterization of Pd-colloids.

The generation of Pd-nanoparticles or colloids in situ was mainly observed and reported using Mizoroki-Heck Cross Coupling (MHCC)

Palladacycles
https://doi.org/10.1016/B978-0-12-815505-9.00009-3
© 2019 Elsevier Inc.
All rights reserved.

reaction as MHCC is often seen as the foundation stone for developing and studying general palladacycle catalyzed reactions. However, a variety of other cross-coupling reactions like Suzuki, Sonogashira, Hiyama, Stille, etc. also followed using palladacycles and the subsequently formed PdNPs or colloids were detected by a variety of techniques, from simpler ones like Hg-Drop Test or more sophisticated methods like X-ray photoelectron spectroscopy (XPS), X-ray diffraction (XRD), transmission electron microscopy (TEM), high resolution-transmission electron microscopy (HR-TEM), etc. All these reactions have been discussed by Didier Astruc in his book *(2)*.

Likewise, various aspects regarding PdNPs and colloids derived from - oxime-based palladacycles have also been discussed by Najera *(13a, b)*, from Pd pincer complexes by Selander and Szabo *(14)*, "homeopathic" PdNPs by Astruc *(15)*, organochalcogen ligands by Singh *(16)*, and most recently, the role of PdNPs in C-C reactions by Augustyniak *(17)* in reviews and perspectives. Hence, later in this book chapter, we discuss developments made in this field and include a detailed discussion of recent reports mostly after 2010 until January 2019.

In 2010, Singh and co-workers *(18)* synthesized new air and moisture stable palladacycles [$PdCl(\{Ph(2\text{-}HOC_6H_4)CHNH(CH_2)_3SePh\}\text{-}H)$] (Fig. 8.1). The synthesized palladacycle exhibited high catalytic activity for Suzuki Miyaura cross-coupling (SMCC) of phenyl boronic acid and bromo and chloroarenes and heteroarenes with TON by chlorides being as high as 9200. The palladacycle complex was converted to a size of ~8 nm NPs of $Pd_{17}Se_{15}$ which could be active catalyst. They showed that on reacting 4-chloronitrobenzene and 4-chloroanisole with phenyl boronic acid in the presence of $Pd_{17}Se_{15}$, the Carbon-Carbon coupling products were not obtained in good yields. They suggest that the agglomeration decreases the catalytic efficiency of $Pd_{17}Se_{15}$NPs in contrast to those generated in situ during the Carbon-Carbon coupling reaction, which converted chloroarenes more competently. Overall this supports their proposition that

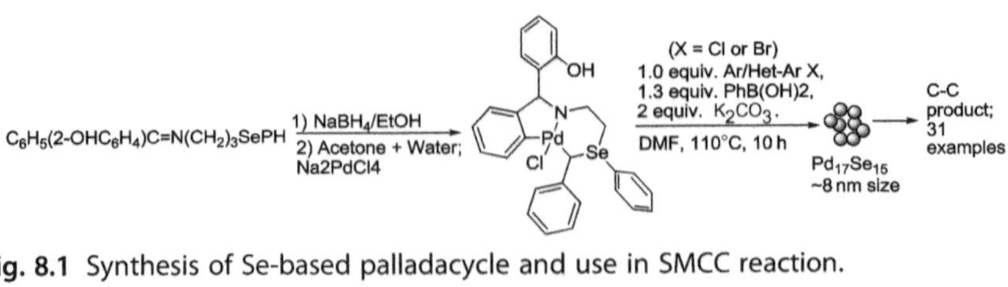

Fig. 8.1 Synthesis of Se-based palladacycle and use in SMCC reaction.

the palladacycle rather acts as a precatalyst which generates nano size $Pd_{17}Se_{15}$.

Continuing on utilizing palladacycles for C-C reactions, Gebbink and co-workers *(19)* in 2010 synthesized novel bis-phosphite palladium complex, palladium complex based on C,P-chelate bonded monophosphite, and PCP-pincer palladium complex. They utilized this pincer Pd complex in diphenyl phosphine borane and a variety of aryl iodides utilizing mild conditions. After carrying out kinetic studies, they proposed a mere precatalyst role of pincer complex and the PdNPs generated were the real or active catalytic species. Although they did use cis-$PdCl_2((o\text{-}Tol)_3P)_2$ and its C-C activity was better, nonetheless, the group continued with the use of the Pincer ligand for the rest of the substrate scope and by conducting kinetic studies, proposed formation of nanoparticles in the reaction conditions and actually catalyzing the reaction (Figs. 8.2 and 8.3).

Najera and co-workers *(20)* in 2011 published their continually developing work on oxime-based palladacycles. They carried out Matsuda-Heck reactions synthesizing close to 20 compounds using both traditionally used $Pd(OAc)_2$ as well as -hydroxyacetophenone oxime-derived palladacycle as a source of PdNPs with extremely low loading of 1 mol% at RT in MeOH without use of an external base. They conclude that the activity of their

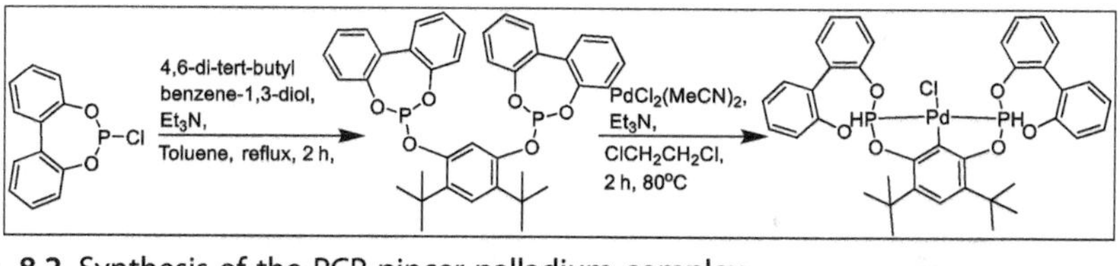

Fig. 8.2 Synthesis of the PCP pincer palladium complex.

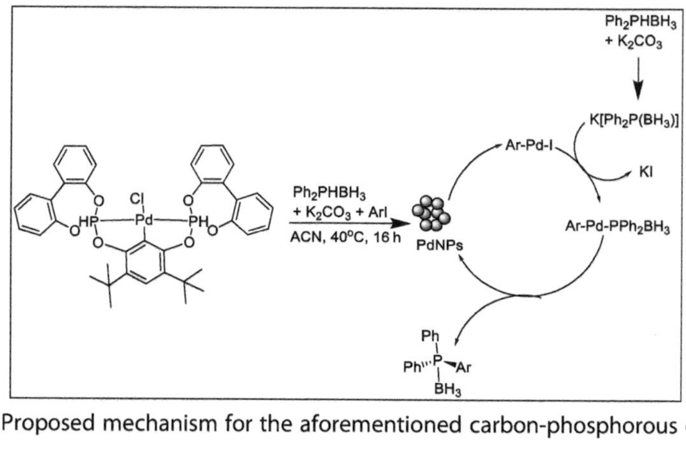

Fig. 8.3 Proposed mechanism for the aforementioned carbon-phosphorous coupling.

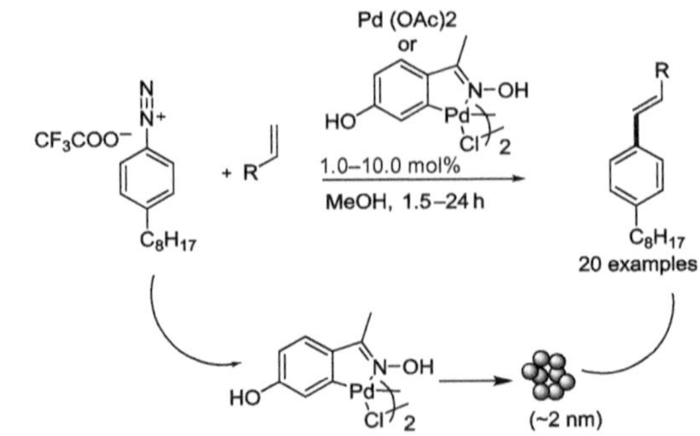

Fig. 8.4 Use of oxime based palladacycle in Matsuda-Heck coupling.

palladacycle highly depends on the type of alkene. They could identify intermediate formed due to the oxidative addition of diazonium salts to the catalyst but not insertion of alkene to metal. TEM images show the presence of PdNPs (~2 nm in diameter in case of the palladacycle) in both the catalysts (for the first time in Matsuda-Heck coupling) (Fig. 8.4).

Ajai Singh and co-workers truly burgeoned the discipline with the synthesis of a series of palladacycles of thioethers- based ligands, S, Se, Te containing amines, pincer ligands as well as 2,3-bis[(phenylthio/-seleno)methyl] quinoxaline *(21–23)* to catalyze SMCC reactions of a variety of boronic acids and aryl/hetero aryl halides. All the palladacycles degraded to give PdNPs of various sizes which were characterized by TEM, XRD, EDS, XPS, etc. They have also carried out experiments and also show that PdNPs worked independently as well with the same or higher activities (Fig. 8.5).

Bergbreiter and co-workers, in 2013, were amongst the very first to have synthesized palladacycle precatalysts based on azo-dyes *(24)* to carry out HMCC and SMCC (yields close to 90%) reactions. They synthesized polyisobutylene (PIB)-bound azo dyes from aryl amines terminated PIB-oligomers. They also employed thermomorphic solutions of heptane-DMF which are monophasic at 80°C and biphasic at RT. This way they were able to recover the catalyst, i.e., Pd colloids formed in situ and they were further reused (~5 cycles without loss of activity). However, some Pd were used to leach into the product solvent layer. They overcame this problem by using a low melting PE (polyethylene) oligomer as a solvent. They were able to reduce the leaching by roughly a magnitude of order 10 as the colloids were dispersed in the polywax (no significant change in yields in this method as well as recyclability). Use of low melting PE oligomer also allowed them to modify the palladacycle precatalyst system by using a PIB bound phosphine ligand. This tuning of the system not only increased the activity and scope

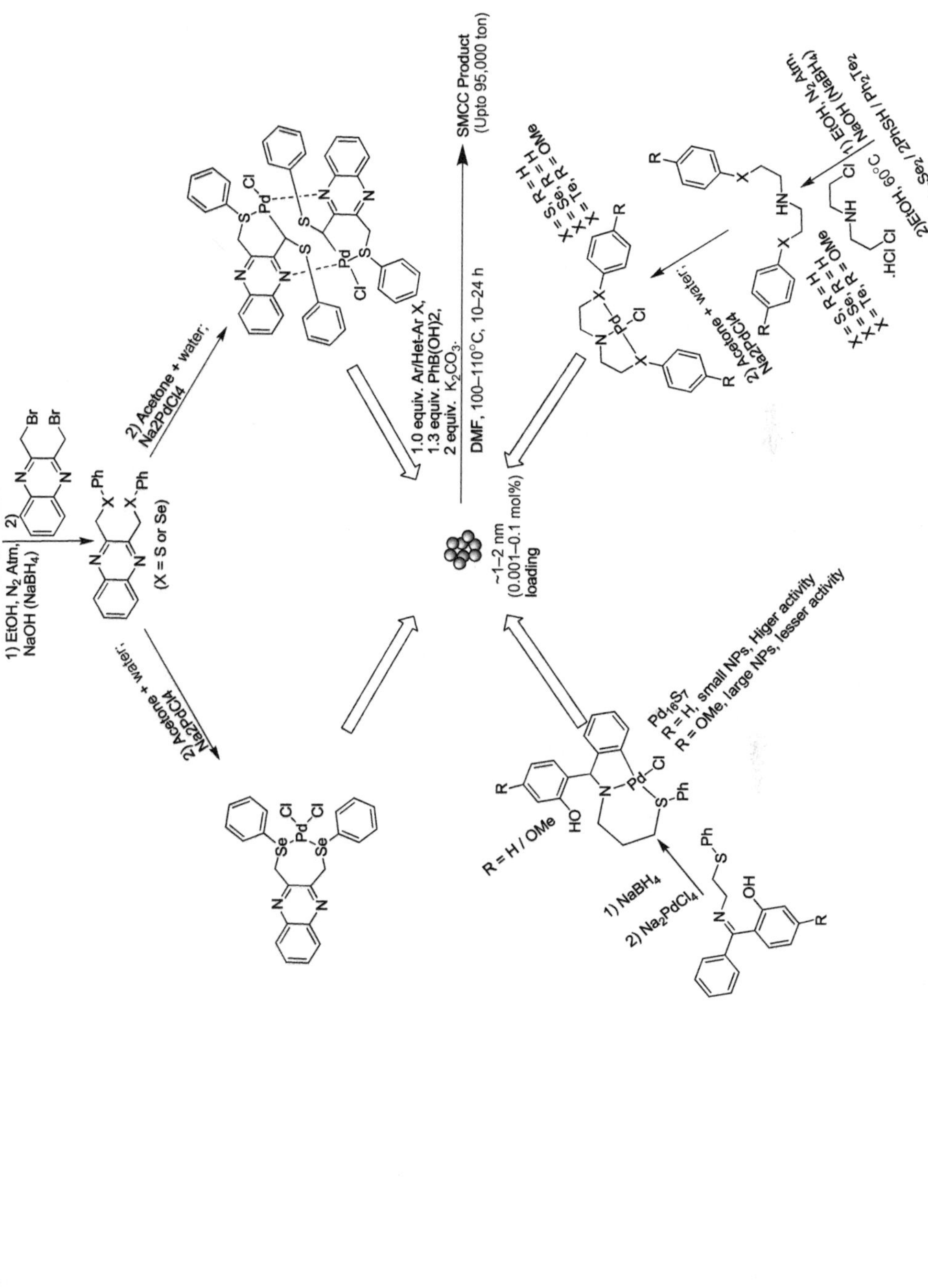

Fig. 8.5 Synthesis of various palladacycles and pincer complexes and subsequent degradation to PdNPs.

of the catalytic system (extended to allylic amination) but also reduced the Pd leaching by 2 orders and they were able to recover almost 99.88% of Pd thereby providing a very sustainable use of Pd colloids in CC reactions (Fig. 8.6).

In the same year, Tamami and co-workers *(25)* synthesized a novel catalytic system of bisphosphinite-based PCP-pincer palladium complex adhered on Merrifield resin. The presence of highly active Pd(0) NPs was proved by XPS and XRD. TEM image of the system exhibited that PdNPs are well distributed through the support in the range of 30–50 nm. This system showed excellent catalytic activities in SMCC, MHCC and SHCC (Sonogashira—Hagihara CC) with different halo arenes including inactive chloroarenes in a short time and reused up to 10 times with no considerable loss in activity. Hot filtration test testified low leaching of the metal into solution from the supported catalyst which showed the catalyst to have a heterogenous activity.

Karami and co-workers *(26)* in 2014 for the first time synthesized PdNPs with narrow size distribution range using PEG (1000 and 15,000) and amine NC palladacycle without using any other chemicals. The Pd-PEG catalysts were characterized completely using XRD, TEM, UV-Vis, FT-IR, etc. The PdNPs of PEG 15,000 showed a very well defined geometry like triangular and rhombohedral of size 2–12 nm. They also carried out MHCC (~12 examples) using both the systems i.e., 1000 PEG-Pd and 15,000 PEG-Pd and found PEG 15,000-Pd to be a better system (yield wise as well as recyclability and Pd leaching wise) which is also shown by the particle size of ~13.2 nm compared to 14.8 nm of PEG 1000 Pd system. The formation of PdNPs in spite of absence of any reducing agent being used shows that the hydroxyl group of PEG gets oxidized to the formyl group which was also confirmed by IR studies.

Continuing their work on degradation of palladacycles to synthesize PdNPs, the group also reported synthesis of an oxime-based palladacyclic complex joisted on Fe_3O_4/oleic acid *(27)*. This catalytic system proved to be an excellent choice for Cu-free SHCC reactions with various halo arenes and heteroarenes with phenylacetylene in air and in EtOH as solvent or varied aqueous mediums. As expected, the reaction was catalyzed by highly active PdNPs with diameters from 5 to 10 nm with an average 9.97 nm. The coupling products were obtained in high yields. There was low loading of Pd and the heterogeneous catalyst can be separated by an external magnet and reused 6 times with no loss of its activity. The catalyst was characterized by XRD, SEM and TEM.

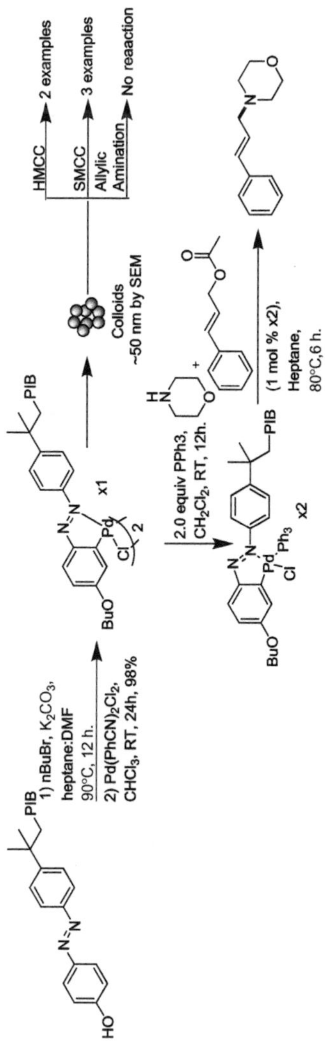

Fig. 8.6 Synthesis of palladacycle based on Azo-compound and its utility in HMCC, SMCC and allylic amination.

In the same year, Holdt and coworkers *(28)* synthesized new *N*-heterocyclic carbene (NHC) complex [$PdCl_2${4,5-dicyano-1,3-dimesitylimidazol-2-ylidene}(PPh_3)] **(a)**) and the NHC palladacycle [PdCl(*N,N*-dimethylbenzylamine){4,5-dicyano-1,3-dimesitylimidazo-l-2-ylidene}] **(b)** by thermolysis of 4,5-dicyano-1,3-dimesityl-2-(pentafluorophenyl)imidazoline in the presence of appropriate palladium(II) precursors. Also a new NHC palladacycle [$PdCl_2${(1-benzyl-4,5-dicyano-3-picolylimidazol-2-ylidene}] **(c)** was prepared by in situ thermolysis of 1-benzyl-4,5-dicyano-2-(pentafluorophenyl)-3-picolylimidazoline in the presence of [$PdCl_2(PhCN)_2$]. Both the palladacycles along with the NHC complex were used to test their catalytic activity in MHCC. They showed high activity with good to great yields and catalyst loading of as low as 0.1 mol%. The nature of the active species was investigated using poisoning tests as well as TEM of the reaction mass. The tests clearly showed the presence and active role of PdNPs (Fig. 8.7).

Ajai Singh and co-workers, in 2015 *(29)*, continued their work on synthesis of newer palladacycles and carrying SMCC reactions (~14 examples). They synthesized palladacycles and Pd(II) complexes of 2-thiophenemethylaminebased Schiff bases. All of the 5 systems containing the same type ligand showed homogenous pathways. However, three of

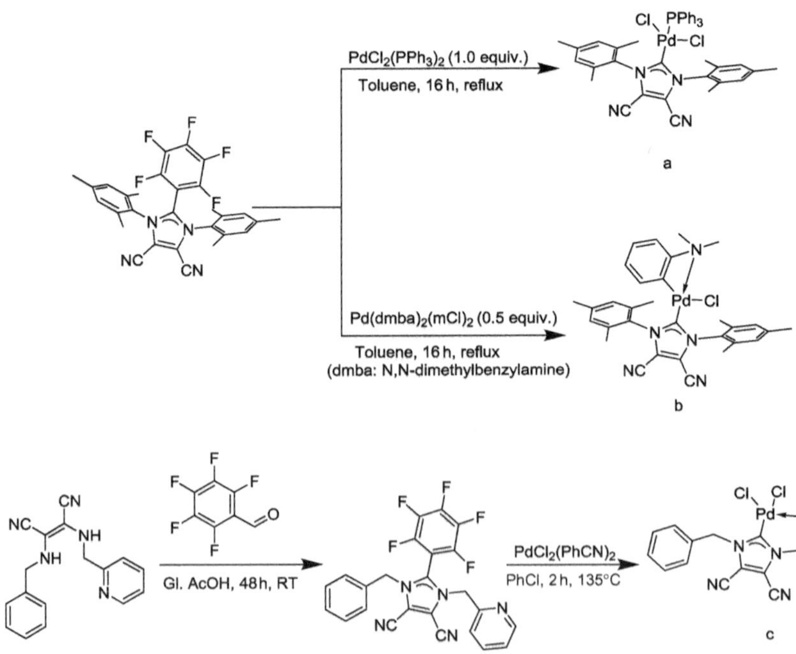

Fig. 8.7 NHC-based palladacycle synthesized by Holdt and co-workers.

them were the catalysts themselves and two palladacycles were precatalysts to give PdNPs of size approximately 2–4 nm mainly being 2.5 nm. The two-phase test carried out by using **d** indicates that the catalytic process is homogeneous, which might be occurring through leaching of Pd(0) from these PdNPs. The catalytic activity of the palladacycles supercedes the Pd complexes by close to 10-fold as the yields obtained by 0.01 mol% loading of the complex were obtained by 0.001 mol% loading of the palladacycle precatalysts. These results are expected as the palladacycles follow the nano particular pathway (Fig. 8.8).

Palladacycles and palladium complexes synthesized using the same ligand coordinating as N,C- and N,O-ligands were discovered to follow distinctive pathways. Thus a direct assessment of contrast of palladacycle with Pd(II) complexes with lesser ligand factors was presented for the first time, which indicated the superiority of the former on the basis of experiments carried out on palladacycle or Pd(II) complexes alone and not taking into consideration differences in ligands.

Cadierno and co-workers *(30)* in the same year also synthesized oxime-based Palladacycle using $PdCl_2$ (1,5 cyclooctadiene) and 1 and 2 equivalents of 2-(diphenylphosphino)benzaldehyde oxime to give **i** and **j** respectively. These compounds showed catalytic prowess in rearrangement reactions of aldoximes to amides in high yields. The reaction was carried in water at 100°C and was supposedly a clean reaction without the use of any co-catalyst. The reaction could also be tuned for nitrile formation by using acetonitrile as solvent. They claim PdNPs to be the active species involved (Fig. 8.9).

Beletskaya and co-workers *(31)*, in early 2015, extensively studied MHCC and SMCC of electronically rich aryl iodides in a catalytic system

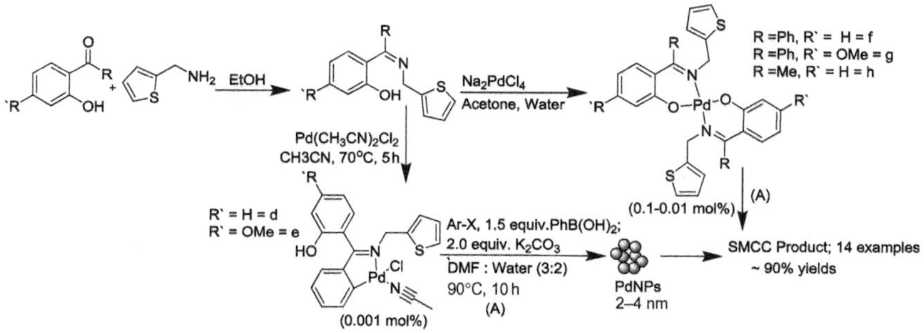

Fig. 8.8 Synthesis of imine-based ligand and its corresponding palladacycles utilizing different Pd salts thereby giving rise to two different pathways to catalyze subsequent SMCC.

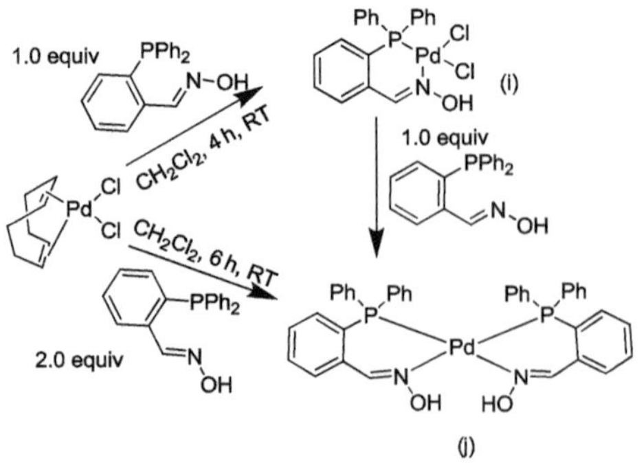

Fig. 8.9 Synthesis of diphenyl phosphino based palladacycles.

free of phosphines and related phosphorous salts. They utilized an extensive range of organo selenium chelators with diverse stability and structure. They employed Hg drop test not only to study the nature of the catalytic pathway but also to understand the role of Hg or lack thereof to establish the critical part of NPs in phosphine free reactions. Their work concludes that there are three stages of the catalytic systems. The first is the slow induction stage—Mononuclear stage characterized by slow release of Pd(O) species from the precatalyst. They are now engaged in nucleation establishment. The second is the fast (or the nanoparticle-enabled) stage. The NPs are formed and forage practically all Pd(0) released up to this time. Lastly is the third decay stage wherein the quality of nanoparticles continuously deteriorates, their number is decreased and size distribution shifts to polydisperse with larger pieces predominating because of the Ostwald ripening.

Karami and Naeini, in 2015 *(32)*, synthesized a new NC palladacycle and supported it on cucurbit *(6)* uril, CB *(6)*, to carry out SMCC (~25 examples) in H_2O-EtOH at RT as well as 40°C. The subsequently obtained NPs showed excellent activity as well as stability confirming it to be a heterogenous catalytic system. Their characterization was done using XRD, TEM and SEM. The recyclability of the catalyst was up to the mark till the 5th cycle yielding 90% of the product. The turnover frequency of 0.055 mol% catalyst was as high as 3920 h^{-1} and TON as high as 2000 (Fig. 8.10).

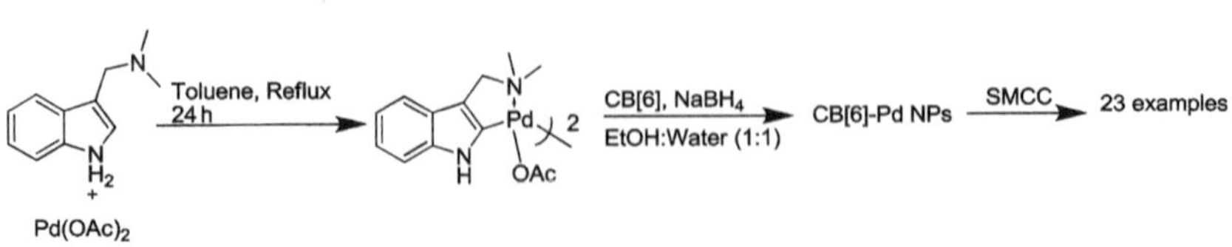

Fig. 8.10 Synthesis of CB *(6)* based Pd complex and use in SMCC.

Nishioka and co-workers, in 2016 *(33)*, synthesized pincer ligands containing Pd complexes. The pincer ligands, which contained one pyridine and two NHC units with acetyl-protected D-glucopyranosyl groups in C-C-N and C-N-C arrangements, were synthesized. Acetyl groups in the ligands were deprotected at RT by K_2CO_3 to afford water soluble complexes with one of the hydroxide groups in the D-glucopyranosyl groups coordinated to the metal ion. The complexes catalyzed SMCC reactions which were carried out in water with TON of 75,000 for the C-C-N and 8900 for C-N-C complexes. C-C-N complex was not "turned off" by addition of mercury suggesting it itself to be the active catalyst. C-N-C, however, was deactivated upon addition of Hg suggesting otherwise and PdNPs to be the probable active species.

Tamami and co-workers *(34)*, continuing their work from 2013, utilized the same modified Merrifield resin-supported PCP pincer palladacycle as precatalyst of PdNPs as a new polymer based catalytic system for cyanation of iodoarenes, thereby advancing their catalytic system.

Holdt and co-workers *(35)* synthesized the phenylidenepyridine (*ppy*) palladacycles [PdCl(*ppy*)(1,3-bis(mesityl)imidazol-2-ylidene (**k**) and [PdCl(*ppy*){4,5-dicyano-1,3-bis(mesityl)imidazol-2-ylidene] (**l**) were prepared. Catalytic experiments using the HMCC reaction of chloroarenes with *n*-butyl acrylate showed that **l** is the superlative performer when compared to **k**. Hence utilizing **l**, 26 substrates were synthesized with high yields. PdNPs were detected to be catalytically active species by Hg poisoning experiments (Fig. 8.11).

In 2016, Hedin and co-workers synthesized a new cyclopalladated porous polymer (cyclo-Pd(II)/PP-2) with up to 20.7 wt% of Pd. It was investigated as a heterogenous catalytic system for carrying out SMCC and HMCC with yields of nearly 90% at 100°C. The palladacycles were formed along the backbone of the azo-linked porous polymer PP-2 with (Pd-N) and (Pd-C) bonds. The palladacycle degraded in Suzuki reaction conditions giving an acyclic Pd/PP-2 and PdNPs bound to PP-2 with size of ~10 nm investigated using HAADF-STEM and TEM.

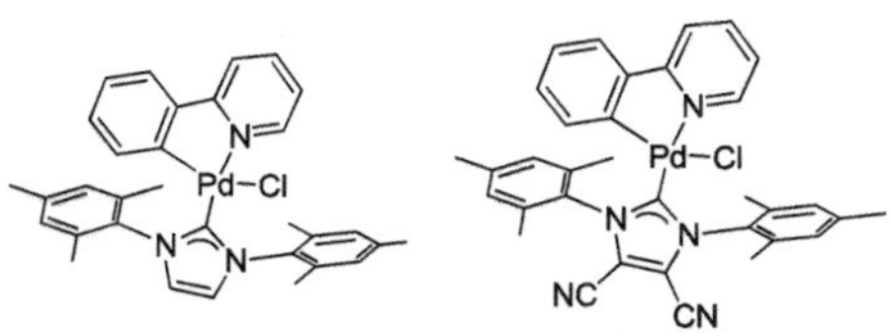

k l

Fig. 8.11 Palladacycles synthesized by Holdt and co-workers.

In 2017, Alonso and co-workers *(36)*, continued their work on oxime-based palladacycles as efficient precatalysts to catalyze SMCC and reported a thorough investigation on the synthesis and characterization of new oxime palladacycle-graphene oxide along with its catalytic activity in the aforementioned SMCC reaction. The catalyst, 1-GO, was fully characterized by Inductively coupled plasma mass spectrometry (ICP-MS), XPS, Thermal Gravimetry Analysis (TGA), and UV-vis analyses and has displayed to be a proficient catalyst for the SMCC between bromo arenes and arylboronic acids and uses very low catalyst loadings (0.002 mol% of Pd) at RT in water as solvent (Fig. 8.12).

Kapdi and co-workers showed that active Pd colloids can be formed upon degradation of a palladacyclic complex (Herrmann-Beller, HB). The Pd colloids have been isolated for the first time and systematically characterized with techniques such as TEM, HR-TEM, XPS, and extended X-ray absorption fine structure spectroscopy. The synthesized palladium colloids, which were detected when the group was carrying out di-aryl methanol synthesis between benzaldehydes and phenyl boronic acids, have been applied for proficient catalysis for the oxidative homocoupling of aryl boronic acids for the first time *(37)* (Figs. 8.13 and 8.14).

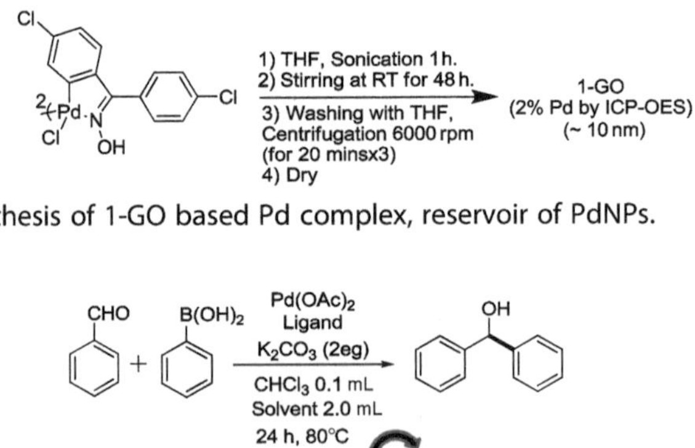

Fig. 8.12 Synthesis of 1-GO based Pd complex, reservoir of PdNPs.

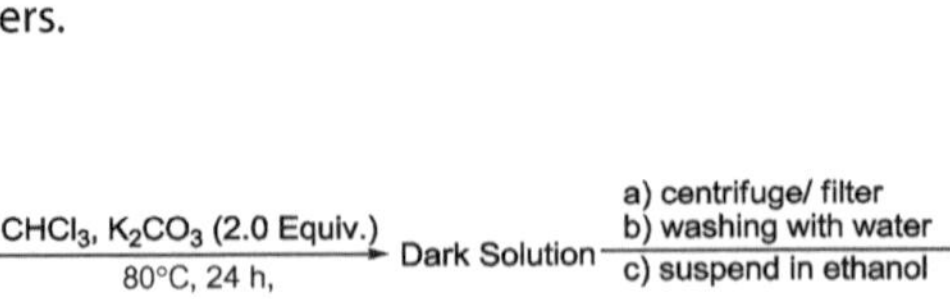

Fig. 8.13 Discovery of Pd colloidal mass while carrying out Diaryl methanol synthesis by Kapdi and co-workers.

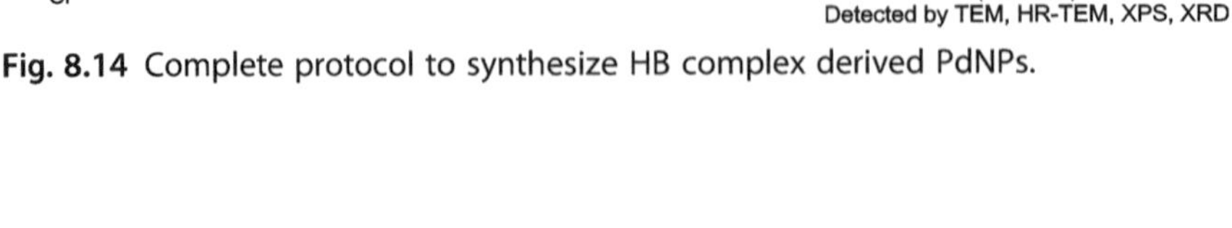

Fig. 8.14 Complete protocol to synthesize HB complex derived PdNPs.

Continuing their work on degradation of HB complex, they fine tuned a new protocol to show a highly selective oxidative heterocoupling of aryl or heteroaryl boronic acids with an active palladium colloidal catalyst. The assorted choice of electronically different aryl boronic acids helped them carry out the synthesis under mild conditions, with the oxidant being air, while extending the substrate scope. This successful approach further allowed the development of a unique one-pot telescoping oxidative heterocoupling/Suzuki-Miyaura cross-coupling tandem process for obtaining substituted terphenyls *(38)* (Fig. 8.15).

In 2017, Ajai Singh and co-workers *(39)* continued their work on ligands based on Chalcogenated Schiff' base (this time based on 1-Naphthaldehyde) and their reduced form to synthesize 4 palladacycles (m,n,o,p) and checked its utility in Cu and amine free SHCC and SMCC. Out of the 4, **m** and **n** had more efficiency to catalyze Cu and amine free SHCC and SMCC reactions. The yields using all the 4 palladacycles were very high 96%–84% and the system showed recyclability up to 8 cycles for both the reactions. The PdNPs were isolated as expected from the reaction mass indicating their active role in the reactions and were characterized by HR-TEM with sizes being **m** and **p** at 10 nm and **n** and **o** at 20 nm (Fig. 8.16).

It was known now that PdNPs can be stabilized onto the surface of Fe_3O_4. In 2018, Zohreh *(40)* and co-workers developed an applicable protocol to oxidize benzyl alcohols and benzoic acids. For this purpose, they stabilized PdNPs onto the surface of Fe_3O_4-entrapped poly(aminoethyl acrylamide) via polymer functionalization with a new class of N,N,N-pincer

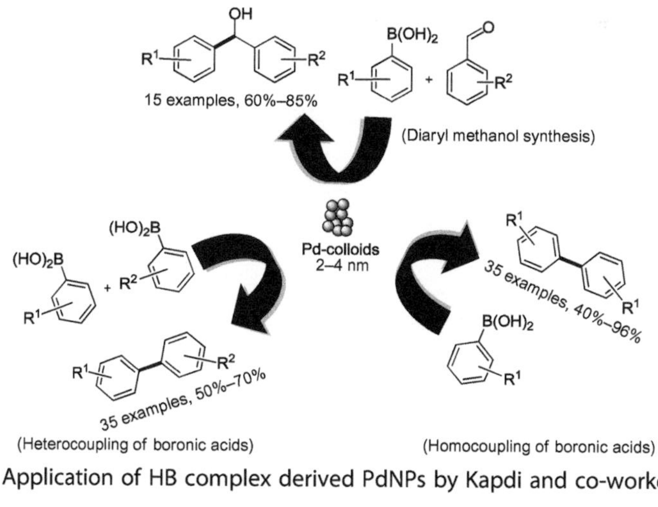

Fig. 8.15 Application of HB complex derived PdNPs by Kapdi and co-workers.

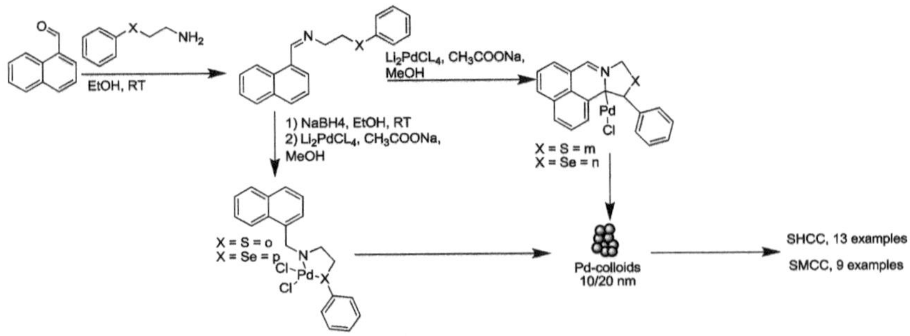

Fig. 8.16 Synthesis of Schiff Based ligand-palladacycle and subsequent application on SHCC and SMCC.

ligands. This provided a high loading or scale up of substrate and stable catalyst for the oxidation of benzyl alcohols. They loaded the catalyst at 0.95 mol% with recyclability up to 5 runs.

Most recently, Bhanage and coworkers *(41)* documented the first ever palladium pincer complex-catalyzed carbonylative SHCC and carbonylative SMCC. They employed their complexes with loading as low as 10^{-4} mol% and 10^{-6} mol% for Carbonylative SMCC and Carbonylative SHCC which in turn resulted in catalytic turnovers of 10^7 and 10^5, respectively. They showed presence of PdNPs (~6.5 nm) formation which were characterized by HR-TEM, XPS, etc. and being the actual catalytic species with the pincer complex being a precatalyst and a reservoir of the PdNPs. Hence, it can be concluded that both the carbonylative cross-coupling reactions proceed through the traditional Pd(0)-Pd(II) pathway (Fig. 8.17).

In conclusion, we have provided a detailed discussion about nanoparticular or colloidal involvement in reactions catalyzed by palladacycles and related pincer ligands. The reactions involving palladacycles are catalyzed by the later generated PdNPs. The fundamental comprehension of the

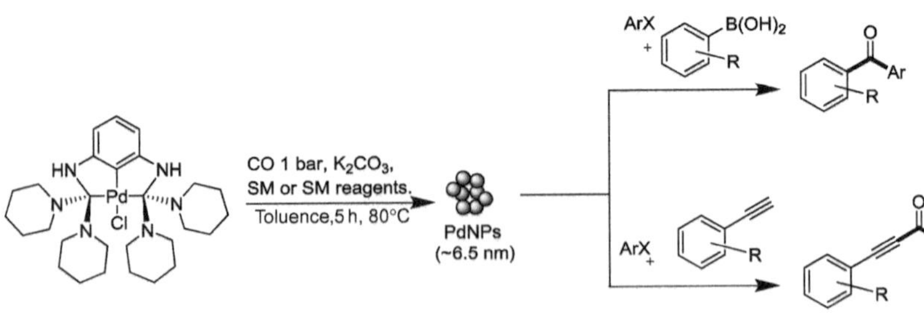

Fig. 8.17 Degradation of pincer complex based Palladacycle in CO atmosphere to give PdNPs and its first ever subsequent use in carbonylative SMCC and SHCC.

accurate role of nanoparticles in Pd-catalyzed reactions is still far from being established to a generally adopted theory. Likewise, the kinetic as well as thermodynamic backing for PdNPs formation from palladacycles have not been studied thoroughly, although, Spencer and Cox *(42)* have taken carried DFT experiments and calculation for validation of the mechanism of in situ Pd(0) formation for cross-coupling reactions from new unsymmetrical pincer palladacycles. Apart from a few generally known conditions like temperatures, solvents, co-ordinating metals, or CO atmosphere, which facilitates NP formation, a solid and fundamental theory (with experimental proof) discussing the actual formation of NPs and establishing some general points involving degradation of all types of palladacycles or related pincer ligands is lacking and is therefore open for further research.

REFERENCES

1. Astruc, D.; Lu, F.; Ruiz Aranzaes, R. J. *Angew. Chem. Int. Ed.* **2005**, *44*, 7852–7872.
2. Astruc, D. *Nanoparticles and Catalysis;* Wiley-VCH: Weinheim, 2008.
3. Thomas, J. M. *Chem. Cat. Chem.* **2010**, *2*, 127–132.
4. Li, Z. X.; Xue, W. B.; Guan, T.; Shi, F. B.; Shi, Z. J.; Jiang, H.; Yan, C. H. *Nanoscale* **2013**, *5*, 1213–1220.
5. Schmid, G. *Nanoparticles: From Theory to Application;* Wiley-VCH: Weinheim, 2012.
6. Reetz, M. T.; Westermann, E. *Angew. Chem. Int. Ed.* **2000**, *39* (1), 165–168.
7. Rocaboy, C.; Gladysz, J. A. Org. Lett. **2002**, *4* (12), 1993–1996.
8. Rocaboy, C.; Gladysz, J. A. New J. Chem. **2003**, *27*, 39–497.
9. Consorti, C. S.; Flores, F. R.; Dupont, J. J. Am. Chem. Soc. **2005**, *127*, 12054–12065.
10. Dupont, J.; Consorti, C. S.; Spencer, J. Chem. Rev. **2005**, *105*, 2527–2571.
11. de Vries, J. G. Dalton Trans. **2006**, 421–429.
12. Alacid, E.; Najera, C. Adv. Synth. Catal. **2007**, *349*, 2572.
13. (a)Alonso, D. A.; Najera, C. Chem. Soc. Rev. **2010**, *39*, 2891–2902; (b)Najera, C. *ChemCatChem* **2016**, *8*, 1865–1881.
14. Selander, N.; Szabo, K. *J.* Chem. Rev. **2011**, *111*, 2048–2076.
15. Deraedt, C.; Astruc, D. Acc. Chem. Res. **2013**, *42* (2), 494–503.
16. Kumar, A.; Rao, G. K.; Singh, A. K. RSC Adv. **2012**, *2*, 12552–12574.
17. Trzeciak, A. M.; Augustyniak, A. W. Coord. Chem. Rev. **2019**, *384*, 1–20.
18. Rao, G. K.; Kumar, A.; Ahmed, J.; Singh, A. K. Chem. Commun. **2010**, *46*, 5954–5956.
19. Li, J.; Lutz, M.; Spek, A. L.; van Klink, G. P. M.; van Koten, G.; Gebbink, R. J. M. K. J. Organomet. Chem. **2010**, *695*, 2618–2628.
20. Salabert, J.; Sebastian, R. M.; Vallribera, A.; Roglans, A.; Najera, C. *Tetrahedron* **2011**, *67*, 8659–8664.
21. Kumar, S.; Rao, G. K.; Kumar, A.; Singh, M. P.; Singh, A. K. Dalton Trans. **2013**, *42*, 16939–16948.
22. Rao, G. K.; Kumar, A.; Kumar, S.; Dupare, U. B.; Singh, A. K. *Organometallics* **2013**, *32*, 2452–2458.
23. Saleem, F.; Kumar Rao, G. K.; Singh, P.; Singh, A. K. *Organometallics* **2013**, *32*, 387–395.
24. Priyadarshani, N.; Suriboot, J.; Bergbreiter, D. E. Green Chem. **2013**, *15*, 1361–1367.

25. Tamami, B.; Nezhad, M. M.; Ghasemi, S.; Farjadian, F. J. Organomet. Chem. **2013**, *743*, 10–16.
26. Karami, K.; Moghadam, Z. K.; Hosseini-Kharat, M. Catal. Commun. **2014**, *43*, 25–28.
27. Karami, K.; Najvani, S. D.; Naeini, N. H.; Pablo Hervés, P. Chin. J. Catal. **2015**, *36*, 1047–1053.
28. Baier, H.; Metzner, P.; Körzdörfer, T.; Kelling, A.; Holdt, H.-J. Eur. J. Inorg. Chem. **2014**, 2952–2960.
29. Joshi, H.; Prakash, O.; Sharma, A. K.; Sharma, K. N.; Sing, A. K. Eur. J. Inorg. Chem. **2015**, 1542–1551.
30. Menéndez-Rodríguez, L.; Tomás-Mendivil, E.; Francos, J.; Najera, C.; Crochet, P.; Cadierno, V. Catal. Sci. Technol. **2015**, *5*, 3754–3761.
31. Sigeev, A. S.; Peregudov, A. S.; Cheprakov, A. V.; Beletskayaa, I. P. Adv. Synth. Catal. **2015**, *357*, 417–429.
32. Karami, K.; Naeini, H. Appl. Organometal. Chem. **2015**, *29*, 33–39.
33. Imanaka, Y.; Shiomoto, N.; Tamaki, M.; Maeda, Y.; Nakajima, H.; Nishioka, T. Bull. Chem. Soc. Jpn. **2017**, *90*, 59–67.
34. Tamami, B.; Nezhad, M. M.; Ghasemi, S.; Farjadian, F. *Phosphorus Sulfur Silicon Relat. Elem.* **2016**, *191*(1), 123–128.
35. Baier, H.; Kelling, A.; Schilde, U.; Holdt, H.-J. Z. Anorg. Allg. Chem. **2016**, *642*, 140–147.
36. Xu, C.; Afewerki, S.; Tai, C.-W.; Cordova, A.; Hedin, N. *ChemistrySelect* **2016**, *1*, 5801–5804.
37. Sable, V.; Maindan, K.; Shejwalkar, P. S.; Hara, K.; Kapdi, A. R. *ACS Omega* **2017**, *2*, 204–217.
38. Sable, V.; Maindan, K.; Bhilare, S.; Chrysochos, N.; Schulzke, C.; Kapdi, A. R. Chem. Asian J. **2018**, *13*, 2489–2498.
39. Bhaskar, R.; Sharma, A. K.; Yadav, M. K.; Sing, A. K. Dalton Trans. **2017**, *46*, 15235–15248.
40. Zohreh, N.; Hosseini, S. H.; Tavakolizadeha, M.; Busuioc, C.; Negrea, R. J. Mol. Liq. **2018**, *216*, 393–404.
41. Gautam, P.; Tiwari, N. J.; Bhanage, B. M. *ACS Omega* **2019**, *4*, 1560–1574.
42. Boonseng, S.; Roffe, G. W.; Targema, M.; Spencer, J.; Cox, H. J. Organomet. Chem. **2017**, *845*, 71–81.

FURTHER READING

43. Blaszczyk, I.; Gniewek, A.; Trzeciak, A. M. J. Organomet. Chem. **2012**, *710*, 44–52.
44. Gómez-Martínez, M.; Baeza, A.; Alonso, D. A. *Catalysts* **2017**, 7, 94.

CHAPTER 9

Palladacycles as Potential Anticancer Agents

Prachi Bangde*, Dharmendra Prajapati†, Prajakta Dandekar*, Ian J.S. Fairlamb‡, Anant R. Kapdi†
*Department of Pharmaceutical Technology, Institute of Chemical Technology, Mumbai, India
†Department of Chemistry, Institute of Chemical Technology, Mumbai, India
‡Department of Chemistry, University of York, York, United Kingdom

Contents

1. INTRODUCTION

Cancer is a multiorgan disease affecting millions of people across the world. According to the WHO, this disease was responsible for an estimated 9.6 million deaths in 2018 alone. The deaths in this year were due to the following types of cancers, listed in descending order according to estimated deaths: lung cancer, breast cancer, colon and rectum cancer, prostate cancer, melanoma of the skin, stomach cancer. Such huge figures indicate that the cancer burden is still a daunting concern and a challenge for public health care systems worldwide. Each type of cancer requires specific treatment regimen, which has led to the development of advanced therapies like immunotherapy, hormonal therapy, targeted therapy etc. These strategies target the disease more effectively by selectively damaging the tumor cells and leaving healthy cells unharmed. However, considering that cancer related deaths mainly occur in low- and middle-income countries, such advanced,

Palladacycles
https://doi.org/10.1016/B978-0-12-815505-9.00010-X
© 2019 Elsevier Inc.
All rights reserved.

expensive therapies add to the economic burden of the existing cancer care expenditure. Thus, presently the mainstay of cancer treatment is chemotherapy, in conjugation with surgery or/and radiotherapy. The chemotherapeutics administered to the cancer patients comprise of powerful medications that exterminate the fastidious cancer cells, with minimal possible dosage. Therefore, this treatment modality remains the highly sought, first-line treatment for cancer, while the other therapies are administered as the second-line therapies, according to need. Cancer chemotherapeutics are mainly dominated by metallodrugs, which are known to have superior anticancer activities simply because of their unusual structural properties due to the coordination of organic molecules with metal atoms, which cannot be achieved by chemical molecules alone.

A remarkable discovery in metallodrugs was made in the 1960s by Barnett Rosenberg, who discovered the anticancer properties of platinum Pt(II) metal, today widely available as cisplatin. This breakthrough further directed the synthesis and biological evaluation of new platinum analogs. However, despite their superior anticancer action, cisplatin and some other platinum-based antitumor drugs, like carboplatin and oxaliplatin, exhibit several disadvantages in the form of serious side-effects, such as nephrotoxicity, neurotoxicity, nausea, vomiting, gastrointestinal toxicity, hearing loss, ototoxicity, elevated blood pressure, allergy etc. It has been observed in rat models and human trials that heavy metals like platinum possess problems of renal clearance, due to their large size, which eventually leads to their accumulation within the renal tissue. The ensuing increase in their onsite concentration leads to their irreversible binding with the protein-bound, sulfur containing biomolecules in the proximal tubules, thus resulting in the inhibition of the renal transport enzymes and eventually renal failure *(1, 2)*. Hence, huge efforts are being made by chemical, biological, and computational scientists, across the world, to explore transition metals as chemotherapeutic moieties, to develop safer therapies for cancer patients by lowering or removing the unpredictable and severe nephrotoxicity associated with heavy metals. Furthermore, the efficacy of platinum-based drugs is hindered by their low solubility in physiological solutions, which offers yet another reason to design complexes with improved solubility. Another major reason for designing newer anticancer drugs is to overcome the acquired tumor resistance to cisplatin and other platinum-based drugs. This resistance is based on the mechanism of binding of platinum molecules to DNA, leading to the formation of DNA adducts and hence resulting in structural damage to the DNA and eventually cell death. Also, the continuous administration of these drugs to disrupt the DNA structure and function can further result in the alteration of cellular DNA and damage its recognition and repair mechanism, which can ultimately result in the continuation of cell viability and hence platinum resistance. These drawbacks of exposure to the platinum drugs have prompted the design of alternative molecules that may have the ability to target other cellular signaling pathways associated with the cancer cells and provide attractive targets for designing alternate anticancer drugs *(3–6)*.

One of the strategies to develop newer anticancer agents is by changing the nature of the central metal ion of the metallodrugs. Here, the Palladium metal, which is above platinum in the periodic table has attracted the attention of researchers all over the world owing to its comparable coordination chemistry, to that of platinum, and also its possibility of yielding water-soluble complexes. However, the Pd(II) complexes are known to be 10^5 times kinetically faster, thus making them more labile than the Pt(II) complexes. Thus, the palladium-based complexes fail to retain their structural integrity in the biological fluids. However, for use as anticancer agents, they can be rendered thermostable by introducing strong metal carbon bonds or multidendate ligands, which prevent untimely hydrolysis of the Pd(II) complexes. Palladium, as the metal center of anticancer complexes, can be stabilized by a judicious choice of ligands having strongly coordinated electron donor groups, along with suitable leaving groups, which may serve as active drugs once present in the cancer cell environment. The thermal and structural stability of palladium complexes, possessing intermolecular coordination chemistry with electron donor ligands, has emerged as an important area of investigation to produce cyclic complexes with enhanced biological activity *(7–9)*. Structurally, for most bidentate or multidendate ligands, a *cis* arrangement of their donor atoms is adopted. In the last few years, palladium complexes having thiosemicarbazones *(10–12)* and phenyl acetaldehyde thiosemicarbazone *(13)* have been reported to be quite stable and good candidates for antitumor therapy *(14)*. Further, the property of the organopalladium compounds to undergo redox swapping between the two stable Pd(II)/Pd(0) oxidation states, is mainly due to their rich chemistry and mainly employed for catalyst applications. Thus, by expanding the knowledge of their properties and exploring their applicability, continuous efforts are being devoted to develop palladium-based complexes for anticancer application. This growing research is based on the understanding that structural properties of palladium complexes are responsible for their mode of action. Some of the important structures related to cyclic palladium-based complexes developed in recent years have been elaborated in this chapter to discuss their structure, biological activity, and stability.

2. PALLADACYCLES/CYCLOPALLADATED COMPLEXES

Palladium compounds are organometallic derivatives, containing at least one metal–carbon bond intra-molecularly stabilized by at least one donor atom (N,S,O,P) from heteromolecules and are termed as cyclopalladated compounds or palladacycles. These types of complexes were characterized in the 1960s, but Hermann discovered their actual utility as catalysts in the 1990s *(15)*. The term cyclopalladation was coined by Trofimenko *(16)*, implying the formation of palladium complexes in which an organic ligand undergoes intramolecular metallation, with the formation of a chelating complex, comprising of at least one metal-carbon bond, along with another donor-metal bond from the ligand. These structures have high thermal stability, but are less inert and thus may be explored for active therapeutic applications *(8)*. In therapeutic applications, the choice of the chelating ligands is mostly related to molecules of natural origin *(17)*. The subtle effects exerted by ligands can be both, steric and electronic in nature, which help to design objective-focused metallo-complexes as per the intended biological use. Such palladacycle structures, featuring functionalized thiosemicarbazone *(18)*, diimine *(19)*, terpyridine *(20)*, and bis(2-pyridylmethyl)amine *(21)* ligands, have been synthesized to contain phosphorous, nitrogen, sulfur, and oxygen donor atoms *(22)*. In the year 2014, an elaborate review article published by Kapdi et al. has described various structures of multidendate ligands complexed with palladium metal to form mono- and multinuclear palladacyclic complexes, as anticancer agents *(23)*. Currently, these complexes have captured the growing interest of researchers for offering unique drug actions, based on the different oxidation states of the palladium metal resulting in different types and number of coordinated ligands and their coordination geometry and hence having diverse anticancer mechanisms *(11)*.

2.1 N,C-Donor Chelating Ligand Complexes

The N-heterocyclic carbene (NHC) moiety is a strong N-donor ligand that can form stable metal carbon bonds with most transition-metal ions *(24–27)*. Recently, the NHC structure has been exploited for anticancer action because of their accessibility and tunable properties. Facile modification of NHC ligands at N1 or N3 positions can significantly endow the ligands with new functions or properties that may be selective for the target biomolecules. This was proven by Jing-Yi Lee et al. who explored the novel tridendate ligands, possessing N-heterocyclic carbene, amidate, and pyridine to prepare normal (complex **1a–2c**; Fig. 9.1) and abnormal NHC (complex **3a–3c**; Fig. 9.1) complexes. These palladium complexes, bearing the "normal" NHC coordination, were found to be more thermally stable than their isomeric analogs with "abnormal" NHC binding. This structural stability of the imidazole-based ring is possibly due to the binding through coordination, via the C4/5 in abnormal NHC binding mode, and C2 atom binding in normal NHC complexes. The potential of these sets of palladium complexes as anticancer drugs was evaluated against human ovarian cancer (TOV21G), colon adenocarcinoma (SW620), and small-cell lung carcinoma cells (NCI-H1688), and exhibited noticeable inhibition activity in a

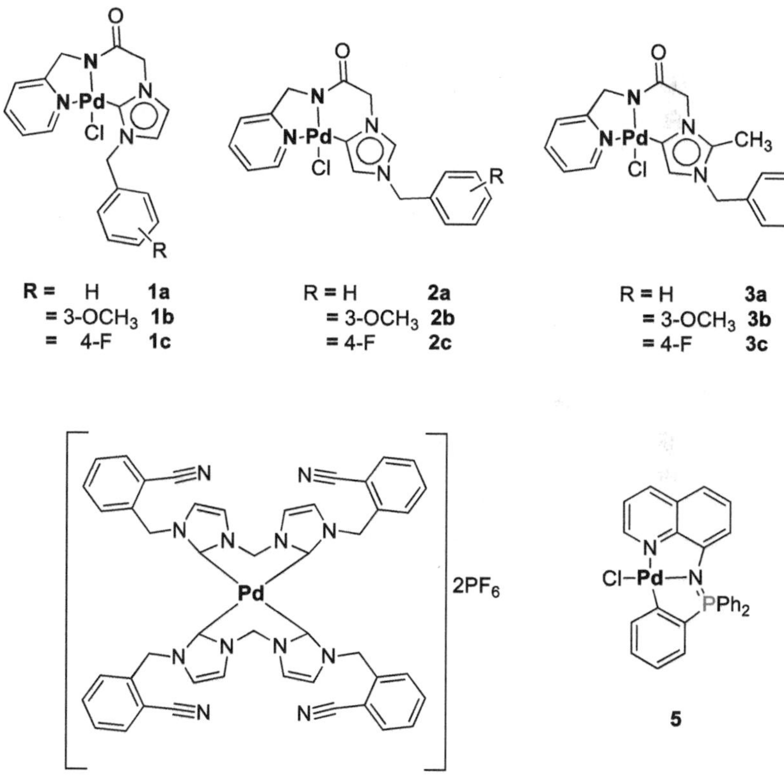

4

Fig. 9.1 Tridentate Pd-NHC and NHC complexes (**1–3**), Bis NHC Pd complex (**4**) and quinoline-based iminophorane Pd(II) complex (**5**).

TOV21G cell line. The presence of an N-3-methoxybenzyl group conferred high inhibition activity on these novel complexes, which exhibited an IC50 of 6.05 μM. On the contrary, their isomeric analogs bearing N-3-methoxy group in 3b, with an H or 4-F atom, led to inferior activities with significantly higher IC50 values of 23.44 and 49.55 μM, respectively. This study implied that the palladium complexes bearing N-4-fluorobenzyl groups in the ligand framework showed superior activity against TOV21G cells *(28)*.

Further, a new Pd(II)-N-heterocyclic carbene (NHC) complex **4** (Fig. 9.1) was derived by chelating with two units of bis-imidazolium ligands, containing two hexafluorophosphate anions, found to possess square-planar geometry. Further, bis-imidazolium ligand and palladium complex **4** were tested against breast cancer cells (MCF-7), wherein the cytotoxicity observed was similar to that due to tamoxifen. Moreover, the ligands alone were found to possess significantly lower antiproliferative activity (IC50 181.0 ± 3.0 μM) in comparison to the metal complexes (2.5 ± 0.2 μM) and tamoxifen (2.4 ± 0.2 μM) *(29)*.

Palladium complex **5** (Fig. 9.1) containing a luminescent iminophosphorane ligand serving as stabilizing (C,N- or N,N), were synthesized and assessed for their anticancer activity in a human ovarian cancer cell line (A2780S), a human lung cancer cell line (A549) and in a nontumorigenic human embryonic kidney cell line (HEK-293T). This study revealed that the palladium complexes were toxic to the ovarian cells (11.0 ± 1.5 μM, 13.2 ± 2.1 μM), compared to the lung (62.5 ± 3.7 μM, 86.5 ± 2.5 μM) and normal cells (53.5 ± 10.4 μM—66.2 ± 5.5 μM). Further, the interaction studies with DNA indicated stronger binding, but of a different nature than that observed with cisplatin. In addition, fluorescence studies with human serum albumin (HSA) revealed higher quenching by the iminophosphorane complex than cisplatin, indicative of faster reactivity of the compounds with HSA than with cisplatin *(30)*.

Other NHC complexes with palladium (II) metal containing different alkyl length chain, Pd2a $PdPPH_3$, Pdiso, and [Pd(C^N^N)Cl] **6–10** (Fig. 9.2) displayed cytotoxic activity toward cervical epithelial cancer cells (HeLa), lung cancer cells (NCI-H1650 and NCIH460), an aggressive triple-negative breast cancer cells (MDAMB-231), and ovarian cancer cells (A2780), as well as its cisplatin resistant clone (A2780cis). All structures exhibited promising antiproliferative activity toward the cancer cell lines, with IC50 values ranging from 0.09 to 2.5 μM, which was 172-fold more cytotoxic than cisplatin. These complexes were also capable of inducing apoptosis, which was confirmed by western blot assay and revealed an increase in the enzymatic activities of caspase-3 and caspase-9 and cleavage of PARP-1. The experiment on interaction of palladium complexes with DNA indicated that DNA was not the major molecular target during cellular destruction. Further, the proteomics data and in vitro biochemical assays suggested that the complexes exerted the anticancer effect via inhibition of epidermal growth factor receptors, induction of mitochondrial dysfunction, and antiangiogenic activity toward the endothelial cells. In vitro stability with the molecules containing thiols groups was also studied to understand reactivity of the metal complexes toward sulfur containing biomolecules like proteins etc., which confirmed inert interactions toward the sulfur molecules. In vivo anticancer studies revealed that the complexes significantly inhibited tumor growth in a nude mice model *(31)*.

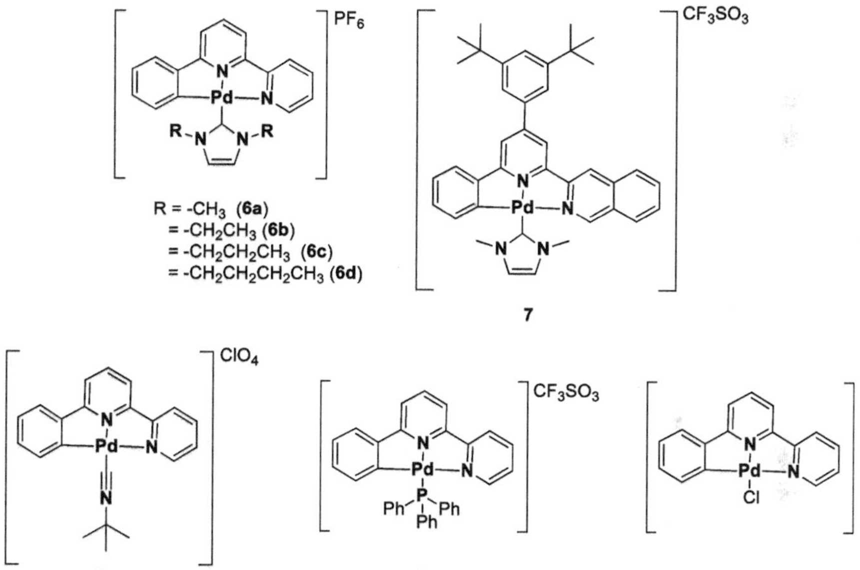

8 9 10

Fig. 9.2 Cyclopalladated Pd(II) complexes containing different alkyl length chain, Pd2a, Pdiso, $PdPPH_3$, and [Pd(C∘N∘N)Cl].

Work by Kazem Karami et al. showed that two newly developed palladium(II) complexes **11 and 12** (Fig. 9.3), containing C,N and N,N chelating ligands, demonstrated interaction with the calf thymus DNA (CT-DNA) by molecular docking method, indicating that the complexes could interact with the DNA structure by engaging in strong partial π-stacking interactions and hydrogen bonding between the coordinated –NH and the functional groups positioned on the edge of the DNA bases. The reactivity toward bovine serum albumin (BSA) revealed that the quenching of BSA fluorescence by the complexes was of static type. The study also revealed that the structures possessed high potency to restrict cancer cells like Jurkat and MCF-7 human (IC50 < 100 μM), with no significant reduction in the viability of normal cells (IC50 > 100 μM) *(32)*.

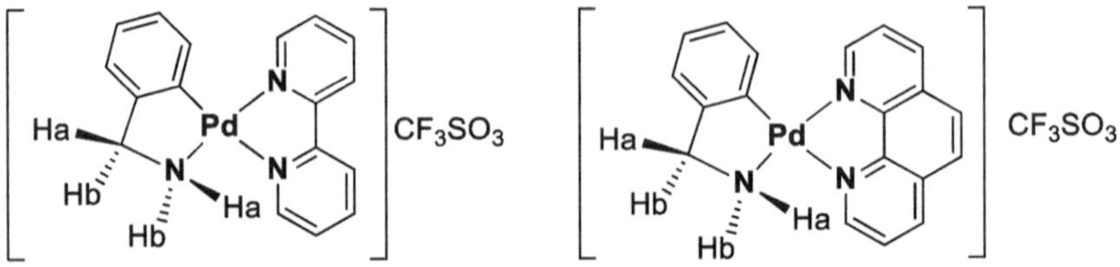

11 **12**

Fig. 9.3 Orthopalladated Palladium complexes based on bipyridine (**11**) and 1,10-phenanthroline (**12**).

The cytotoxic nature of novel palladacycles complexes **13–14** (Fig. 9.4), based on substituted benzylidene-2,6-diisopropylphenylamine ligands in conjunction with 1,3,5-triaz-7-phosphaadamantane as the auxiliary ligand, were studied against the breast cancer cell lines MCF and MDA-MB-231, as well as against the melanoma cell-line ME1402. The complexes C2 and C3 exhibited reasonable activity compared to that of cisplatin. These complexes were also relatively more active than analogous orthopalladated complexes based on aryl imines/amines, in which other tertiary phosphines were employed as the auxiliary ligands. Such complexes existed as neutral palladacycles and their activity was due to the aquation or solvolysis of the Pd-Cl bond. Western blot assays demonstrated that C2 induced apoptosis, which was confirmed by the detection of the proteinγ-H2AX and the cleavage of PARP. Further, it was observed that this apoptosis may not have been due to

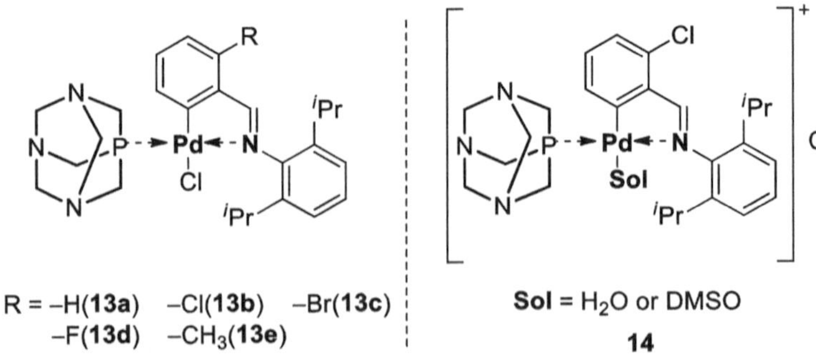

Fig. 9.4 Phosphaadamantane (PTA) ligand-based cyclopalladated complexes.

DNA damage as C2's mode action with DNA did not involve intercalation *(33)*.

A series of palladium complexes **15** (Fig. 9.5), containing metformin and benzylamine, were designed, developed, and synthesized. These complexes were investigated for their anticancer properties against Hela (cervical cancer), MCF7 (breast cancer), and A549 (lung cancer) cell lines and demonstrated comparable activity to cisplatin. Furthermore, detailed investigations of complexes **15** and metformin with CT-DNA showed that the interaction was of intercalation type for metformin and through the groove binding mode for complexes **15**. In addition, the interaction studies involving BSA, using UV–Vis and fluorescence spectroscopy, as well as determination of the binding constant values indicated good interaction with BSA *(34)*.

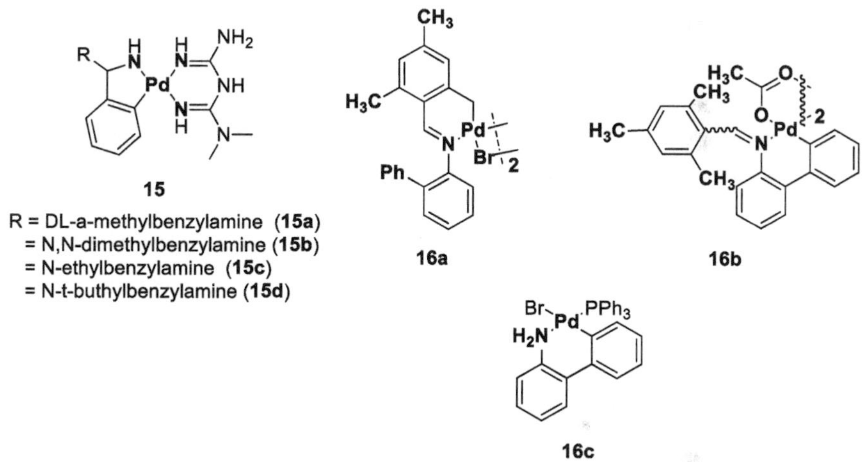

Fig. 9.5 Cyclopalladated Pd complex containing metamorphine (**15**) and mesitylmethanamine (**16a, 16b, 16c**).

Two novel structural mononuclear cyclopalladated isomers of imine a (E)-N-([1,1′-biphenyl]-2-yl)-1-mesitylmethanimine, termed as 3a-endo and 3a-exo complex **16a–c** (Fig. 9.5), were demonstrated for their anticancer properties imparted due to their structural features. Compound 3a-exo, containing an exo imine functional group, presented cytotoxicity in a low micromolar range (5–20 μM) toward the MDA-MB-231 and MCF-7 human breast cancer cell lines and the HCT-116 human colon cancer cell line. Interestingly, the cytotoxicity of complex 3a-exo was at least seven times lower than cisplatin, toward the normal BJ cells. These results contrasted with those of compound **16c**, which contained an amino function in its structural formula that was highly toxic (1–5 μM), and those of compound 3a-endo **(16a),** which was a structural isomer of compound 3a-exo **(16b)**. Compound **16b** was more cytotoxic than compound **16a** toward the cancer cell lines, but was not cytotoxic toward the normal BJ cells. On the other hand, the biological activity displayed by the compound 3a-endo was very weak as compared to the compound 3a-exo. Finally, it was observed that the compounds **16b** and **16c** were more cytotoxic toward cancer cells than imine but were not efficient in altering the DNA tertiary structure or inhibiting cathepsin B activity *(35)*.

2.2 N,N-Donor Chelating Ligand Complexes

The nitrogen-based donor ligands provide several distinct geometrical and physiochemical features that have resulted in various structures to suit the industrial applications. Different types of modes of coordination chemistry can be achieved by using bidendate amine groups like ethylenediaminoalkanes, diaminocyclohexane, alkylaminophosphine oxides mercaptoimidazoles, pyridine, and pyrimidine derivatives which have been proven to improve palladium metal stability and antitumoral activity *(14, 36)*. A rigid framework designed by coordination of palladium complexes with two terminal imino-quinolyl units bridged by a phenylene ring complex **17** (Fig. 9.6) was developed by Motswainyana et al. The antitumor activity presented by these compounds examined against breast cancer MCF-7 and human colon HT-29 cancer cell lines showed activity of IC50 60 μM and IC50 46 μM, respectively, and were found to be highly toxic when compared to cisplatin complex (IC50 > 100 μM). This activity suggests that binuclear complexes are more toxic owing to the synergistic effect of two metal centers *(37)*.

Pincer type ligands having a specific tridentate monoanionic framework were used to form N-metallated palladacycles complex **18** (Fig. 9.5) which were tested for antitumor properties and were found to possess pronounced cytotoxic properties. The resulting palladacycles complexes containing sulfur ancillary donor group were found to be highly cytotoxic (0.45–22 μM) on human colon cancer HCT116, human breast cancer MCF7, and human prostate cancer PC3 cell lines compared to cells treated with cisplatin (IC50 > 25 μM). The presence of ancillary group on palladium complexes was found to be essential in imparting cytotoxic effect by increased bioavailability as a result of preventing interaction with sulfur containing

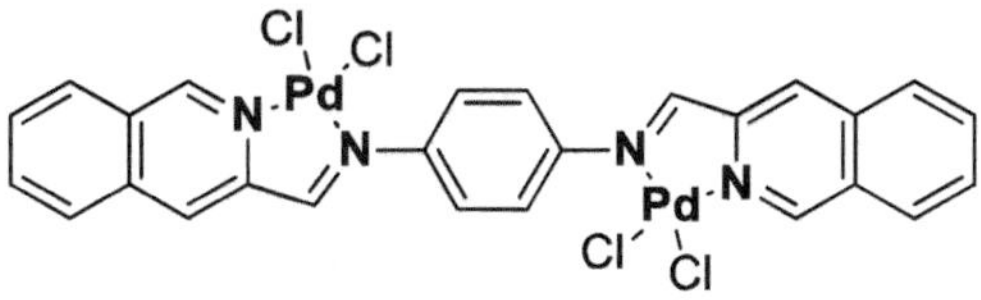

17

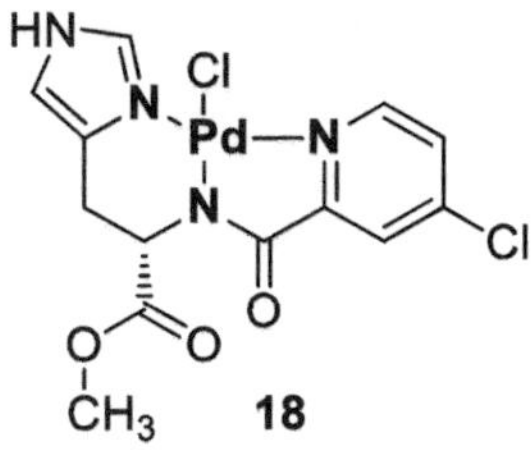

18

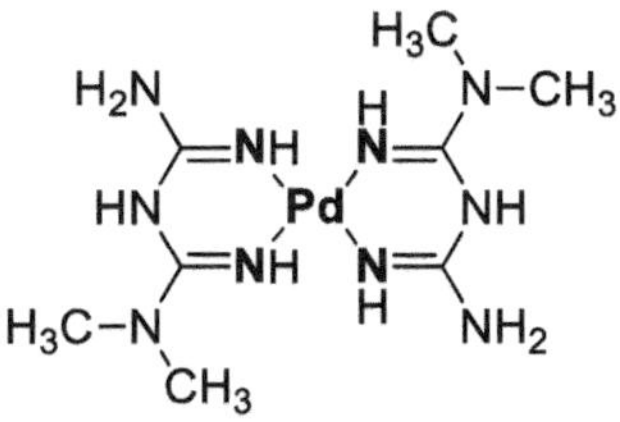

19

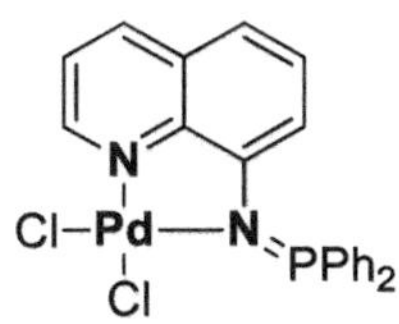

20

R = CH_3 (**21a**)
= CH_2CH_3 (**21b**)
= $CH_2CH_2CH_3$(**21c**)

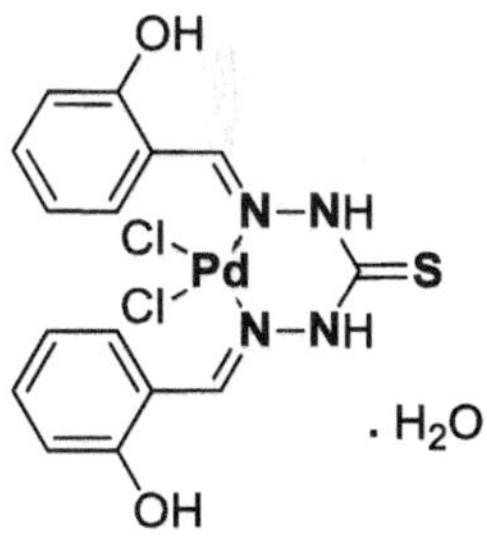

22

Fig. 9.6 Palladacycles (**17–22**) bridged by a phenylene ring complex **17**, Pincer type palladium complex **18**, palladium complex benzylamine ligands **19**, iminophosphorane ligand based **20**, R2edda-type ligands **21**, thiocarbohydrazide with salicylaldehyde **22**.

biomolecules. The stability of these pincer palladium molecules was confirmed by UV-Vis studies, which indicated no decomposition of pincer structure either in DMSO-water and DMSO-PBS solutions. Further understanding of palladium complexes interaction with DNA was carried out by gel electrophoresis to study electrophoretic mobility of plasmid DNA. This study resulted in the decrease of DNA mobility with increasing concentration of the complexes evidently signifying their binding nature with DNA, where methionine-based derivative show stronger interactions and hence uncoiling of plasmid DNA even at low concentration. Further, its superior inhibition activity of complexes on DNA topoisomerase I enzyme was confirmed by testing its ability to convert supercoiled DNA into relaxed DNA strand *(22)*.

Biochemical studies for palladium benzylamine based complex **19** (Fig. 9.6) developed by Karami et al. were studied on Hela (cervical cancer), MCF7 (breast cancer) and A549 (lung cancer) cancer cells, which revealed that complex **19** is highly active on MCF-7 cells, and weakly toxic on other cells, as compared to cisplatin. UV–Vis and fluorescence spectroscopy and competitive studies with methylene blue displayed that complex **19** interact with DNA through the groove binding mode. In addition, the interaction studies with BSA using UV–Vis and fluorescence spectroscopy show good binding affinity with BSA *(34)*.

Palladium complexes with a luminescent iminophosphorane ligand as in complex **20** (Fig. 9.6) were synthesized and assayed for their anticancer activity over a human ovarian cancer cell line (A2780S), in human lung cancer cells (A549) and in a nontumorigenic human embryonic kidney cell line (HEK-293T). This study reveals that palladium complexes are toxic to ovarian cells (11.0 ± 1.5 μM, 13.2 ± 2.1 μM) compared to lung (62.5 ± 3.7 μM, 86.5 ± 2.5 μM) and normal cells (53.5 ± 10.4 μM—66.2 ± 5.5 μM). Further interaction studies with DNA have stronger binding but of a different nature than those exerted by cisplatin. In addition, interaction with human serum albumin (HSA) is observed to be faster than cisplatin *(30)*.

Bidentate chelating N,N-ligands like O,O-dialkyl esters of (*S,S*)-ethylenediamine-*N,N*-di-2-propanoic,(*S,S*)-ethylenediamine-*N,N*-di-2-(3-cyclohexyl)propanoic and (*S,S*)-propylenediamine-*N,N*-di-2-(3-cyclohexyl)propanoic acids represent a "chiral subgroup" of the class of R2edda-type ligands in which nitrogen becomes chiral by coordination with a metal ion. These structure ligands **21a–c** (Fig. 9.6) were used to form coordinated structure with palladium as a metal center. Cytotoxicity of these palladium(II) complexes was investigated against a panel of three cancer cell lines viz. human cervical adenocarcinoma (HeLa), human alveolar basal adenocarcinoma (A549) and noncancerous human fetal lung fibroblast (MRC-5) cell lines using colorimetric MTT assay. However, cytotoxicity values obtained of these structures did not exceed that of cisplatin *(14)*. New palladium complexes complex **22** (Fig. 9.6) was formed using bis derivative for thiocarbohydrazide with salicylaldehyde which includes multidonor centers thus enriching the coordination toward palladium metal. Antiproliferative activity was determined for compounds against hepatocellular carcinoma cells, showing mild activity toward the inhibition of all in carcinoma cell line investigation *(38)*. Antitumor efficacy of palladium(II) (Pd)-saccharinate complex with terpyridine ligand complex **23** (Fig. 9.7) was tested against nine different cancerous cell lines from breast (MDA-MB-231, MDA-MB-435), lung (A-549, LLC), cervix (HeLa), prostate (PC-3), neuroblastoma (SH-SY5Y], and neural (C6) cancer cell lines and normal cells (primary human aortic smooth muscle cells: HASMC-1 and HASMC-2 as well as noncancer transformed cells (CHO-K1). Palladium complexes exhibited cytotoxicity to several cells when treated for a period of 72 h showing different efficiencies ranging between 2 and 42 μM. These complexes displayed apoptotic potential confirmed by microscopic and flow-cytometry studies. These complexes were evaluated for their clinical potential for drug-induced liver toxicity on healthy C57BL/6 mice, which showed no significant change in hepatic enzymes (ALT and AST), which are responsible for producing side effects, when compared to elevated response in mice treated with cisplatin. Further in vivo safety of the Pd(II) complex were examined by analyzing reduction in tumor size, changes in body and organ weight, histopathological analysis of liver, kidney, and tumor sections, and biochemical analysis of serum in mice. The study results displayed that the Pd(II) complex was more cytotoxic to cancer cells than noncancer cell lines and caused cell death through apoptotic pathways. The treatment of the Pd(II) complex in tumor-bearing mice effectively reduced the tumor size at half the dose used for cisplatin. The Pd(II) complex appeared to exert less liver damage than the cisplatin-based complex on changes in the hepatic enzyme levels in the serum. Hence, the complex appears to be a potential chemotherapeutic *(39, 40)*. The benzothiazole (BTA) and its analogs offer a high degree of structural tunability which was exploited for its potential anticancer properties by Rubino, et al. These heterocyclic compounds form a comparative structure as present in natural products thus qualifying to show a variety of pharmacological properties. The BTA is a heterocyclic compound in which benzene ring is fused to 4,5-positions of thiazole ring which results in complete planar structure. The ambidentate BTA ligands with N and S donor groups coordinates with metal ion to form stable complexes. Such ligands were metallated with palladium complex **24** (Fig. 9.7) to afford $[Pd_2(m\text{-}Cl)_2(DTBTA)_2]Cl_2$ and were evaluated of their in vitro antitumor activity against two human tumor cell lines—human breast cancer (MCF-7) and hepatocellular carcinoma (HepG2)—using cisplatin as positive control. However, this structure exhibited very poor cytotoxicity activity against tumor cell lines *(41)*.

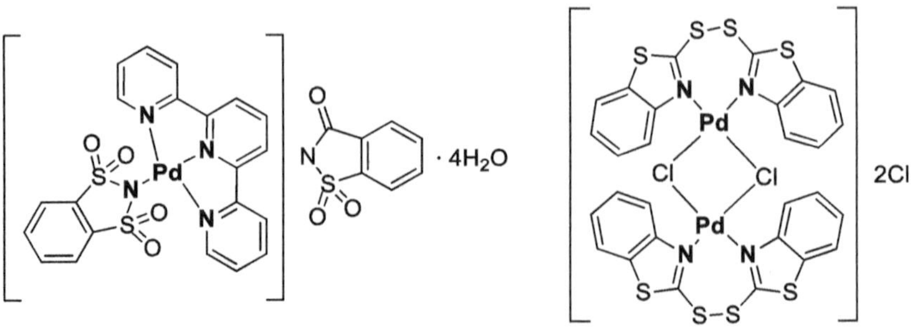

23 **24**

Fig. 9.7 Pd(II) sachrinate complex of terpyridine (**23**) and dithiobis(benzothiazole)-based Pd(II) complexes (**24**).

Formamidine ligands were coordinated with palladium metal formed palladium complexes **25–28** as represented in Fig. 9.8 forming square planar structures having neutral, stable and water soluble properties. These complexes revealed good anticancer properties against MCF-7-human breast adenocarcinoma (0.0740 μM), HEP-2- human laryngeal (0.0074–0.0082 μM), and HepG-2-human liver (0.0119–0.0144 μM) cell lines *(42)*.

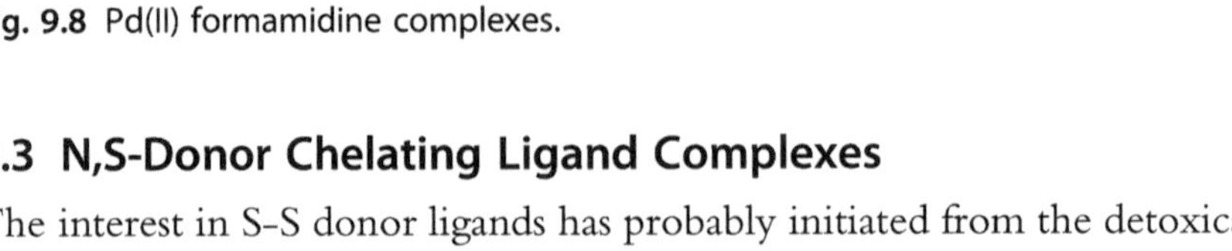

25 **26** **27** **28**

Fig. 9.8 Pd(II) formamidine complexes.

2.3 N,S-Donor Chelating Ligand Complexes

The interest in S-S donor ligands has probably initiated from the detoxication properties of the sulfur-containing ligands against heavy metal intoxication *(43)*. Palladium complexes having *N*- and *S*-donor ligands have an exceptionally high structural stability. However, the sulfur containing molecules not only obstruct them from reaching their pharmacological targets, but also significantly convert them into inactive *trans* isomers that lower anticancer activity. This problem can be overcome by the use of appropriate ligands that may turn down the negative lability by reducing reactivity toward biomolecules. Various literature suggests use of sulfur containing ligands like dithiocarbamate to complex with palladium metal ion, through sulfur and nitrogen atoms. Such bulky ligand provides stability to the complex structure and facilitates the movement of the intact complex through physiological solutions, to the target cells *(44)*.

The amino acid methionine and cysteine based S-coordination ligands were complexed with palladium metal ion complexes **29–30** (Fig. 9.9), which improved the cytotoxic properties of the N,N,S-derivatives. The cytotoxic activities of the sulfide-based complexes against human colon cancer cells HCT116, human breast cancer cells MCF7, and human prostate cancer cells PC3 demonstrated higher IC50 (0.3–22 μM) than that of cisplatin (IC50 > 25 μM). The complexes also showed strong DNA binding, implying a strong intercalation mode of binding, and inhibited the activity of DNA topoisomerase I *(22)*.

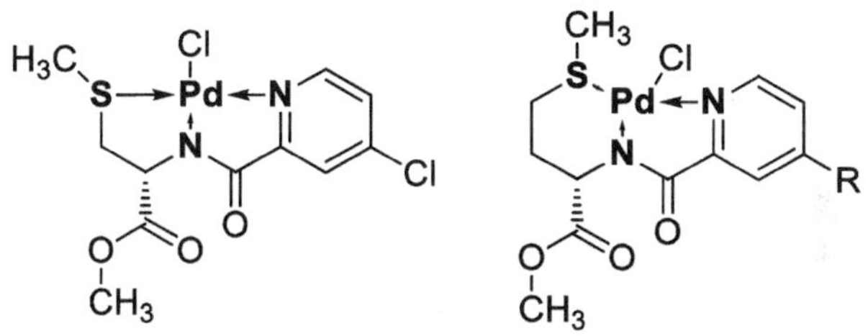

29 R = H **(30a)** -Cl **(30b)**

Fig. 9.9 Pd(II) complexes of Picolinylamides functionalized with S-donor group.

Shahraki et al. developed water-soluble complexes having a framework of pyrrolidindithiocarbamate ligand to yield palladium(II)-dithiocarbamate, $[Pd(phen)(pyr\text{-}dtc)]NO_3$ (phen = 1,10-phenanthroline and pyr-dtc = pyrrolidinedithiocarbamate) complex **31** (Fig. 9.10). In these complexes, the dithiocarbamato ligand coordinated the Pd(II) center with two sulfur atoms, as a bidentate ligand. They were examined for their biological activity against a chronic myelocytic leukemia K562 cell line. These complexes exhibited significant cytoxicity, as compared to cisplatin. Further, these complexes were found to bind by intercalation mode and were found to denature the DNA, even at low concentrations *(45)*. A palladium(II) complex with dithiocarbamate moiety was examined for its cytotoxic activity against a leukemia K562 cell line and resulted in an IC50 value of 53.06 μM. However, a combination of this complex with vitamin K3 complex **32a and 32b** (Fig. 9.10) resulted in a dramatic decrease in the IC50 value (32.95 μM). The change in palladium structure in the absence and presence of vitamin K3 was examined for its interactions with CT-DNA, which showed intercalation mode of binding. Further, the UV–Vis spectroscopy indicated that in the presence of vitamin K3, the complex surprisingly denatured the CT-DNA at very low concentrations. Fluorescence spectroscopy results indicated that the Pd(II) complex possessed a static quenching mechanism, in the presence and absence of the vitamin *(46)*.

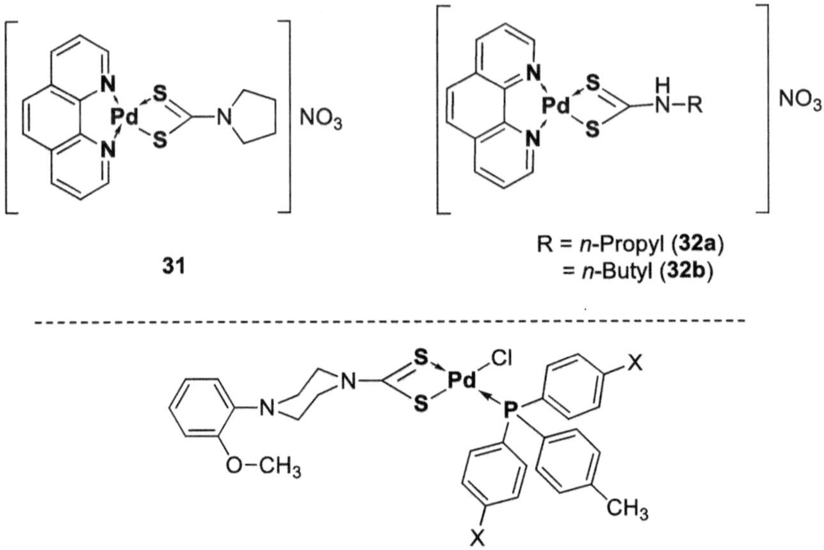

31

R = *n*-Propyl (**32a**)
= *n*-Butyl (**32b**)

X = H (**33a**)
= CH_3 (**33b**)

Fig. 9.10 Pd(II) complexes of 1,10-Phnanthroline and dithiocarbamate (**31–32**) and heteroleptic Pd(II) complexes based on dithiocarbamate (**33**).

Ligands like sodium 4-(2-methoxyphenyl)piperazine-1-carbodithioate and diphenyl-*p*-tolylphosphine (1) or tri-*p*-tolylphosphine (2) were used to form palladium complexes **33a and 33b** (Fig. 9.10) and were analyzed for their cytotoxic activity against cancer cell lines (LU—human lung carcinoma, MCF7—human breast adenocarcinoma, MDA-MB-231—human breast adenocarcinoma, Hepa-1c1c7—mouse liver hepatoma, PC-3—human prostate adenocarcinoma). Here, the complex **33b** demonstrated higher activity as compared to complex **33a**. Also, studies concerning interaction with DNA revealed that the complexes demonstrated a good binding affinity. Theoretical studies revealed that complex **33b** was thermally more stable than complex **33a** *(47)*.

Thiosemicarbazones, a type of N, S donor ligands having pharmacological properties, were complexed with the palladium metal center to understand the therapeutic properties of the resulting complexes. Pd(II) complexes, [PdCl(L)(PPh3)] (1–5) and [Pd(L)2] (6 and 7) (HL = indole-3-carbaldehyde thiosemicarbazones, HL1–HL5) complex **34–35** (Fig. 9.11) were tested for their in vitro anticancer activity via the MTT assay in three cancer cell lines (HepG-2, A549 and MCF7) and one normal cell line (L929). Here, the triphenylphosphine showed better activity in HepG-2, with an IC50 value of 22.8 μM as compared to the activity of cisplatin, which demonstrated an IC50 value of 67.1 μM. However, the developed complexes showed moderate anticancer activity against A549 and MCF7 cancer cell lines and less toxicity toward the normal cell line. The cell death occurred by apoptosis and was confirmed from the morphological changes and DNA fragmentation assay. Additional studies with CT-DNA demonstrated that the complexes could cleave the pDNA. The interaction of the complexes with BSA showed static quenching. Further docking studies revealed efficient binding with DNA, BSA and DNA topoisomerase I, which was responsible for the cell death *(48)*.

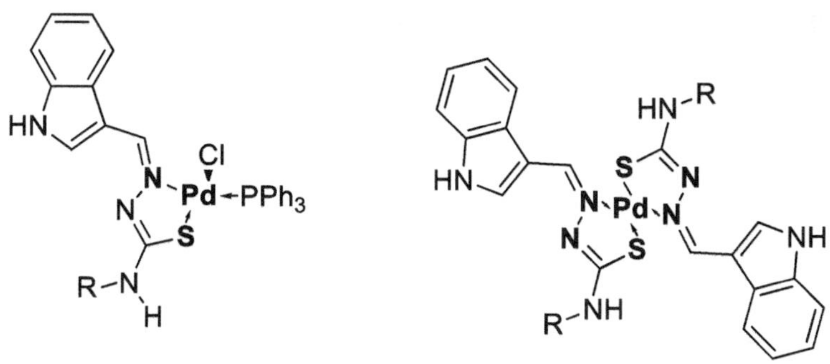

R= -H(**34a**) -CH_3(**34b**) -C_2H_5(**34c**) -C_6H_5(**34d**)

R = -CH_3 (**35a**) -C_6H_{11} (**35b**)

Fig. 9.11 Indole-3-carbaldehyde thiosemicabazone based Pd(II) complexes (**34–35**).

2.4 N,O-Donor Chelating Ligand Complexes

A number of palladium complexes containing N,O-chelating ligands have shown potential antitumor activities. The square-planar and octahedral configurations provided by the N_2O_2 donors to the geometries of the Pd(II) complexes afforded them with excellent inhibitory activity toward the cancer cell line *(38)*.

The quinolin-8-ol (8-HQ) derivatives, 5-chloro-7-iodo-quinolin-8-ol (CQ), 5,7-dichloro-quinolin-8-ol (dClQ) and 5,7-dibromo-quinolin-8-ol(dBrQ), were used to form palladium complexes **36** (Fig. 9.12).

These ligands possess antifungal, antibacterial, and antiviral activities by themselves. The complexes were prepared in the form of ions, using cations, like Na^+, K^+ or Cs^+, in order to improve their anticancer efficacy. Interaction of these complexes with CT-DNA revealed intercalation as the probable mode of their interaction with the DNA. The in vitro antitumor properties of the complexes were studied using the mouse leukemic cell line L1210, the ovarian cancer cell line A2780 and the noncancerous cell line HEK293, revealing superior cytotoxicity as compared to cisplatin *(49)*.

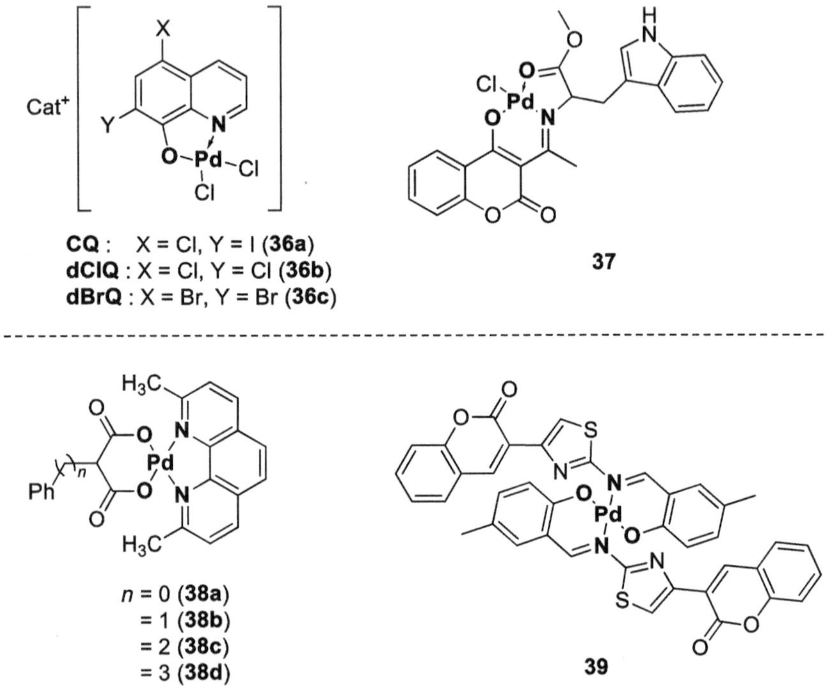

Fig. 9.12 Pd(II) complexes of dihaloquinolinolato (**36**), coumarin and typtophan (**37**), benzenealkyldicarboxylate (**38**), coumarin derived thiazole (**39**).

Palladium structure **37** (Fig. 9.12), containing 3-acetyl-4-hydroxy coumarin and methyl esters of tryptophane, were examined for their toxicity toward two leukemia cell lines (JVM-13 and MOLT-4) and against primary leukemic cells isolated from chronic lymphocytic leukemia (CLL) patients. Biological studies revealed that the ligands had no effect on viability of the cancer cells (442.7 ± 77.6 μM, 286.3 ± 65.2 μM and 423.9 ± 68.2 μM), in vitro, whereas the IC50 values for the complex 3c were

73.3 ± 12.7 μM, 5.6 ± 3.4 μM and 76.7 ± 18.1 μM for JVM-13, MOLT-4 and CLL cells, respectively *(50)*.

A series of Pd(II) complexes derived from benzenealkyldicarboxylate ligands, [Pd(Ln)(phen)] (phen = 2,9-dimethyl-1,10-phenanthroline, complex 1: L1 = phenylmalonate; complex 2: L2 = benzylmalonate; complex 3: L3 = (2-phenylethyl)malonate; complex 4: L4 = (3-phenylpropyl)malonate) **38** (Fig. 9.12) were studied for understanding the relationship between the carbon chain length and the biological activity. The palladium complexes showed significant inhibitory action against HeLa and HL-60 cancer cells. Further, these complexes demonstrated the ability to cleave the pBR322 plasmid DNA and induced apoptosis in the HL-60 tumor cell line. Further, the molecular dynamic simulations and docking studies revealed their binding affinity toward DNA and DNA-topoisomerase I, which indicated the mechanism of their cytotoxicity *(51)*.

Coordination through oxygen and nitrogen atoms provided by coumarin-thiazole with palladium metal yielded a new complex **39** (Fig. 9.12). The anticancer activity of ligand and its Pd(II) complexes were evaluated in the human cancer lines like MCF-7 (human breast adenocarcinoma), LS174T (human colon carcinoma), and LNCAP (human prostate adenocarcinoma), wherein the complexes resulted in significant anticancer action *(52)*.

Oxalate and formamide were used to synthesize a series of palladium complexes **40–42** (Fig. 9.13) that were thermally stable, as reveled from their relatively high, overall activation energies (441–688 kJ mol $-^{1}$). Further, these complexes were toxic toward cancer cells, with IC50 in the range of 0.011–0.168 μM against MCF-7 (human breast adenocarcinoma), 0.012–0.150 μM against HCT-116 (human colon), 0.042–0.094 μM against PC-3 and 0.006–0.222 μM against HepG-2 (human liver)cell lines, which suggested their applicability as anticancer agents *(53)*.

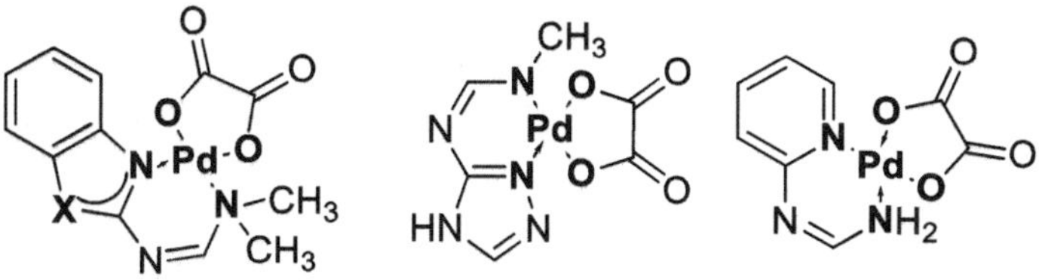

X -= N(**40a**) S(**40b**) **41** **42**

Fig. 9.13 Pd(II) oxalate complexes based on formamidine ligands.

2.5 Miscellaneous

Tris(2-carboxyethyl)phosphine $P(CH_2CH_2COOH)_3$ (TCEP) is a reducing agent which is known to break the disulfide bonds in peptides, proteins, and other compounds. Therefore, water soluble and air stable palladium(II) complexes having TCEP framework were developed. Cytotoxic studies were performed in melanoma cells: SK-mel, SH-4, Colo-829, C-32 and breast cancer cells, MCF7,T-47D and MDA-MB-231 cell lines. The results indicated that the complexes were highly active against SH-4, Colo-829 and MDA-MB-231 cells, as compared to cisplatin *(54)*. Schiff bases are formed by the condensation of primary amines with either ketones or aldehydes. These are versatile compounds, which have the ability to chelate with metals, through imino nitrogen, another oxygen or sulfur donor atom. These ligands form palladium metal complexes to yield N-(2-oxidophenyl) salicylideneiminatotriphenylphosphine palladium(II) (**43**); Fig. 9.14 and 14 N-(2-sulfidophenyl)salicylideneiminatotriphenylphosphine palladium(II) (**44**); Fig. 9.14. These were found to have inhibitory effect on alkaline phosphatase (ALP) enzyme activity. The biological activity of these compounds were tested by brine shrimp and potato disc antitumor assay, which reflected their activity, probably due to the presence of oxygen moiety in ortho position of the ring. Further, DNA protection assay revealed that they possessed moderate protective activity against hydroxyl free radicals. The cyclic voltammetric method ascertained the susceptibility of the compounds through intercalative mode of DNA-drug interactions *(55)*.

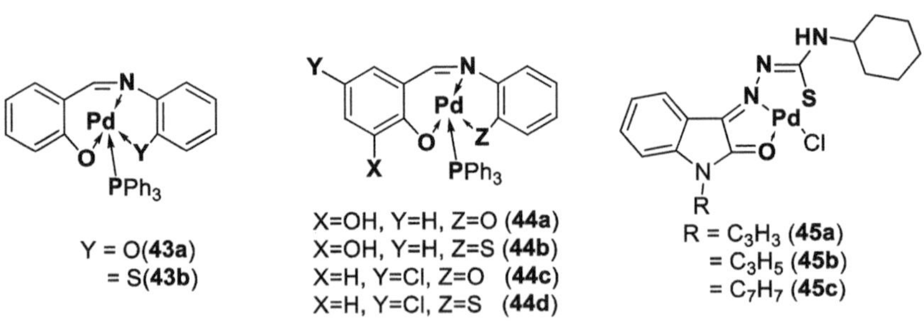

Fig. 9.14 Schiff base triphenylphosphine Pd(II) complexes (**43–44**) and Pd(II) complexes with N-substituted Isatinthiosemicarbazone (**45**).

Three pincer type palladium(II) complexes (1–3) were synthesized from the reactions between $PdCl_2$ with thiosemicarbazone ligands (L1–L3). These complexes **45** (Fig. 9.14) had significant cytotoxicity against human breast cancer (MCF7) and lung cancer (A549) cell lines. The interactions studies of the palladium(II) complexes toward CT-DNA revealed an intercalative mode of DNA binding. DNA binding studies revealed intercalative-, nongroove binding mode, which resulted in cleavage of the pUC19 plasmid. Also, these complexes demonstrated good binding affinity for proteins *(56)*.

Boron dipyrromethane (BODIPY) was used as a ligand to form new palladium metal supramolecules **46** (Fig. 9.15), with triangular/square architectures. These supramolecules were found to be more cytotoxic toward brain cancer (glioblastoma) cells than the normal lung fibroblasts, as compared to cisplatin. The green fluorescence, the inherent property of the BODIPY ligand in these complexes, was explored for bioimaging, using confocal microscopy, which showed the accumulation of the complexes in the cytoplasm and on the plasma membrane of the cells *(57)*.

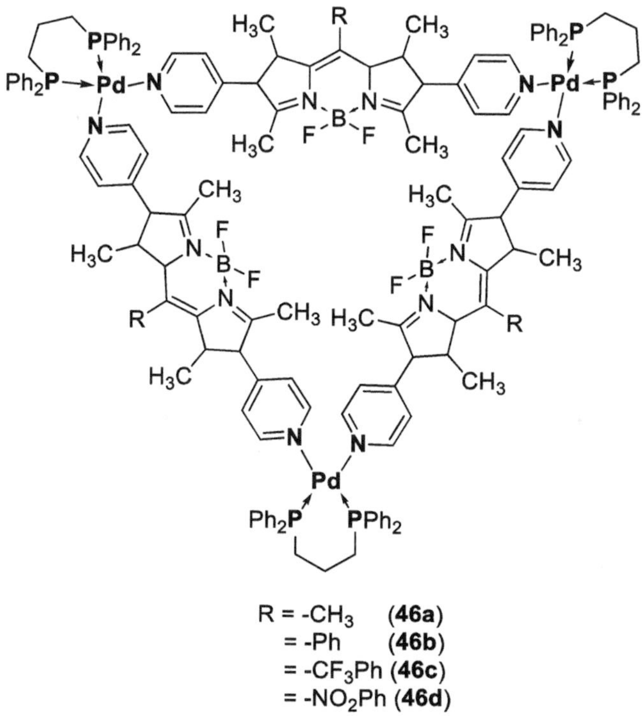

Fig. 9.15 BODIPY based Pd(II) supramolecules.

Palladacycles **47–48** (Fig. 9.16) were prepared by the activation of nonaromatic C—H bonds and the displacement of the hydrogen atom using ligand derived from cinnamaldehyde and o-amino phenol. In vitro cytotoxicity revealed that the palladacycles exhibited growth inhibitory activity against the human breast cancer (MCF7) cell line *(58)*.

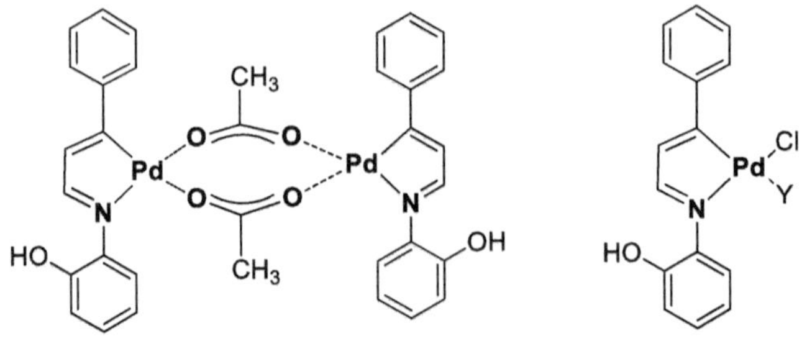

47

Y = Py (**48a**)
= PPh_3 (**48b**)

Fig. 9.16 Schiff base2-[(3-phenylallylidene)amino]phenol-based palladacycle.

Biphosphinic Palladacycle Complex (BPC) was produced using *N,N*-dimethyl-1-phenethylamine and the coordinating ligand 1,10-bis(diphenylphosphine)ferrocene **49** (Fig. 9.17). This complex induced cell death in drug-resistant osteosarcoma Saos-2 cells, mainly by apoptosis. Further investigations revealed that the apoptosis was associated by the effectors of caspase-3 activation; this event were mostly an outcome of pro-apoptotic protein Bax translocation, and increased cytosolic calcium mobilization from intracellular compartments. Lysosomal Membrane Permeabilisation (LMP) was also observed after BPC exposure which revealed that significant decrease in BAPTA-AM (a calcium chelator) and CA074 (cathepsin B inhibitor), suggesting that both calcium and cathepsin B were required for the antitumor activity of BPC. Preclinical studies were performed in mice that were intravenously administered with murine melanoma B16F10-Nex2 cells which were treated with BPC intraperitoneally (i.p.) for ten consecutive days, after which the lung metastatic nodules were counted. This study demonstrated that BPC protected the mice against murine metastatic melanoma. In conclusion, the BPC complex was found to be an effective anticancer compound against metastatic murine melanoma which presents high chemotherapy resistant property by inducing apoptosis triggered by calcium signaling and a lysosomal-dependent pathway *(59)*.

Cyclopalladated compounds bearing biphosphinic ligand **50** (Fig. 9.17) were explored for their anticancer properties, against human cervix carcinoma (Hela), colon cancer (HT-29), leukemia cancer (K562), and human

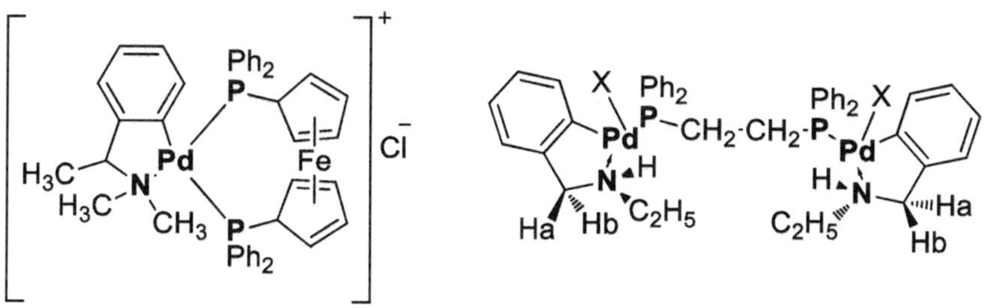

49

X = Cl (**50a**)
X = Br (**50b**)

Fig. 9.17 Cyclopalladated complexes based on bisphosphinic ligand (**49**) and diphosphine bridge (**50**).

breast carcinoma (MCF-7) tumor cell lines which revealed that the metal complexes were more effective in conjunction with bromide instead of chloride as the leaving group. DNA binding assay suggested that the complexes interacted with FS-DNA, through the intercalation mode. The UV absorption and tryptophan quenching studies revealed reactivity of these complexes with BSA suggested static quenching and complexes exhibited good binding affinity for BSA *(60)*.

Saccharin anions have various poly-functional ligands due to the presence of several potential donor sites such as nitrogen, one carbonyl and two sulfonyl oxygen atoms. A strong in vitro anticancer activity was shown by the [Pd(sac)(tpy)](sac)_4H_2O (sac: a saccharinate and tpy: 2,2′:60,200-terpyridine) palladium(II) complex **51** (Fig. 9.18) against the MCF-7 and

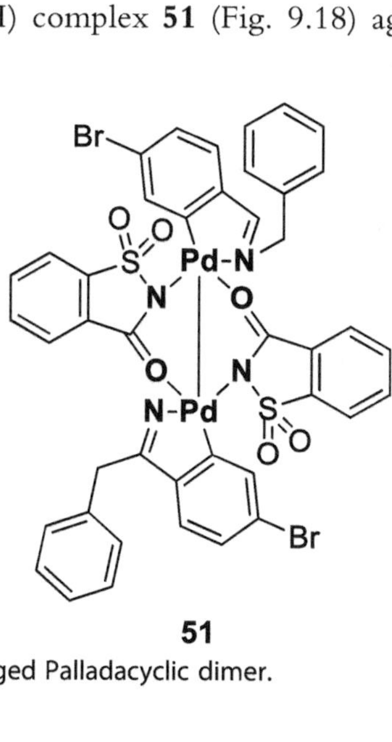

51

Fig. 9.18 Sachrinate bridged Palladacyclic dimer.

MDA-MB-231 human breast cancer cell lines, through the induction of apoptosis via cell death receptors. Therefore, saccharine-based Pd(II) complex was synthesized. The DNA binding properties of this complex suggested its interaction with the CT-DNA is via groove binding mode. Interaction of complex with BSA demonstrated a static type of binding, resulting due to quenching of BSA fluorescence. In addition, a strong in vitro anticancer activity was observed against MCF-7 and A549 cell lines, however no significant viability reduction was observed in normal cells *(40, 61, 62)*.

Palladium structures **52** (Fig. 9.19) containing 3-acetyl-4-hydroxy coumarine and methyl esters of methionine were examined for their cytotoxicity against two leukemia cell lines (JVM-13 and MOLT-4) and primary leukemic cells from leukemia (CLL) patients. Studies revealed that the complexes had no effect on the viability of all the three types of cancer cells *(47)*.

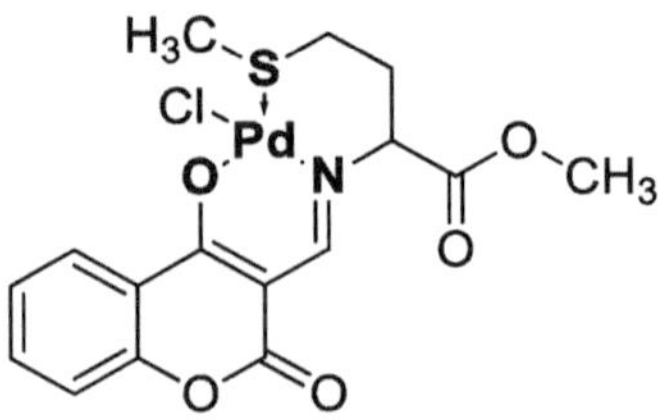

52

Fig. 9.19 Pd(II) complexes with coumarine and methionine.

3. CONCLUSION

Numerous research investigations have explored the influence of the structural properties of palladium complexes over their biological activities, which have suggested the potential of palladium metal as an alternative anticancer moiety. Design and synthesis of thermally stable palladium complexes highly depends on a judicious choice of ligands, having good electron donating groups and leading to stronger coordination bonds, thus improving their chances to reach the target cells. Additionally, the bulkiness of ligands is another factor to be considered, which prevents their premature hydrolysis before entering the cells, by protecting the metal center, thereby permitting its interactions with DNA strands. However, the choice of bulky ligands needs to be balanced with their water solubility. Further, it is studied that sulfur containing complexes are more potent in their anticancer activities.

Thus, it is imperative to design palladium complexes having bulky electron donating groups, yet possessing water-solubility and inertness toward sulfur containing biomolecules. The greater part of investigations, to date, have been directed toward the synthesis of different palladium-based therapeutic agents, having different structural properties, which in turn has produced an enormous amount of research on palladium coordination compounds. However, to date no palladium drug is qualified for clinical use. However, various studies suggest the possibility of using palladacycles as anticancer agents to circumvent the problems caused by the kinetic liability of Pd(II). This chapter highlights the structural advantages enjoyed by palladium complexes to improve their stability, hence rendering them as suitable anticancer drugs. Various studies highlight the need to understand the interaction of palladium complexes with various bio molecules containing sulfur groups, so as to predict their efficacy and ability to undergo renal clearance. The choice of the ligand structure and its coordination with the metal center yields an array of complexes with diverse conformational and geometrical properties, which can be exploited to target cellular signaling pathways other than binding to the cellular DNA. Also, the higher stability and solubility of complexes relative to cisplatin allows more time for the drug to reach the target site. The high stability of palladacyclic compounds in physiological media and the resultant low toxicity to normal cells make them promising candidates as future anticancer agents. The greatest hope lies, however, in finding a more powerful relationship between their chemical structures and antitumor activities, to directly exterminate the cancer cells.

REFERENCES

1. Borch, R. F.; Pleasants, M. E. Inhibition of cis-platinum nephrotoxicity by diethyldithiocarbamate rescue in a rat model. *Proc. Natl. Acad. Sci.* **1979**, *76*(12), 6611–6614.
2. Madias, N. E.; Harrington, J. T. Platinum nephrotoxicity. *Am. J. Med.* **1978**, *65*(2), 307–314.
3. Che, C.-M.; Siu, F.-M. Metal complexes in medicine with a focus on enzyme inhibition. *Curr. Opin. Chem. Biol.* **2010**, *14*(2), 255–261.
4. van Rijt, S. H.; Sadler, P. J. Current applications and future potential for bioinorganic chemistry in the development of anti-cancer drugs. *Drug Discov. Today* **2009**, *14*(23–24), 1089–1097.
5. Kelland, L. The resurgence of platinum-based cancer chemotherapy. *Nat. Rev. Cancer* **2007**, 7(8), 573.
6. Martin, L. P.; Hamilton, T. C.; Schilder, R. J. Platinum resistance: the role of DNA repair pathways. *Clin. Cancer Res.* **2008**, *14*(5), 1291–1295.
7. Pranczk, J.; et al. Platinum (II) and palladium (II) complex compounds as anti-cancer drugs. Methods of cytotoxicity determination. *Curr. Pharm. Anal.* **2014**, *10*(1), 2–9.

8. Dupont, J.; Pfeffer, M. *Palladacycles: Synthesis, Characterization and Applications;* John Wiley & Sons: Weinheim, Germany, 2008.
9. Newkome, G.; et al. Square-planar cis-and trans-C-palladium (II) complexes of N-electron-deficient heteroaromatic ligands. Ligand synthesis, complexation, spectral analyses, and complex interaction with phage PM2 DNA+. *Inorg. Chem.* **1985**, *24*(6), 811–826.
10. Quiroga, A. G.; Ranninger, C. N. Contribution to the SAR field of metallated and coordination complexes: studies of the palladium and platinum derivatives with selected thiosemicarbazones as anti-tumoral drugs. *Coord. Chem. Rev.* **2004**, *248*(1-2), 119–133.
11. Dupont, J.; Consorti, C. S.; Spencer, J. The potential of palladacycles: more than just precatalysts. *Chem. Rev.* **2005**, *105*(6), 2527–2572.
12. Quiroga, A. G.; et al. Novel tetranuclear orthometalated complexes of Pd (II) and Pt (II) derived from p-isopropylbenzaldehyde thiosemicarbazone with cytotoxic activity in cis-DDP resistant tumor cell lines. Interaction of these complexes with DNA. *J. Med. Chem.* **1998**, *41*(9), 1399–1408.
13. Quiroga, A. G.; et al. Palladated and platinated complexes derived from phenylacetaldehyde thiosemicarbazone with cytotoxic activity in cis-DDP resistant tumor cells. Formation of DNA interstrand cross-links by these complexes. *J. Inorg. Biochem.* **1998**, *70*(2), 117–123.
14. Zmejkovski, B. B.; et al. Synthesis, characterization and in vitro anti-tumor activity of new palladium (II) complexes with (S, S)-R2edda-type esters. *Polyhedron* **2014**, *80*, 106–111.
15. Herrmann, W. A.; et al. Palladacycles as structurally defined catalysts for the Heck olefination of chloro-and bromoarenes. *Angew. Chem. Int. Ed. Engl.* **1995**, *34*(17), 1844–1848.
16. Trofimenko, S. Cyclopalladation reaction. *Inorg. Chem.* **1973**, *12*(6), 1215–1221.
17. Albéniz, A. C.; Espinet, P. Palladium: Inorganic & Coordination Chemistry. In *Encyclopedia of Inorganic and Bioinorganic Chemistry;* John Wiley & Sons, Ltd., 2006, John Wiley & Sons, Ltd., https://doi.org/10.1002/0470862106.ia178.
18. Ramachandran, E.; et al. Role of substitution at terminal nitrogen of 2-oxo-1, 2-dihydroquinoline-3-carbaldehyde thiosemicarbazones on the coordination behavior and structure and biological properties of their palladium (II) complexes. *Inorg. Chem.* **2013**, *52*(3), 1504–1514.
19. Cruz, S.; et al. Synthesis, characterization and anti-cancer activity of new palladacycles derived from chiral α-diimines. *Appl. Organomet. Chem.* **2010**, *24*(1), 8–11.
20. Icsel, C.; et al. Cationic Pd (II)/Pt (II) 5, 5-diethylbarbiturate complexes with bis (2-pyridylmethyl) amine and terpyridine: Synthesis, structures, DNA/BSA interactions, intracellular distribution, cytotoxic activity and induction of apoptosis. *J. Inorg. Biochem.* **2015**, *152*, 38–52.
21. Ari, F.; et al. Palladium (II) saccharinate complexes with bis (2-pyridylmethyl) amine induce cell death by apoptosis in human breast cancer cells in vitro. *Bioorg. Med. Chem.* **2013**, *21*(11), 3016–3021.
22. Churusova, S. G.; et al. Highly cytotoxic palladium (II) pincer complexes based on picolinylamides functionalized with amino acids bearing ancillary S-donor groups. *Inorg. Chem.* **2017**, *56*(16), 9834–9850.
23. Kapdi, A. R.; Fairlamb, I. J. Anti-cancer palladium complexes: a focus on PdX 2 L 2, palladacycles and related complexes. *Chem. Soc. Rev.* **2014**, *43*(13), 4751–4777.
24. Teyssot, M.-L.; et al. Metal-NHC complexes: a survey of anti-cancer properties. *Dalton Trans.* **2009**, *35*, 6894–6902.
25. Liu, W.; Gust, R. Metal N-heterocyclic carbene complexes as potential anti-tumor metallodrugs. *Chem. Soc. Rev.* **2013**, *42*(2), 755–773.
26. Zou, T.; et al. Anti-cancer metal-N-heterocyclic carbene complexes of gold, platinum and palladium. *Curr. Opin. Chem. Biol.* **2018**, *43*, 30–36.

27. Gründemann, S.; et al. Bis-carbene complexes from oxidative addition of imidazolium C–H bonds to palladium(0). *J. Chem. Soc. Dalton Trans.* **2002**, *10*, 2163–2167.
28. Lee, J.-Y.; et al. Palladium complexes with tridentate n-heterocyclic carbene ligands: selective "normal" and "abnormal" bindings and their anti-cancer activities. *Organometallics* **2015**, *34*(17), 4359–4368.
29. Haque, R. A.; et al. Bis-N-heterocyclic carbene silver (I) and palladium (II) complexes: efficient antiproliferative agents against breast cancer cells. *Inorg. Nano-Met. Chem.* **2017**, *47*(1), 131–137.
30. Frik, M.; et al. Luminescent iminophosphorane gold, palladium and platinum complexes as potential anti-cancer agents. *Inorg. Chem. Front.* **2014**, *1*(3), 231–241.
31. Fong, T. T. H.; et al. Cyclometalated palladium (II) N-heterocyclic carbene complexes: anti-cancer agents for potent in vitro cytotoxicity and in vivo tumor growth suppression. *Angew. Chem. Int. Ed.* **2016**, *55*(39), 11935–11939.
32. Karami, K.; et al. Synthesis, characterization and biological activities of two novel ortho-palladated complexes: Interactions with DNA and bovine serum albumin, antitumour activity and molecular docking studies. *Appl. Organomet. Chem.* **2017**, *31*(10), e3740.
33. Blanckenberg, A.; et al. Preparation, characterization and evaluation of novel 1, 3, 5-triaza-7-phosphaadamantane (PTA)-based palladacycles as anti-cancer agents. *J. Organomet. Chem.* **2017**, *851*, 68–78.
34. Karami, K.; et al. Cyclopalladated complexes containing metformin and benzylamine derivatives: synthesis, characterization, binding interactions with DNA and BSA, in vitro cytotoxicity studies. *Inorg. Chim. Acta* **2017**, *467*, 46–55.
35. Albert, J.; et al. Endo and exo cyclopalladated (E)-N-([1,1'-biphenyl]-2-yl)-1-mesityl-methanimines: anticancer, antibacterial and antioxidant activities. *J. Organomet. Chem.* **2017**, *839*, 116–125.
36. Jurisevic, M.; et al. Platinum complexes with eDDa (ethyleneDiamine-N, N-Diacetate) liganDs as potential anti-cancer agents. *Serb. J. Exp. Clin. Res.* **2016**, *17*(4), 285–296.
37. Motswainyana, W. M.; et al. Anti-tumor activity of phenylene bridged binuclear bis (imino-quinolyl) palladium (II) and platinum (II) complexes. *Bioorg. Med. Chem. Lett.* **2014**, *24*(7), 1692–1694.
38. El-Metwaly, N. M.; El-Ghalban, M. G. Hartree-Fock, molecular docking, spectral, kinetic and antitumor considerations for cobalt, nickel, palladium and platinum (II)-bis carbothiohydrazide complexes. *J. Mol. Liq.* **2016**, *220*, 265–276.
39. Cetin, Y.; et al. A palladium (II)–saccharinate complex of terpyridine exerts higher anti-cancer potency and less toxicity than cisplatin in a mouse allograft model. *Anti-Cancer Drugs* **2017**, *28*(8), 898–910.
40. Ulukaya, E.; et al. Anti-cancer activity of a novel palladium (II) complex on human breast cancer cells in vitro and in vivo. *Eur. J. Med. Chem.* **2011**, *46*(10), 4957–4963.
41. Rubino, S.; et al. Synthesis, properties, anti-tumor and antibacterial activity of new Pt (II) and Pd (II) complexes with 2, 2′-dithiobis (benzothiazole) ligand. *Bioorg. Med. Chem.* **2017**, *25*(8), 2378–2386.
42. Soliman, A. A.; et al. New palladium (II) formamidine complexes: preparation, characterization, theoretical calculations and cytotoxic activity. *J. Mol. Struct.* **2017**, *1137*, 453–460.
43. Makedonas, C.; Mitsopoulou, C. A. An investigation of the reactivity of [(diimine) (dithiolato) M] complexes using the Fukui functions concept. *Eur. J. Inorg. Chem.* **2006**, *2006*(3), 590–598.
44. Amir, M. K.; et al. Anti-cancer activity, DNA-binding and DNA-denaturing aptitude of palladium (II) dithiocarbamates. *Inorg. Chim. Acta* **2016**, *451*, 31–40.
45. Shahraki, S.; et al. Platinum (II) and Palladium (II) complexes with 1, 10-phenanthroline and pyrrolidinedithiocarbamato ligands: synthesis, DNA-binding and anti-tumor activity in leukemia K562 cell lines. *Iran. J. Sci. Technol. (Sciences)* **2015**, *39*(2), 187–198.

46. Shahraki, S.; et al. Effects of vitamin K3 on Cytotoxicity and DNA-binding properties of a Palladium (II) complex as a potential anti-tumor agent. *Biomacromol. J.* **2017**, *3*(1), 60–74.
47. Khan, S. Z.; et al. New heteroleptic palladium (II) dithiocarbamates: synthesis, characterization, packing and anti-cancer activity against five different cancer cell lines. *Appl. Organomet. Chem.* **2016**, *30*(6), 392–398.
48. Haribabu, J.; et al. Synthesis, structures and mechanistic pathways of anti-cancer activity of palladium (ii) complexes with indole-3-carbaldehyde thiosemicarbazones. *New J. Chem.* **2018**.
49. Farkasová, V.; et al. Low-dimensional compounds containing bioactive ligands. Part VIII: DNA interaction, antimicrobial and anti-tumor activities of ionic 5, 7-dihalo-8-quinolinolato palladium (II) complexes with K+ and Cs+ cations. *J. Inorg. Biochem.* **2017**, *167*, 80–88.
50. El-Metwally, N. M.; et al. Spectral, magnetic, electrical and thermal studies on malonyl bis (thiosemicarbazide) complexes. *Transit. Met. Chem.* **2006**, *31*(1), 71–78.
51. Zhu, M.; et al. The structures, cytotoxicity, apoptosis and molecular docking controlled by the aliphatic chain of palladium (II) complexes. *J. Inorg. Biochem.* **2016**, *157*, 34–45.
52. Şahin, Ö.; et al. New platinum (II) and palladium (II) complexes of coumarin-thiazole Schiff base with a fluorescent chemosensor properties: synthesis, spectroscopic characterization, X-ray structure determination, in vitro anti-cancer activity on various human carcinoma cell lines and computational studies. *J. Photochem. Photobiol. B Biol.* **2018**, *178*, 428–439.
53. Soliman, A. A.; et al. New formamidine ligands and their mixed ligand palladium (II) oxalate complexes: Synthesis, characterization, DFT calculations and in vitro cytotoxicity. *Spectrochim. Acta A Mol. Biomol. Spectrosc.* **2016**, *152*, 358–369.
54. Pruchnik, H.; et al. Palladium (II) complexes with tris (2-carboxyethyl) phosphine, structure, reactions and cytostatic activity. *J. Inorg. Biochem.* **2016**, *156*, 14–21.
55. Shabbir, M.; et al. Schiff base triphenylphosphine palladium (II) complexes: synthesis, structural elucidation, electrochemical and biological evaluation. *J. Mol. Struct.* **2016**, *1118*, 250–258.
56. Muralisankar, M.; et al. An investigation on the DNA/protein binding, DNA cleavage and in vitro anti-cancer properties of SNO pincer type palladium (II) complexes with N-substituted isatin thiosemicarbazone ligands. *Inorg. Chim. Acta* **2017**, *466*, 61–70.
57. Gupta, G.; et al. Self-assembled novel bodipy-based palladium supramolecules and their cellular localization. *Inorg. Chem.* **2017**, *56*(8), 4615–4621.
58. Nassar, A.; Hassan, A.; Alabd, S. Anti-tumor and antimicrobial activities of novel palladacycles with abnormal aliphatic CH activation of Schiff Base 2-[(3-phenylallylidene) amino] phenol. *Synth. React. Inorg. Met.-Org. Nano-Met. Chem.* **2015**, *45*(2), 256–270.
59. Bechara, A.; et al. Palladacycle (BPC) antitumour activity against resistant and metastatic cell lines: the relationship with cytosolic calcium mobilisation and cathepsin B activity. *Eur. J. Med. Chem.* **2014**, *79*, 24–33.
60. Karami, K.; et al. In vitro cytotoxicity studies of palladacyclic complexes containing the symmetric diphosphine bridging ligand. Studies of their interactions with DNA and BSA. *Eur. J. Med. Chem.* **2014**, *73*, 8–17.
61. Karami, K.; et al. A saccharinate-bridged palladacyclic dimer with a Pd–Pd bond: experimental and molecular docking studies of the interaction with DNA and BSA and in vitro cytotoxicity against human cancer cell lines. *New J. Chem.* **2018**, *42*(1), 574–586.
62. Ari, F.; et al. Synthesis, structural characterization and cell death-inducing effect of novel palladium (II) and platinum (II) saccharinate complexes with 2-(hydroxymethyl) pyridine and 2-(2-hydroxyethyl) pyridine on cancer cells in vitro. *Bioorg. Med. Chem.* **2013**, *21*(21), 6427–6434.

CHAPTER 10

Miscellaneous Applications of Palladacycles

Gaurav R. Gupta*, Shatrughn A. Bhilare*, Debabrata Maiti†, Anant R. Kapdi*
*Department of Chemistry, Institute of Chemical Technology, Mumbai, India
†Department of Chemistry, Indian Institute of Technology, Mumbai, India

Contents

1. INTRODUCTION

The impact of CO on human health accentuates the scientific community to consider the harm caused by CO as a principle subject of interest. Carbon monoxide is a member of the gasotransmitter family, a crucial gas message molecule, which plays an important role in the regulation of both physiological and pathological processes. Hypoxia-induced CO is involved in modulating various cellular activities, including signal transduction, proliferation, and apoptosis *(1–6)*. The main clinical manifestations of CO poisoning, such as increase in blood pressure, and adverse effects on the central nervous system, causing headache, dizziness, myasthenia, paralysis, convulsions, and changes in perception, including changes in the visual and auditory system, were skillfully reviewed by Lewin *(7)*. The past decade has seen an unprecedented rise in the number of attempts made by the scientific community to develop subtle techniques and systems for monitoring, or the utilization

Palladacycles
https://doi.org/10.1016/B978-0-12-815505-9.00011-1
© 2019 Elsevier Inc.
All rights reserved.

of CO, thus reducing the hazards associated with CO up to a certain extent. Although exogenous therapeutic additions of CO to tissues and whole animals have been studied thoroughly *(8)*, the real-time spatial and temporal tracking of CO at the cellular level still has not been achieved thus far.

Developments in science and technology offer highly efficient sensing devices for the detection of chemical constituents that could be in the solid, liquid, or in gaseous form, and even the presence of trace amounts could be detected. In this regard, chemical sensors have in recent years been developed in such a way that a chemical (or even biological) molecular process could be tuned very sophisticatedly to detect or quantify a certain chemical species in a gas, liquid, or solid sample. The presence of the chemical constituent could be analyzed and their response documented in terms of current, voltage, mass change, altered optical property, or temperature *(9)*.

2. LUMINESCENCE PROPERTIES OF PALLADACYCLES

Spectroscopic examination of organometallic complexes evidenced that luminescence from low-lying MLCT excited states offer a new type of spectroscopic probe; the metal complexes exhibit a number of key features that, taken together, provide advantages over conventional probe molecules *(10–12)*. Investigations related to the luminescence properties of palladacycles and the effect of the surrounding ligands and ligand systems have been systematically scrutinized by Gonzalez et al. *(13)* by developing novel palladacyclic systems (Scheme 10.1) (**1a–c**). Authors very systematically examined the photo-optical properties of the synthesized palladacycles **(1a–c)**. In addition to that, the effects induced by:

(a) the nature of the substituent and

(b) the placement of the phenyl rings with 2-phenylindole unit on the potential luminescence of the ligands and their palladacycles have also been studied by the authors.

The UV-visible spectra of the free ligands and the palladacycles **1a–c** in dichloromethane at 298 K (Table 10.1) showed two bands for free ligands and three bands for the palladacycles **1a–c**, respectively. Amongst these, the band which appeared at higher energies when compared with the bands of free ligands are attributed to a metal perturbed intra-ligand transition (MPILT). An extra band found in the case of the palladacycle depends on the polarity of the solvent and is assigned to the metal to ligand charge transfer transition (MLCT). Authors also explored the potential emissive

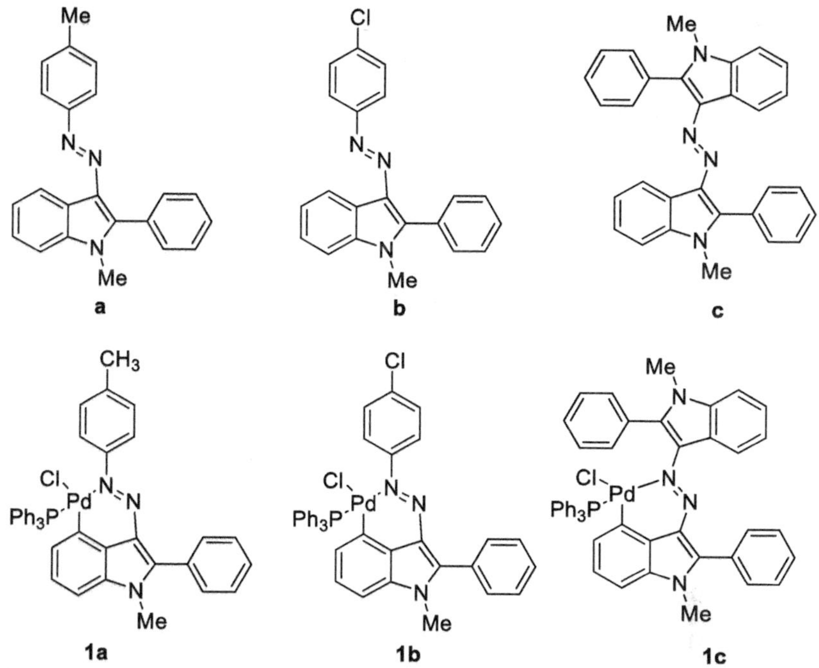

Scheme 10.1 Structure of ligands (a–c) and corresponding palladacycles 1a–c.

properties of the ligands and the palladacycles **1a–3c** in the solid state and in dichloromethane solution at 298 K. The study revealed that the free ligands were not emissive in the solid state, but in dichloromethane they became luminescent upon excitation at $\lambda_{exc} = 280$ nm (for **a** and **b**) or 300 nm (for **c**). The emission spectra of solid samples of the palladacycles **1a** and **1b** (at $\lambda_{exc} = 480$ nm) at 298 K showed a broad emission band in the range of 590–595 nm; while for **1c** a stronger emission (at $\lambda_{exc} = 630$ nm) was observed. No significant emission was found when palladacycles **1a** and **1b** (at $\lambda_{exc} = 480$ nm) were excited in dichloromethane solution (Table 10.1). In contrast with the results obtained for **1a** and **1b,** complex **1c** was strongly luminescent under the same conditions and the emission spectra obtained when excited at 520 nm showed a band at low energies ($\lambda_{max} = 610$ nm). Attempts have been made by the authors to measure total luminescence quantum yields at 298 K for ligands **a–c**, and palladacycle **1a** and **1b** against quinine sulfate in 1N H_2SO_4 ($\Phi = 0.54$) as the standard reference and for **1c**, $[Ru(bipy)_3]Cl_2$ in water ($\Phi = 0.042$) as the standard reference system. The study reveals that the quantum yield of **c** is significantly higher than those of **a** and **b**.

Table 10.1 Absorption and emission properties of the free ligands **a–c** and the cyclopalladated complexes **1a–c** in dichloromethane at 298 K (wavelengths, λ_i (i = 1,2 or 3) excitation wavelength λ_{exc}. And λ_{max} in nm and extinction coefficients ε_i (in M^{-1} cm^{-1}) and quantum yields, Φ)

		Absorption spectroscopic data			Emission data λ_{max}			
Entry		λ_1 (ε_1)	λ_2 (ε_2)	λ_3 (ε_3)	λ_{exc}	**Solid state**	**In solution**	Φ
Ligands								
1	**A**	381 (27.0×10^3)	275 (45.6×10^3)	–	280	–	430	0.0015
2	**B**	386 (11.8×10^3)	273 (101×10^3)	–	280	–	396	0.0012
3	**C**	301 (6.9×10^3)	232 (10.8×10^3)	–	300	–	415, 379, 363 and 349	0.17
Palladacycles								
4	**1a**	468 (20.6×10^3)	326 (22.3×10^4)	269 (18.4×10^4)	480 275	594 –	NE 408	– 0.0018
5	**1b**	472 (16.6×10^3)	326 (62.7×10^3)	255 (49.1×10^3)	480 275	590 –	NE 395	– 0.0016
6	**1c**	511 (20.0×10^3)	326 (92.4×10^4)	234 (66.3×10^3)	520	630	610	0.0053

3. PALLADACYCLES IN CO SENSING

One of the earliest examples of carbon monoxide sensing comes from the pioneering work by Cinellu et al. *(14)* in 1989 in the form of carbonylation of palladacycle (**2b**) under high pressure, opening up an avenue to the scientific community for developing chemical sensors based on the palladacyclic unit as a core for the estimation of CO (Scheme 10.2). The authors developed a general protocol for the carbonylation of the palladacycle **2b**. The utilization of dichloromethane as the solvent under 70–90 atmospheric pressure enabled a mixture of three different products which were contaminated by a small amount of metal. Thereafter, the treatment of ether allowed the release of orange crystals of a major product **2a** which exhibited a strong absorption at 1675 cm^{-1} in the infrared spectrum. The second product was very difficult to isolate from the reaction mixture as it got converted rapidly into compound **2c**. Finally, the third compound **2c** was isolated in the form of a hydrochloride. After successful isolation of the products, the attempt was made by the authors to generalize the mechanism of carbonylation reaction of the palladacycle, and it was observed that the reaction initiated the unstable six-membered acyl complex through the insertion of carbon monoxide into the palladium-carbon bond. The labile behavior of the palladacycle toward CO resulted in a number of palladacycles for the quantitative detection of CO and has been of keen interest to researchers.

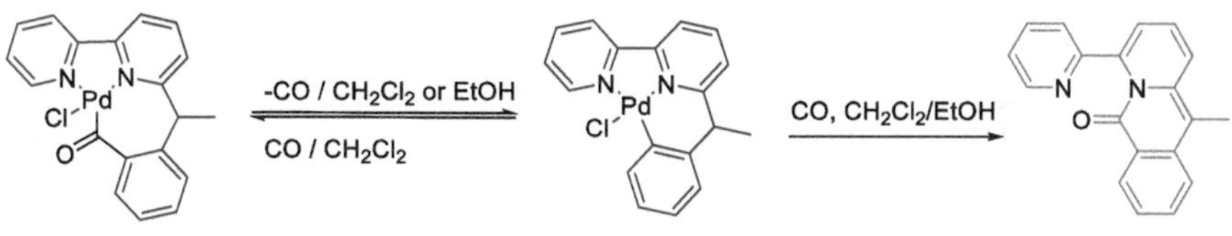

2a 2b 2c

Scheme 10.2 Carbonylation of palladacycle 2b.

To have a better understanding of the phenomenon of carbonylation of palladacycles, Michel et al. *(15)* very meticulously developed an efficient turn-on fluorescent probe for selective CO detection (Scheme 10.3). Investigations related to the fluorescence properties and CO reactivity of the synthesized palladacycle **3a** in aqueous solution buffered to physiological pH have also been reported by the authors. The authors employed

[Ru(CO)$_3$Cl(glycinate)] as an easy-to-handle CO source to allow the in vitro analysis of CO. It was noted that the synthesized palladacycle **3a** exhibits very weak fluorescence in Dulbecco's phosphate buffered saline (DPBS) buffered to pH 7.4 ($\lambda_{em} = 503$ nm, $\Phi = 0.01$). While on the addition of 50 μM [Ru(CO)$_3$Cl(glycinate)] to a solution of palladacycle **3a** at 310 K, carbonylation product **3b** was formed which was found to be responsible for the robust fluorescent turn-on response ($\lambda_{max} = 499$ nm, $\varepsilon = 23\ 000\ L\ M^{-1}\ cm^{-1}$, $\lambda_{em} = 507$ nm, $\Phi = 0.44$). A ten-fold increase in the fluorescence was observed within 60 min after initiating the reaction. Furthermore, good to excellent selectivity for palladacycle **3a** as the fluorescence turn-on probe was scrutinized over other biologically relevant reactive oxygen, nitrogen, and sulfur species, including H_2O_2, *tert*-butylhydroperoxide (*t*-BuOOH), hypochlorite (OCl^-), superoxide (O^{2-}), NO, peroxynitrite ($ONOO^-$), and H_2S. This study revealed that exposure of palladacycle **3a** to these molecules failed to trigger any fluorescence response as compared to that observed with CO. Interestingly, the use of water as solvent for such reactions provided acid as the major product, while in organic solvent formation of dealkylative amide as a product was observed.

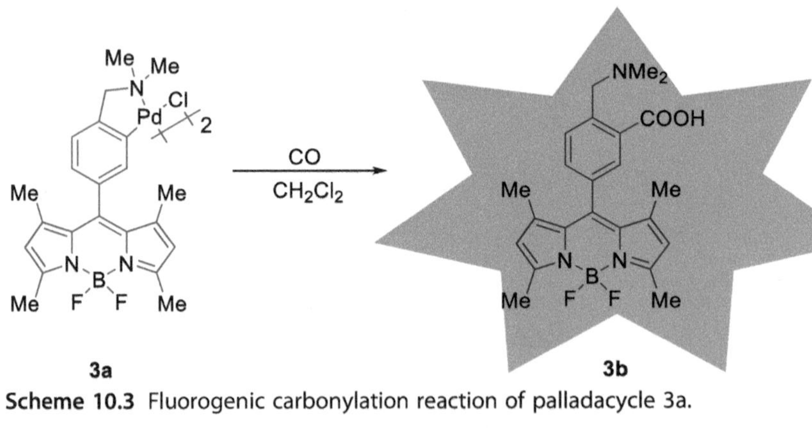

3a **3b**

Scheme 10.3 Fluorogenic carbonylation reaction of palladacycle 3a.

A comprehensive review by Bumbrah et al. *(16)* unfolding the utility of Raman spectroscopy in the forensic analysis of different types of inks, characterization of drugs of abuse, and other illicit substances on different matrices, such as cloth, currency notes, fiber etc., without extensive sample preparation in a nondestructive manner, was recently published highlighting

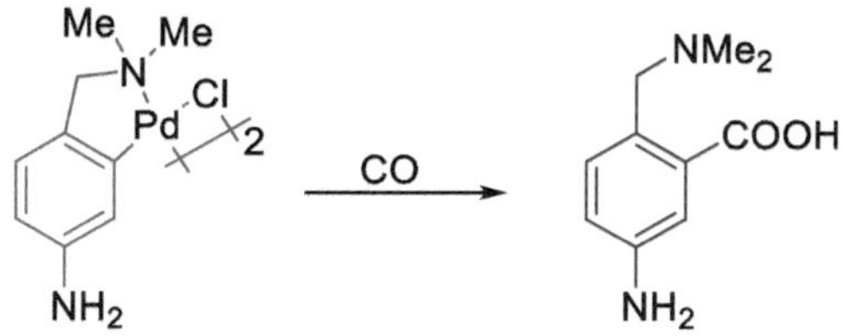

4a **4b**

Scheme 10.4 Carbonylation reaction of palladacycle 4a on the surface of AuNP.

all these aspects. Cao et al. *(17)* very competently explored the affinity of palladacycle **4a** in carbonylation processes for the development of highly efficient and robust nano-sensor models, which exhibits high selectivity toward the detection of intracellular carbon monoxide (CO) by the employment of surface-enhanced Raman spectroscopy (SERS) (Scheme 10.4). The study has been divided into three parts:

(1) SERS observation of palladacycle carbonylation on AuNP/PC nano-sensors

(2) SERS detection of CO in solution using AuNP/PC nano-sensors and

(3) detection of CO in living cells.

In the first part of their experiment, the process of carbonylation of palladacycle with the prepared AuNP/PC (palladacycle accumulated on gold nanoparticles) nano-sensors was monitored by SERS technique. The results obtained for the initial phase of the reaction were monitored with the SERS spectrum of the palladacycle **4a** as a self-assembled monolayer (SAM) on the surface of AuNPs. The resulting spectrum exhibits a Raman band at 1173 cm^{-1} due to the symmetric stretching vibration of C−C, whereas the peak at 1209 cm^{-1} is attributed to the phenyl ring modes. As the reaction commences, and after 10 min of time interval of the reaction, the intensity of the palladacycle bands at 1173 and 1209 cm^{-1} decrease significantly, while a simultaneous formation of a new set of vibrational Raman bands at 1032, 1118, and 1217 cm^{-1} was observed in the spectrum. The formation of new vibrational Raman bands has been strongly attributed to the coordination of CO to palladium center. The newly developed spectral bands at 1032, 1118, and 1217 cm^{-1} have been assigned more precisely to the contribution of R-NH_2 rocking, $\nu_{(C\text{-}X)}$ stretch, and $\nu_{(C\text{-}O)}$ stretch respectively. As the reaction progresses steadily few more Raman bands became much stronger at 1071, 1242, and 1288 cm^{-1} in the spectrum, which might be attributed to the presence of —COOH functionality in the product **4b**.

Secondly, the CO sensing potential of the developed AuNP/PC nano-sensors in solution was achieved by monitoring the reaction under pH 7.4 at temperature 310 K. The study was performed by using biologically compatible and SERS inactive tricarbonylchloro(glycinato)ruthenium(II) as a CO source. Moreover, when the concentration of tricarbonylchloro(glycinato) ruthenium(II) was varied, significant impact on the SERS spectra of AuNP/PC was observed, e.g., when 0.5 μM tricarbonylchloro(glycinato) ruthenium(II) was responsible for the instant exclusion of the bands at 1173 and 1209 cm^{-1}. On the other hand, bands at 1032, 1118, 1242, 1288, and 1358 cm^{-1} appeared in the spectrum and their intensity was found to be enhanced with the increasing concentration of tricarbonylchloro(glycinato)ruthenium(II).

The high selectivity of the developed SERS nano-sensors was advanced by screening the nano-sensors against other biologically active species. The study revealed that no significant change in the SERS spectra of AuNP/PC was recorded, even in the presence of other gaseous molecules (Na_2S, NO), reactive oxygen species, and prevalent biological molecules (GSH, CN^-) *(17)*. In addition, detection of CO in living cells could be afforded very selectively using the developed nano-sensor. The SERS nano-sensors, when they entered the cells and via endocytosis, were accumulated in the cytosol, dark-field microscope (DFM) images, and were then taken and utilized for the visualization of the positions of the SERS nano-sensors; thereafter the corresponding SERS spectrum was recorded. It has been well understood that the SERS response of the palladacycle and the compositions of the cells were too weak to record, and therefore no SERS spectra was recorded without the addition of exogenous CO. The presence of tricarbonylchloro(glycinato)ruthenium(II) (0.5 μM) was responsible for the absorption at 1032 cm^{-1} and more SERS bands at 1118 and 1242 cm^{-1} appeared and became more intense when higher concentrations of tricarbonylchloro (glycinato)ruthenium(II) (5.0 μM, 30 μM) were introduced. The present protocol based on the chemical reactivity of palladacycle monitored with SERS technique affords a novel and promising approach for the analysis of CO functions in the systems which are actively involved in life governing processes where highly selective and sensitive detection of intra- as well as intercellular CO is required.

Li et al. *(18)* also developed a protocol (Scheme 10.5) for the carbonylative sensing of CO using azobenzene-cyclopalladium-based fluorescent probes **5a–b**. The authors explored commercially available dichlorotricarbonylruthenium(II) dimer as a CO source in a pH range of 3-9. In the

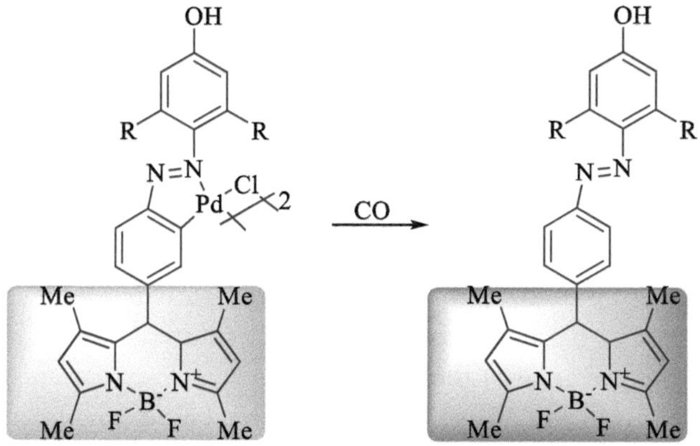

5a (R = H)
5b (R = CH_3)

5c (R = H)
5d (R = CH_3)

Scheme 10.5 Carbonylation reaction of palladacycle (5a–b).

beginning, excitation/emission spectra of the probes were recorded and analyzed, with the results of the study highlighting that the synthesized palladacycles **5a** and **5b** exhibited excitation/emission maxima at 498/512 nm. A very strong signal-to-noise ratio and an exceptional fluorescent enhancement was recorded when treated with 20 equiv of dichlorotricarbonylruthenium(II) dimer with palladacycle **5b.** The selectivity and sensitivity trend for the palladacycles also compared well with each other. On the basis of the result obtained, it should be noted that palladacycle **5b** was found to be quite promising and exhibited high sensitivity toward CO as compared to palladacycle **5a**. Table 10.2 summarizes the molar extinction coefficients (ε) and fluorescence quantum yield (Φ_f) for the palladacycles **5a** and **5b** as well as their carbonylation reaction products when treated with CO gas (**5c** and **5d**). On closer scrutiny, Table 10.2 indicates that these four compounds have almost the same ε values, while comparable difference was displayed for the Φ_f values for the palladacycles (**5a and 5b**) and the products of carbonylation reaction (**5c and 5d**).

Table 10.2 Molar attenuation coefficients (ε) and fluorescence quantum yield (Φ_f) of **5a, 5b, 5c and 5d**

Entry	Compounds	5a	5b	5c	5d
1	ε (L mol^{-1} cm^{-1})	10200	10100	9800	9900
2	Φ_f	0.039	0.009	0.155	0.176

Attempts have also been made by the authors to understand the detailed mechanism for the carbonylation reaction of palladacycles **5a** and **5b**. The spectroscopic evidence confirms that the reaction progresses through the protonolysis of the Pd—C bond in the presence of CO. Next, the specificity of **5b** toward CO was evaluated by performing fluorescence measurement of **5b** in the presence of a series of reactive oxygen species or reactive nitrogen species as well as reactive sulfide species, and other reducing species including glutathione, cysteine, homocysteine, hydrogen sulfide, ascorbic acid, and ferrous ion. It has been observed that no fluorescence increment was recorded in the presence of these species.

Encouraged by the above results, studies based on the fluorescent imaging of CO in the living cells were carried out by the authors by employing a laser-scanning confocal microscope. Human hepatocellular carcinoma (HepG2) cells were used throughout the experiment. Preliminary results revealed that palladacycle **5b** is a potential candidate for the specific monitoring of intracellular CO production. The present protocol offers a wide variety of fluorescent probes for CO detection with exceptionally high order of selectivity as well as sensitivity through the installation of fluorophores over the palladacycle backbone for bio-analytical applications.

Recently, Sun et al. *(19)* synthesized a novel benzimidazole-based palladacycle **6a** framework as a fluorescence turn-on probe that exhibited high selectivity and quantitative detection of carbon monoxide (Scheme 10.6). The recent investigations of the feasibility of benzimidazole-based palladacycle **6a** for the quantitative detection of CO have been advanced by the authors. Carboxylic acid functionalized benzimidazole was used to enhance water solubility of the synthesized palladacycle **6a**. The nonfluorescent behavior of the synthesized palladacycle **6a** was attributed to the heavy-atom

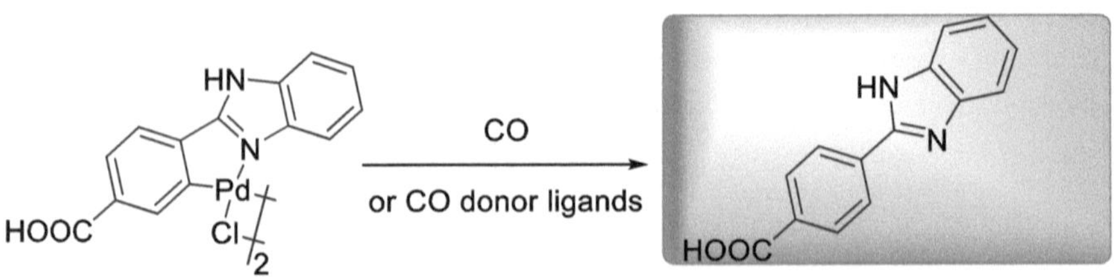

6a **6b**

Scheme 10.6 Degradation of palladacycle (6a) in presence of CO or CO donor ligands.

electronic effects exerted by palladium. The synthesized palladacycle **6a** triggered high contrast of the fluorescence intensity while interacting with CO. During the course of the studies, the authors investigated the spectral properties of palladacycle **6a** in aqueous solution buffered at physiological pH (20 mM 4-(2-hydroxyethyl)-1-piperazineethanesulfonic acid (HEPES) buffer, pH 7.4, 50% EtOH). The solution of palladacycle **6a** in HEPES buffer displayed weak fluorescence with maximum emission at 415 nm ($\phi_f = 0.009$). A dramatic fluorescence turn-on at 415 nm was observed after exposure of palladacycle **6a** (1 μM in 20 mM HEPES buffer, pH 7.4) to the CO gas. After 20 min of exposure of palladacycle **6a** to excess amounts of CO, the fluorescence intensity enhanced by 20-fold, while the fluorescence quantum yield of the probe improved more than 20-fold (QY = 0.28). For highly precise quantification, carbon monoxide donors, such as dichlorotricarbonylruthenium(II) dimer, were selected as a CO source as one can easily control its concentration. The present protocol could tolerate a variety of biologically relevant reactive oxygen or nitrogen species including hydrogen peroxide, superoxide anions, *tert*-butylperoxyl radical, *tert*-butylhydroperoxide, nitric oxide, peroxynitrite, hydrogen sulfide, andhydroxyl radical species, and afford high CO sensing. Besides this, other species such as nitrogen containing bases such as tryptophan, serine, sulfur containing species, and ions like fluoride, iodide, and nitrate have also been examined and no substantial effect was encountered during the analysis. The mechanism of action of the developed fluorescence probe originated via the carbonylation of palladacycle **6a**, and offers a fluorescent turn-on probe in which a weakly fluorescent motif very quickly and in a highly selective manner reacts with carbon monoxide with the reaction yielding an extremely fluorescent benzimidazole moiety **6b**. It has been observed that the synthesized palladacycle probe was innocent toward fluorescence and could release the fluorescence only in the presence of CO or CO donor ligands. This ability of palladacycle **6a** was extensively exploited by the authors for quantitative as well as qualitative detection of CO with high sensitivity. Intracellular imaging results indicate that the developed fluorescent probe was having cell permeability, posing low toxicity, and exhibiting the capability to perform in vivo analysis of CO. The probe based on palladacycle **6a** also exhibits some unique features such as being easy to synthesize, quick responding, and possessing high order of selectivity.

4. PALLADACYCLES AS THIOCYANATE SENSORS

The exploration of palladacycles as fluorescent probes for the quantitative detection of carbon monoxide in solution as well as in living cells facilitates the formation of a bridge between the science of molecules and the science of materials. This could be well illustrated by the following example, wherein Hassan et al. *(20)* broke new ground for the exploration of palladacycles (**7a**) as anion exchangers or iono-phores in potentiometric sensors (Scheme 10.7). On closer inspection of the results described in this study, the palladacycle **7a** exhibits high exchange capacity toward thiocyanate ion.

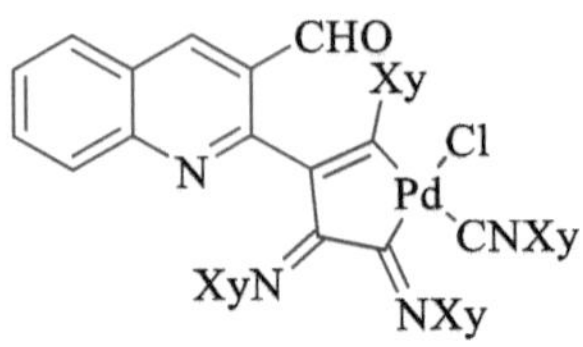

7a

Scheme 10.7 Structure of palladium ion exchanger complex (palladacycle 7a).

Initially, the ion exchange experiments were carried out through the plasticized PVC-based membrane sensors with and without Pd-complex for the determination of the thiocyanate ions. During the course of the studies, it has been observed that the sensor without palladacycle **7a** displayed no substantial response for thiocyanate or other common anions. On the other hand, the sensor developed with palladacycle **7a** as an ingredient displayed a linear Nernstian response toward SCN ions. The effect of solvent polarities on the sensor response has also been reported with different solvents (e.g., dioctylphthalate (DOP, $\varepsilon = 7$), dioctylsebacate (DOS, $\varepsilon = 3.9$) and *o-nitro*-phenyloctylether (*o*-NPOE, $\varepsilon = 24$)) as sensor membrane plasticizers with the best performance of the sensor recorded with 60–67 wt% *o*-NPOE. On the addition of tridodecylmethylammonium chloride (TDMAC) as cationic excluders (50 mol% of Pd-complex) no significant effect on the efficacy of the sensor was visualized, while the addition of sodium tetraphenylborate (NaTPB) as anionic additive (30 mol% of Pd-complex) enhanced both the response and selectivity of the sensor (Table 10.3).

Closer scrutiny of Table 10.3 also reveals that a sensor with membrane consisting of ionophore (1.5 wt%), *o*-NPOE as plasticizer (64.5 wt%) and PVC (34 wt%) produces a linear response over the concentration range of $1.0 \times 10^6 - 1.0 \times 10^2$ mol L^{-1} SCN with a slope value of 57.8 ± 0.2 mV·decade^{-1}. In addition to that, the lower limit of detection was 6.3×10^{-7} mol L^{-1}. At such dilute concentration level, the

Table 10.3 Potentiometric response characteristics of palladium complex based thiocyanate PVC membrane sensors using different plasticizers

Entry	Parameter	DOP	DOS	NPOE	NPOE + 50 mol % TDMAC	NPOE + 30 mol % Na TPB
1	Slope ($mV \cdot decade^{-1}$)	37.4 ± 0.1	41.5 ± 0.1	57.8 ± 0.2	57.8 ± 0.2	59.2 ± 0.1
2	Correlation coefficient (r^2)	0.9748	0.9936	0.9859	0.9929	0.9926
3	Intercept (mV)	59	68	112	59	82
4	Linear range ($mol\ L^{-1}$)	6.3×10^{-5}–1.0×10^{-2}	3.1×10^{-5}–1.0×10^{-2}	1.0×10^{-6}–1.0×10^{-2}	1.3×10^{-6}–1.0×10^{-2}	8.0×10^{-7}–1.0×10^{-2}
5	Detection limit ($mol\ L^{-1}$)	3.1×10^{-5}	1.2×10^{-5}	6.3×10^{-7}	6.3×10^{-7}	3.6×10^{-7}
6	Working range pH	2–10	2–10	2–10	2–10	2–10
7	Response time (s)	≤ 30	≤ 30	≤ 20	≤ 20	≤ 20
8	Precision (%)	98	98.3	99	99	99
9	Within-day reproducibility (mV)	0.5	0.5	0.3	0.3	0.3
10	Between-day variability (mV)	0.6	0.6	0.6	0.8	0.8
11	Accuracy (%)	98.0	98.5	99.0	99.1	99.0
12	Standard deviation (%)	0.4	0.4	0.3	0.3	0.3

reproducibility of the potential response detected was much better (0.7 mV). Sensors developed with membranes incorporating DOS and DOP as solvent mediators furnishes narrower response ranges as shown in Table 10.3.

Furthermore, authors also reported the sensor response toward other anions which has been summarized in Table 10.4 with the focus of the study on the selectivity of the developed sensor for thiocyanate ion over several other common inorganic and organic anions. High affinity of palladacycle **7a** toward SCN ion has been attributed to the high nucleophilic character of SCN ion which played a key role in the displacement of chloride ion. The development of a new characteristics absorption band at 2069 cm^{-1} in the infrared absorption spectrum of the complex after treating with thiocyanate solution, confirms the presence of ν_{SCN} group.

Table 10.4 Potentiometric selectivity coefficients of palladium complex based thiocyante PVC membrane sensor with *o*-NPOE plasticizer

Entry	Interferent, *J*	$K^{pot}_{SCN^-}$ without NaTPB	$K^{pot}_{SCN^-}$ with 30 mol% NaTPB
1	Cl^-	2.4×10^{-4}	4.6×10^{-5}
2	I^-	3.3×10^{-3}	5.3×10^{-4}
3	F^-	7.1×10^{-4}	8.8×10^{-5}
4	NO_3^-	1.4×10^{-3}	3.9×10^{-4}
5	NO_2^-	8.7×10^{-4}	6.3×10^{-5}
6	CN^-	4.2×10^{-3}	6.7×10^{-4}
7	SO_4^{2-}	6.5×10^{-4}	7.8×10^{-5}
8	$S_2O_3^{2-}$	4.9×10^{-4}	6.9×10^{-5}
9	SO_3^{2-}	5.6×10^{-4}	7.6×10^{-5}
10	PO_4^{3-}	1.6×10^{-4}	3.8×10^{-5}
11	ClO_4^-	1.0×10^{-2}	1.2×10^{-3}
12	$(Citrate)^{3-}$	5.4×10^{-4}	6.5×10^{-5}
13	$(Acetae)^-$	2.3×10^{-4}	4.1×10^{-5}
14	$(Oxalate)^{2-}$	1.9×10^{-4}	3.6×10^{-5}

The data tabulated in Table 10.5 provides further evidence that the proposed sensor not only has a superior selectivity behavior (even in very dilute concentration range) over many of the previously suggested sensors *(21–32)* but also displays better linear response range than those already reported in the literature *(21–32)*.

After developing good working conditions for the palladacycle based sensor, authors fine-tuned the feasibility of the sensor for the determination of thiocyanate concentration in human urine and saliva samples. The data obtained from the developed palladacycle-based sensor was found to be in good agreement with the data obtained by other authors using spectrophotometric techniques and the results have been summarized in Table 10.6.

Table 10.5 General performance characteristics of some potentiometric SCN^- sensors

Entry	Ionophore	Slope (mV decade^{-1})	Linear range (mol L^{-1})	Limit of detection (mol L^{-1})	Interference
1	AgX/SCN solid state *(21)*	−58.5	1.0×10^{-6}–1.0×10^{-1}	–	I^-, S^{2-}, X^-
2	Di-,tetra- and hexa-imidepyridinederivatives *(22)*	−55.6 −58.3	9.0×10^{-6}–1.0×10^{-2}	1×10^{-6}	IO_4^-
3	Cadmium-Schiff's complex *(23)*	−59.0 ± 0.2	1.0×10^{-6}–1.0×10^{-1}	1×10^{-6}	Salicylate, I^-, MnO_4^-
4	Mn(II) complex of N,N-bis-(4-phenylazosalicylidene)ophenylenediamine *(24)*	−57	7.0×10^{-6}–1.0×10^{-1}	1×10^{-6}	I^-, ClO_4^-, salicylate
5	Bis[N-(2-hydroxyethyl)salicylaldimino]copper (II) *(25)*	−59	6.0×10^{-6}–1.0×10^{-1}	1×10^{-6}	I^-, ClO_4^-, salicylate
6	Urea-functionalized porphyrin *(26)*	−58	5.0×10^{-5}–1.0×10^{-2}	1×10^{-6}	NO_3^-, NO_2^-, I^-, Cl^-, Br^-, ClO_4^-
7	Cobalt and Mn-phthalocyanine *(27)*	−57.5	1.0×10^{-6}–1.0×10^{-1}	1×10^{-6}	I^-, N_3^-, salicylate
8	Dibenzyl-hexaazacyclo-tetradecane-Ni(II)-perchlorate *(28)*	−58.0 ± 0.4	3.3×10^{-6}–1.0×10^{-1}	1×10^{-6}	I^-, $S_2O_3^{2-}$, CrO_4^{2-}, MnO_4^-
9	Tetrakis(2,4,6-triphenylphenyl) porphyrinato)-Mn(III)-chloride *(29)*	−57	1.0×10^{-5}–1.0×10^{-2}	1×10^{-6}	Salicylate
10	Butane-2,3-dione bissalicylhydrazonate-zinc *(30)*	−56.0 ± 0.1	1.0×10^{-6}–1.0×10^{-1}	1×10^{-6}	–
11	Tetrakis(pentafluorophenyl)porp-hyrin-Mn(III)-chloride *(31)*	−59.5	1.0×10^{-7}–1.0×10^{-1}	1×10^{-6}	N_3^-, ClO_4^-, CN^-
12	Bis(2-mercaptobezoxazolato)-Mercury *(32)*	−60.6 ± 0.8	1×10^{-6}–0.1	1×10^{-6}	I^-
13	OrganoPd-complex *(20)*	−57.8 ± 0.2	1.0×10^{-6}–1.0×10^{-1}	1×10^{-6}	–

Table 10.6 Flow injection potentiometric determination of thiocyanate in some biological samples using palladium complex based thiocyanate PVC membrane sensor and spectrophotometry

Entry	No of smoked cigarettes/day	Age	Spectrophotometry[a] [Saliva-SCN^-] mol L^{-1}	[Urine-SCN^-] mol L^{-1}	Potentiometry[a] [Saliva-SCN^-] mol L^{-1}	[Urine-SCN^-] mol L^{-1}
1	0	30–33	$(6.8–7.6 \pm 0.3) \times 10^{-5}$	$(4.3–4.8 \pm 0.1) \times 10^{-5}$	$(6.6–7.5 \pm 0.2) \times 10^{-5}$	$(4.1–4.7 \pm 0.1) \times 10^{-5}$
2	1–5	30.42	$(2.7–3.8 \pm 0.1) \times 10^{-4}$	$(1.7–2.7 \pm 0.1) \times 10^{-4}$	$(2.8–3.7 \pm 0.1) \times 10^{-4}$	$(1.7–2.5 \pm 0.1) \times 10^{-4}$
3	6–10	38–42	$(4.2–6.2 \pm 0.2) \times 10^{-4}$	$(2.8–3.6 \pm 0.1) \times 10^{-4}$	$(4.1–6.1 \pm 0.2) \times 10^{-4}$	$(2.6–3.5 \pm 0.1) \times 10^{-4}$
4	11–20	36–45	$(2.9–3.8 \pm 0.1) \times 10^{-3}$	$(1.8–2.6 \pm 0.1) \times 10^{-3}$	$(2.8–3.7 \pm 0.1) \times 10^{-3}$	$(1.7–2.5 \pm 0.1) \times 10^{-3}$
5	>20	46–54	$(4.6–5.5 \pm 0.1) \times 10^{-3}$	$(2.5–3.5 \pm 0.1) \times 10^{-3}$	$(4.5–5.4 \pm 0.1) \times 10^{-3}$	$(2.4–3.4 \pm 0.1) \times 10^{-3}$

[a] Average of five measurements from each individual.

The sensor was tuned by the authors in a very elegant way providing satisfactory results with greater precision and high order of reproducibility for both the manual and flow injection potentiometric determination of SCN^- in the samples of saliva and urine of the cigarette smokers and the nonsmokers.

Positive results obtained as a part of the experimentation with the biological samples of urine and saliva, and with living cell models for the sensing of SCN ion and CO, respectively allowed the authors to further explore the possibility of examining the toxicity of palladacycles. The results of the study confirmed that the palladacycles were nontoxic to the living cells. The positive influence of the palladacycles on the living cells prompted the scientific community to further investigate the biological potential of palladacycles *(15, 17–20)*. To date, palladacycles have been explored for their anticancer ability *(33)* and a few of the palladacycles are under clinical trials as anticancer agents (the details have already been discussed in the earlier chapter). Besides anticancer ability, several other biological applications of palladacycles have also been examined by the organometallic chemists. These have been detailed below as a part of the exploration of palladacycles as bio-active agents.

5. BIOLOGICAL APPLICATIONS OF PALLADACYCLES

At the beginning of the 20th century, the discovery of metal-based pharmaceutically useful compounds boosted the importance of organometallics in the field of medicines. Organometallic compounds of a variety of transition metals have in the past shown potential drug activities *(34–37)*. However, so far only compounds of platinum have attained any real importance in terms of clinical usage. The serendipitous discovery that *cis*-$[Pt(NH_3)_2Cl_2]$ could suppress cell division led ultimately to its clinical use in the form of the commercially relevant drug *cis*-platin. Since then much work has been devoted to comprehending the possible mechanism of the action of *cis*-platin, and developing new drugs in this class (metallo-drugs) with reduced side effects, and a wider spectrum of anticancer activity. As for the current situation in metallo-drugs development, certain transition metals such as palladium, ruthenium, and rhodium complexes have shown promising results and are the subject of much current research and testing in model systems. *(38, 39)*

Palladacycles are unique because of their outstanding photo-optical properties, which can be exploited for both sensing and therapeutic applications. An extensive review of this area was reported by Elgazwy et al. *(40)* discussing the application of palladacycles on the basis of structure-activity-relationship (SAR) in particular for biomedical applications. From the biological standpoint, the types of antibacterial, antifungal, antimycobacterial, and antiprotozoal (antiamoebic and antitrypanosomal) activities will vary considerably from one geographic region to another throughout the world. To assess the importance of using palladacycles in diagnostics and their potential applications as antibacterial, antifungal, antimycobacterial, and antiprotozoal agents, the following sections are organized according to the biological potentials exhibited by palladacycles.

5.1 Antibacterial Activity

One of the earliest examples of palladacycles exhibiting antibacterial activity was given by Ma et al. for a Pd(II) complex **8b** of fluorine-containing Schiff base as a ligand *(41)*. Complex **8b** (Scheme 10.8) is a μ-chloro-bridged dinuclear cyclometallated Pd(II) complex and the thermal profile of the complex **8b** revealed that the complex exhibited high thermal stability.

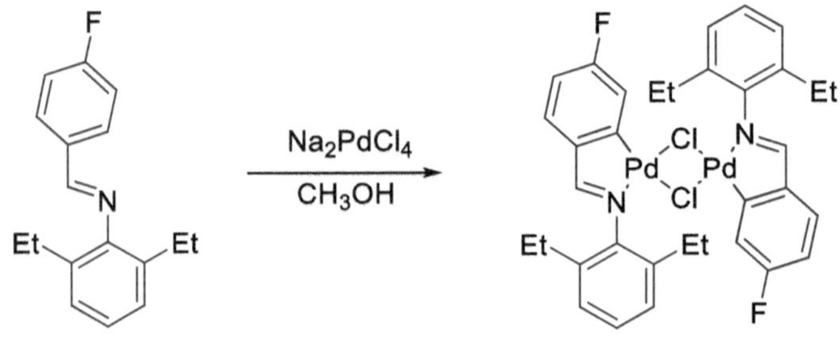

8a **8b**

Scheme 10.8 Structure of fluorine-containing Schiff base containing cyclopalladated complex 8b.

Antimicrobial activities of compound **8a** and palladacyclic complex **8b** was examined toward the commercially available strains of bacterium such as *Staphylococcus aureus, Bacillus cereus, Rhizopus,* and *Escherichia coli* by the authors and the results have been represented in Table 10.7. The data shown in Table 10.7 clearly suggest that the effect of **8a** and **8b** against *S. aureus* was found in very close agreement with the data obtained for sodium penicillinate as a reference. The inhibition diameters for the compounds **8a** and **8b** were found to be larger than those of sodium penicillinate against *B. cereus*. Both the compounds **8a** and **8b** were also found to be superior to penicillin against *rhizopus* and *E. coli*. Overall, the data reported in Table 10.7 indicates that the presence of Pd^{2+} activates compound **8b** in such a way that **8b** exhibits exceptionally superior activities as compared with compound **8a**.

Table 10.7 Antimicrobial activities of **8a** and **8b**

		Zone of inhibition (mm)			
Entry	**(μg/disk)**	***S. aureus***	***B. cereus***	***Rhizopus***	***E. coli***
8a	20	30.2	20.4	21.2	23.8
	10	27.8	18.7	19.4	22.1
	2	20.3	16.5	17.1	20.4
8b	20	32.7	23.4	25.5	26.7
	10	31.9	21.7	23.1	24.5
	2	24.6	20.1	20.1	21.2
Peniciline	20	34.7	12.2	0.0	0.0
	10	14.8	0.0	0.0	0.0
	2	0.0	0.0	0.0	0.0
DMF	20	–	–	–	–
	10	–	–	–	–
	2	–	–	–	–

5.2 Antimycobacterial Activity

Recently, Moro et al. *(42)* synthesized and investigated the thermal behavior of palladacycle (**9a–d**) (Scheme 10.9). The studies undertaken by the authors suggest minimum inhibitory concentration (MIC) values against *M. tuberculosis* for cyclopalladated complexes **9a, 9b, 9c** and **9d** were 31.2, 156, 15.6 and 15.6 mg mL^{-1} respectively. The principle of hard acid and hard base was found to play an important role in deciding the antimycobacterial potential of the synthesized complexes **9a–d**. Complex **9a**, for example, is the result of an interaction between a soft Lewis acid and a hard Lewis base and is responsible for the lability of complex **9a**, while the remaining three complexes, **9c–d**, are associated with soft Lewis bases which are responsible for the reduction of the lability of the Pd(II) complexes. The complexes, when compared with Isoniazid (one of the first-line antitubercular drugs), were found to be less active than isoniazid (MIC value of 0.030 mg mL^{-1}).

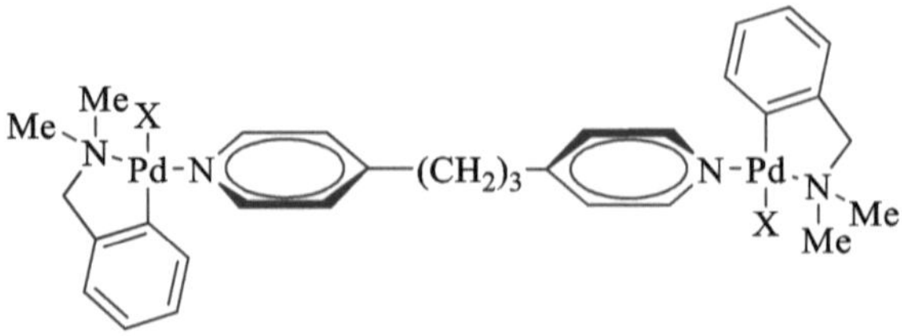

Scheme 10.9 Structures of cyclopalladated complex (9a–d).

5.3 Anticonvulsant Activity

Barros et al. *(43)* recently synthesized two organometallic diazepam–palladium(II) derivatives by C-H activation of diazepam (DZP) with palladium salts, i.e., $PdCl_2$ and $Pd(OAc)_2$ (Scheme 10.10). Two well-established animal models: pentylenetetrazole (PTZ)- and picrotoxin (PTX) -induced convulsions were employed for the assessment of the anti-convulsant potential of the synthesized complexes, which were labeled as $[(DZP)PdOAc]_2$ (**10a**) and $[(DZP)PdCl]_2$ (**10b**). The results have been discussed in terms of the latencies when the animals were protected against convulsions induced by PTZ and PTX by the authors and it has been found that the organometallic DZP-palladium(II) acetate complex, $[(DZP)PdOAc]_2$ (**10a**), in comparison with the chloro derivative, $[(DZP)PdCl]_2$ (**10b**), was far more effective in both the PTX and PTZ modes of action, while the chloro derivative was suitable only in the PTZ model. Attempts have been made by the authors to explain the possible mechanism of action of the DZP-palladium(II) complexes (**10a–b**), and the mechanism was also confirmed with the use of flumazenil (FLU), a GABAA-benzodiazepine receptor complex site antagonist. For the first time authors have assessed the anticonvulsant properties of organometallic DZP-palladium(II) complexes (**10a–b**) as well as suggesting that these compounds have potential to play a subtle role in the study related to the discovery of new drugs molecules for the treatment of epilepsy. Authors have also emphasized the significant protection provided by the complexes against convulsions induced by PTX, which in the case of $[(DZP)PdOAc]_2$ (**10a**) was attributed to the possibility of $[(DZP)PdOAc]_2$ (**10b**) acting as GABAA receptor agonist by increasing the chloride influx via brain chloride channels.

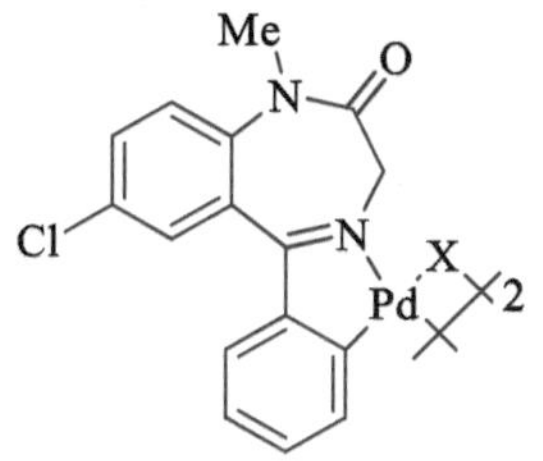

10a (X = OAc)
10b (X = Cl)

Scheme 10.10 Structures of diazepam (DZP) based cyclopalladated complexes (10a–b).

6. CONCLUSION AND PERSPECTIVES

The importance for high living standards and hygienic disciplines presents great challenges for the flourishing development of some innovative, trustworthy, but highly effective biologically active motifs or sensors that should be environmentally benign and exceptionally safe for human use. As cited above in this chapter, quite a large number of palladacycles are currently available with flourishing assortments of innovative structural features having their use as sensing probes, as catalysts for organic transformations, and as molecules for biological examinations. Various applications are cited which range from sensing of carbon monoxide to thiocyanate ion, antimalarial, antimycobacterial, and anticonvulsant agents. It has been encouraged that the ligands on these species have been designed and developed in such a way that they advocate an appropriate balance between electronic and steric properties of the metal center. Thus, the major problem is not how to develop a palladacycle, but how to design the performance of the palladacycle. In spite of the potential exhibited toward the large variety of applications, palladacycles as sensors or bio-active agents remains a challenging prospect given the high cost of the metal and therefore it is important for researchers to focus their efforts in making these species more efficient at concentrations (preferably nano molar concentrations) that are attractive and cost effective.

REFERENCES

1. Tsuji, J. *Palladium Reagents and Catalysts, New Perspectives for the 21st Century*. John Wiley & Sons: England, 2004.
2. Malleron, J. L.; Fiaud, J. C.; Legros, J. Y. *Handbook of Palladium-Catalyzed Organic Reactions Synthetic Aspects and Catalytic Cycles*. Academic Press: London, 1997.
3. Lorion, M. M.; Maindan, K.; Kapdi, A. R.; Ackermann, L. *Chem. Soc. Rev.* **2017**, *46*, 7399.

4. Jairton, D.; Michel, P. *Palladacycles, Synthesis, Characterization and Applications.* Wiley-VCH: Weinheim, 2008.
5. Dupont, J.; Pfeffer, M. *Palladacycles, Synthesis, Characterization and Applications.* Wiley-VCH: Weinheim, 2008.
6. Sable, V.; Maindan, K.; Shejwalkar, P.; Hara, K.; Kapdi, A. R. *ACS Omega* **2017**, *2*, 204.
7. Penney, D. G. *Carbon monoxide toxicity.* Taylor & Francis Group, CRC Press: Florida, 2000.
8. Penney, D. G. *Carbon Monoxide Poisoning.* Taylor & Francis Group, CRC Press: Florida, 2008.
9. Jürgen Buschow, K. H.; Cahn, R. W.; Flemings, M. C.; Ilschner, B.; Kramer, E. J.; Mahajan, S.; Veyssière, P. *Encyclopedia of Materials: Science and Technology.* Elsevier Ltd, 2007.
10. Ramamurthy, V.; Schanze, K. S. *Molecular and Supramolecular Photochemistry, Optical sensors and switches.* Marcel Dekker: New York, 2001.
11. Zhang, X.; Hou, L.; Samori, P. *Nat. Commun.* **2016**, *711118.*
12. (a)Yamauchi, S. *Chemical Sensing Technology.* Elsevier: London, 1992; (b)Gründler, P. *Chemical Sensors: An Introduction for Scientists and Engineers.* Springer-Verlag: Berlin, 2007.
13. Gonzalez, A.; Granell, J.; Lopez, C.; Bosque, R.; Rodriguez, L.; Font-Bardia, M.; Calvet, T.; Solans, X. *J. Organomet. Chem.* **2013**, *726*, 21.
14. Cinellu, M. A.; Gladiali, S.; Minghetti, G. J. *Organomet. Chem.* **1989**, *363*, 401.
15. Michel, B. W.; Lippet, A. R.; Chang, C. J. J. Am. Chem. Soc. **2012**, *134*, 15668.
16. Bumbrah, G. S.; Sharma, R. M. *Egypt J. Forensic Sci.* **2016**, *6*, 209.
17. Cao, Y.; Li, D. W.; Zhao, L. J.; Liu, X. Y.; Cao, X. M.; Long, Y. T. Anal. Chem. **2015**, *87*, 9696.
18. Li, Y.; Wang, X.; Yang, J.; Xie, X.; Li, M.; Niu, J.; Tong, L.; Tang, B. Anal. Chem. **2016**, *88*, 11154.
19. Sun, M.; Yu, H.; Zhang, K.; Wang, S.; Hayat, T.; Alsaedi, A.; Huang, D. *ACS Sens.* **2018**, *3*, 285.
20. Hassan, S. S. M.; Mohmoud, W. H.; Elgazwy, H.; Abdel-Sattar, S.; Badawy, N. M. *Electroanalysis* **2006**, *18*, 2070.
21. Umezawa, Y. *CRC Handbook of Ion-Selective Electrodes Selectivity Coefficients.* CRC Press: Boca Raton, 1990.
22. Hassan, S. S. M.; Abou Ghalia, M. H.; Amr, A. G. E.; Mohamed, A. H. K. *Anal. Chim. Acta* **2003**, *482*, 9.
23. Ganjali, M. R.; Poursaberi, T.; Basiripour, F.; Salavati-Niasari, M.; Yousefi, M.; Shamsipur, M. *Fresenius J. Anal. Chem.* **2001**, *370*, 1091.
24. Li, Z. Q.; Wu, Z. Y.; Yuan, R.; Ying, M.; Shen, G. L.; Yu, R. Q. *Electrochim. Acta* **1999**, *44*, 2543.
25. Huang, X.; Chai, Y.; Yuan, R.; Wang, X.; Li, Q. *Anal. Sci.* **2004**, (20), 1185.
26. Taylor, J. K. *Quality Assurance of Chemical Measurements.* CRC Press: Boca Raton, 1987.
27. Parrish, M. E.; Hart, J. L. L.; Shafer, K. H. *Vib. Spectrosc.* **2001**, *27*, 29.
28. Galanti, L. M. *Clin. Chem.* **1997**, *43*, 184.
29. Brown, D. V.; Chaniotakis, N. A.; Lee, I. H.; Ma, S. C.; Park, S. B.; Meyerhoff, M. E.; Nick, R. J.; Groves, J. T. *Electroanalysis* **1989**, *1*, 477.
30. Sun, B.; Noller, B. N. *Water Res.* **1998**, *32*, 3698.
31. Khorasani, J. H.; Amini, M. K.; Motaghi, H.; Tangestaninejad, S.; Moghadam, M. *Sens. Actuators B* **2002**, *87*, 448.
32. Amin, M. K.; Rafi, A.; Ghaedi, M.; Habibi, M. H.; Zahory, M. M. *Microchem. J.* **2003**, *75*, 143.
33. Bangde, P. S.; Prajapati, D. S.; Dandekar, P. P.; Kapdi, A. R. *ChemistrySelect* **2018**, *3*, 5709.

34. Kaim, W.; Schwederski, B.; Klein, A. *Bioinorganic Chemistry: Inorganic Elements in the Chemistry of Life.* John Wiley & Sons: United Kingdom, 2013.
35. Crichton, R. R. *Biological Inorganic Chemistry An Introduction.* Elsevier: United Kingdom, 2008.
36. Sutton, R.; Rockett, B.; Swindells, P. G. *Chemistry for the Life Sciences.* Taylor & Francis Group: Boca Raton, 2009.
37. Fraústo da Silva, J. J. R.; Williams, R. J. P. *The Biological Chemistry of the Elements The Inorganic Chemistry of Life.* Oxford University Press: London, 2001.
38. Jones, C.; Thornback, J. *Medicinal Applications of Coordination Chemistry.* The Royal Society of Chemistry: United Kingdom, 2007.
39. Kessissoglou, D. P. *Bioinorganic Chemistry. An Inorganic Perspective of Life.* Kluwer Academic Publishers: Greece, 1995.
40. Elgazwy, A.-S. S. H.; Ismail, N. S. M.; Atta-Allah, S. R.; Sarg, M. T.; Soliman, D. H. S.; Zaki, M. Y.; Elgamas, M. A. Palladacycles as Antimicrobial Agents. *Curr. Med. Chem.* **2012**, *19*, 3967–3981.
41. Ma, D. Y.; Zhang, L. X.; Rao, X. Y.; Wu, T. L.; Li, D. H.; Xie, X. Q. *J. Coord. Chem.* **1486**, *2013*, 66.
42. Moro, A. C.; Urbaczek, A. C.; De Almeida, E. T.; Pavan, F. R.; Leite, C. Q. F.; Netto, A. V. G.; Mauro, A. E. *J. Coord. Chem.* **2012**, *65*, 1434.
43. Galvao Barros, W. B. Z.; Queiroz da Silva, A. H.; Lima Barbosa, A. S.; Nunes, A. M.; Machado Reys, J. R.; de Araujo-Filho, H. G.; de Souza Siqueira Quintan, J.; Quintans-Junior, L. J.; Pfeffer, M.; Rodrigues dos Santos Malta, V.; Meneghetti, M. R. J. Inorg. Biochem. **2016**, *155*, 129.

INDEX

Note: Page numbers followed by *f* indicate figures, *t* indicate tables, and *s* indicate schemes.

A

B

D

E

F

I

K

L

M

N

Q

S

T

FSC
www.fsc.org
MIX
Papier | Fördert
gute Waldnutzung
FSC® C083411

Kolibri 360 GmbH
Ferdinand-Jühlke-Straße 7
99095 Erfurt, Deutschland
produktsicherheit@kolibri360.de